D1036888

Structural Steel Design

LRFD Approach

Second Edition

Structural Steel Design

LRFD Approach

Second Edition

J. C. Smith
North Carolina State University

John Wiley & Sons, Inc.
New York Chichester Brisbane Toronto Singapore

ACQUISITIONS EDITOR Cliff Robichaud
ASSISTANT EDITOR Catherine Beckham
MARKETING MANAGER Debra Reigert
PRODUCTION EDITOR Ken Santor
COVER DESIGNER Harry Nolan
INTERIOR DESIGN Michael Jung
MANUFACTURING MANAGER Dorothy Sinclair

This book was set in 10/12 Palatino by John Wiley & Sons, Inc., and printed and bound by Hamilton Printing. The cover was printed by Hamilton Printing.

Recognizing the importance of preserving what has been written, it is a policy of John Wiley & Sons, Inc. to have books of enduring value published in the United States printed on acid-free paper, and we exert our best efforts to that end.

The paper on this book was manufactured by a mill whose forest management programs include sustained yield harvesting of its timberlands. Sustained yield harvesting principles ensure that the number of trees cut each year does not exceed the amount of new growth.

Library of Congress Cataloging-in-Publication Data:

Smith, J.C., 1933-
Structural steel design : LRFD approach / J.C. Smith.—2nd ed.
p. cm.
Includes bibliographical references and index.
ISBN 0-471-10693-3 (cloth: alk. paper)
1. Building, Iron and steel. 2. Steel, Structural. 3. Load factor design. I. Title.
TA684.S584 1996
624.1'821—dc20 95-36503
CIP

Printed in the United States of America

10 9 8 7 6 5 4 3 2 1

Preface

This book has been written to serve as the undergraduate-level textbook for the first two structural steel design courses in Civil Engineering.

In this edition, each chapter was modified to reflect the changes made in the 1993 AISC LRFD Specification for Structural Steel Buildings and the 1994 LRFD Manual of Steel Construction, Second Edition, which consists of

Volume I: Structural Members, Specifications, & Codes

Volume II: Connections

The chapter on the behavior and design of tension members is located before the chapter on connections for tension members, which is separated from the chapter on other types of connections. Bolted connections for tension members are discussed before welded connections. The long examples in the first edition have been replaced by shorter ones.

Each professor has particular course constraints and preferences of what to present in each course. Chapters 1 to 6 probably contain most of the material that is taught in the first steel design course. Chapters 7 to 11 contain material to meet the other needs of each professor. Appendix B gives the review material needed for a thorough understanding of principal axes involved in column and beam behavior. Appendix C provides some formulas for the warping and torsional constants of open sections.

The LRFD Specification requires a factored load analysis and permits either an elastic analysis or a plastic analysis. In our capstone structural design course, the students are required to design a steel-framed building and a reinforced-concrete-framed building. Since the ACI Code permits only an elastic analysis due to factored loads, I use only the elastic analysis approach in the capstone structural design course. Consequently, Chapter 6 and Appendix A give students a brief but realistic introduction to elastic analysis and design of unbraced frames in the LRFD approach. Chapter 11 should be adequate for those who wish to discuss plastic analysis and design. Appendix D provides some handbook information pertaining to plastic analysis.

I use the textual material associated with Appendix A in the classroom whenever appropriate.

The reviewers of this edition were:

P. R. Chakrabarti, California State University - Fullerton; W. S. Easterling, Virginia Tech; S. C. Goel, University of Michigan; R. B. McPherson, New Mexico State University; and A. C. Singhal, Arizona State University.

I am appreciative of their comments, suggestions for improvement, constructive criticisms, and identified errors.

J. C. Smith

Contents

Chapter 2 Tension Members 43

Chapter 3 Connections for Tension Members 83

Chapter 4 Columns 118

Chapter 8 Connections 312

Chapter 9 Plate Girders 366

Introduction

1.1 STRUCTURAL STEEL

Steel is extensively used for the frameworks of bridges, buildings, buses, cars, conveyors, cranes, pipelines, ships, storage tanks, towers, trucks, and other structures.

1.1.1 Composition and Types

Yield strength is the term used to denote the *yield point* (see Figure 1.1) of the common structural steels or the stress at a certain offset strain for steels not having a well-defined yield point. Prior to about 1960, steel used in building frameworks was ASTM (American Society for Testing and Materials) designation A7 with a yield strength of 33 ksi. Today, there are a variety of ASTM designations available with yield strengths ranging from 24 to 100 ksi.

Steel is composed almost entirely of iron, but contains small amounts of other chemical elements to produce desired physical properties such as *strength, hardness, ductility, toughness,* and *corrosion resistance*. Carbon is the most important of the other elements. Increasing the carbon content produces an increase in strength and hardness, but decreases the ductility and toughness. Manganese, silicon, copper, chromium, columbium, molybdenum, nickel, phosphorus, vanadium, zirconium, and aluminum are some of the other elements that may be added to structural steel. Hot-rolled structural steels may be classified as *carbon steels, high-strength low-alloy steels,* and *alloy steels.*

Carbon steels contain the following maximum percentages of elements other than iron: 1.7% carbon, 1.65% manganese, 0.60% silicon, and 0.60% copper. Carbon and manganese are added to increase the strength of the pure iron. Carbon steels are divided into four categories: (1) *low carbon* (less than 0.15%); (2) *mild carbon* (0.15–0.29%); (3) *medium carbon* (0.30–0.59%); and, (4) *high carbon* (0.60–1.70%). Structural carbon steels are of the mild-carbon category and have a distinct yield point [see curve (a) of Figures 1.1 and 1.2]. The most common structural steel is A36, which has

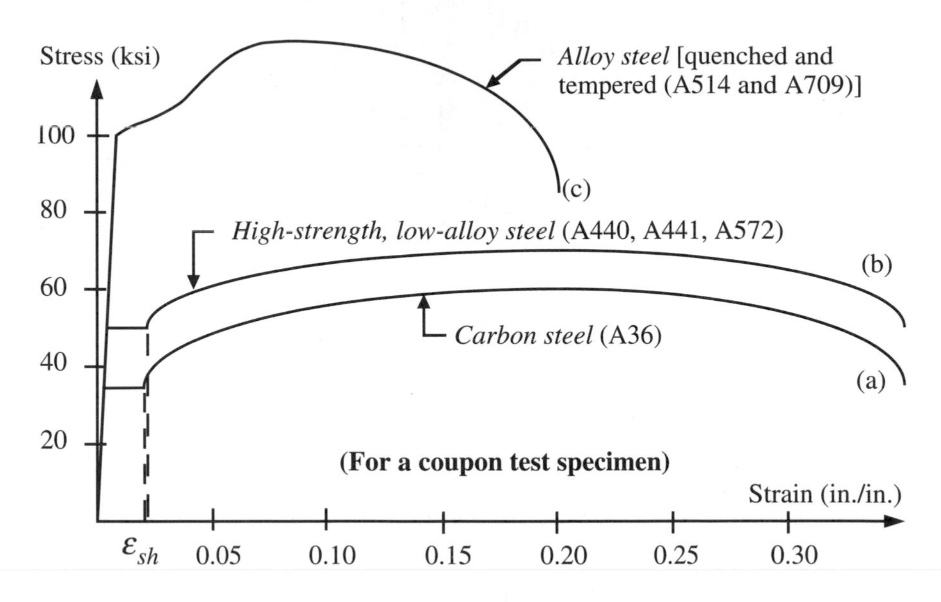

FIGURE 1.1 Typical stress–strain curves for steel.

a maximum carbon content of 0.25 to 0.29%, depending on the thickness, and a yield strength of 36 ksi. The carbon steels of Table 1.1 are A36, A53, A500, A501, A529, A570, and A709 (grade 36); their yield strengths range from 25 to 100 ksi.

High-strength low-alloy steels [see curve (b) of Figures 1.1 and 1.2] have a distinct yield point ranging from 40 to 70 ksi. Alloy elements such as chromium, columbium,

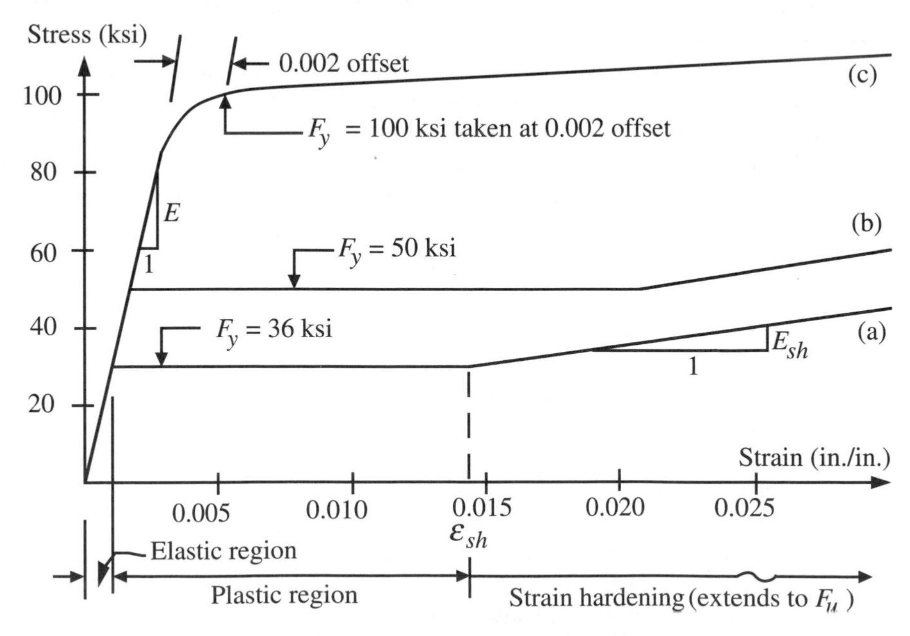

FIGURE 1.2 Enlargement of Figure 1.1 in vicinity of yield point.

Table 1.1 Steels Used for Buildings and Bridges

Steel Type	ASTM Designation	F_y (ksi)	F_u (ksi)	Thickness (in.)	Common Usage
Carbon	A36	32	58–80	Over 8	General; buildings
		36	58–80	To 8	General; buildings
	A529 Grade 42	42	60–85	To 0.5	Metal building systems
	Grade 50	50	70-100	To 1.5	Metal building systems
High-strength low-alloy	A441	40	60	4-8	Welded construction
		42	63	1.5-4	Welded construction
		46	67	0.75-1.5	Welded construction
		50	70	To 0.75	Welded construction
	A572 Grade 42	42	60	To 6	Buildings; bridges
	Grade 50	50	65	To 2	Buildings; bridges
	Grade 60	60	75	To 1.25	Buildings; bridges
	Grade 65	65	80	To 1.25	Buildings; bridges
Corrosion resistant	A242	42	63	1.5-4	Bridges
		46	67	0.75-1.5	Bridges
High-strength, low-alloy		50	70	To 0.75	Bridges
	A588	42	63	5-8	Weathering steel
		46	67	4-5	Weathering steel
		50	70	To 4	Weathering steel bridges
Quenched & tempered low-alloy	A514	90	100-130	2.5-6	Plates for welding
		100	110-130	To 2.5	Plates for welding
Quenched and tempered alloy	A852	70	110-190	To 4	Plates for welding

copper, manganese, molybdenum, nickel, phosphorus, vanadium, and zirconium are added to improve some of the mechanical properties of steel by producing a fine instead of a coarse microstructure obtained during cooling of the steel. The high-strength low-alloy steels of Table 1.1 are A242, A441, A572, A588, A606, A607, A618, and A709 (grades 50 and 50W).

Alloy steels [see curve (c) of Figures 1.1 and 1.2] do not have a distinct yield point. Their yield strength is defined as the stress at an offset strain of 0.002 with yield strengths ranging from 80 to 110 ksi. These steels generally have a maximum carbon content of about 0.20% to limit the hardness that may occur during heat treating and welding. Heat treating consists of quenching (rapid cooling with water or oil from 1650°F to about 300°F) and tempering (reheating to 1150°F and cooling to room temperature). Tempering somewhat reduces the strength and hardness of the quenched material, but significantly improves the ductility and toughness. The quenched and tempered alloy steels of Table 1.1 are A514 and A709 (grades 100 and 100W).

Bolts and *threaded fasteners* are classified as:

1. A307 (low-carbon) bolts, usually referred to as common or machine or unfinished bolts, do not have a distinct yield point (minimum yield strength of 60 ksi is taken at a strain of 0.002). Consequently, the Load and Resistance

Factor Design (LRFD) Specification [2][1] does not permit these bolts to be used in a slip-critical connection [see LRFD J1.11 (p. 6-72), J3.1(p. 6-79), and Table J3.2(p. 6-81)]. However, they may be used in a bearing-type connection.

2. A325 (medium-carbon; quenched and tempered with not more than 0.30% carbon) bolts have a 0.2% offset minimum yield strength of 92 ksi (0.5–1 in.-diameter bolts) and 81 ksi (1.125–1.5 in.-diameter bolts) and an ultimate strength of 105 to 120 ksi.
3. A449 bolts have tensile strengths and yield strengths similar to A325 bolts, have longer thread lengths, and are available up to 3 in. in diameter. A449 bolts and threaded rods are permitted only where greater than 1.5-in. diameter is needed.
4. A490 bolts are quenched and tempered, have alloy elements in amounts similar to A514 steels, have up to 0.53% carbon, and a 0.2% offset minimum yield strength of 115 ksi (2.5–4-in. diameter) and 130 ksi (less than 2.5–in. diameter).

Weld electrodes are classified as E60XX, E70XX, E80XX, E90XX, E100XX, and E110XX where E denotes electrodes, the digits denote the tensile strength in ksi, and XX represents characters indicating the usage of the electrode.

1.1.2 Manufacturing Process

At the steel mill, the manufacturing process begins at the blast furnace where iron ore, limestone, and coke are dumped in at the top and molten pig iron comes out at the bottom. Then the pig iron is converted into steel in basic oxygen furnaces. Oxygen is essential to oxidize the excess of carbon and other elements and must be highly controlled to avoid gas pockets in the steel ingots since gas pockets will become defects in the final rolled steel product. Silicon and aluminum are deoxidizers used to control the dissolved oxygen content. Steels are classified by the degree of deoxidation: (1) *killed steel* (highest); (2) *semikilled steel* (intermediate); and (3) *rimmed steel* (lowest).

Potential mechanical properties of steel are dictated by the chemical content, the rolling process, finishing temperature, cooling rate, and any subsequent heat treatment. In the rolling process, material is squeezed between two rollers revolving at the same speed in opposite directions. Thus, rolling produces the steel shape, reduces it in cross section, elongates it, and increases its strength. Ordinarily, ingots are poured from the basic oxygen furnaces, reheated in a soaking pit, rolled into slabs, billets, or blooms in the bloom mill, and then rolled into shapes, bars, and plates in the breakdown mill and finishing mill. If the continuous casting process is used, the ingot stage is bypassed.

A chemical analysis, also known as the heat or ladle analysis, is made on samples of the molten metal and is reported on the mill test certificate for the heat or lot (50–300 tons) of steel taken from each steel-making unit. One to 8 hours are required to produce a heat of steel depending on the type of furnace being used.

[1]We assume that each reader has a copy of the LRFD Manual[2]. Throughout this text, each applicable specification and design aid in the LRFD Manual is cited. Also, to enable the reader to quickly locate these items, the corresponding page numbers are given.

Mechanical properties (*modulus of elasticity, yield strength, tensile strength,* and *elongation* to determine the *degree of ductility*) of steel are determined from tensile tests of specimens taken from the final rolled product. These mechanical properties listed on the mill test certificate normally exceed the specified properties by a significant amount and merely certify that the test certificate meets prescribed steel-making specifications. Each piece of steel made from the heat of steel covered by the mill test certificate does not have precisely the properties listed on the mill test certificate. Therefore, structural designers do not use the mill test certificate properties for design purposes. The minimum specified properties listed in the design specifications are used by the structural designer.

1.1.3 Strength and Ductility

Strength and *ductility* are important characteristics of structural steel in the structural design process. Suppose identical members (same length and same cross-sectional area) are made of wood, reinforced concrete, and steel. The steel member has the greatest strength and stiffness, which permit designers to use fewer columns in long clear spans of relatively small members to produce steel structures with minimum dead weight.

Ductility, the ability of a material to undergo large deformations without fracture, permits a steel member to yield when overloaded and redistribute some of its load to other adjoining members in the structure. Without adequate ductility, (1) there is a greater possibility of a fatigue failure due to repeated loading and unloading of a member; and (2) a brittle fracture can occur.

Strength and ductility are determined from data taken during a standard, tensile, load–elongation test. (We contend that more appropriately for a member subjected to bending, the area under the moment–curvature curve is a better measure of ductility due to bending.) A stress–strain curve such as Figure 1.1 can be drawn using the load–elongation test data. On the stress–strain curve, after the peak or *ultimate strength* F_u, is reached, a descending branch of the curve occurs for two reasons:

1. *Stress* is defined as the applied load divided by the original, unloaded, cross-sectional area. However, the actual cross-sectional area reduces rapidly after the ultimate strength is reached.
2. The load is hydraulically applied in the lab. If the load were applied by pouring beads of lead into a bucket, for example, no decrease in load would occur from the time the ultimate strength was obtained until the specimen fractured and a horizontal, straight line would occur on the usual stress–strain curve from the ultimate strength point to the fracture point.

1.1.4 Properties and Behavorial Characteristics of Steel

For purposes of most structural design calculations, the following values are used for steel:

1. Weight = 490 lb/ft^3.
2. Coefficient of thermal expansion, CTE = 0.0000065 strain/°F).
3. Poisson's ratio $v = 0.3$.

The stress–strain curves shown in Figure 1.1 are for room-temperature conditions. As shown in Figure 1.3, after steel reaches a temperature of about 200°F, the yield strength, tensile strength, and modulus of elasticity are significantly influenced by the temperature of the steel. Also, at high temperatures, steel creeps (deformations increase with respect to time under a constant load). Temperatures in the range shown in Figure 1.3 can occur in members of a building in case of a fire, in the vicinity of welds, and in members over an open flame in a foundry, for example.

Temperature and *prior straining into the strain-hardening region* have an adverse effect on ductility. Fractures at temperatures significantly below room temperature are brittle instead of ductile. *Toughness* (ability to absorb a large amount of energy prior to fracture) is related to ductility. Toughness usually is measured in the lab by a Charpy V-notch impact test in which a standard notched specimen chilled or heated to a specified temperature is struck by a swinging pendulum. Toughness, as implied by the type of test for toughness, is important for structures subjected to impact loads (earthquakes, vertical motion of trucks on bridges, and on elevator cables if an elevator suddenly stops). Killed steels and heat-treated steels have the most toughness. As

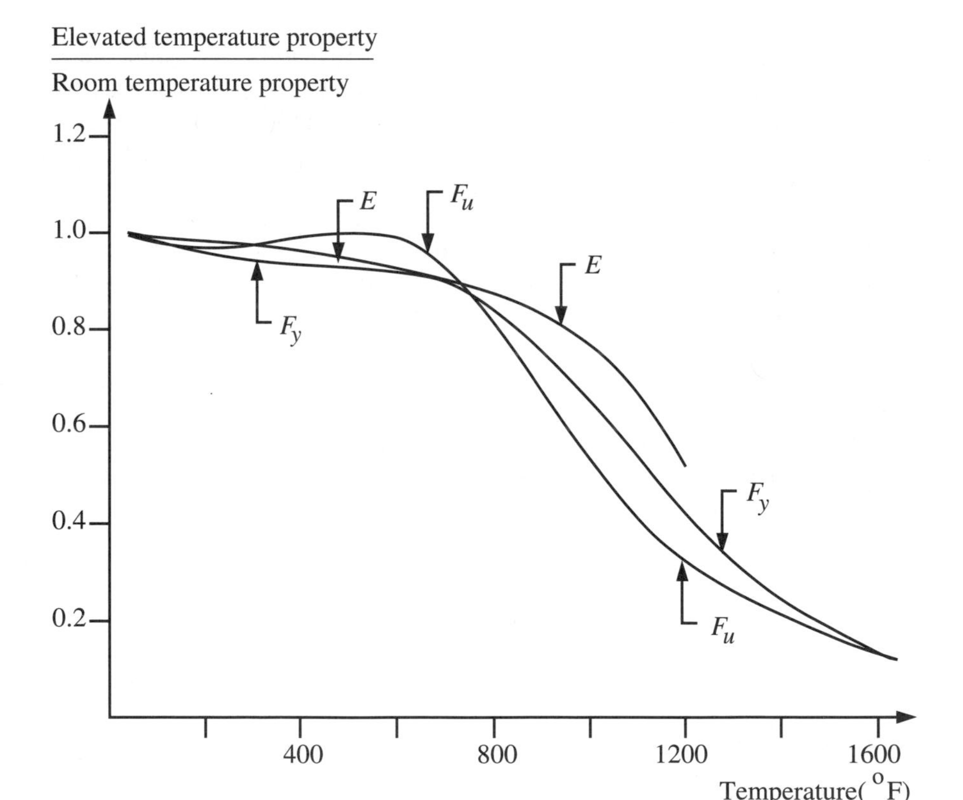

FIGURE 1.3 W section.

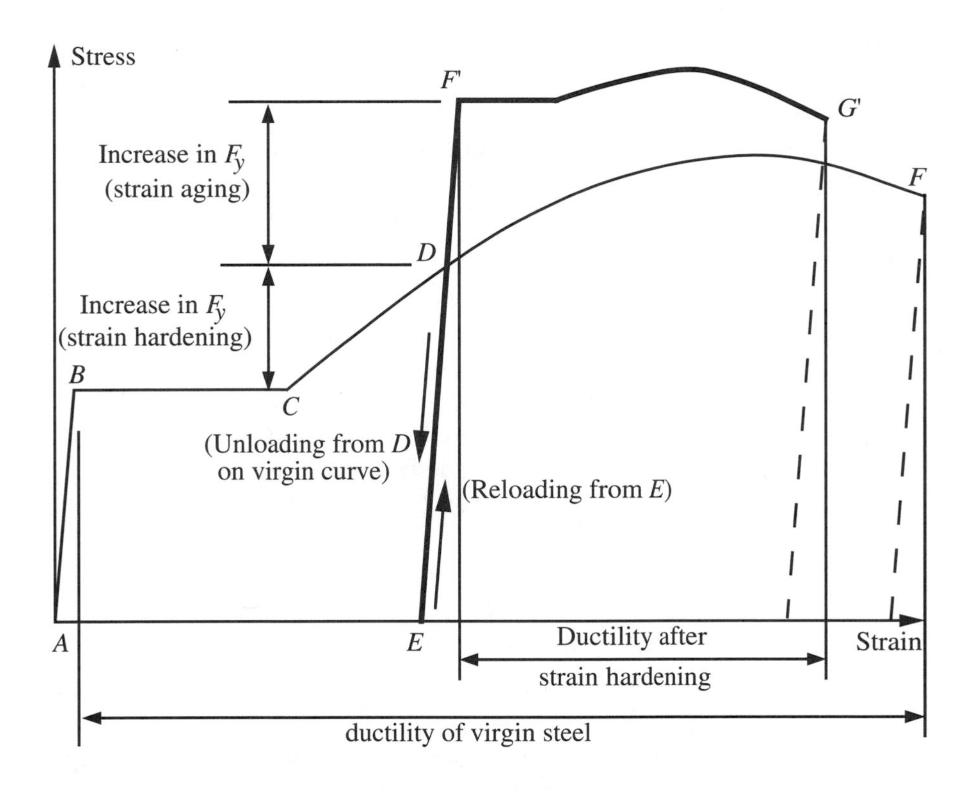

FIGURE 1.4 Tensile test of a W section member.

shown in Figure 1.4, ductility is significantly reduced after a structure has been overloaded into the strain-hardening region. Overloaded was chosen by the author as the descriptor since a building framework does not experience strains in the strain-hardening region under normal service condition loads except for severe earthquakes, for example. However, corners (bends of 90°or more at room temperature) in cold-formed steel sections are strained into the strain-hardening range.

Corrosion resistance increases as the temperature increases up to about 1000°F. In the welding process, a temperature of about 6500°F occurs at the electric arc tip of a welding electrode. Thus, high temperatures due to welding occur and subsequently dissipate in a member in the vicinity of welds. High-strength low-alloy steels have several times more resistance to rusting than carbon steels. Weathering steels form a crust of rust that protects the structure from further exposure to oxidation.

Weldability (relative ease of producing a satisfactory, crack-free, structurally sound joint) is an important factor in structural steel design since most connections in the fabrication shop are made by welding using automated, high-speed welding procedures wherever possible. The temperature of the electric arc increases as the speed of welding increases, and more of the structural steel mixes with the weld. Steels with a carbon content ≤ 0.30% are well suited to high-speed welding. Steels with a carbon content > 0.35% require special care during welding.

Members and their connections in a highway or railway bridge truss, for example, may be repeatedly loaded and unloaded millions of times during the life of the bridge. Some of the diagonal truss members may be in tension and later on in compression as a truck traverses the bridge. Even if the yield point of the steel in a member or its connections is never exceeded during the repeated loading and unloading occurrences, a fracture can occur and is called a *fatigue fracture*. Anything that reduces the ductility of the steel in a member or its connections increases the chances of a brittle, fatigue fracture. Thus, *fatigue strength* may dictate the definition of nominal strength of members and connections that are repeatedly loaded and unloaded a very large number of times during the life of the structure. Indeed, the life of a repeatedly loaded and unloaded structure may be primarily dependent on the fatigue strength of its members and connections.

1.1.5 Residual Stresses

Residual stresses exist in a member due to:

1. the *uneven cooling* to room temperature of a hot-rolled steel product,
2. *cold bending* (process used in straightening a crooked member and in making cold-formed steel sections), and
3. *welding* two or more sections or plates together to form a built-up section (e.g., four plates interconnected to form a box section).

Figure 1.5 shows a cross section of a steel rolled shape designated as a W section, which is the most common shape used in structural steel design as a *beam* (bending member), a *column* (axially-loaded compression member), and a *beam-column* (bending plus axial compression member).

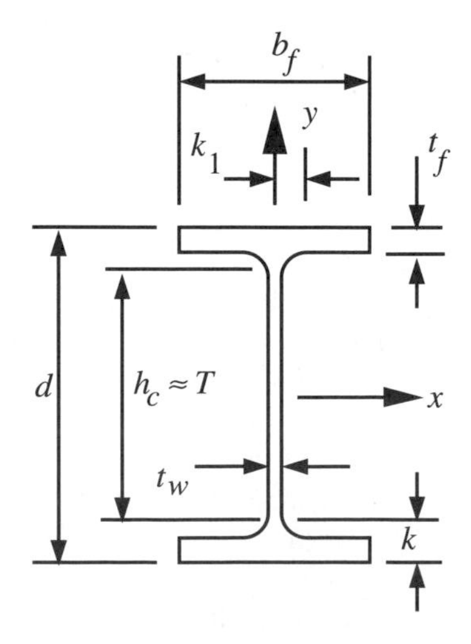

FIGURE 1.5 Local buckling and column buckling.

Consider a hot-rolled W section after it leaves the rollers for the last time. Consider any cross section along the length of the W section product. The flange tips and the middle of the web cool under room-temperature conditions at a faster rate than the junction regions of the flanges and the web. Steel shrinks as it cools. The flange tips and the middle of the web shrink freely when they cool since the other regions of the cross section have yet to cool. When the junction regions of the flanges and the web shrink, they are not completely free to shrink since they are interconnected to the flange tips and the middle of the web regions, which have already cooled. Thus, the last-to-cool regions of the cross section contain residual tensile stresses, whereas the first-to-cool regions of the cross section contain residual compressive stresses. These residual stresses, caused by shrinkage of the last-to-cool portions of the cross section and their being interconnected to regions that are already cool, have a symmetrical pattern with respect to the principal axes of the cross section of the W section. Therefore, the residual stresses are self-equilibrating and do not cause any bending about either principal axis of a cross section at any point along the length direction of the member. Residual stresses in a W section are in the range of 10 to 20 ksi, regardless of the yield strength of the steel.

1.1.6 Effect of Residual Stresses on Tension Member Strength

Consider a laboratory tension test of a particular W section member. Some W sections have the residual stress pattern shown in Figure 1.6(a), which illustrates that:

1. The *maximum residual compressive stress* f_{rc} occurs at the flange tips and at midheight of the web.
2. the *maximum residual tensile stress* f_{rt} occurs at the junction of the flanges and web.

The residual stresses vary through the thickness of the flanges and web. Cross-sectional geometry (flange thickness and width, web thickness and depth) influences the cooling rate and residual stress pattern. Some W sections are configured to be efficient as axial compression members, and other W sections are configured to be efficient as bending members. Depending on the cross-sectional geometry, some W sections have only residual tensile stresses in the web, with the maximum value occurring at the junction of the flanges and web. Furthermore, the magnitude of the residual stresses is smaller for quenched and tempered members. Thus, the residual stress pattern as well as average values of f_{rc} and f_{rt} through the thickness are dependent on several variables. Residual stress magnitudes on the order of 10 to 15 ksi or more occur if the member is not quenched and tempered.

As shown in Figure 1.6(c), in a tensile test of a W section member:

1. Fibers in the cross section begin to yield when $(f_{rt} + T/A_g) = F_y$; that is, at locations where the residual tensile stress and applied tensile stress add up to the yield stress.
2. All fibers in the cross section yield before the first-to-yield fibers begin to strain harden.

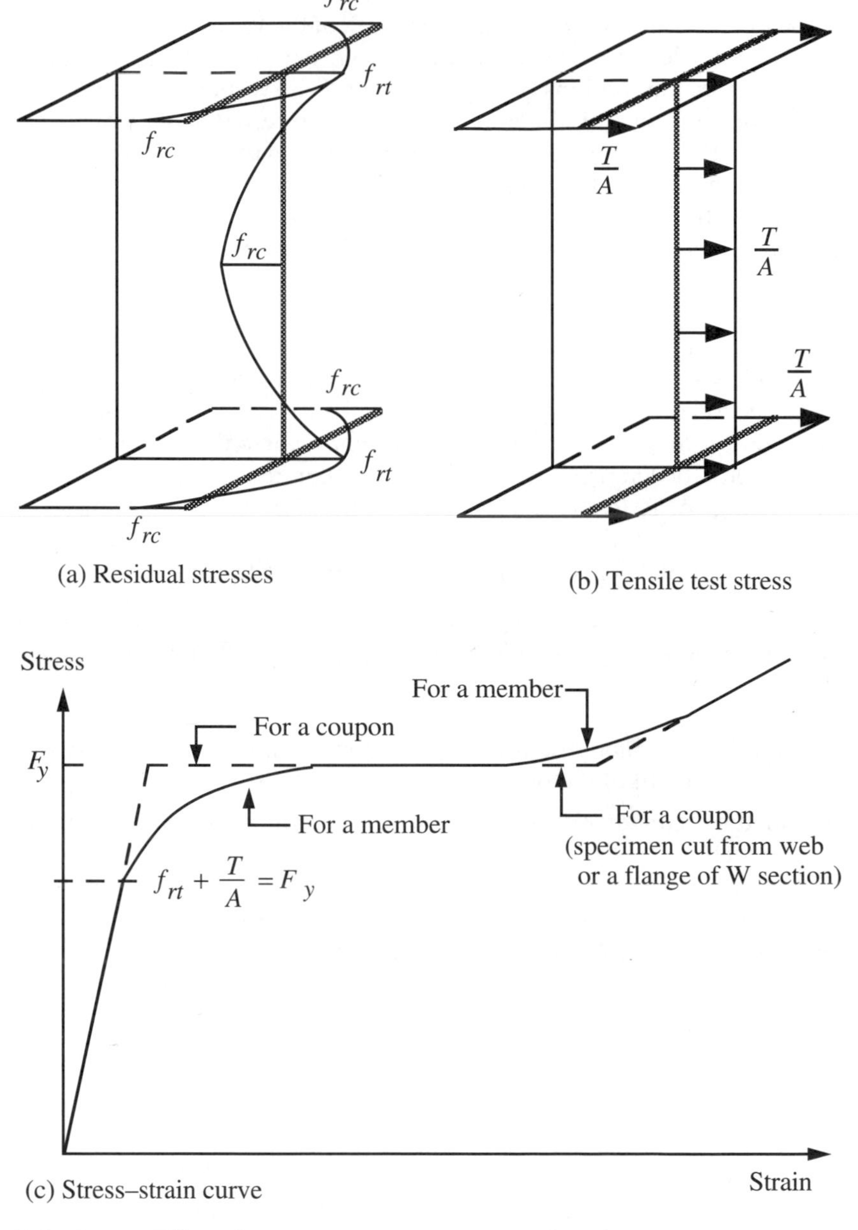

FIGURE 1.6 Effect of temperature on properties of steel.

Thus, the second phenomenon is the same condition that occurs in a coupon test specimen torch-cut from a flange or the web of a W section, properly machined, and laboratory tested to determine the stress-strain curve. Cutting the coupon specimen from the member's flange or web completely removes the residual stresses from the coupon specimen. From a comparison of the tension tests on the coupon and the member in Figure 1.6 (c), the yield strength is identical. Therefore, the tension member strength is not affected by the presence of residual stresses.

The fatigue strength of a tension member is determined by alternately loading (stretching) and unloading the member repeatedly until fracture occurs at the cross-sectional location where the tensile stress $(f_{rt} + T/A_g)$ is maximum during the loading cycle. Fatigue fractures can occur when the maximum tensile stress $(f_{rt} + T/A_g)$ is much less than F_y. Therefore, the fatigue strength of a tension member is affected by the presence of residual stresses.

1.1.7 Effect of Residual Stresses on Column Strength

In Figure 1.5, we see that a W section is I-shaped and composed of five elements (one vertical element and four horizontal elements). Each pair of horizontal elements is called a *flange* and the vertical element is called the *web*.

Suppose Figure 1.5 is the cross section of a column (an axially loaded compression member), of length L. Let A_g denote the cross-sectional area, and let P denote the axial compression force applied at each end of the column [see Figure 1.7(a)]. If a W section is used as a column, the cross section is composed of five *compression elements,* each of which is subjected to a uniform compressive stress of P/A_g. Each compression element in a cross section is classified as being either *stiffened* or *unstiffened* (projecting). A *stiffened compression element* is attached on both ends to other cross-sectional elements. An *unstiffened compression element* is not attached to anything on one end and is attached to another cross-sectional element on the other end. When a W section is used as a column, the web is a stiffened compression element and each flange is composed of two unstiffened compression elements.

Each of the five compression elements of the cross section in Figure 1.5 essentially is a rectangle. The longer side of the rectangle is the *width* and the other side is the *thickness*. Each compression element has a property known as the *width–thickness ratio* or b/t in mathematical terms. For each of the four unstiffened elements in Figure 1.5, $b = 0.5b_f$ and $t = t_f$, where b_f is the overall or total width of each top and bottom flange and t_f is the flange thickness. For the stiffened element (the web) in Figure 1.5, $b = h_c$ and $t = t_w$, where t_w is the thickness of the web and, for a W section, h_c is the clear height of the web.

If b/t does not exceed the limiting value stipulated in LRFD B5 (p. 6-36) for each compression element in a W section, local buckling does not occur before column buckling occurs. If b/t of a compression element exceeds the stipulated limiting value, local buckling of the compression element occurs as shown in Figure 1.7 before the column buckles and affects the column buckling strength.

Column buckling strength is affected by the presence of *residual stresses*. If a W section column buckles inelastically, the first-to-cool regions of the cross section yield in compression when $(f_{rc} + P/A_g) = F_y$. However, the last-to-cool regions of the cross section contain residual tension stresses and the applied compressive stress (P/A_g). Consequently, some portions of these last-to-cool regions of the cross section are still elastic when inelastic column buckling occurs; that is, $(-f_{rt} + P/A_g) < F_y$, where the negative sign indicates a tension stress and the compression stresses are positive.

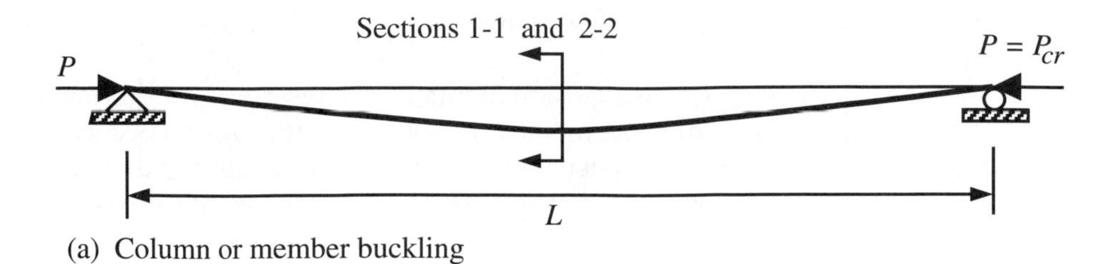

(a) Column or member buckling

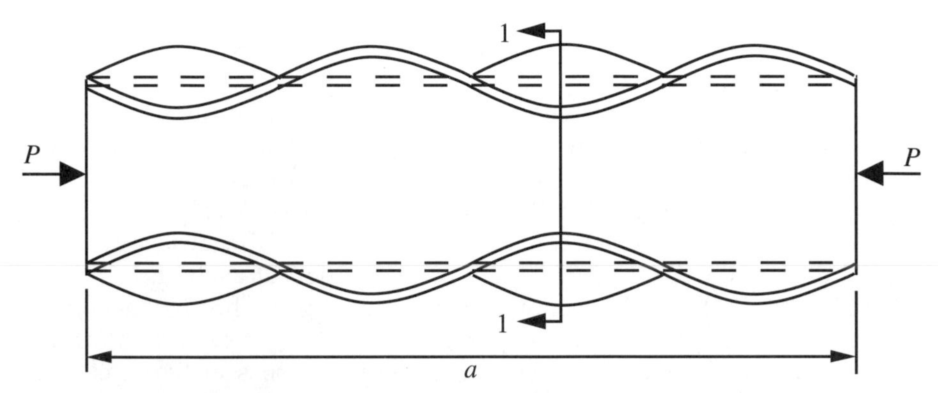

Number of half sine waves is a function of a/b and b/t of flange.

(b) Flange local buckling of a W section column

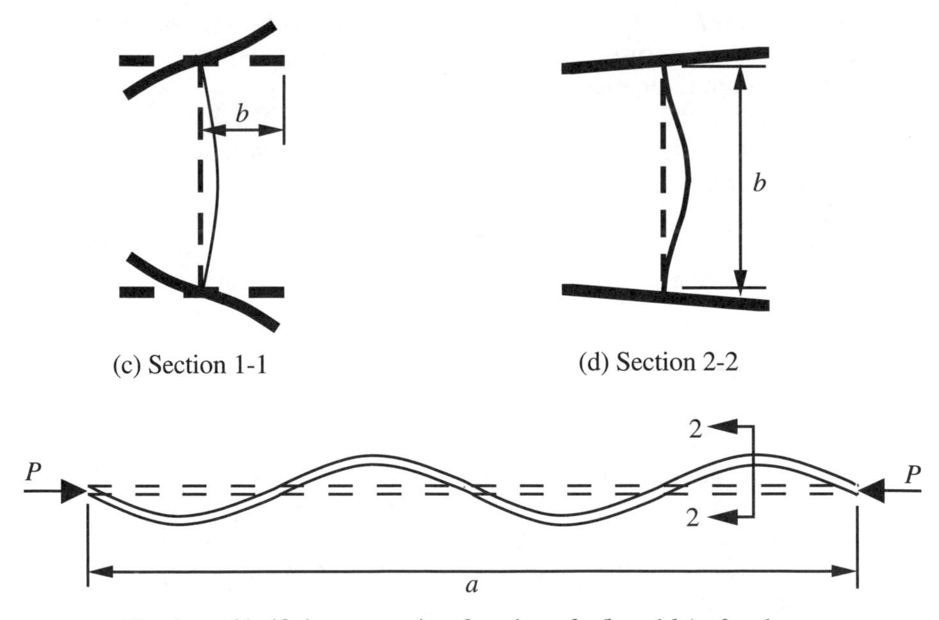

(c) Section 1-1

(d) Section 2-2

Number of half sine waves is a function of a/b and b/t of web.

(e) Web local buckling of a W section column

FIGURE 1.7 Stress–strain curves.

1.2 STRUCTURAL BEHAVIOR, ANALYSIS, AND DESIGN

A *structure* is an assembly of members interconnected by joints. A *member* spans between two joints. The points at which two or more members of a structure are connected are called *joints*. Each support for the structure is a boundary joint that is prevented from moving in certain directions as defined by the structural designer.

Structural behavior is the response of a structure to applied loads and environmental effects (wind, earthquakes, temperature changes, snow, ice, rain).

Structural analysis is the determination of the reactions, member forces, and deformations of the structure due to applied loads and environmental effects.

Structural design involves:

1. Arranging the general layout of the structure to satisfy the owner's functional requirements (for nonindustrial buildings, an architect usually does this part)
2. Conducting preliminary cost studies of alternative structural framing schemes and/or materials of construction
3. Performing preliminary analyses and designs for one or more of the possible alternatives studied in item 2
4. Choosing the alternative to be used in the final design
5. Performing the final design, which involves the following:
   (a) Choosing the analytical model to use in the analyses
   (b) Determining the loads
   (c) Performing the analyses using assumed member sizes that were obtained in the preliminary design phase
   (d) Using the analysis results to determine if the trial member sizes satisfy the design code requirements
   (e) Resizing the members, if necessary, and repeating items (c) and (d) if necessary
6. Checking the steel fabricator's shop drawings to ensure that the fabricated pieces will fit together properly and behave properly after they are assembled
7. Inspecting the structure as construction progresses to ensure that the erected structure conforms to the structural design drawings and specifications

Structural analysis is performed for structural design purposes. In the design process, members must be chosen such that design specifications for deflection, shear, bending moment, and axial force are not violated. Design specifications are written in such a manner that separate analyses are needed for *dead loads* (permanent loads), *live loads* (position and/or magnitude vary with time), *snow loads*, and effects due to *wind* and *earthquakes*. Influence lines may be needed for positioning live loads to cause their maximum effect. In addition, the structural designer may need to consider the effects due to fabrication and construction tolerances being exceeded, temperature changes, and differential settlement of supports. Numerical values of E and I must be known to perform continuous-beam analyses due to differential settlement of supports, but only relative values of EI are needed to perform analyses due to loads.

Structural engineers deal with the analysis and design of buildings, bridges, conveyor support structures, cranes, dams, offshore oil platforms, pipelines, stadiums, transmission towers, storage tanks, tunnels, pavement slabs for airports and highways, and structural components of airplanes, spacecraft, automobiles, buses, and ships. The same basic principles of analysis are applicable to each of these structures.

Architectural, heating, air conditioning, and other requirements by the owner impose constraints on the structural designer's choice of the structural system for a building. The owner wants a durable, serviceable, and low-maintenance structure, and possibly a structure that can be easily remodeled. The structural designer's choice of the structural framing scheme and the structural material are influenced by these factors. Sometimes, a special architectural effect dictates the choice of the material and framing scheme.

The engineer in charge of the structural design must

1. Decide how the structure is to behave when it is subjected to applied loads and environmental effects.
2. Ensure that the structure is designed to behave that way. Otherwise, a designed structure must be studied to determine how it responds to applied loads and environmental effects. These studies may involve making and testing a small-scale model of the actual structure to determine the structural behavior (this approach is warranted for a uniquely designed structure—no one has ever designed one like it before). Full-scale tests to collapse are not economically feasible for one-of-a-kind structures. For mass-produced structures such as airplanes, automobiles, and multiple-unit (repetitive) construction, the optimum design is needed, and full-scale tests are routinely made to gather valuable data that are used in defining the analytical model employed in computerized solutions.

Analytical models (some analysts prefer to call them mathematical models) are studied to determine which analytical model best predicts the desired behavior of the structure due to applied loads and environmental effects. Determination of the applied loads and the effects due to the environment is a function of the structural behavior, any available experimental data, and the designer's judgment based on experience.

A properly designed structure must have adequate *strength, stiffness, stability,* and *durability*. The applicable structural design code is used to determine if a structural component has adequate *strength* to resist the forces required of it, based on the results obtained from structural analyses. Adequate *stiffness* is required, for example, to prevent excessive deflections and undesirable structural vibrations. There are two types of possible *instability*:

1. A structure may not be adequately configured either externally or internally to resist a completely general set of applied loads.
2. A structure may buckle due to excessive compressive axial forces in one or more members.

Overall internal structural *stability* of determinate frames may be achieved by designing either truss-type bracing schemes or shear walls to resist the applied lateral loads. In the truss-type bracing schemes, members that are required to resist axial compression forces must be adequately designed to prevent buckling; otherwise, the integrity of the bracing scheme is destroyed. Indeterminate structural frames do not need shearwalls or truss-type bracing schemes to provide the lateral stability resistance required to resist the applied lateral loads. However, indeterminate frames can become unstable due to sidesway buckling of the structure.

In the course work that an aspiring structural engineer takes, the traditional approach has been to teach at least one course in structural analysis and to require that course as a prerequisite for the first course in structural member behavior and design. This traditional approach of separately teaching analysis and design is the proper one in our opinion, but in this approach, the student is not exposed to the true role of a structural engineer unless the student takes a structural design course that deals with the design of an entire structure. In the design of an entire structure, it becomes obvious that structural behavior, analysis, and design are interrelated. A bothersome thing to the student in the first design of an entire structure using plane frame analyses is the determination of the loads and how they are transferred from floor slab to beams, from beams to girders, from girders to columns, and from columns to supports. Transferral of the loads is dependent on the analytical models that are deemed to best represent the behavior of the structure. Consequently, in the first structural design courses, the analytical model and the applied loads are given information, and the focus is on structural behavior and learning how to obtain member sizes that satisfy the design specifications.

1.3 IDEALIZED ANALYTICAL MODELS

Structural analyses are conducted on an analytical model that is an idealization of the actual structure. Engineering judgment must be used in defining the idealized structure such that it represents the actual structural behavior as accurately as is practically possible. Certain assumptions have to be made for practical reasons: Idealized material properties are used, estimations of the effects of boundary conditions must be considered, and complex structural details that have little effect on the overall structural behavior can be ignored (or studied later as a localized effect after the overall structural analysis is obtained).

All structures are three-dimensional, but in many cases it is possible to analyze the structure as being two-dimensional in two mutually perpendicular directions. This text deals only with truss and frame structures. If a structure must be treated as being three-dimensional, in this text it is classified as being either a space truss or space frame. If all members of a structure lie in the same plane, the structure is a two-dimensional or planar structure. Examples of planar structures shown in Figure 1.8 are a plane truss, a beam, plane frames, and a plane grid. In Figure 1.8 each member is represented by only one straight line between two joints. Each joint is assumed to be a point that has no size. Members have dimensions of depth and width, but a single line is chosen for graphical convenience to represent the member spanning between two joints. Thus, the idealized structure is a line diagram configuration. The length of each line

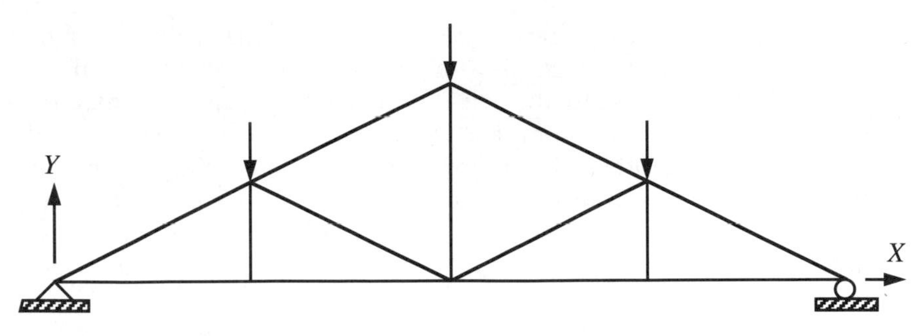

(a) Plane truss lying in XY plane

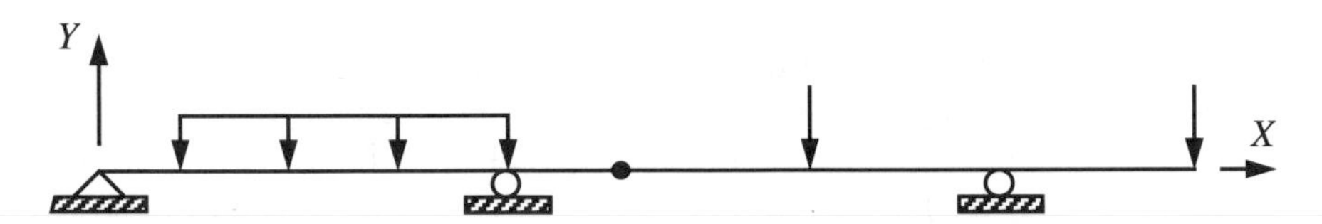

(b) Beam lying in *XY* plane

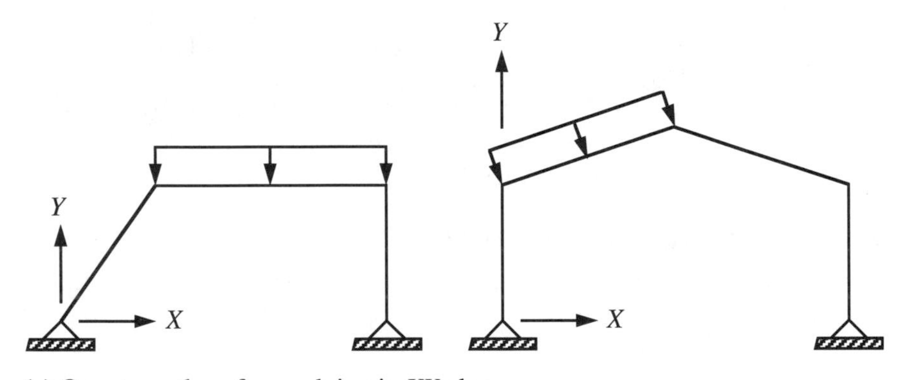

(c) One story plane frames lying in *XY* plane

FIGURE 1.8 Examples of planar structures.

defines the span length of a member, and usually each line is the trace along the member's length of the intersecting point of the centroidal axes of the member's cross section.

A *plane truss* [see Figure 1.8(a)] is a structural system of members lying in one plane that are assumed to be pin-connected at their ends. Truss members are designed to resist only axial forces and truss joints are designed to simulate a no moment resistance capacity.

A *plane frame* [see Figures 1.8 (b–d)] is a structural system of members lying in one plane. Each member end is connected to a joint capable of receiving member end moments and capable of transferring member end moments between two or more member-ends at a common point.

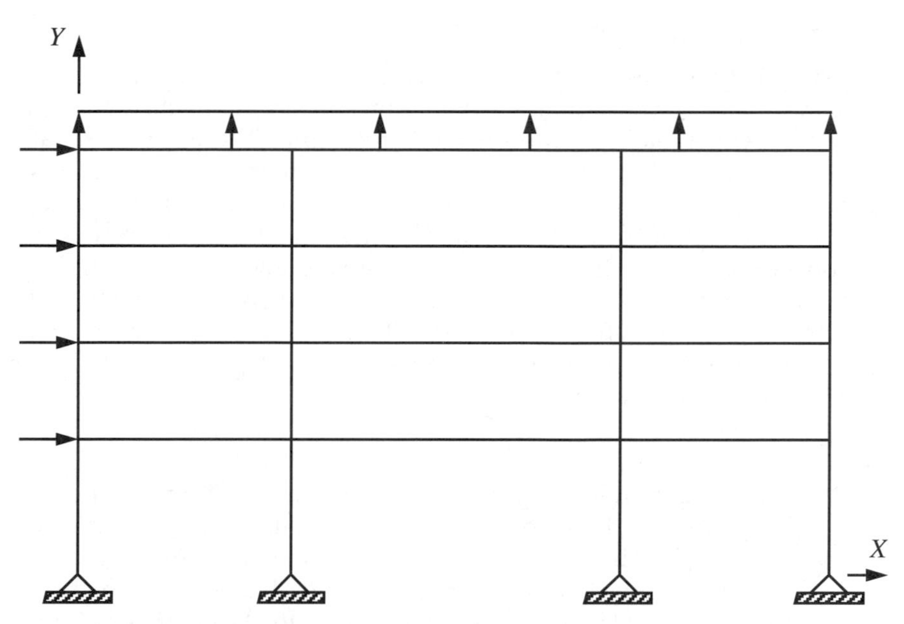

(d) Multistory, multibay plane frame lying in XY plane

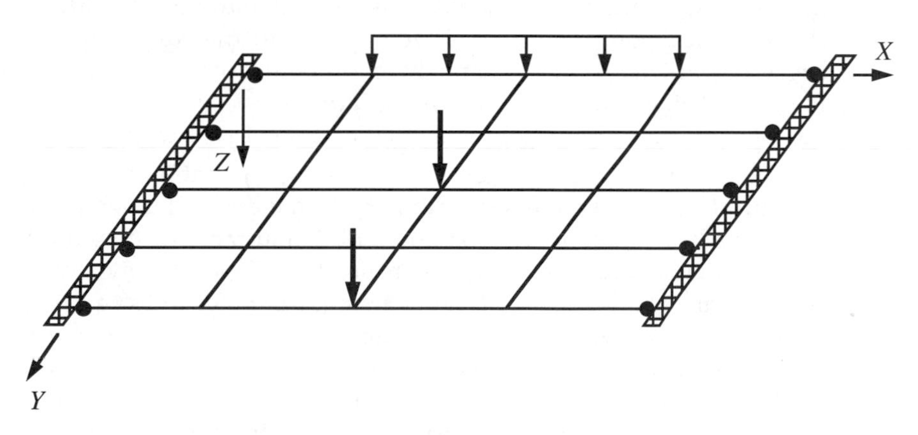

(e) Plane grid lying in XY plane—all loads in Z direction

FIGURE 1.8 (continued)

A *plane grid* [see Figure 1.8(e)] is a structural system of members lying in one plane that are connected at their ends to joints capable of receiving and transferring member-end moments and torques between two or more member ends at a common point.

Note that all members of a plane grid lie in the same plane, but all loads are applied perpendicular to that plane. For all other planar structures in Figure 1.8, all applied loads and all members of the structure lie in the same plane.

1.4 BOUNDARY CONDITIONS

For simplicity purposes in the following discussion, the structure is assumed to be a plane frame. At one or more points on the structure, the structure must be connected either to a foundation or another structure. These points are called *support joints* (or boundary joints, or exterior joints). The manner in which the structure is connected to the foundation and the behavior of the foundation influence the number and type of restraints provided by the support joints. Since the support joints are on the boundary of a structure and special conditions can exist at the support joint locations, the term *boundary conditions* is used for brevity to embody the special conditions that exist at the support joints. The various idealized boundary condition symbols for the line diagram structure are shown in Figure 1.9 and discussed in the following paragraphs.

A *hinge* [Figure 1.9(a)] represents a structural part that is pin-connected to a foundation that does not allow translational movements in two mutually perpendicular directions. The pin connection is assumed to be frictionless. Therefore, the attached structural part is completely free to rotate with respect to the foundation. Since many of the applied loads on the structure are caused by and act in the direction of gravity, one of the two mutually perpendicular support directions is chosen to be parallel to the gravity direction. In conducting a structural analysis, the analyst assumes that the correct direction of this support force component is either opposite to the direction of the forces caused by gravity or in the same direction as the forces caused by gravity. In Figure 1.9, the reaction components are shown as vectors whose arrow indicates our choice for the assumed direction of each vector.

A *roller* [Figure 1.9(b)] represents a foundation that permits the attached structural part to rotate freely with respect to the foundation and to translate freely in the direction parallel to the foundation surface, but does not permit any translational movement in any other direction. To avoid any ambiguity for a roller on an inclined surface [Figure 1.9(c)], we prefer to use a different roller symbol than used on a horizontal surface. A *link* is defined as being a fictitious, weightless, nondeformable, pinned-ended member that never has any loads applied to it except at the ends of the member. Some analysts prefer to use a link [Figure 1.9(d)] instead of a roller to represent the boundary condition described at the beginning of this paragraph.

A *fixed support* [Figure 1.9(e)] represents a bedrock type of foundation that does not deform in any manner whatsoever, and the structural part is attached to the foundation such that no relative movements can occur between the foundation and the attached structural part.

A *translational spring* [Figure 1.9(f)] is a link that can deform only along its length. This symbol is used to represent either a joint in another structure or a foundation resting on a deformable soil.

A *rotational spring* [Figure 1.9(g)] represents a support that provides some rotational restraint for the attached structural part, but does not provide any translational restraint. The support can be either a joint in another structure or

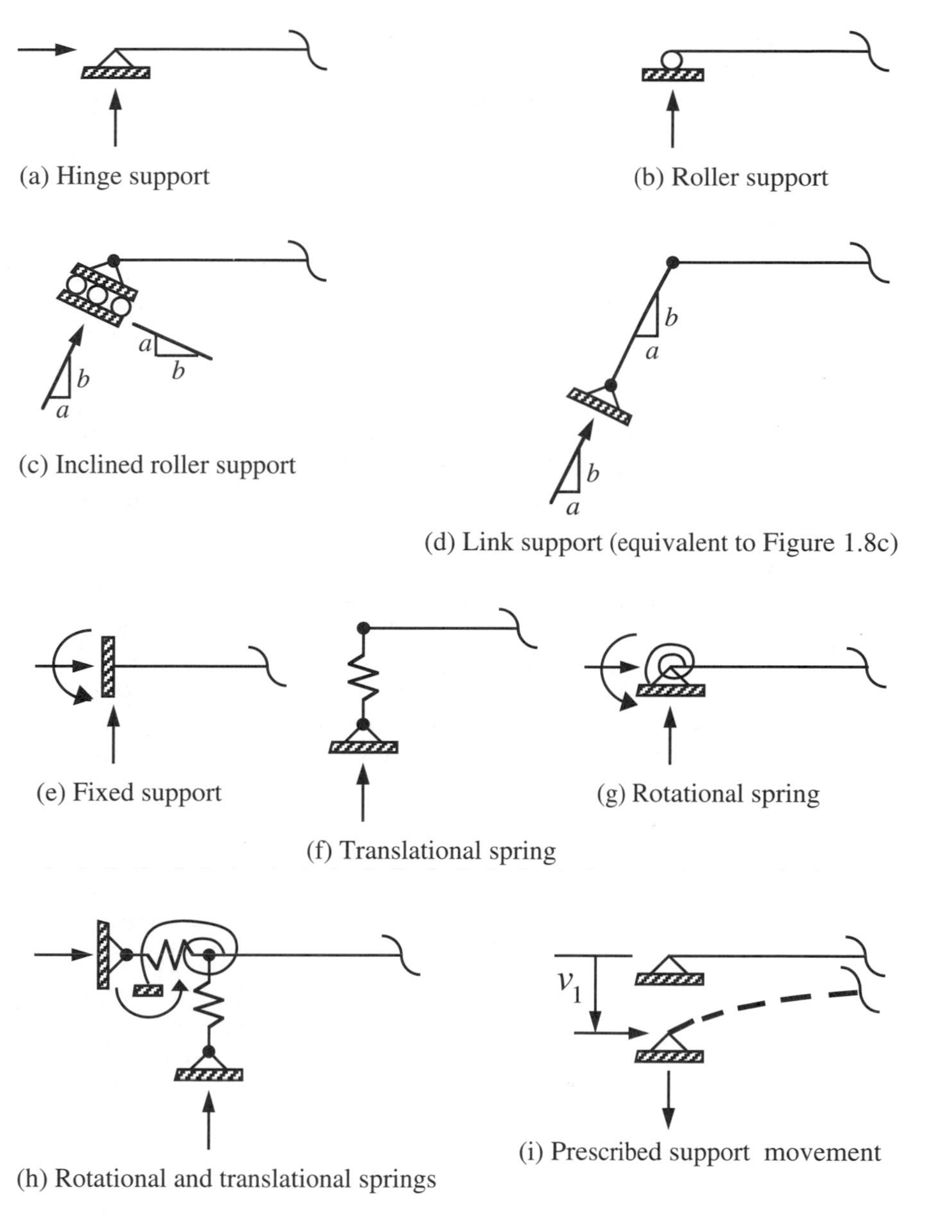

FIGURE 1.9 Boundary condition symbols and reaction components.

a foundation resting on a deformable soil. Generally, as shown in Figures 1.9(g) and (h), a rotational spring is used in conjunction with either a hinge, or a roller, or a roller plus a translational spring, or a translational spring, or two mutually perpendicular translational springs.

The soil beneath each individual foundation is compressed by the weight of the structure. Soil conditions beneath all individual foundations are not identical. The weights acting on the foundations are not identical and vary with respect to time. Therefore, nonuniform or differential settlement of the structure occurs at the support joints. Estimated differential settlements of the supports are made by the foundation engineer and treated as prescribed support movements by the structural engineer. Figure 1.9(i) shows a *prescribed support movement.*

1.5 INTERIOR JOINTS

For simplicity and generality purposes in the following discussion, the structure is assumed to be a plane frame. On a line diagram structure, an *interior joint* is a point at which two or more member length axes intersect. For example, in Figure 1.10, points 2, 4, 5, 7, 8, and 10 are interior joints, whereas points 1, 3, 6, and 9 are support joints (or exterior joints, or boundary joints).

The manner in which the member ends are connected at an interior joint must be accounted for on the line diagram. The types of connections for a structure composed of steel members can be broadly categorized as being one of the following types:

1. A *shear connection* develops no appreciable moment. If the connection at joint 10 of Figure 1.10 is as shown in Figure 1.11, it is classified by designers as being a shear connection. Thus, an internal hinge is shown on the line diagram at joint 10 of Figure 1.10 to indicate that no moment can be transferred between the ends of members 2 and 10 at joint 10. However, the internal hinge is capable of transferring translational-type member-end forces (axial forces and shears) between the ends of members 2 and 10 at joint 10. Note that this type of connection can transfer a small amount of moment, but the moment is small and can be ignored in design.
2. A *rigid connection* fully transfers all member-end forces. If the connection at joint 7 of Figure 1.10 is as shown in Figure 1.12, it is classified by designers as a joint that behaves like a rigid (nondeformable) body. Thus, if joint 7 of Figure 1.10 rotates 5° in the counterclockwise direction, the ends of members 1, 2, 8, and 9 at joint 7 also rotate 5° in the counterclockwise direction.
3. A *semirigid connection* is a partial member-end moment transferral connection. If the beam-to-column connection at joint 4 of Figure 1.10 is as shown in

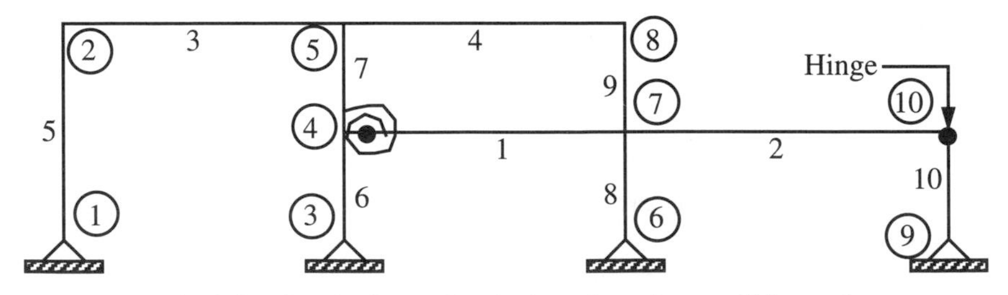

At the joint 4 end of member 1, there is an internal hinge plus a rotational spring spanning across the hinge.

FIGURE 1.10 Idealized interior joint conditions.

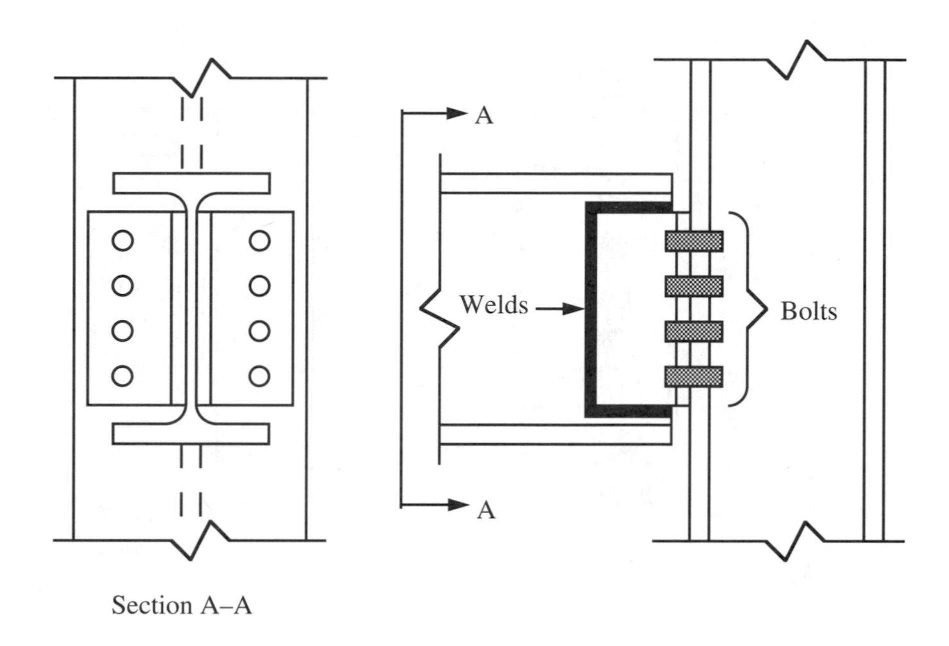

FIGURE 1.11 Web connection (shear connection).

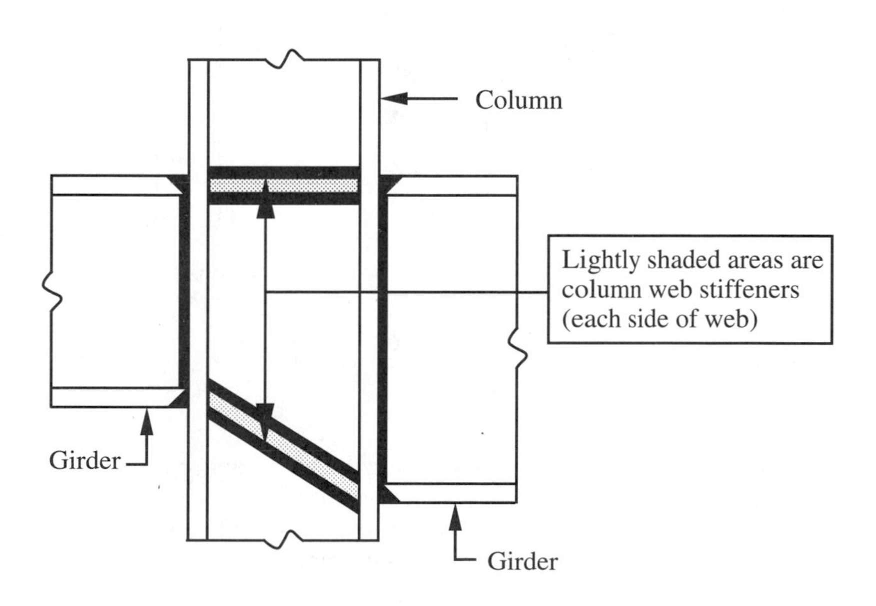

FIGURE 1.12 Rigid connection: fully welded plus web stiffeners.

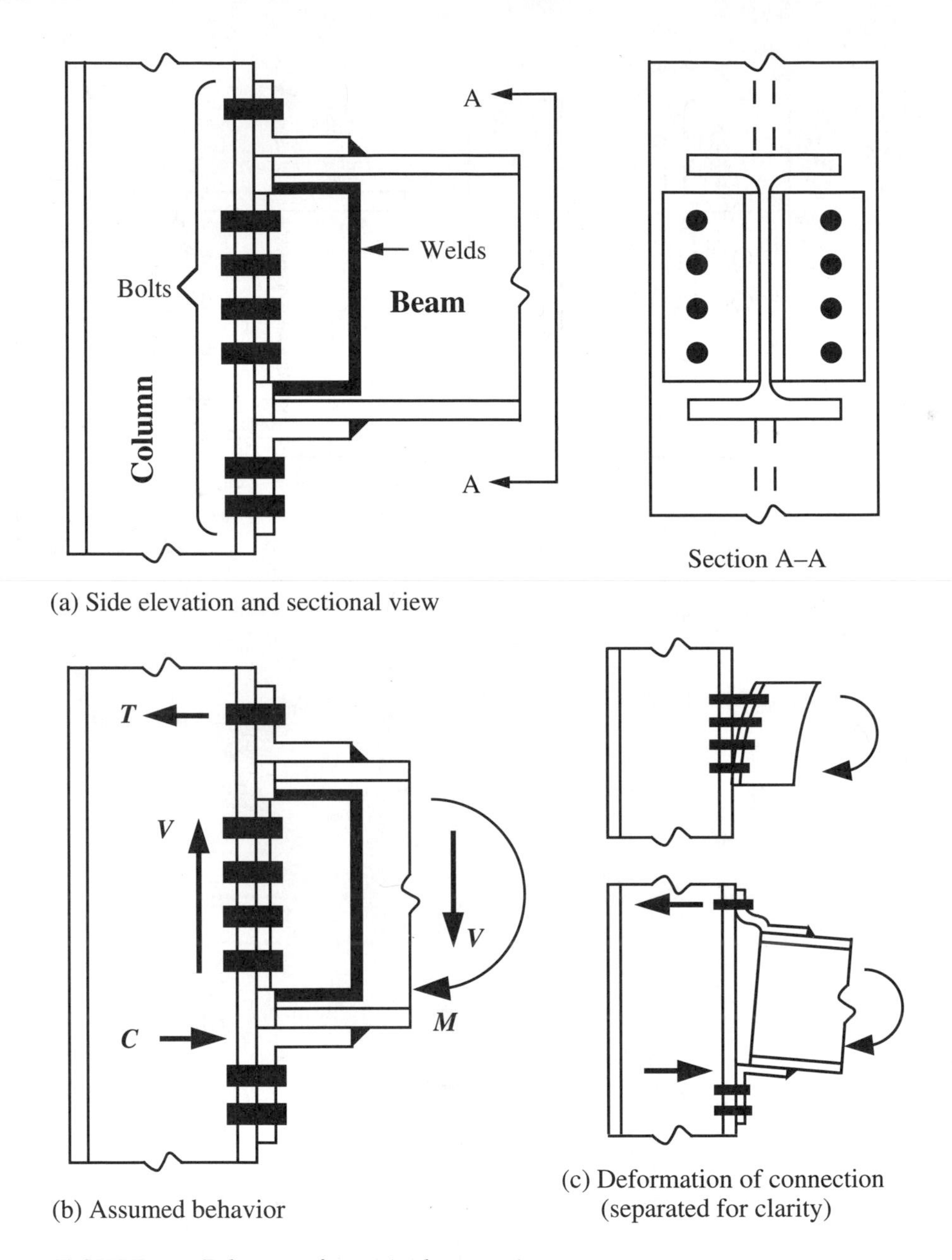

FIGURE 1.13 Behavior of semirigid connection.

Figure 1.13, it is classified by designers as being a semirigid connection. (Webster's dictionary definition of *semirigid* is "rigid to some degree or in some parts.") The top and bottom flange angles in Figure 1.13 transfer almost all of the beam-end moment to the column. The web angles in Figure 1.13 transfer almost all the beam-end shear to the column flange and fully ensure that the Y direction displacement at the end of member 1 is identical to the Y

> direction displacement of joint 4. (On the line diagram structure in Figure 1.10, joints 3, 4, and 5 lie on the same straight line that is the longitudinal axis of members 6 and 7. Thus, joint 4 is located at the point where the longitudinal axes of members 1, 6, and 7 intersect.) Consequently, joint 4 is treated as being rigid in the Y direction. However, the top and bottom flange angles in Figure 1.13 are not flexurally stiff enough to ensure that the flanges of member 1 always remain completely in contact with the flanges of members 6 and 7. Thus, joint 4 cannot be treated as being completely rigid. Therefore, at the left end of member 1 in Figure 1.10, a rotational (spiral) spring is shown to denote that a rotational deformation occurs between joint 4 and the end of member 1. It should be obvious that a semirigid connection is capable of developing more moment than a web connection can develop, but not as much moment as a rigid connection can develop.

In Figure 1.13, the angles are welded to the beam and bolted to the column. M effectively is transferred to the top and bottom flange angles. Consequently, due to the action of M, the top flange angle and the web angles flexurally deform, allowing the top beam flange to translate a finite amount [see Figures 1.13(c) and (d)]. However, the bottom flange angle remains in contact with the column flange. Thus, the gap between the end of the beam and the column flange is trapezoidal after the angle deformations occur. The bolts resist V and ensure that the beam end does not translate in the Y direction.

1.6 LOADS AND ENVIRONMENTAL EFFECTS

In structural analysis courses, the analytical model and the applied loads are given information, and the focus is on the applicable analysis techniques. In structural design, the loads that are to be applied to the analytical model of the structure must be established by the structural designer.

In this country, each state has a building code mandated by law that must be used in the design of an engineered structure. The building code gives minimum design loads that must be used in the design of a building to ensure a desired level of public safety unless the structural designer decides that higher design loads should be used. Coping with building codes and determining the applied loads are topics covered in a structural design course dealing with the design of an entire building. We choose to give only a brief description of loads and environmental effects. However, the terminology used in the discussion conforms to the terminology in the building code definitions for the loads and environmental effects.

All loads are treated as being statically applied to the structure, and the load classifications are dead loads, live loads, and impact loads. *Environmental effects* due to snow and ice, rain, wind, earthquakes, temperature changes, differential settlement of supports, misfit of members, construction tolerances, soil pressures, and hydrostatic pressures are converted into statically equivalent applied live loads.

There are three different types of loads: concentrated loads, line loads, and surface loads. *Concentrated loads* are applied on a relatively small surface area; examples are wheel loads of cranes, forklifts, and traffic vehicles (particularly on bridges). A *line load* is confined to a rather narrow strip in the structure; examples are member weights and partition wall weights. As the name implies, *surface loads* are distributed over a large area; examples are the weight of a floor slab or a roof, wind pressure on an exterior wall, and snow on a roof.

1.6.1 Dead Loads

Dead loads do not vary with time in regard to position and weight. Thus, they are not moved once they are in place and, therefore, are called dead loads. A worn floor or roof cover is removed and replaced with a new one in a matter of days. A load that is not there for only an interval of a few days in the 50-year life of a structure is considered to be a permanent load and is classified as a dead load. Examples are the weight of the structure; heating and air-conditioning ducts; plumbing; electrical conduits, wires, and fixtures; floor and roof covers; and ceilings. Since the weights of the indicated items are provided by their manufacturer, dead loads can be estimated with only a small margin of error.

1.6.2 Live Loads

Gravity loads that vary with time in regard to magnitude and/or position are called *live loads*. Examples of live loads are people, furniture, movable equipment, movable partition walls, file cabinets, and stored goods in general. Forklifts and other types of slow-moving vehicles (cranes in an industrial building and traffic vehicles in a parking garage, e.g.) are treated as live loads. An estimated maximum expected value of a live load contains a much larger margin of error than an estimated dead load.

Occupancy Loads for Buildings

Building codes specify minimum values that must be used for this classification of loads in the design of a building. Each designer must use at least the minimum values stated in the applicable building code. Some representative values of uniformly distributed live loads for this classification of loads are 40 psf for apartments, hotel rooms, and school rooms; 50 psf for offices in a professional building; 75 to 100 psf for retail stores; 100 psf for corridors on the exit floor level of public buildings (80 psf for corridors on other floor levels) and for bleachers in a sports arena; 150 psf for library stacks; and 250 psf for warehouses (floors and loading docks).

Traffic Loads for Bridges

Minimum loads for highway bridges are given in the *Standard Specifications for Highway Bridges* [12]. Designers usually refer to them as the AASHTO specs since they are published by the American Association of State Highway and Transportation Officials. A lane loading with a roving concentrated load as well as wheel loads for a standardized van and for a semitrailer truck are given in these specifications.

Minimum loads for railroad bridges are given in the *Specifications for Steel Railway Bridges* [13]. Designers usually refer to them as the AREA specs since they are published by the American Railway Engineering Association.

1.6.3 Roof Loads

In some of the loading combinations listed in LRFD A4.1 (p. 6-30), one of the independent loadings is shown as L_r or S or R, where L_r is roof live load, S is snow load, and R is load due to initial rainwater or ice exclusive of the ponding contribution. Some state building codes give minimum load values that must be used for each of these variables. Other state building codes give only a single minimum load value that must be used on the roof. For example, except for counties in either the coastal region or the mountainous region of North Carolina, 20 psf is given as the minimum load value that must be used on the roof. The coastal region counties are subject to hurricane rains and the mountainous region counties are subject to deeper snow accumulations. For each county in these regions, a minimum value greater than 20 psf is listed for either rain or snow.

Snow Loads

Snow loads corresponding to a 50-year mean recurrence interval are specified in most building codes. The minimum snow load value that must be used is either listed for each county or shown on a map with varying color shades and corresponding minimum snow load values for a group of counties. A 1 in. snow accumulation on a flat surface weighs about 0.5 psf at mountain elevations and weighs more at lower elevations. Snow loads in the range of 20 to 40 psf are commonly found as the minimum snow load value listed in building codes.

If the roof surface is not flat, a reduction factor that is a function of the roof slope may be given to convert the snow load specified for a flat roof to a value for a pitched roof. However, the snow load specified for a pitched roof is given as acting on a horizontal projection of the roof surface. Depending on the profile shape of the roof, the snow depth may not be constant over the entire roof surface. The deepest accumulations can be expected to occur in the roof valleys. Also, snow drifts can occur on a flat roof. If either a flat or sloped roof is below a higher roof on the same building or close enough to a roof on an adjacent taller building, snow can either blow off or slide off the higher roof onto the lower roof. Thus, snow drifts can be expected to occur on some roofs. A structural designer should account for these variations in the snow depth on the roof surface, even if the applicable building code does not explicitly state that such variations must be considered.

Rain or Ice Loads

Some building codes group ice loads with snow loads, but LRFD A4.1 (p. 6-30) groups ice loads with rain. Ice can accumulate on members in an exposed structure (bridges and signs, e.g.). An ice coating on such members increases the structural area exposed to wind. Thus, icing in such cases increases the wind-induced loads as well as the gravity direction loads.

If the drains for a flat roof become clogged or if rainwater accumulates faster than the drains can remove the water, ponding occurs, causing the roof to sag and to accumulate more water. Thus, rainwater on a flat roof causes more serious problems than snow. A slope of at least 0.25 in./ft is needed on the top surface of a flat roof for rainwater to drain properly. Furthermore, in hurricane-prone regions 120-mph winds occur with the heaviest rainfalls, push the rainwater on a flat roof to one side of the roof, and cause ponding. For these conditions, in addition to the primary roof drainage system, a secondary drainage system (*scuppers*, large holes in parapet walls) located above the primary drainage system can be installed to prevent water from accumulating above a certain level. These roofs are usually designed to resist rainwater loads for the rainwater elevation being at the elevation of the secondary drainage system plus 5 psf.

Roof Live Loads

Mobile equipment may be used on the roof either during construction or when the roof needs repair. Installation or replacement of an air-conditioning unit housed on the roof may require a portable crane to be hoisted to a flat roof and used to lift the unit into place. A flat roof may be used as an outdoor setting for a restaurant or as a helicopter port. These are possible sources of the L_r variable in LRFD A4.1 (p. 6-30).

1.6.4 Wind Loads

Wind on an enclosed building causes a pressure to occur on the windward vertical surface and a suction on the leeward vertical surface. Suction is actually an outward pressure—the atmospheric pressure inside the building is greater than the pressure on the outside of the leeward wall. Wind causes a suction (uplift) on flat ($\theta \leq 15°$) roof surfaces of an enclosed building. On a sloping roof with a mean-height/width ≤ 0.3 and $\theta > 15°$, wind causes pressure on the windward slope and a suction on the leeward slope.

Maximum wind speeds vary with geographical location (mountain tops and coastal regions prone to hurricanes may experience 120-mph winds), types of terrain (open, wooded, urban, proximity and shapes of nearby structures), height above the ground, air density, and other factors. Wind speed data are collected by the weather bureau at an elevation of 10 m (32.8 ft) above ground level. Formerly, a recorded wind speed was the speed for a mile of wind flowing past the recording device. Now, wind speeds are being recorded for a 3-sec-duration gust of wind, which is the familiar type of information given in the local TV weather news.

The *effects due to wind* are converted into an equivalent static pressure acting on the structure. Wind pressures based on the maximum wind speed for a 50-year mean recurrence interval are specified in most building codes. A *basic wind pressure* (function of the mass density of air and the wind velocity) is given in the building code either as a formula or in tabular form (pounds per square foot along the height direction of the building). Wind velocity is least at ground level and increases along the height direction of the building. *Shape factors* are given for buildings and components of buildings. The basic wind pressure is multiplied by the *building shape factor* and possibly other given factors to obtain a *design wind pressure* that is applied to the structure. For example, for an enclosed rectangular-shaped building, a shape factor of 1.3 (+0.8 on the windward surface and -0.5 on the leeward surface) is not uncommon. The design wind pressure distribution up the side of the building is determined and converted to wind loads acting on the structural framework accounting for the way the cladding is supported. In most cases, the wind loads are applied joint loads. Half of the wind load on a wall segment located between two adjacent floor levels and two adjacent column lines goes to the floor slab at the top of this wall segment, and the other half of the load goes to the floor slab at the bottom of this wall segment. The floor slab surrounds the columns and delivers the wind loads as concentrated loads on the columns at the floor levels (at the joints of the framework).

1.6.5 Earthquake Loads

The *effect of an earthquake* on a building is similar to the effect of a football player being clipped. For our purposes, say a clip is a hit around or below the knees and from the blind side. The football player is unaware that he is going to be hit. Consequently, his feet must go in the direction of the person who hits him, but his upper body does not want to move in that direction until the momentum of his lower body tends to drag the upper body in that direction. An earthquake consists of horizontal and vertical ground motions. The horizontal ground motion effect on a structure is similar to the football player being clipped. It is this type of motion that is converted into an equivalent static loading to simulate the effect of an earthquake on a building. An equivalent static loading (essentially a force $F = ma$ with modification factors accounting for seismic zone, type of occupancy, structural load-resisting characteristic, and soil–structure interaction conditions) is applied at all story levels and in the opposite direction of the ground motion since the foundation of the structure remains stationary in a static analysis. *All dynamic loads cannot always be replaced by equivalent static loads,* and a dynamic analysis of the structure subjected to time-dependent motions induced by an earthquake or rotating machinery should be conducted in such cases.

1.6.6 Impact Loads

An *impact load* is a live load that is increased to account for the dynamic effect associated with a suddenly applied load. Impact loads are applicable for cranes, elevators, reciprocating machinery, and vehicular traffic on highway or railroad bridges. LRFD A4.2 (p. 6-30) stipulates the percentage of increase in live loads to account for impact. LRFD A4.3 (p. 6-31) stipulates the horizontal and longitudinal crane forces that must be applied to the crane support beam to account for the effect of moving crane trolleys and lifted loads. Similar longitudinal forces are applied to highway and railway bridges to account for sudden stops of vehicles on a bridge.

1.6.7 Water and Earth Pressure Loads

If a structure has walls (or portions thereof) below the ground level, the active earth pressure must be applied to these walls. If a portion of a structure extends below the water table, water pressure must be applied thereon. Also, water pressure must be applied to dams and flumes.

1.6.8 Induced Loads

The effects due to *temperature changes, shrinkage, differential settlement of supports,* and *misfit of members* [1] are also converted into equivalent static loadings.

1.7 CONSTRUCTION PROCESS

If the framework of the structure is made of steel, the construction process involves the *fabrication, field erection,* and *inspection* of the erected structural steel. The general contractor chooses the shop to fabricate the steel and the subcontractor to do the field erection of the steel (in some cases, the general contractor erects the steel framework). Field inspection is done by an employee hired by the structural engineer and/or the architect. *Field inspection* is an integral part of the construction process and the final phase of the design process.

Fabrication involves interpreting design drawings and specifications, preparing shop fabrication and field erection drawings, obtaining the material from a steel mill if the needed material is not in the stockpile, cutting, forming, assembling the material into shippable units, and shipping the fabricated units to the construction site.

The fabricator cuts the main members to the correct length, cuts the connection pieces from larger pieces including steel plates, and either punches or drills the holes wherever bolted field connections are specified. A shearing machine is used to cut thin material, and a gas flame torch is used to cut thick material and main members unless extreme precision or a smooth surface is required, in which case the cut is made with a saw. If the design specifications do not tolerate as much crookedness in a member as the allowed steel mill tolerances, the fabricator reduces the amount of crookedness by using presses or sometimes by applying heat to localized regions of the member.

Bolt holes are made by punching, if possible, or drilling. The holemaking process may cause minute cracks or may make the material brittle in a very narrow rim around the hole. The LRFD B2 (p. 6-34) requires the structural designer to assume that the bolt hole diameter is 1/16 in. larger than the actual hole in order to account for the material that was "damaged" by the hole-making process.

The steel *field erection* contractor uses ingenuity and experience to devise an erection plan that involves lifting the fabricated units into place with a crane. Without a proper plan, lifting operations may cause compression forces to occur in members of a truss that were designed to resist only tension, for example. Also, improperly lifting a plate girder could cause local buckling to occur. Temporary bracing generally must be provided by the erection contractor to avoid construction failures due to the lack of three-dimensional or space frame stability. After permanent bracing designed by the structural designer, the roof, and the walls are in place, the structure has considerably more resistance to wind loads. Consequently, more failures due to wind loads occur during construction due to the lack of an adequately designed temporary bracing scheme by the erection contractor.

1.8 LOAD AND RESISTANCE FACTOR DESIGN

A building code for a state is prepared by a committee of experienced structural engineers and is mandated by law to be used in the design of a public building. The state building code defines minimum loads (live, snow, wind) for which the structure must be designed, but the structural designer may use larger loads if they are deemed to be more appropriate. These *service condition loads* are called *nominal loads* that are code-specified loads. In the LRFD approach, each nominal load is multiplied by a *load factor*. The factored loads are applied to the structure before performing structural strength analyses needed in the design process. Either an elastic analysis or a plastic analysis due to the factored loads is permitted. LRFD A4.1 (p. 6-30) requires the following load combinations to be investigated to find the critical combination of factored loads:

1. $1.4D$
2. $1.2D + 1.6L + 0.5\ (L_r \text{ or } S \text{ or } R)$
3. $1.2D + 1.6\ (L_r \text{ or } S \text{ or } R) + (0.5L \text{ or } 0.8W)$
4. $1.2D + 1.3W + 0.5L + 0.5\ (L_r \text{ or } S \text{ or } R)$
5. $1.2D \pm 1.0E + 0.5L + 0.2S$
6. $0.9D \pm (1.3W \text{ or } 1.0E)$

where

The numerical values are load factors.

D, L, W, L_r, S, and R are *nominal loads* (code-specified loads).

D is *dead load* due to the weight of the structural elements and permanent features on the structure.

L is *live load* due to occupancy and movable equipment.

W is *wind load*.

L_r is *roof live load*.

S is *snow load*.

R is load due to initial *rainwater* or *ice* exclusive of the ponding contribution.

Cross-sectional properties listed in the LRFD Manual for rolled sections are nominal values. Steel mills have + and - tolerances (see LRFD, p. 1–188) for the cross-sectional dimensions of a rolled shape. The permissible variation in area and weight is ±2.5% (see LRFD, p. 1–189). For a rolled shape that is used as a tension member with its ends welded to connections, for example, the limit of internal *resistance* (nominal strength) is the cross-sectional area times the yield strength of the steel. If bolted connections are used, fracture of the member in the connection region may govern the limit of internal resistance. To account for the uncertainty in the cross-sectional area and the steel properties, the *nominal strength* (resistance) is multiplied by a *resistance* (strength reduction) *factor* to obtain the *design strength* of a tension member. Since a mathematical statement of the design requirement for a tension member is more convenient than words, let:

1. ϕ = *resistance factor* (strength reduction factor)
2. P_n = *nominal strength* (resistance) for a tension member
3. P_u = *required tensile strength* (maximum axial tension force obtained from an elastic factored load analysis)

The LRFD Specification requires that $\phi P_n \geq P_u$.
Some examples of the *strength reduction factor* (resistance factor), ϕ, are:

1. $\phi_c = 0.85$ for axial compression
2. $\phi_v = 0.90$ for shear
3. $\phi_b = 0.90$ for flexure (bending moment)
4. $\phi_t = 0.90$ for yielding in a tension member
5. $\phi_t = 0.75$ for fracture in a tension member

The load and resistance factors in the LRFD Specification were developed using a probabilistic approach to ensure with a reasonable margin of safety that the maximum strength of each member and each connection in a structure is not less than the maximum load imposed on each of them. A portion of the margin of safety is in the load factors and the other portion is in the resistance factors.

In addition to being adequately designed for strength requirements, the structure must perform satisfactorily under nominal or service load conditions. Deflections of floor and roof beams must not be excessive. In the direction of wind, relative deflections of the column ends or story drift due to wind load must be controlled. Excessive vibrations cannot be tolerated. Thus, the structural designer must provide a structure that satisfies the owner's performance requirements and the safety requirement on strength as imposed by the applicable building code and LRFD Specification.

After the structure has been adequately *designed for strength*, the structural designer investigates the performance of the structure under service conditions. In addition to adequate *strength*, a member and the entire structure must have adequate *stiffness* for serviceability reasons. Many of the owner's serviceability requirements can be met by ensuring that deflections do not exceed acceptable limits. Some of the common serviceability problems are [3]:

1. Local damage of nonstructural elements (e.g., windows, ceilings, partitions, walls) occurs due to displacements caused by loads, temperature changes, moisture, shrinkage, and creep.
2. Equipment (e.g., an elevator) does not function normally due to excessive displacements.
3. Drift and/or gravity direction deflections are so noticeable that occupants become alarmed.
4. Extensive nonstructural damage occurs due to a tornado or a hurricane.
5. Structural deterioration occurs due to age and usage (e.g., deterioration of bridges and parking decks due to deicing salt).
6. Motion sickness of the occupants occurs due to excessive vibrations caused by
   (a) Routine occupant activities (floor vibrations).
   (b) Lateral vibrations due to the effects of wind or an earthquake.

In Table 1.2, these serviceability problems are categorized as a function of either the gravity-direction deflection or the lateral deflection.

It is customary steel design practice to limit the *deflection index* to:

1. $L/360$ *due to live load* on a floor or snow load on a roof when the beam supports a plastered ceiling.
2. $L/240$ *due to live load* or *snow load* if the ceiling is not plastered.
3. $h/667$ to $h/200$ for each story *due to the effects of wind* or *earthquakes*—only a range of limiting values can be given for many reasons (type of facade, activity of the occupants, routine design, innovative design, structural designer's judgment and experience).
4. $H/715$ to $H/250$ for entire building height H *due to the effects of wind* or *earthquakes*—comment in item 3 applies here too.

The *deflection index* limits for drift are about the same as the accuracy that can be achieved in the erection of the structure. The largest tolerable deflection due to live load is 0.5% of the member length. Consequently, deflected structure sketches are grossly exaggerated for clarity in textbooks.

To aid in the discussion of how the required strength of a member in a structure is determined from a factored load analysis, we choose to use the plane frame structure shown in Figures 1.14 and 1.15 (see Appendix A for the results obtained from a factored load analysis). This structure is a roof truss supported by two beam-columns (members 1 to 4 in Figure 1.15). A *beam-column* is a member that is subjected to axial compression plus bending. Behavior and design of beam-columns are discussed in Chapter 6. In Figure 1.15, members 1 to 4 and the roof truss ends are interconnected to provide resistance due to wind, as well as overall lateral stability of the structure for the gravity direction loads. In Figure 1.15, note the moment springs at the foundation ends of the columns. In the factored load analysis given in Appendix A, we assumed that the moment springs represented a boundary condition of half-way-fixed ($G = 2$ as explained in Chapter 6) due to gravity loads. To provide resistance due to wind perpendicular to the plane of Figure 1.15, some bracing scheme (see Figure 1.16 for an acceptable scheme) must be devised and designed.

For Figure 1.14, the nominal loads are:

1. *Dead*

   Built-up roof on metal decking = 8 psf

   Purlins = 20 lb/ft

   Truss = 0.15 kips at each interior joint

   Columns = 40 lb/ft

2. *Live* (crane loads)

   8.0 kips at joints 6 and 18

   16.0 kips at joint 12

Table 1.2 Deflection Index and Serviceability Behavior

Deflection Index	Typical Serviceability Behavior
$h/1000$	No visible cracking of brickwork.
$h/500$	No visible cracking of partition walls.
$h/300$ or $L/300$	Visible architectural damage. Visible cracks in reinforced walls. Visible ceiling and floor damage. Leaks in structural facade.
$L/200$ to $L/300$	Cracks are visually annoying.
$h/200$ to $h/300$	Visible damage to partitions and large. plate-glass windows.
$L/100$ to $L/200$	Visible damage to structural finishes.
$h/100$ to $h/200$	Doors, windows, sliding partitions, and elevators do not function properly.

Note:
L = span length of a floor or roof member,
h = story height.

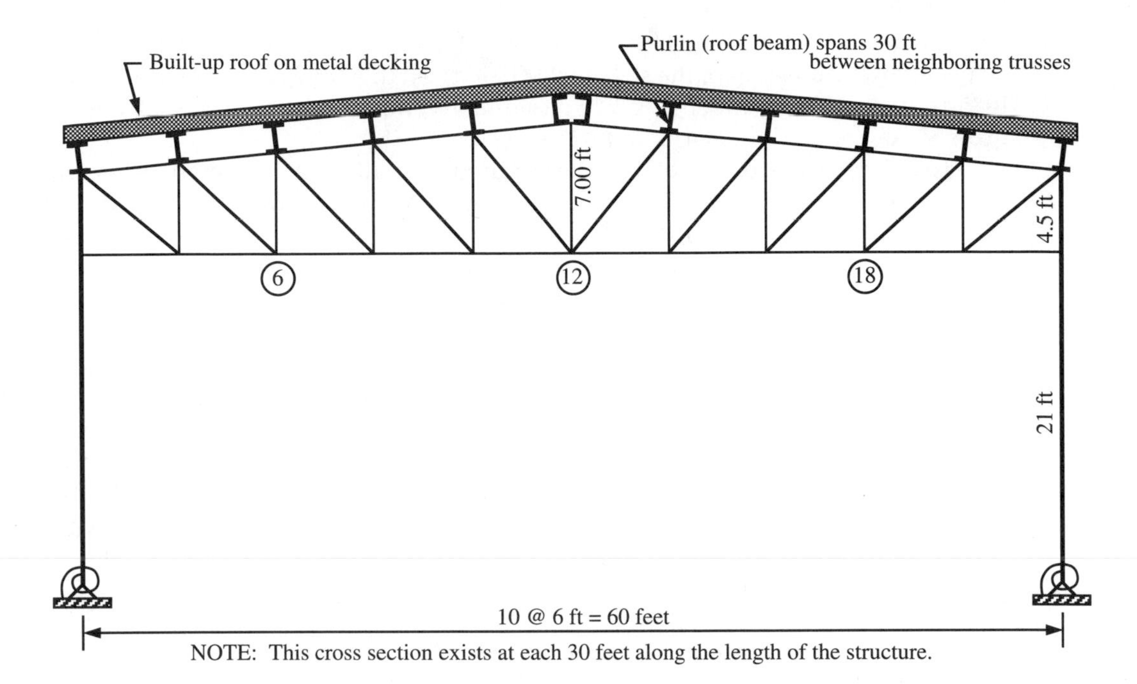

FIGURE 1.14 Cross section of an industrial building.

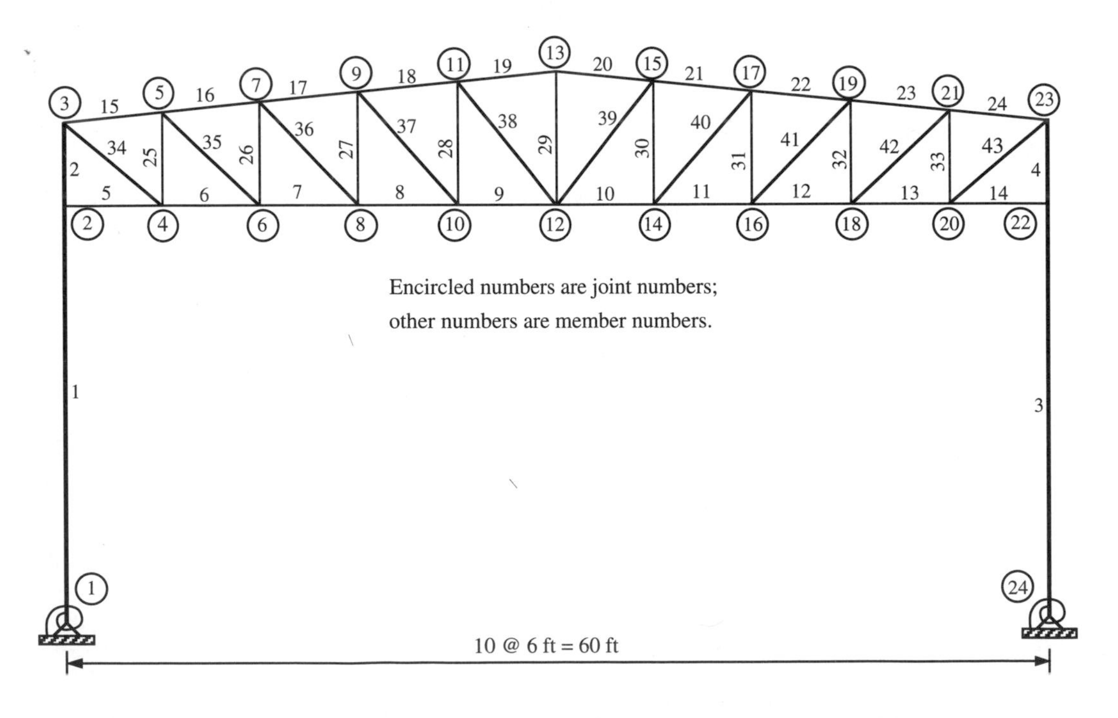

FIGURE 1.15 Joint numbers and member numbers for the structure in Figure 1.14.

3. *Snow*

 20 psf (perpendicular to horizontal surface)

4. *Wind*

 12 psf (pressure) on windward surface

 7.5 psf (suction) on leeward surface

 11 psf (suction) on roof surface

For Figure 1.15, the joint loads are given in Appendix A.

LRFD A4.1 (p. 6-30) load combinations that must be considered are:

$$1.4D$$

$$1.2D + 1.6L + 0.5\,(L_r \text{ or } S \text{ or } R)$$

$$1.2D + 1.6\,(L_r \text{ or } S \text{ or } R) + (0.5L \text{ or } 0.8W)$$

$$1.2D + 1.3W + 0.5L + 0.5\,(L_r \text{ or } S \text{ or } R)$$

$$0.9D + 1.3W$$

Discussions of the structure in Figures 1.14 to 1.16 are made in some other chapters of the text. Since the discussions will be related to the required strength of

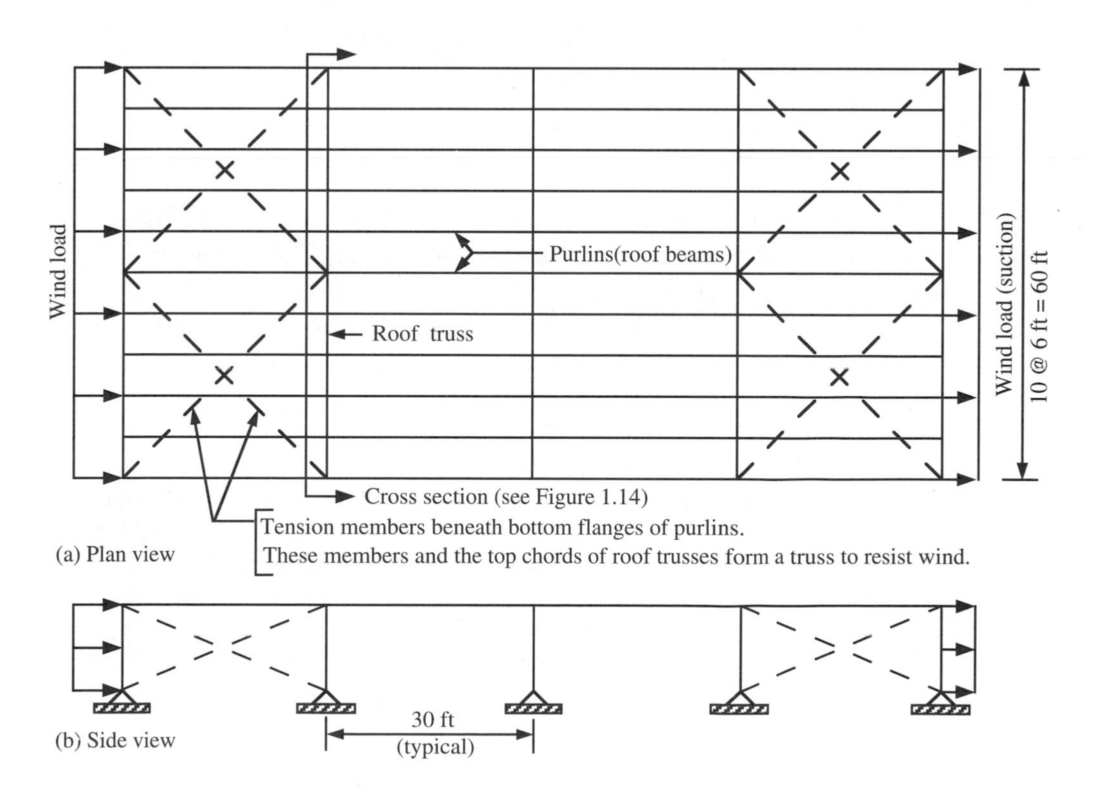

FIGURE 1.16 Side elevation view and plan view of building for Figure 1.14.

connections and members, the following examination of displacements for serviceability purposes is presented now. From Appendix A, due to nominal loads:

1. At joint 2 due to $0.9D + W$, $\Delta_x = 1.024$ in. $= 0.0853$ ft $= (h/246)$, where $h = 21$ ft. From item 3 of Table 1.2, a story-drift index of $h/246$ will be acceptable since $h/246$ lies in the range of $h/667$ to $h/200$.
2. Due to snow plus the crane loads, the vertical deflection at joint 12 is 1.041 in. $= 0.08675$ ft $= (L/692)$, where $L = 60$ ft. According to item 2 of Table 1.2, the *live-load deflection* should not exceed $(L/240) = 0.25$ ft and 0.087 ft is less than 0.25 ft. Consequently, based on the member properties used for the analysis of Figure 1.15, the truss has more than adequate stiffness for gravity loads.

1.9 STRUCTURAL SAFETY

The structural designer must provide a structure that satisfies the owner's performance requirements and the strength requirements stipulated by the applicable building code and LRFD Specification. Safety, serviceability, and economy are accounted for in designing a structure to fulfill the intended usage during the expected lifetime. A safe structure must perform satisfactorily under the expected loads with little or no damage and without injury to the occupants due to any structural malfunctions. For a properly designed structure, the probability of a partial or total collapse due to extreme accidental overloads must be very small. Since forecasting the future always involves some uncertainty, an absolutely safe structure during its expected lifetime cannot be designed. For example, a record-breaking rainfall, snowfall, windstorm, or earthquake may occur for the locale of the building. Thus, the actual loads on the building may exceed the maximum expected loads used in the design of the structure.

The first paragraph on LRFD Commentary, p. 6-169, is:

The LRFD Specification is based on (1) probabilistic models of loads and resistance, (2) a calibration of the LRFD criteria to the 1978 edition of the AISC ASD Specification for selected members, and (3) the evaluation of the resulting criteria by judgment and past experience aided by comparative design office studies of representative structures.

A brief discussion of the probabilistic model and calibration is presented later. However, *calibration to the 1978 AISC ASD Specification* is related to our discussion of structural safety. Hereafter, the 1978 AISC ASD Specification is referred to as the ASD (Allowable Stress Design) Specification.

Consider a plane truss whose member ends are welded to gusset plates (joints in the truss). Structural safety of a member in the truss is to be discussed in regard to the ASD and LRFD Specifications. Strength terminology in the LRFD Specification is in terms of forces, but the strength terminology in the ASD Specification is in terms of stresses. We choose to discuss the strength requirements of both specifications in terms of forces.

For a tension member in our plane truss, the ASD requirement for strength is

$$P_a \geq P_s$$

where

P_a = allowable tension force
P_s = maximum tension force

(P_s is determined from a truss analysis for the required ASD loading combinations of service loads applied at the truss joints.)

Let

$$P_y = A_g F_y$$

where

$$A_g = \text{gross cross-sectional area}$$

$$F_y = \text{yield stress of steel}$$

If the force in our tension member reaches P_y due to an extreme accidental overload, this is classified as a "failure" condition (excessively large deflections certainly will occur even though collapse may not occur). To ensure an adequate margin of safety against this failure condition, the ASD requirement for strength is

$$(P_a = 0.6P_y) \geq P_s$$

which can also be written as

$$\left(P_a = \frac{P_y}{\text{FS}} \right) \geq P_s$$

where

$$\text{FS (factor of safety)} = \frac{1}{0.6} = \frac{10}{6} = 1.67$$

The ASD Specifications from 1924 to 1989 did not give the basis for choosing the indicated FS = 1.67, but this choice can be rationalized as follows. $P_y = A_g F_y$ is a nominal value since neither A_g nor F_y is a perfect parameter. For example, the dimensions of a steel section can be manufactured only within acceptable tolerances (+ and -). Therefore, we should assume that the minimum failure strength of our truss tension member is less than the nominal value by an amount ΔP_y, which means that the minimum failure strength is $P_y - \Delta P_y$. We should assume that an extreme accidental overload of ΔP_s occurs that causes the force in our member to be $P_s + \Delta P_s$. The failure condition occurs when

$$(P_s + \Delta P_s) = (P_y - \Delta P_y)$$

$$P_s \left(1 + \frac{\Delta P_s}{P_s} \right) = P_y \left(1 - \frac{\Delta P_y}{P_y} \right)$$

and we obtain

$$FS = \frac{P_y}{P_s} = \frac{\left(1 + \dfrac{\Delta P_s}{P_s}\right)}{\left(1 - \dfrac{\Delta P_y}{P_y}\right)}$$

For

$$\frac{\Delta P_y}{P_y} = 0.1 \quad \text{and} \quad \frac{\Delta P_s}{P_s} = 0.5$$

we obtain

$$FS = \frac{1 + 0.5}{1 - 0.1} = \frac{1.5}{0.9} = 1.67$$

Values of FS > 1.67 were appropriately chosen by the ASD Specification writers for other failure conditions. Two examples where FS > 1.67 was chosen are:

1. For a tension member with bolted member-end connections, FS = 1/0.5 = 2.00 since the ASD Specification gives $P_a = 0.5A_eF_u$, where A_e is the effective net area and F_u is the specified minimum tensile strength of steel.
2. For elastic column buckling,

$$FS = \frac{23}{12} = 1.92$$

since the ASD Specification gives

$$P_n = \frac{P_{cr}}{\left(\dfrac{23}{12}\right)}$$

where P_{cr} is the elastic column buckling load.

The preceding discussion illustrates that structural design involves a forecast of the actual loads and the actual member strengths by estimating in some way that the chance of high loads and low strengths will occur. The main variables involved in our discussion of structural safety for a proposed steel structure are:

1. Strength
   (a) Stress–strain characteristics
   (b) Cross-sectional properties
   (c) Workmanship in the fabrication shop and in field erection
   (d) Structural deterioration due to repetitive loads (unloading and reloading) and corrosion, for example, particularly at the connection locations
   (e) Field inspection and quality control
   (f) Accuracy of the analysis and design calculations

2. Loads
   (a) Magnitude
   (b) Position
   (c) Duration
   (d) Load combinations
3. Consequences of a collapse
   (a) Loss of life
   (b) Property damage
   (c) Lawsuits and legal fees

The LRFD Specification accounts for the factors that influence strength and loads by using a probabilistic basis (involves probability theory and statistical methods). This probabilistic basis ensures a more consistent margin of safety than was the case in the ASD Specification. In order to discuss how structural safety is achieved in the LRFD Specification, we must define the pertinent LRFD terminology.

Each failure condition is referred to as a "limit state." A *limit state* is a condition at which a member, a connection, or the entire structure ceases to fufill the intended function. There are two kinds of limit states: *serviceability* and *strength*. Serviceability limit states deal with the functional requirements of the structure and involve the control of deflections, vibrations, and permanent deformations. Examples of strength limit states are yielding of a tension member, fracture of a tension member end, formation of a plastic hinge, formation of a plastic mechanism, overall frame instability, member instability, local buckling, and lateral-torsional buckling. Our discussion of structural safety is related to the strength limit states.

Again, consider our truss tension member; yielding of the member is the LRFD limit state chosen for our discussion. If we use a generalized notation, the discussion can be extended to a limit state of bending in a beam, for example. Let R denote the resistance (yield strength P_y) of our truss tension member. Let Q denote the axial force in the member due to a factored loading combination (for convenience, assume that only dead and live loads are involved). Structural safety is a function of R and Q, which are random variables. R is random due to acceptable tolerances (+ and -), which are necessary in the steel mill (steel-making and rolling processes), fabrication shop, and field erection. Dead load is random due to the uncertainty in the weight of the nonstructural elements (HVAC, sprinkler systems, electrical and communication systems, insulation, partitions, and ceilings) and the acceptable tolerances in the dimensions of steel members and concrete slab thicknesses, for example. Live load varies from structure to structure within a group of supposedly identical structures and varies as a function of time for each of these structures. Strength test data can be plotted as a histogram or frequency distribution of R (see Figure 1.17). For the data used to plot the histogram, statistical definitions can be used to obtain the *mean* (average) R_m and the *standard deviation* σ_R, which is a measure of the dispersion. Probability theory can be used to fit a continuous theoretical curve [*probability density function* (PDF) of R] to the histogram. Similarly, by using field measurement data of dead and live loads, a PDF of Q can be obtained.

If the PDF of R is normal or Gaussian, then the analytical form of the PDF of R is

$$f_R(x) = \frac{1}{\sqrt{2\pi}\,\sigma_R} \exp\left[-\frac{1}{2}\left(\frac{x - R_m}{\sigma_R}\right)^2\right]$$

where

$$R_m = \int_{-\infty}^{+\infty} x\, f_R(x)\, dx$$

$$\sigma_R^2 = \int_{-\infty}^{+\infty} (x - R_m)^2 f_R(x)\, dx$$

and the constant $1/\sqrt{2\pi}$ is used so that the normalized PDF of R encloses a unit area:

$$\int_{-\infty}^{+\infty} f_R(x)\, dx = 1$$

Between $(R_m - \sigma_R)$ and $(R_m + \sigma_R)$, the area under the PDF is 0.6827; that is, the probability of an occurrence within this range is 68.27%. The probability of an occurrence between $(R_m - 2\sigma_R)$ and $(R_m + 2\sigma_R)$ is 95.45%. The probability of an occurrence between $(R_m - 3\sigma_R)$ and $(R_m + 3\sigma_R)$ is 99.865%.

If the PDF of Q is normal or Gaussian, then the previous paragraph is applicable, provided that we replace each R with Q. If R and Q are normal random variables and if we define $Z = R - Q$ as the safety margin, then Z is a normal random variable. The probability of $Z < 0$ is the probability of failure (achievement of the limit state of

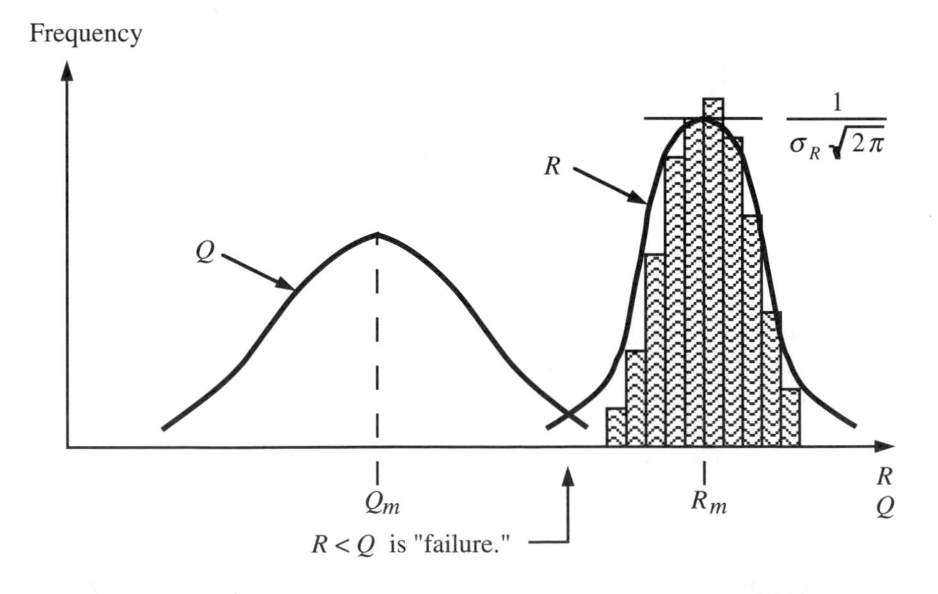

FIGURE 1.17 Frequency distribution of load Q and resistance R.

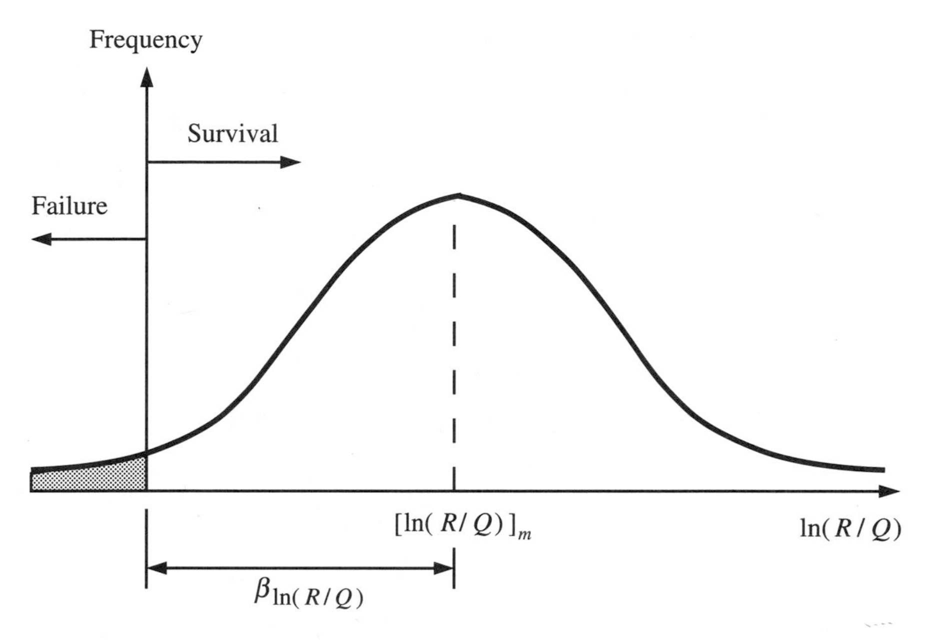

FIGURE 1.18 Reliability index *b*.

yielding) for our truss tension member. The equivalent representation of structural safety shown in Figure 1.18 was used in the LRFD probabilistic model. Figure 1.18 is a schematic plot of the PDF of ln (R/Q) and the shaded area is the probability of failure for our truss tension member. In Figure 1.18, $\beta\sigma_{\ln(R/Q)}$ is the distance from the origin to the mean [ln $(R/Q)]_m$. Using

$$\left[\beta\sigma_{\ln(R/Q)} \approx \beta\sqrt{V_R^2 + V_Q^2}\right] \approx \ln(R_m / Q_m)$$

where

$$V_R = \frac{\sigma_R}{R_m}$$

are the coefficients of variation, we obtain:

$$\beta \approx \frac{\ln(R_m / Q_m)}{\sqrt{V_R^2 + V_Q^2}}$$

and *b* is called the *reliability* (safety) *index*. Thus, in obtaining *b* for each strength limit state, a *first-order, second-moment* (FOSM) probabilistic model was used. This model only uses the mean values and coefficients of variation for the random variables involved in a particular strength limit state to obtain the reliability index *b*. For increasing values of β, the probability of failure (achievement of the limit state of yielding) for our truss tension member decreases.

The reliability index β is only a relative measure of safety and must be chosen to give the desired degree of reliability. For each LRFD strength criterion, β was selected by requiring that the LRFD criterion produce the same design as the corresponding ASD requirement for an "average" design situation. For example, for the LRFD loading combination consisting of gravity loads only, a live-load-to-dead-load ratio of 3 was used. This procedure of selecting b is called "calibration" to an existing design criterion (1978 AISC ASD Specification in this case). Therefore, the actual distribution shape of the PDF of R/Q was not required and the resistance factors were chosen such that for $D + L$ or S, targeted values of $b = 2.6$ for members and $b = 4.0$ for members were achieved. Thus, for our truss tension member example, the LRFD design requirement for the limit state of gross section yielding of the member became

$$\left(\phi P_n = 0.9 A_g F_y\right) \geq P_u$$

where

$$\phi P_n = \text{design tensile strength}$$

$$P_u = \text{required tensile strength}$$

(P_u is the maximum value of the tension member force obtained from a factored load analysis for all required LRFD load combinations.)

1.10 SIGNIFICANT DIGITS AND COMPUTATIONAL PRECISION

Estimated dead loads can be revised after all member sizes are finalized. However, estimated live loads and equivalent static loads for the effects due to wind and earthquakes are much more uncertain than the estimated dead loads. Perfection is impossible to achieve in the fabrication and erection procedures of the structure. Also, certain simplifying assumptions have to be made by the analyst to obtain practical solutions. For example, joint sizes are usually assumed to be infinitesimal, whereas they really are finite. Interior joints and boundary joints may be assumed to be either rigid or pinned, whereas they really are somewhere between rigid and pinned. Thus, the final structure is never identical to the one that the structural engineer designed, but the differences between the final structure and the designed structure are within certain tolerable limits.

A digit in a measurement is a significant digit if the uncertainty in the digit is less than 10 units. Standard steel mill tolerance for areas and weights is $\pm 2.5\%$ variation. Consider a rolled shape in the LRFD Manual with a weight of 100 lb/ft and a cross-sectional area of 29.4 in.^{2}. The weight variation tolerance is $\pm 0.025(100) = \pm 2.5$, and the actual weight lies between 97.5 and 102.5 lb/ft. Since the third digit in 100 is uncertain by only 5 units, the 100 lb/ft value is valid to 3 significant digits. However, the area variation tolerance is $\pm 0.025(29.4) = \pm 0.735$, and the actual area lies between 28.665 and 30.135 in.^{2}. There are 14.7 units of uncertainty in the third digit of 29.4, and ordinarily only the first 2 digits in the value 29.4 would be significant in engineering calculations, which means that 29.0 would be appropriate. In the LRFD approach, however, the uncertainty in the 29.4 in.^{2} area is accounted for in the resistance factor and it is appropriate to use the 29.4 in.^{2} value recorded in the LRFD Manual.

Most computers will accept arithmetic constants having an absolute value in the range 1×10^{-38} to 1×10^{38}. Some computers will accept a much wider range. A computer holds numeric values only to a fixed number of digits, usually the equivalent of between 6 and 16 decimal (or base 10) digits. The number of decimal digits held is called the precision of the arithmetic constant.

For discussion purposes, suppose that the loads and structural properties are accurate to only 3 significant digits. Most commercially available structural analysis software appropriately use at least 16-digit precision in the solution of the set of simultaneous equations. In a multistory building, there may be thousands of equations in a set. If only 3-digit precision were used in computerized solutions, the truncation and round-off errors in the mathematics would in some cases contribute as much uncertainty in the computed results as there is in the structural properties and loads.

Reconciliation of the actual number of significant digits in the computed results should be made by the structural designer after all of the computed results are available. We always make electronic calculator computations using the maximum available precision in the same manner that a computer would make them in floating-point form. We record our computed results with at least three-digit precision and rounded in the last digit.

PROBLEMS

Modify Figure 1.15 by deleting the columns (members 1 and 3) to obtain a determinate truss supported on a fixed hinge at joint 2 and supported on a roller at joint 22. Use the nominal loads given in Appendix A for Loadings 1 to 4 at joints 3 to 21 and 23. For the member specified in each of the following problems, from a truss analysis find the axial force for each nominal load case (D, L, S, and W). Then, see the factored load combinations required by LRFD(A4-1) to (A4-6) on LRFD p. 6-30. For the member specified in each of the following problems, use the previously found axial forces due to nominal loads and find the axial force for each LRFD load combination. Since a nominal load was not given for E(earthquake), ignore any load combination which is a function of E.

1.1 Find the axial force in member 33. Indicate which LRFD load combination gives P_u (the maximum axial tension and/or compression). Compare P_u to the corresponding P_u given in Appendix A.

1.2 Find the axial force in member 43. Indicate which LRFD load combination gives P_u (the maximum axial tension and/or compression). Compare P_u to the corresponding P_u given in Appendix A.

1.3 Find the axial force in member 10. Indicate which LRFD load combination gives P_u (the maximum axial tension and/or compression). Compare P_u to the corresponding P_u given in Appendix A.

1.4 Find the axial force in member 20. Indicate which LRFD load combination gives P_u (the maximum axial tension and/or compression). Compare P_u to the corresponding P_u given in Appendix A.

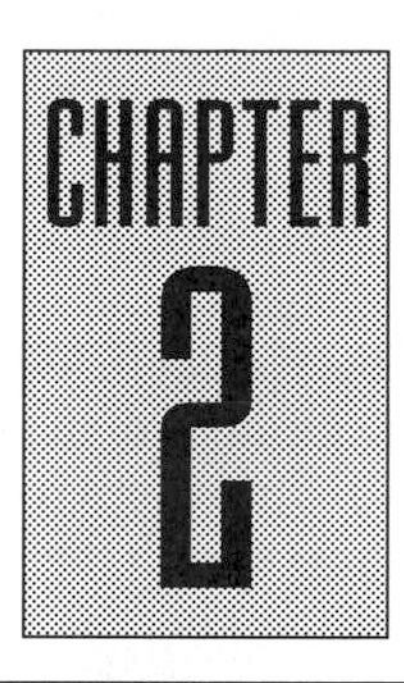

Tension Members

2.1 INTRODUCTION

A tension member is designed on the assumption that the member has to provide only axial tensile strength. Cables or guy wires are used as tension members to stabilize wood poles that support telephone and electricity transmission lines. Steel cables (wire ropes) and very slender rods ($L/d \leq 500$) have negligible bending stiffness. Thus, the assumption that a cable or very slender rod only provides tensile strength is indeed very reasonable. A tension member in a truss is fastened by welds or bolts at the member ends to either other members or connection plates (gusset plates). Truss members do not necessarily have negligible bending stiffness. Therefore, if the structural designer wants a structure to behave like a truss (all joints assumed to be pins and no bending occurs in any member), the design details must be chosen such that negligible bending occurs in each member. This means that the design details must (1) provide for all loads except the self-weight of the members to actually occur only at truss joints, and (2) ensure that the joints do not cause appreciable member-end moments to occur. If a structural designer wants a structure to behave in a certain manner, the structural details must be carefully chosen such that the desired structural behavior is closely approximated.

The LRFD definition of design strength is a *resistance factor* times the *nominal strength*. Since the resistance factor is less than unity, a resistance factor is a strength reduction factor. Separate LRFD design strength definitions are given for members, connectors (bolts, welds), and joints (angles, brackets, gusset plates, splice plates, stiffeners). The definitions of design strength for connection plates, fillet welds, and shear of bolts are given in Chapter 3.

In the LRFD approach, the members and connections of a structure are designed to have adequate strength to resist the factored loads imposed on the structure. For a tension member, the *design strength* is ϕP_n, where ϕ is a *resistance factor* (strength reduction factor), and P_n is the *nominal strength* (resistance) to be defined in the following discussion. For a tension member, the LRFD *design requirement for strength* is $\phi P_n \geq P_u$, where P_u is the *required tensile strength,* which is the maximum axial tension force obtained from an elastic factored load analysis.

After the structure has been adequately designed for strength, the structural designer investigates the performance of the structure under service conditions. If a tension member is too flexible (does not have enough bending stiffness), (1) special handling in the fabrication shop and in the field erection stages may be necessary, resulting in extra costs; (2) the member may sag excessively due to its own weight; and (3) in a building containing large machines with rotating parts or in a bridge truss exposed to wind, the member may vibrate too much. Thus, in addition to adequate *strength,* a member and the entire structure must have adequate *stiffness* for *serviceability* reasons. Many of the owner's serviceability requirements can be met by ensuring that deflections do not exceed acceptable limits (see Section 1.8).

2.2 STRENGTH OF A TENSION MEMBER WITH BOLTED-END CONNECTIONS

Consider Figure 2.1, which shows a tension member fastened by bolts to a gusset plate. Along the member at some finite distance from the bolts, all cross-sectional fibers of the member on Section 2-2 in Figure 2.1(d) can attain the yield strength, when the bolts and member-end connection are stronger than the member. If all cross-sectional fibers of a member yield in tension, the member elongates excessively, which can precipitate failure somewhere in the structural system, of which the tension member is a part. Consequently, *yielding on* A_g of the member is classified as a failure condition.

The member-end connection in Figure 2.1 is called a *bearing-type connection* since the transfer of axial force in the member to the gusset plate is made by bearing of the bolts at each bolt hole in the member end and gusset plate. Failure due to *shear of the bolts* and *bearing at the bolt holes* is discussed in Chapter 3, where we find that when failure is due either to shear of the bolts or to bearing at the bolt holes, each bolt in Figure 2.1 is assumed to transfer $P_u/3$ from the member to the gusset. Bearing at the bolt holes in the member end is shown in Figure 2.1(c), where P denotes the bearing (or pushing) force provided by each bolt.

In the connection region of Figure 2.1, the stress distribution due to the applied load is not uniform in the member since some of the cross-sectional elements of the member are not bolted to the gusset plate. Hence, a transition region exists in the member from the connection region to some finite distance from the connection where the stress distribution in the member becomes uniform when yielding occurs. Thus, before yielding occurs in the member, the regions of the member end and the gusset end containing the bolt holes usually experience strain-hardening, and fracture can occur through the bolt holes either in the member end or in the gusset plate. In Figure 2.1, fracture of the member end occurs on Section 3-3 that has full P_u (or all of P_u) on it. Fracture of the gusset end is discussed in Chapter 3.

In Figure 2.1(e), the cross section of the member is a pair of angles. Each angle section consists of two elements, and their centerlines form a capital "L". Sometimes in the LRFD Specification, the elements of an angle section are called legs. When some of the cross-sectional elements do not have bolts in them as in Figure 2.1, the force in the elements having no bolts must be shunted to the elements where bolts exist in order for the bolts to transfer their force to the gusset plate through bearing at the bolt holes. Removal of the force from an element that has no bolt in it is made

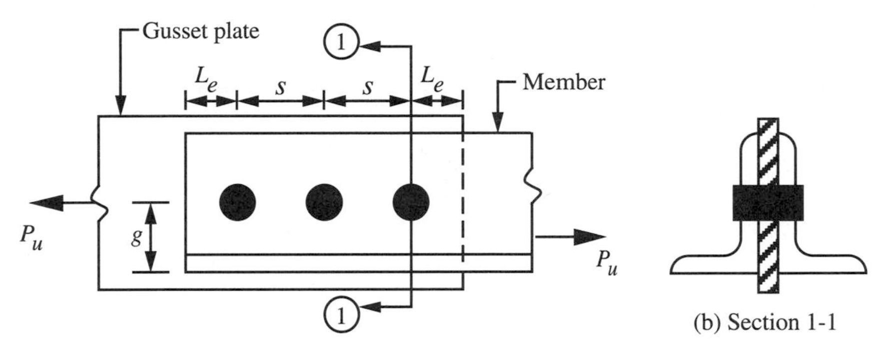

(a) Member end bolted to gusset plate

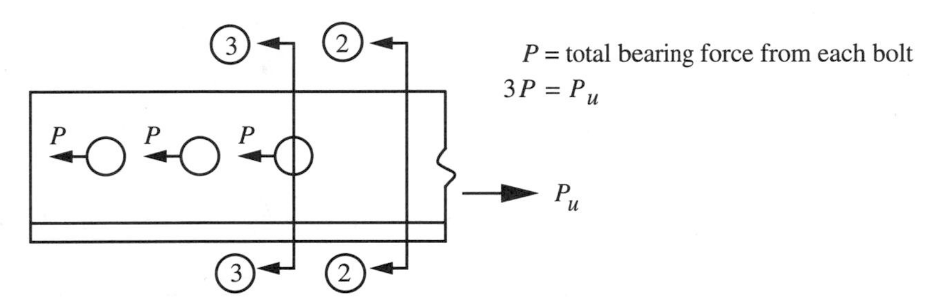

(c) FBD of member (pair of angles)

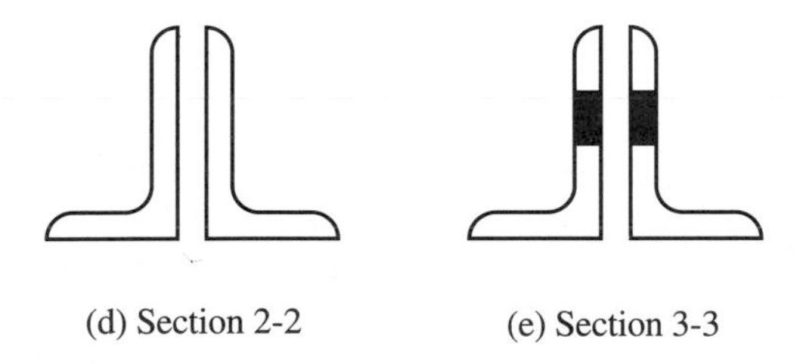

(d) Section 2-2 (e) Section 3-3

FIGURE 2.1 Tension member bolted to a gusset plate

through shear to an element that has a bolt in it, and the force removal process is called *shear lag*. When the shear lag has to occur in too short a distance on the member end, the fracture strength is smaller than when no shear lag exists. The LRFD definition of *fracture on* A_e (effective net area) accounts for any effect due to shear lag.

As shown in Figure 2.2(b), another failure mode of the member end can occur and is called *block shear rupture*. Note that a displaced block is pushed out of each angle leg that has bolts in it. When block shear rupture occurs, a portion of the resistance is due to shear P_v, and the other portion of the resistance is due to tension P_t. Fracture occurs on the plane of the block where the larger resistance exists and yielding is conservatively assumed to occur on the other plane of the block.

All bolt holes are standard holes except when the structural engineer specifies an oversized hole. In the fabrication shop, standard bolt holes are punched in the

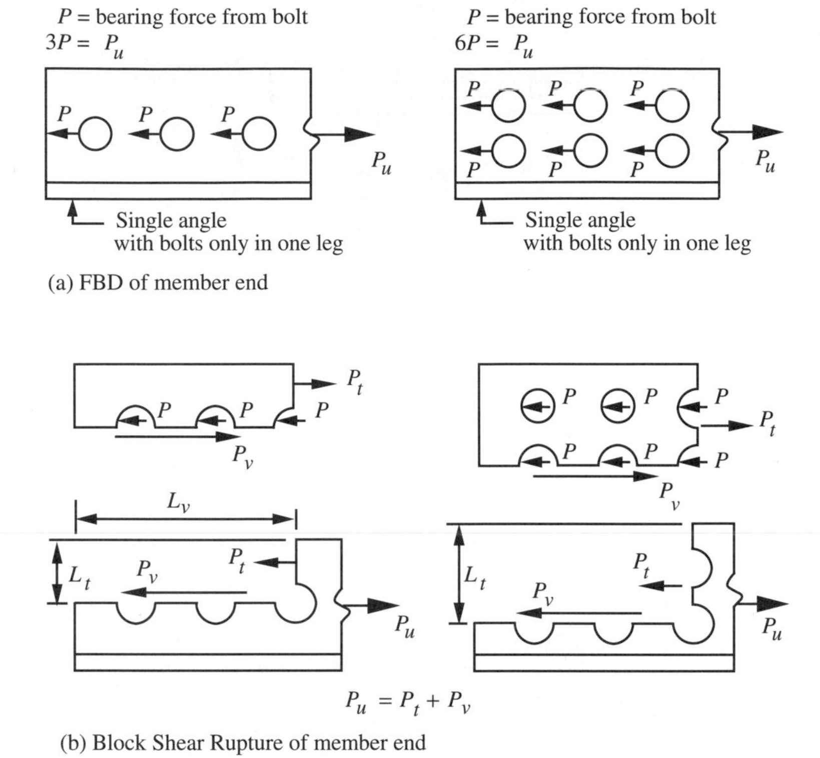

FIGURE 2.2 Examples of block shear rupture

member, when the material thickness does not exceed the hole diameter. The nominal diameter of a standard bolt hole (LRFD Table J3.3, p. 6-82) is the bolt diameter plus 1/16 in. In *strength* calculations, the hole diameter is defined (LRFD B2, p. 6-34) as the nominal diameter of the hole plus 1/16 in.

For each cited LRFD Specification, the reader should look in the LRFD Commentary to see if there is any information that may be helpful with the interpretation of that specification.

The LRFD design strength definitions of a tension member with bolted connections are:

1. *Yielding on* A_g [see Figures 2.1(c) and (d)] (LRFD D1, p. 6-44)

$$\phi P_n = 0.90 F_y A_g$$

where

$$F_y = \text{specified minimum yield stress (ksi)}$$

$$A_g = \text{gross area of member (in.}^2\text{)}$$

2. *Fracture on* A_e [see Figures 2.1(c) and (e)] (LRFD D1, p. 6-44, and B3, p. 6-34)

$$\phi P_n = 0.75 F_u A_e$$

where

F_u = specified minimum tensile stress (ksi)

A_e = effective net area of the critical section (in.^{2})

$A_e = A_n U$ when not all cross-sectional elements transfer some of P_u

$A_e = A_n$ when all cross-sectional elements transfer some of P_u

$A_n = A_g - \Sigma(t_h d_h)$ = net area of member (in.^{2})

t_h = thickness of hole (in.)

d_h = diameter of hole (in.)

$d_h = (d + 1/8$ in.) for a standard bolt hole

d = diameter of bolt (in.)

$U = (1 - \bar{x}/L_c) \leq 0.9$ = strength reduction factor

L_c = length of connection parallel to P_u (in.)

$\bar{x}$ = connection eccentricity (in.) (see Figure 2.3)

$U < 1$ when all cross-sectional elements of the member do not participate in transferring the tension force at the member ends through bearing to the bolts and then the bolts transfer their forces through bearing to the gusset plate. Figure 2.3 (adapted from LRFD Figures C-B3.1 and C-B3.2) illustrates how the connection eccentricity $\bar{x}$ is to be computed. For the supposed angle(s) to be used in Figures 2.3 (b–d) for computing $\bar{x}$ in the vertical direction, the implications are that the vertical leg of the supposed angle section is measured from the center of the indicated bolt hole to the free edge of the actual section. $\bar{x}$ in the vertical direction is measured from the center of the indicated bolt hole to the horizontal axis through the centroid of the supposed angle section. The strength-reduction factor for shear lag is based on the research conducted by Munse and Chesson [15] and Easterling and Giroux[16].

3. *Block shear rupture*, abbreviated BSR (LRFD J4.3, pp. 6-87 and 6-228) [see Figure 2.2(b) for some examples] When $F_u A_{nt} \geq 0.6 F_u A_{nv}$,

$$\phi R_n = 0.75(F_u A_{nt} + 0.6 F_y A_{gv})$$

When $0.6 F_u A_{nv} \geq F_u A_{nt}$,

$$\phi R_n = 0.75(0.6 F_u A_{nv} + F_y A_{gt})$$

where

A_{gv} = gross area on BSR shear plane(s), in.^{2}

A_{gt} = gross area on BSR tension plane, in.^{2}

$A_{nv} = A_{gv}$ - [A_{holes} on BSR shear plane(s)]

$A_{nt} = A_{gt}$ - [A_{holes} on BSR tension plane]

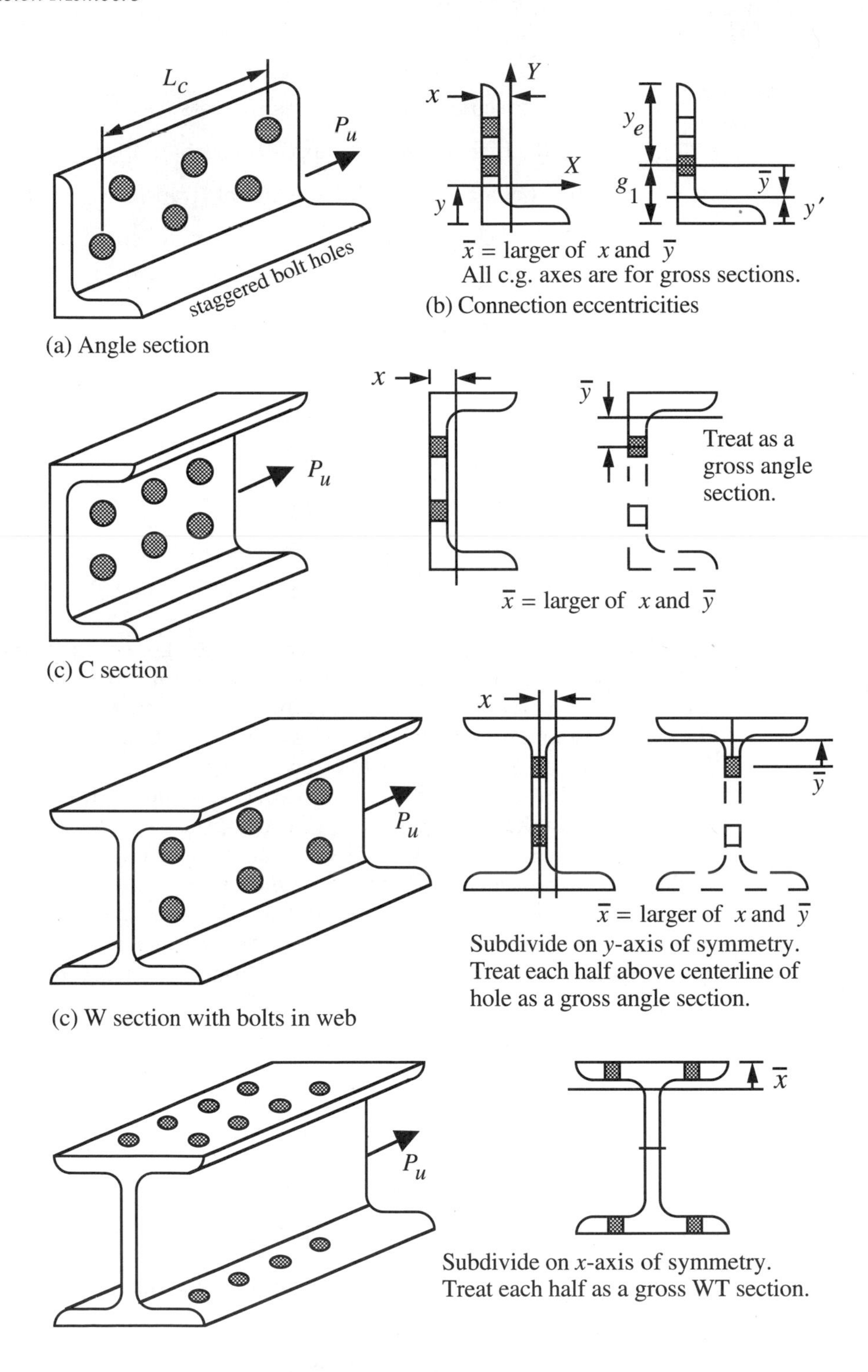

FIGURE 2.3 Connection eccentricity in shear lag reduction coefficient

Example 2.1

The A36 steel member in Figure 2.1(a) is a pair of angles, L3.5×3×0.25, with the long legs fastened to a gusset plate by 0.75-in.-diameter bolts. Use $L_e = 2$ in., $s = 3$ in., and $g = 2$ in. The gusset plate thickness is 0.5 in. Find the governing design strength of the tension member for a bearing-type bolted connection.

Solution

LRFD, p. 1-62: For a single L3.5 x 3 x 0.25,

$A = 1.56$ in.^{2} $x = 0.785$ in.; $y = 1.04$ in.

The governing design strength of the tension member is the least ϕP_n value obtained from:

1. *Yielding on* $A_g = 2(1.56) = 3.12$ in.^{2} [see Figures 2.1(c) and (d)]

$$\phi P_n = 0.90 F_y A_g = 0.90(36)(3.12) = 101 \text{ kips}$$

2. *Fracture on* $A_e = A_n U$ [see Figures 2.1(c) and (e)]

$$d_h = d + 1/8 \text{ in.} = 0.75 + 0.125 = 0.875 \text{ in.}$$

$$t_h = 2(0.25) = 0.500 \text{ in.}$$

$$d_h t_h = 0.875(0.500) = 0.4375 \text{ in.}^2$$

$$A_n = A_g - d_h t_h = 3.13 - 0.4375 = 2.69 \text{ in.}^2$$

From Figure 2.3(b), we see that:

(a) the connection eccentricity from the face of the gusset plate is $x = 0.785$ in.
(b) the other connection eccentricity for a supposed long leg $= g$ must be computed.

$$y_e = (leg - g) = (3.5 - 2) = 1.50 \text{ in.}$$

$$A' = A - y_e t = 1.56 - (1.50)(1/4) = 1.185 \text{ in.}^2$$

$$y' = \frac{Ay - y_e t(leg - 0.5y_e)}{A'} = \frac{(1.56)(1.04) - 1.50(0.25)(3.5 - 1.50/2)}{1.1875} = 0.499 \text{ in.}$$

$$\bar{x} = \text{larger of} \begin{cases} x = 0.661 \text{ in.} \\ g - y' = 2 - 0.499 = 1.50 \text{ in.} \end{cases}$$

$$U = \text{smaller of} \begin{cases} 1 - \bar{x}/L_c = 1 - 1.50/6 = 0.750 \\ 0.9 \end{cases}$$

$$A_e = A_n U = 2.69(0.750) = 2.02 \text{ in.}^2$$

$$\phi P_n = 0.75 F_u A_e = 0.75(58 \text{ ksi})(2.02 \text{ in.}^2) = 87.8 \text{ kips}$$

3. *Block shear rupture* [see Figure 2.2(b), one-row-of-bolts case]

$$L_v = L_e + 2s = 2 + 2(3) = 8 \text{ in.}$$

$$A_{gv} = L_v t = 8(2)(0.25) = 4.00 \text{ in.}^2$$

$$A_{nv} = A_{gv} - A_{\text{holes}} = 4.00 - 2.5(0.4375) = 2.906 \text{ in.}^2$$

$$L_t = leg - g = 3.5 - 2 = 1.5 \text{ in.}$$

$$A_{gt} = L_t t = 1.5(2)(0.25) = 0.75 \text{ in.}^2$$

$$A_{nt} = A_{gt} - A_{\text{holes}} = 0.75 - 0.5(0.4375) = 0.53125 \text{ in.}^2$$

$$F_u A_{nt} = 58(0.53125) = 30.8 \text{ kips}$$

$$0.6 F_u A_{nv} = 0.6(58)(2.906) = 101.14 \text{ kips}$$

$$\phi P_n = 0.75(0.6 F_u A_{nv} + F_y A_{gt})$$

$$\phi P_n = 0.75[101.14 + 36(0.75)] = 96.1 \text{ kips}$$

$\phi P_n = 87.8$ kips, due to fracture on A_e, is the governing design strength of the tension member.

Example 2.2

The A36 steel member in Figure 2.4 is a pair of angles, L6 × 4 × 0.375, with the long legs fastened to a gusset plate by 0.75-in.-diameter bolts. Use $L_e = 2$ in., $s = 3$ in., $g_1 = 2.25$ in., and $g_2 = 2.5$ in. The gusset plate thickness is 0.5 in. Find the governing design strength of the tension member for a bearing-type bolted connection.

Solution

LRFD, p. 1-58: For a single L6 x 4 x 0.375
$A = 3.61$ in.^{2} $\quad x = 0.941$ in. $\quad y = 1.94$ in.

The governing design strength of the tension member is the least ϕP_n value obtained from:

1. *Yielding on* A_g

$$\phi P_n = 0.90 F_y A_g = 0.90(36)(7.22) = 233.9 \text{ kips}$$

2. *Fracture on* $A_e = A_n U$

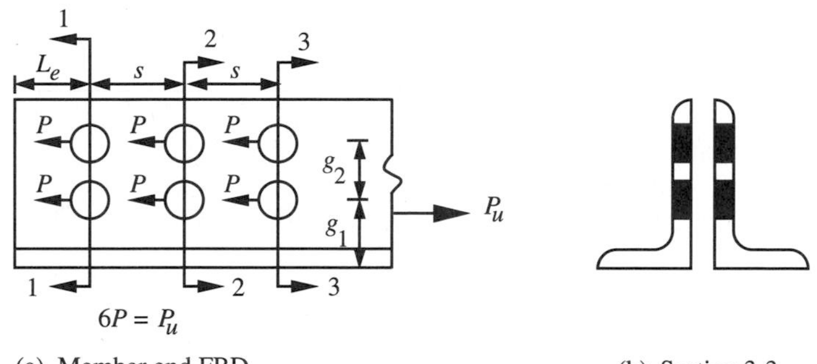

(a) Member end FBD

(b) Section 3-3

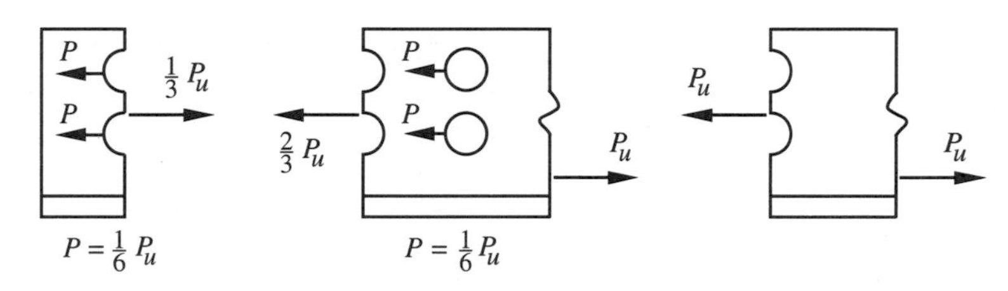

(c) FBD Section 1-1 (d) FBD Section 2-2 (e) FBD Section 3-3

FIGURE 2.4 Tension member with two lines of bolts

$$d_h = d + 1/8 \text{ in.} = 0.75 + 0.125 = 0.875 \text{ in.}$$

$$t_h = 2(0.375) = 0.750 \text{ in.}$$

$$d_h t_h = 0.875(0.750) = 0.656 \text{ in.}^2$$

$$A_n = A_g - \Sigma(d_h t_h) = 7.22 - 2(0.656) = 5.908 \text{ in.}^2$$

From Figure 2.3(b), we see that:

(a) The connection eccentricity from the face of the gusset plate is $x = 0.941$ in.
(b) The other connection eccentricity for a supposed long leg $= g$ must be computed:

$$y_e = (leg - g) = (6 - 2.25) = 3.75 \text{ in.}$$

$$A' = A - y_e t = 3.61 - (3.75)(0.375) = 2.20 \text{ in.}^2$$

$$y' = \frac{Ay - y_e t(leg - 0.5y_e)}{A'} = \frac{(3.61)(1.94) - 3.75(0.375)(6 - 3.75/2)}{2.20} = 0.540 \text{ in.}$$

$$\bar{x} = \text{larger of} \begin{cases} x = 0.661 \text{ in.} \\ g - y' = 2.25 - 0.540 = 1.71 \text{ in.} \end{cases}$$

$$U = \text{smaller of} \begin{cases} 1 - \bar{x}/L_c = 1 - 1.71/6 = 0.715 \\ 0.9 \end{cases}$$

$$A_e = A_n U = 5.908(0.715) = 4.22 \text{ in.}^2$$

$$\phi P_n = 0.75 F_u A_e = 0.75(58 \text{ ksi})(4.22 \text{ in.}^2) = 184 \text{ kips}$$

3. *Block shear rupture* [see Figure 2.2(b), two-row-of-bolts case]

$$L_v = L_e + 2s = 2 + 2(3) = 8 \text{ in.}$$

$$A_{gv} = L_v t = 8(2)(0.375) = 6.00 \text{ in.}^2$$

$$A_{nv} = A_{gv} - A_{\text{holes}} = 6.00 - 2.5(0.656) = 4.688 \text{ in.}^2$$

$$L_t = leg - g_1 = 6 - 2.25 = 3.75 \text{ in.}$$

$$A_{gt} = L_t t = 3.75(2)(0.375) = 2.81 \text{ in.}^2$$

$$A_{nt} = A_{gt} - A_{\text{holes}} = 2.81 - 1.5(0.656) = 1.83 \text{ in.}^2$$

$$F_u A_{nt} = 58(1.83) = 106 \text{ kips}$$

$$0.6 F_u A_{nv} = 0.6(58)(4.688) = 163 \text{ kips}$$

$$\phi P_n = 0.75(0.6 F_u A_{nv} + F_y A_{gt})$$

$$\phi P_n = 0.75[163 + 36(2.81)] = 198 \text{ kips}$$

$\phi P_n = 184$ kips, due to fracture on A_e, is the governing design strength of the tension member.

2.3 EFFECT OF STAGGERED BOLT HOLES ON NET AREA

Figure 2.4(a) is the FBD of a tension member separated from a bearing-type bolted connection. To ensure that the socket can be fitted on each nut of all bolts in a bolt group, LRFD J3.3 stipulates that the minimum spacing center to center between two adjacent bolt holes is $2.67d$, where d is the nominal diameter of the bolts. To speed up the installation of the nuts, the preferred minimum spacing is $3d$. In Figure 2.4(a), s and g_2 are the parameters for which the preferred minimum spacing is $3d$. In Example 2.2, a pair of L6 × 4 × 0.375 was used as the tension member in Figure 2.4(a). The usual gages for the long leg of a L6 × 4 × 0.375 are $g_1 = 2.25$ in. and $g_2 = 2.5$ in. Since $g_2/3 = 2.50/3 = 0.833$ in., we cannot use $d > 0.75$ in. if we want to use the preferred minimum spacing of $3d$ at g_2. Suppose that we want to use d = 1-in.-diameter bolts and the preferred minimum spacing of $3d$. Then, as shown in Figure 2.5(a), the bolt holes must be staggered to ensure that $C \geq 3d$ and $2s \geq 3d$.

When a staggered bolt hole pattern is used in the connection of a tension member, two or more net sections must be investigated to find the critical net section for fracture on A_e. If fracture occurs on the net section shown in Figures

2.5(c) and (d), the fracture surface lies in one plane. However, if fracture occurs on the net section of Figures 2.5(e) and (f), the fracture surface does not lie in one plane. Figures 2.5(d) and (h) have the same net area and the same fracture design strength. However, the internal force in Figure 2.5(h) is smaller than in Figure 2.5(d). Furthermore, the internal force on all net sections at and to the left of Section 3-3 in Figure 2.5(g) is less than P_u. Therefore, the critical net section (where the fracture on A_e would occur) is either Figure 2.5(c) or (e). The design strength due to tensile fracture in the net section is the smaller value obtained for the net sections in Figures 2.5(c) and (e).

In Figure 2.5(e), Section 2-2 shows a potential tensile fracture path across the member, and this path is not a straight line. Such a fracture path is called a staggered path. If a line segment on the staggered path is not perpendicular to P_u, we call that line segment a stagger. As shown in Figure 2.5(e), a stagger has a longitudinal component called s and a transverse component called g. If fracture occurs on Section 2-2 of Figure 2.5(e), a tensile fracture force exists perpendicular to each line segment of Section 2-2. On each stagger, a shear force must exist in addition to the tension force. This shear force is required to equilibrate the transverse component of the tension force on the stagger. Therefore, a combined state of failure stress exists on each stagger and we only need the sum of all fracture force components parallel to P_u. LRFD B2 (p. 6-34) gives an empirical definition of the net area based on the research of Cochrane [14], which when multiplied times F_u gives the desired tensile fracture force component parallel to P_u. We prefer to show this definition as a formula. Let A_{gs} denote the equivalent gross area for a staggered path:

$$A_{gs} = A_g + \sum \frac{s^2 t}{4g}$$

where

A_g = planar gross section area

s = longitudinal component of a stagger

g = transverse component of a stagger

Then the empirical definition of the net area given in LRFD B2 for a staggered path can be written as

$$A_n = A_{gs} - A_{holes}$$

This definition of A_n is applicable for Figures 2.5(e) and (f).

Example 2.3

The A36 steel member in Figure 2.5 is a pair of angles, L6 × 4 × 0.375, with the long legs fastened to a gusset plate by 1-in.-diameter bolts. Use L_e = 2 in., s = 2 in., g_1 = 2.25 in., and g_2 = 2.5 in. The gusset plate thickness is 0.5 in. Find the governing design strength of the tension member for a bearing-type bolted connection.

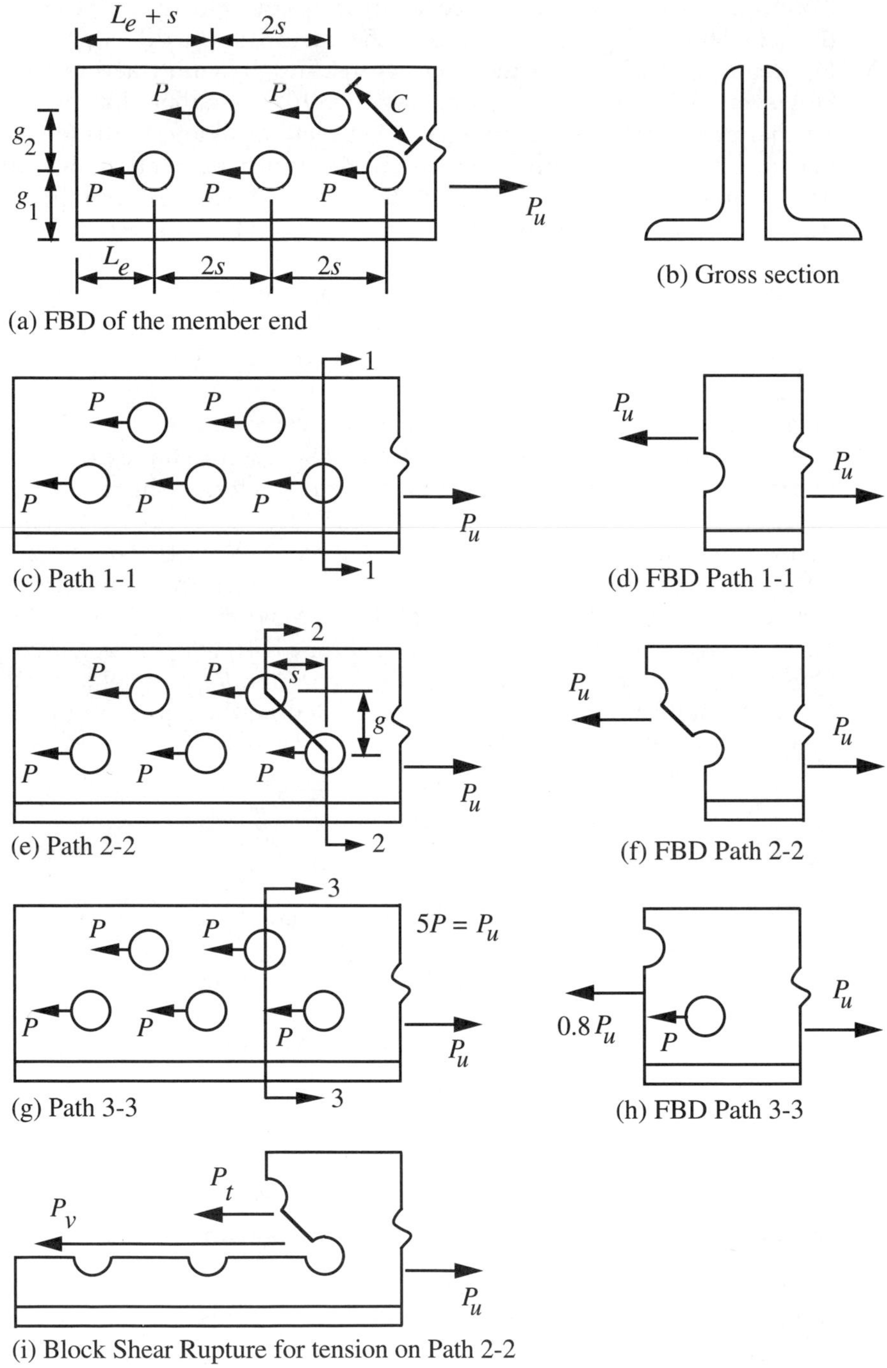

$L_e + s$
$2s$
P
C
g_2
g_1
P_u
L_e
$2s$
$2s$
(a) FBD of the member end
(b) Gross section
1
(c) Path 1-1
(d) FBD Path 1-1
2
s
g
(e) Path 2-2
(f) FBD Path 2-2
3
$5P = P_u$
$0.8P_u$
(g) Path 3-3
(h) FBD Path 3-3
P_t
P_v
(i) Block Shear Rupture for tension on Path 2-2

FIGURE 2.5 Staggered bolt configuration in a tension member

Solution

LRFD, p. 1-96: For a pair of L6 × 4 × 0.375, $A_g = 7.22$ in.^{2}
LRFD, p. 1-59: For L6 × 4 × 0.375, $\bar{x} = (x = 0.941$ in.)
The governing design strength of the tension member is the least ϕP_n value obtained from:

1. *Yielding on A_g*

$$\phi P_n = 0.90 F_y A_g = 0.90(36)(7.22) = 234 \text{ kips}$$

2. *Fracture on $A_e = A_n U$*

$$U = \text{smaller of } \begin{cases} 1 - \bar{x}/L_c = 1 - 0.941/6 = 0.843 \\ 0.9 \end{cases}$$

$$d_h = d + 1/8 \text{ in.} = 1 + 0.125 = 1.125 \text{ in.}$$

$$t_h = 2(0.375) = 0.750 \text{ in.}$$

$$d_h t_h = 1.125(0.750) = 0.844 \text{ in.}^2$$

We must compute A_n for Sections 1-1 and 2-2 in order to determine the critical net section. Both sections have full P_u on them, but we cannot tell by inspection which section has the smaller A_n.

Section 1-1

$$A_n = A_g - \Sigma(d_h t_h) = 7.22 - 0.844 = 6.376 \text{ in.}^2$$

Section 2-2

$$s = 2 \text{ in.}$$

$$g = (g_2 = 2.5 \text{ in.})$$

$$A_{gs} = A_g + \sum\left(\frac{s^2 t}{4g}\right) = 7.22 + 2\left[\frac{(2)^2\,(2)(0.375)}{4(2.5)}\right] = 7.82 \text{ in.}^2$$

$$A_n = A_{gs} - \Sigma(d_h t_h) = 7.82 - 2(0.844) = 6.13 \text{ in.}^2$$

The critical section for *fracture on A_e* is Section 2-2:

$$A_e = A_n U = 6.13(0.843) = 5.17 \text{ in.}^2$$

$$\phi P_n = 0.75 F_u A_e = 0.75(58 \text{ ksi})(5.17 \text{ in.}^2) = 225 \text{ kips}$$

3. *Block shear rupture* [see Figure 2.5(i)]

$$L_v = L_e + 2(2s) = 2 + 2(2)(2) = 10 \text{ in.}$$

$$A_{gv} = L_v t = 10(2)(0.375) = 7.50 \text{ in.}^2$$

$$A_{nv} = A_{gv} - A_{holes} = 7.50 - 2.5(0.844) = 5.39 \text{ in.}^2$$

$$L_t = leg - g_1 = 6 - 2.25 = 3.75 \text{ in.}$$

$$A_{gt} = L_t t + \sum\left(\frac{s^2 t}{4g}\right) = 3.75(2)(0.375) + \frac{(2)^2(2)(0.375)}{4(2.5)} = 3.11 \text{ in.}^2$$

$$A_{nt} = A_{gt} - A_{holes} = 3.11 - 1.5(0.844) = 1.85 \text{ in.}^2$$

$$F_u A_{nt} = 58(1.85) = 107.3 \text{ kips}$$

$$0.6F_u A_{nv} = 0.6(58)(5.39) = 185.6 \text{ kips}$$

$$\phi P_n = 0.75(0.6F_u A_{nv} + F_y A_{gt}) = 0.75[185.6 + 36(3.11)] = 223 \text{ kips}$$

$\phi P_n = 223$ kips, due to BSR, is the governing design strength of the tension member.

Example 2.4

The A36 steel member in Figure 2.6 is a pair of angles, L6×3.5×0.5, with the long legs fastened to a gusset plate by 0.75-in.-diameter bolts. Use $L_e = 2$ in., $s = 1.5$ in., $g_1 = 2.25$ in., $g_2 = 2.5$ in., and $g = 2$ in. The gusset plate thickness is 0.5 in. All cross-sectional elements of the member contain bolts; therefore, $A_e = A_n$. For visualization purposes in computing A_n, one angle is shown flattened out in Figure 2.6(c). The design strength of the tension member is for a bearing-type bolted connection. Is the design satisfactory for $P_u = 275$ kips?

Solution

LRFD, p. 1-58: For a pair of L6 × 3.5 × 0.5, $A_g = 2(4.50) = 9.00$ in.²

1. *Yielding on A_g*

$$\phi P_n = 0.90 F_y A_g = 0.90(36)(9.00) = 291.6 \text{ kips}$$

$$(\phi P_n = 291.6 \text{ kips}) \geq (P_u = 275) \quad \text{as required}$$

2. *Fracture on $A_e = A_n$*

$$d_h = d + 1/8 \text{ in.} = 0.75 + 0.125 = 0.875 \text{ in.}$$

$$t_h = 2(0.5) = 1.00 \text{ in.}$$

$$d_h t_h = 0.875(1.00) = 0.875 \text{ in.}^2$$

$$A_n = A_g - \Sigma(d_h t_h) = 7.22 - 0.844 = 6.376 \text{ in.}^2$$

For the net section that passes through only hole B,

$$A_n = A_g - \Sigma(d_h t_h) = 9.00 - 0.875 = 8.125 \text{ in.}^2$$

$$\phi P_n = 0.75 F_u A_e = 0.75(58)(8.125) = 353.4 \text{ kips}$$

$$(\phi P_n = 353.4 \text{ kips}) \geq (P_u = 275) \quad \text{as required}$$

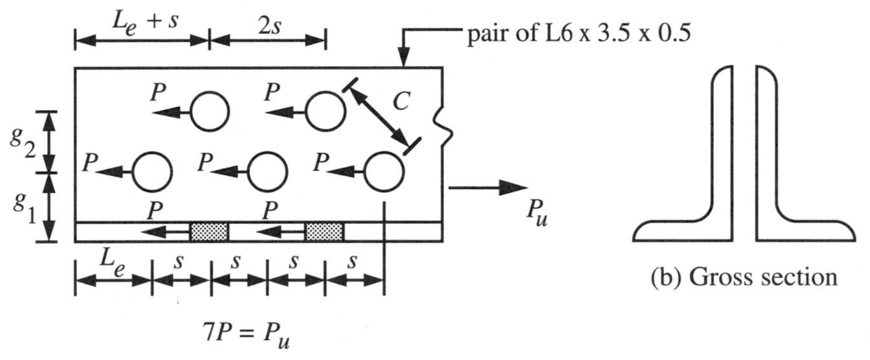

(a) Member end FBD

(b) Gross section

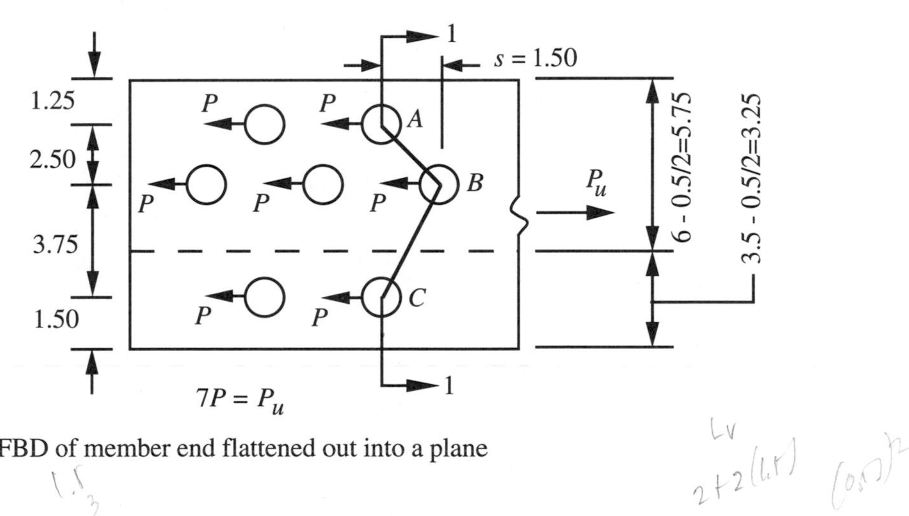

(c) FBD of member end flattened out into a plane

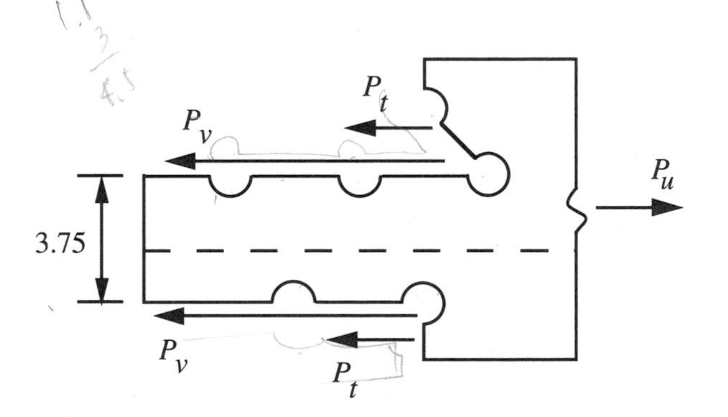

(d) BSR of member end flattened out into a plane

FIGURE 2.6 Bolts in all cross-sectional elements

For the net section that passes through holes A and C,

$$A_n = A_g - S(d_h t_h) = 9.00 - 2(0.875) = 7.25 \text{ in.}^2$$

$$\phi P_n = 0.75 F_u A_e = 0.75(58)(7.25) = 315.4 \text{ kips}$$

$$(\phi P_n = 315.4 \text{ kips}) \geq [(6/7)P_u = 6(275)/7 = 235.7 \text{ kips}] \quad \text{as required}$$

For the net section that passes through holes A and B,

$$A_n = A_g + \sum\left(\frac{s^2 t}{4g}\right) - \sum(d_h t_h)$$

$$A_n = 9.00 + \frac{(1.5)^2(2)(0.5)}{4(2.5)} - 2(1.00) = 7.225 \text{ in.}^2$$

$$\phi P_n = 0.75 F_u A_e = 0.75(58)(7.225) = 314.3 \text{ kips}$$

$$(\phi P_n = 314.3 \text{ kips}) \geq (P_u = 275) \quad \text{as required}$$

For the net section that passes through holes A, B, and C [see Figure 2.6(c)],

$$A_n = A_g + \sum\left(\frac{s^2 t}{4g}\right) - \sum(d_h t_h)$$

$$A_n = 9.00 + \frac{(1.5)^2(2)(0.5)}{4(2.5)} + \frac{(1.5)^2(2)(0.5)}{4(3.75)} - 3(1.00) = 6.375 \text{ in.}^2$$

$$\phi P_n = 0.75 F_u A_e = 0.75(58)(6.375) = 277.3 \text{ kips}$$

$$(\phi P_n = 277.3 \text{ kips}) \geq (P_u = 275) \quad \text{as required}$$

3. *Block shear rupture* [see Figure 2.6(d)]

$$A_{gv} = \Sigma L_v t = (8 + 6.5)(2)(0.5) = 14.50 \text{ in.}^2$$

$$A_{nv} = A_{gv} - A_{holes} = 14.50 - (1.5 + 2.5)(0.875) = 11.0 \text{ in.}^2$$

$$A_{gt} = \sum(L_t t) + \sum\left(\frac{s^2 t}{4g}\right) = (3.75 + 1.50)(2)(0.5) + \frac{(1.50)^2(2)(0.5)}{4(2.5)} = 5.475 \text{ in.}^2$$

$$A_{nt} = A_{gt} - A_{holes} = 5.475 - (1.5 + 0.5)(0.875) = 3.725 \text{ in.}^2$$

$$F_u A_{nt} = 58(3.725) = 216.05 \text{ kips}$$

$$0.6 F_u A_{nv} = 0.6(58)(5.39) = 382.8 \text{ kips}$$

$$\phi P_n = 0.75(0.6 F_u A_{nv} + F_y A_{gt}) = 0.75[382.8 + 36(5.475)] = 434.9 \text{ kips}$$

$$(\phi P_n = 434.9 \text{ kips}) \geq (P_u = 275) \quad \text{as required}$$

The design is satisfactory for $P_u = 275$.

2.4 DESIGN OF A TENSION MEMBER WITH BOLTED-END CONNECTIONS

For a tension member with bearing-type bolted connections at the member ends, the design requirement for strength is

$$\phi P_n \geq P_u$$

and the design strengths ϕP_n that must be considered are yielding on A_g; fracture on A_e; BSR; and bearing at the bolt holes. The latter is accounted for when the connection design strengths are computed in Chapter 3. In Examples 2.1 to 2.4, we found in three of the four examples that ϕP_n due to BSR governed and was only slightly smaller than ϕP_n due to fracture on A_e. If BSR governs the design, we usually can easily make some changes in the connection layout to increase the design strength to a satisfactory level. Consequently, in the design of a tension member with bearing-type bolted connections at the member ends, we recommend the following itemized approach:

1. Tentatively choose the bolt diameter, number of bolts, spacing center to center between the bolt holes, and the end edge distance. (*Note*: This step is done by the author for the reader since concepts in Chapter 3 must be used in making the tentative choices.)
2. Assume that ϕP_n due to BSR does not govern.
3. Using the information given in item 1, pick a trial section that has adequate ϕP_n for both of the following cases:

   a. *Yielding on* A_g

   $$(\phi P_n = 0.90 F_y A_g) \geq P_u$$

   $$A_g \geq P_u / (0.90 F_y)$$

   b. *Fracture on* $A_e = A_n U$ (illustrated for the most general case)

   $$(\phi P_n = 0.75 F_u A_e) \geq P_u$$

   $$A_e \geq P_u / (0.75 F_u)$$

   $$A_n = A_g - \Sigma d_h t_h$$

   $$A_g \geq \frac{P_u}{0.75 F_u U} + \sum (d_h t_h)$$

   In $U = (1 - \bar{x} / L_c) \leq 0.9$, $\bar{x}$ is unknown before the section has been chosen and L_c is tentative until all details of the bolt group have been finalized. For the design of tension members, the following U values (from LRFD C-B3, p. 6-172) are good first estimates:
   (i) $U = 0.85$ when longer line of bolts parallel to P_u has ≥ three bolts.
   (ii) $U = 0.75$ when longer line of bolts parallel to P_u has only two bolts.
4. Compute ϕP_n due to BSR to check our assumption.

   a. If $\phi P_n \geq P_u$, the trial section is satisfactory and it becomes the chosen section. Exit the design process.

b. If $\phi P_n < P_u$, increase ϕP_n by either changing some of the previously chosen values in item 1 (to increase the length of the shear plane, e.g.) or by choosing a section whose elements in which *BSR* occurs are thicker. Now, compute $U = (1 - \bar{x} / L_c) \leq 0.9$ and compare it to the assumed *U* value. If they differ by more than 2%, compute ϕP_n due to *fracture on* A_e and check the strength design requirement for *fracture on* A_e. If necessary, repeat step 4 until the design is satisfactory.

Examples 2.5, 2.7, and 2.8 discuss the design of tension members in the "truss region" of the structure shown in Figure 1.15. Stability of this structure in the horizontal direction and resistance due to wind loads are ensured by bending of the columns (members 1 to 4). The truss-to-column connections at joints 2, 3, 22, and 23 will necessarily have to account for bending of the columns, which causes some bending in the "truss members" attached to those joints. Consequently, in the plane frame analysis of this structure (see Appendix A), the author did not release the moment at the end of any member. Therefore, for some of the members in the "truss region" of this structure, the bending effects cannot be ignored. In the final design check of such members, P_u and M_u must be accounted for simultaneously as discussed in Chapter 6. In the preliminary design phase (selecting trial member sizes) of such members, we account for M_u by using an *equivalent* P_u.

Example 2.5

For members 34 and 43 in Figure 1.15, select the lightest available double-angle section with long legs back to back (see LRFD, p. 1-98) of A36 steel that can be used in a bearing-type bolted connection.

For the connection layout shown in Figure 2.1, the author has determined that the following values are required for the connection design to be satisfactory with standard-size bolt holes: d = 0.75-in.-diameter bolts; s = 3-in. bolt spacing; L_e = 1.5-in. end edge distance, and each angle thickness $t \geq 3/16$ in. for strength due to bearing at the bolt holes of the member. Note that the length of the connection is $L_c = L_e + 2s$ = 1.5 +2(3) = 7.5 in.

From Appendix A, for members 34 and 43 due to loading 8, we find P_u = 66.4 kips (tension) and M_u = 0.16 ft-kips. In the final design check of these members, P_u and M_u must be accounted for simultaneously as discussed in Chapter 6. In the preliminary design phase of a tension-plus-bending member, we account for M_u by using an *equivalent* P_u; in this case, we know (from Chapter 6) that a 10% increase in P_u is adequate. Try *equivalent* P_u = 1.10(66.3) = 72.9 kips.

Solution

Assumptions

1. Try U = 0.85 since there are three bolts in the longer line of bolts.
2. ϕP_n due to BSR does not govern.

For *yielding on* A_g, the design requirement is

$$(\phi P_n = 0.90F_yA_g) \geq (equivalent\ P_u = 72.9 \text{ kips})$$

$$A_g \geq \left[\frac{P_u}{0.9F_y} = \frac{72.9}{0.9(36)} = 2.25 \text{ in.}^2 \right]$$

For *fracture on* $A_e = A_nU$, the design requirement is

$$[\phi P_n = 0.75F_u(A_nU)] \geq (equivalent\ P_u = 72.9 \text{ kips})$$

$$A_n \geq \left[\frac{P_u}{0.75F_uU} = \frac{72.9}{0.75(58)(0.85)} = 1.97 \text{ in.}^2 \right]$$

$$d_ht_h = (0.75 + 0.125)t = 0.875t$$

$$A_n = A_g - 2(0.875)t = A_g - 1.75t$$

$$A_g \geq (A_n + 1.75t = 1.97 \text{ in.}^2 + 1.75t)$$

Summary of the design requirements

1. $t \geq 3/16$ in. for each angle due to bearing at the bolt holes.
2. $A_g \geq 2.25$ in.^{2} for *yielding on* A_g.
3. $A_g \geq (1.97$ in.$^2 + 1.75t)$ for *fracture on* A_e.

   (a) For $t = 3/16$ in., we need:

   $$A_g \geq [1.97 + 1.75(0.1875) = 2.30 \text{ in.}^2] \geq 2.22 \text{ in.}^2$$

   Since a pair of angles with long legs back to back is to be chosen, pick the trial section from LRFD, pp. 1-96 to 1-99.
   No available section for $t = 3/16$ in. has $A_g \geq 2.30$ in.^{2}

   (b) For $t = 1/4$ in., we need:

   $$A_g \geq [1.97 + 1.75(0.25) = 2.41 \text{ in.}^2] \geq 2.22 \text{ in.}^2$$

   Try a pair of L3 x 2.5 x 1/4, ($A_g = 2.63$ in.^{2}) ≥ 2.41 in.^{2} $U = 0.85$ was assumed; check this assumption.

From Table 2.1 for *leg* = 3 in., we find $g = 1.75$ in. From LRFD, pp.1- 64 and 1-65, for a single L3 x 2.5 x 1/4, we find $x = 0.661$ in. and $y = 0.911$ in. that are needed to determine the governing connection eccentricity for each angle. From Figure 2.3(b), we see that (1) the connection eccentricity from the face of the gusset plate is $x = 0.661$ in.; and (2) the other connection eccentricity for a supposed long leg = g must be computed. The y-edge distance is

$$y_e = (leg - g) = (3 - 1.75) = 1.25 \text{ in.}$$

For simplicity in the following computations, a prime is placed on all variables for the supposed-angle section.

$$A' = A - y_et = 2.63/2 - (1.25)(1/4) = 1.0025 \text{ in.}^2$$

$$y' = \frac{Ay - y_e t(leg - 0.5y_e)}{A'} = \frac{(1.31)(0.911) - 1.25(0.25)(3 - 1.25/2)}{1.00} = 0.456 \text{ in.}$$

$$\bar{x} = \text{larger of} \begin{cases} x = 0.661 \text{ in.} \\ g - y' = 1.75 - 0.456 = 1.29 \text{ in.} \end{cases}$$

$$U = \text{smaller of} \begin{cases} 1 - \bar{x}/L_c = 1 - 1.29/7.5 = 0.828 \\ 0.9 \end{cases}$$

Since ($U = 0.828$) < (assumed $U = 0.85$), revise A_g required for *fracture on* A_e to

$$A_g \geq [1.97(0.85)/0.828 + 1.75(0.25) = 2.46 \text{ in.}^2]$$

Try a pair of L3 × 2.5 × 1/4, ($A_g = 2.63$ in.^{2}) ≥ 2.46 in.^{2}.
This is the original trial section. We cannot find any lighter pair of angles for $t = 1/4$ in. Weight = 9.0 lb/ft.

(c) For $t = 5/16$ in. and $U = 0.85$, the A_g requirement is

$$A_g \geq [1.97 + 1.75(0.3125) = 2.52 \text{ in.}^2] \geq (2.22 \text{ in.}^2).$$

Try a pair of L2.5 × 2 × 5/16, ($A_g = 2.62$ in.^{2}) ≥ 2.52 in.^{2}
Check our assumption that $U = 0.85$:

$$y_e = (\text{leg} - g) = (2.5 - 1.375) = 1.125 \text{ in.}$$

$$A' = A - y_e t = 2.62/2 - (1.125)(5/16) = 0.958 \text{ in.}^2$$

TABLE 2.1 Usual gages in angle legs, inches

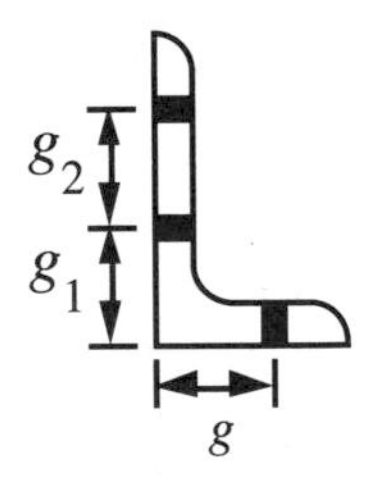

Leg	8	7	6	5	4	$3\frac{1}{2}$	3	$2\frac{1}{2}$	2	$1\frac{3}{4}$	$1\frac{1}{2}$	$1\frac{3}{8}$	$1\frac{1}{4}$	1
g	$4\frac{1}{2}$	4	$3\frac{1}{2}$	3	$2\frac{1}{2}$	2	$1\frac{3}{4}$	$1\frac{3}{8}$	$1\frac{1}{8}$	1	$\frac{7}{8}$	$\frac{7}{8}$	$\frac{3}{4}$	$\frac{5}{8}$
g_1	3	$2\frac{1}{2}$	$2\frac{1}{4}$	2										
g_2	3	3	$2\frac{1}{2}$	$1\frac{3}{4}$										

From LRFD, p. 9–13.

$$y' = \frac{Ay - y_e t(leg - 0.5y_e)}{A'} = \frac{(1.31)(0.809) - 1.125(0.3125)(2.5 - 1.125/2)}{0.958} = 0.395 \text{ in.}$$

$$\bar{x} = \text{larger of} \begin{cases} x = 0.559 \text{ in.} \\ g - y' = 1.375 - 0.395 = 0.980 \text{ in.} \end{cases}$$

$$U = \text{smaller of} \begin{cases} 1 - \bar{x}/L_c = 1 - 0.980/7.5 = 0.869 \\ 0.9 \end{cases}$$

$(U = 0.869) > (\text{assumed } U = 0.85)$. Revise A_g required for *fracture on* A_e to

$$A_g \geq [1.97(0.85)/0.869 + 1.75(0.3125) = 2.47 \text{ in.}^2]$$

Try L2.5 x 2 x 5/16, $(A_g = 2.62 \text{ in.}^2) \geq 2.47 \text{ in.}^2$. This is the original trial section. We cannot find any lighter pair of angles for $t = 5/16$ in. Weight = 9.0 lb/ft.

Conclusion:

Two acceptable sections of equal weight have been found that satisfy the design requirements thus far. Our preference is a pair of $L3 \times 2.5 \times 1/4$ with long legs back-to-back. Now we must compute ϕP_n due to BSR for the trial section to check our assumption that BSR does not govern the design selection. See Figure 2.7:

$$d_h t_h = (0.75 + 0.125)(2)(0.25) = 0.4375 \text{ in.}^2$$

$$L_v = 1.5 + 2(3) = 7.5 \text{ in.}$$

$$A_{gv} = (7.5)(2)(0.25) = 3.75 \text{ in.}^2$$

$$A_{nv} = 3.75 - 2.5(0.4375) = 2.656 \text{ in.}^2$$

$$A_{gt} = (1.25)(2)(0.25) = 0.625 \text{ in.}^2$$

$$A_{nt} = 0.625 - 0.5(0.4375) = 0.406 \text{ in.}^2$$

$$F_u A_{nt} = 58(0.406) = 23.55 \text{ kips}$$

$$0.6F_u A_{nv} = 0.6(58)(2.656) = 92.43 \text{ kips}$$

$$\phi P_n = (0.6F_u A_{nv} + F_y A_{gt}) = 0.75[92.43 + 36(0.625)] = 86.2 \text{ kips}$$

$$(\phi P_n = 86.2 \text{ kips}) \geq (P_u = 66.3 \text{ kips})$$

As we assumed, BSR does not govern our choice of the section.

Conclusion:

Use a pair of $L3 \times 2.5 \times 1/4$; weight = 9.0 lb/ft.

2.5 STRENGTH OF A TENSION MEMBER WITH WELDED-END CONNECTIONS

Consider Figure 2.8, which shows a tension member fastened by fillet welds to a gusset plate. Along the member at some finite distance from the welds, as shown

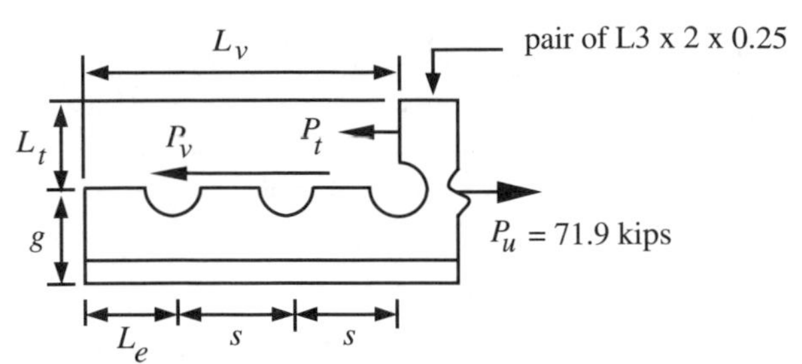

FIGURE 2.7 Block shear rupture of bolted member end

in Section 1.1.4 and Figure 1.4(c), all cross-sectional fibers can attain the yield strength, when the welds and gusset plate are stronger than the member. In the region of the member-end connection, an edge on one leg of each angle in Figure 2.8(c) is not welded to the gusset plate. Therefore, the stress distribution due to the applied load is not uniform in the member end. However, the LRFD definitions account for a *shear lag* effect in the member end when only transverse welds are used at the member end to transfer the force in the member to the gusset plate. A transition region exists from the connection region to some finite distance from the connection where the stress distribution in the member becomes uniform when yielding occurs in the member. Thus, before yielding occurs in the member, the connection region of the member end usually experiences strain-hardening, and fracture can possibly occur in the region where strain-hardening occurs due to shear lag.

The LRFD design strength definitions of a tension member without any holes in it and with fillet-welded end connections are:

1. *Yielding on* A_g [see Figure 2.8(b)] (LRFD D1, p. 6-44)

$$\phi P_n = 0.90 F_y A_g$$

2. *Fracture on* A_e (LRFD D1 p. 6-44)

$$\phi P_n = 0.75 F_u A_e$$

where A_e, the effective area, is to be determined from:

(a) When holes exist in the member (LRFD D1, p. 6-44),

$$A_e = A_n$$

(b) When holes do not exist in the member (LRFD B3, p. 6-34),

$$A_e = AU$$

where A and U are to be determined from:

(i) When the weld group consists only of transverse welds,

$$A = \text{area of the directly connected elements}$$

$$U = 1.0$$

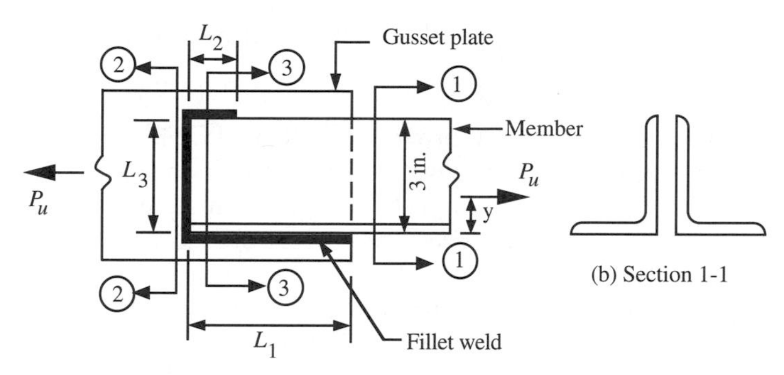

Gusset plate: 5 x 5/8; **Member:** pair of L3 x 2 x 5/16

Fillet welds: 0.25 in.; L_1 = 4.50 in.; L_2 = 1.50 in.; L_3 = 3.00 in.

(a) Member end connection detail

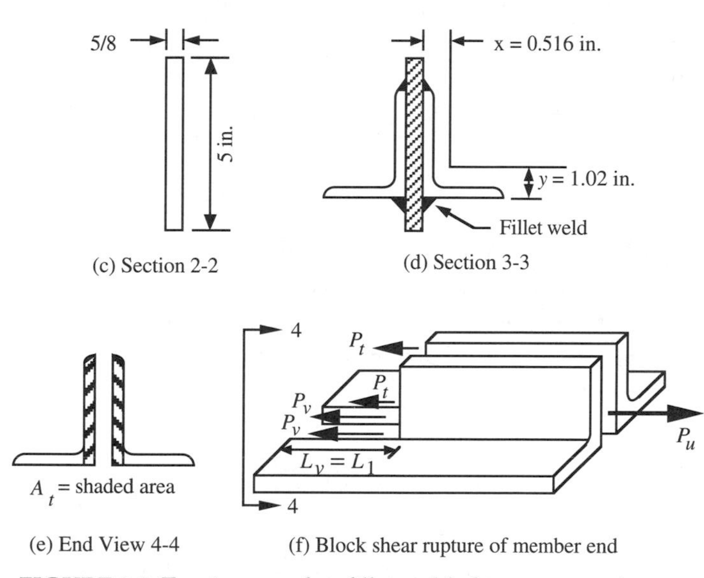

FIGURE 2.8 Tension member fillet welded to a gusset plate

(ii) When the member is a single plate fastened by only longitudinal welds for (L_c = length of longitudinal weld) ≥ (w = width of plate),

$$A = A_g$$

$$U = 1.00 \quad \text{when } L_c \geq 2w$$

$$U = 0.87 \quad \text{when } 2w > L_c \geq 1.5w$$

$$U = 0.75 \quad \text{when } 1.5w > L_c \geq w$$

(iii) When preceding item (i) or (ii) is not applicable,

$$A = A_g$$

$$U = (1 - \bar{x} / L_c) \leq 0.9$$

L_c = length of longest longitudinal weld

$\bar{x}$ = connection eccentricity

3. *Block shear rupture*, abbreviated BSR (LRFD J4.3, p. 6-87)[see Figure 2.8(e) for an example] BSR design strength is a function of the weld group arrangement in the welded-end connections and the thickness of the block in the member end. Welded-end connections are discussed in Chapter 3.

   Since all cases we will discuss have no holes in the members, we can simplify the BSR definitions as shown below:

   A_v = gross area on BSR shear plane(s) (in.^{2})

   A_t = gross area on BSR tension plane (in.^{2})

   When $F_u A_t \geq 0.6 F_u A_v$,

   $$\phi P_n = 0.75(F_u A_t + 0.6 F_y A_v)$$

   When $0.6 F_u A_v \geq F_u A_t$,

   $$\phi P_n = 0.75(0.6 F_u A_v + F_y A_t)$$

Example 2.6

A fillet weld group arranged as shown in Figure 2.8(a) is used on each long leg of a pair of L3 × 2 × 0.3125 of A36 steel to fasten this tension member to a 0.625 in. thick gusset plate. Find the governing design strength of the tension member with welded-end connections.

Solution

LRFD, p. 1-98: For a pair of L3 x 2 x 0.3125,

$$A_g = 2.93 \text{ in.}^2$$

The governing design strength of the tension member is the least ϕP_n value obtained from:

1. *Yielding on* A_g = 2.93 in.^{2}

   $$\phi P_n = 0.90 F_y A_g = 0.90(36)(2.93) = 94.9 \text{ kips}$$

2. *Fracture on* $A_e = A_g U$

   $$\phi P_n = 0.75 F_u A_e$$

   $$U = \text{smaller of } \begin{cases} 1 - \bar{x}/L_c = 1 - 0.516/4.50 = 0.885 \\ 0.9 \end{cases}$$

   $$\phi P_n = 0.75(58)(2.93)(0.885) = 114.0 \text{ kips}$$

3. *Block shear rupture* of the member end [see Figure 2.8(e)]

   $$A_v = L_v t = 4.5(2)(0.3125) = 2.8125 \text{ in.}^2$$

$$A_t = L_t t = 3(2)(0.3125) = 1.875 \text{ in.}^2$$

$$F_u A_t = 58(1.875) = 108.75 \text{ kips}$$

$$0.6F_u A_v = 0.6(58)(2.8125) = 97.875 \text{ kips}$$

$$\phi P_n = 0.75(F_u A_t + 0.6F_y A_v)$$

$$\phi P_n = 0.75[108.75 + 0.6(36)(2.8125)] = 127.1 \text{ kips}$$

For the tension member shown in Figure 2.8, the governing design strength is $\phi P_n =$ 94.9 kips.

2.6 DESIGN OF A TENSION MEMBER WITH WELDED-END CONNECTIONS

For a tension member with welded-end connections, the design requirement for strength is

$$\phi P_n \geq P_u$$

and the applicable design strengths are *yielding on* A_g, *fracture on* A_e, and *BSR*. However, if *yielding on* A_g is not the governing case, we usually can easily make some changes in the connection layout to increase the other design strengths to a satisfactory level. Consequently, in the design of a tension member with welded-end connections, we recommend the following itemized approach:

1. Assume that *yielding on* A_g governs; pick a trial section that satisfies:

$$(\phi P_n = 0.90F_y A_g) \geq P_u$$

$$A_g \geq P_u/(0.90F_y)$$

2. Tentatively choose the weld size and weld lengths. (*Note*: This step is done by the author for the reader since concepts explained in Chapter 3 must be used in making the tentative choices.)
3. Compute ϕP_n due to *fracture on* A_e and *BSR:*
   (a) If $\phi P_n \geq P_u$, our trial section is satisfactory, it now becomes our chosen section, and we exit the design process.
   (b) If $\phi P_n < P_u$, we must increase ϕP_n by either changing some of our previously chosen values in item 2 (to increase the length of the shear plane, e.g.) or choosing a section with thicker elements in those elements in which *BSR* occurs. Repeat step 3 until our design is satisfactory.

Example 2.7

Repeat Example 2.5 except, as shown in Figure 2.9, the long legs of each angle section are to be welded to a gusset plate. The longer longitudinal weld length is 4.50 in. and 3/16-in. fillet welds are used.

From Example 2.5, for members 34 and 43 in Figure 1.15,

$$\textit{Equivalent } P_u = 72.9 \text{ kips}$$

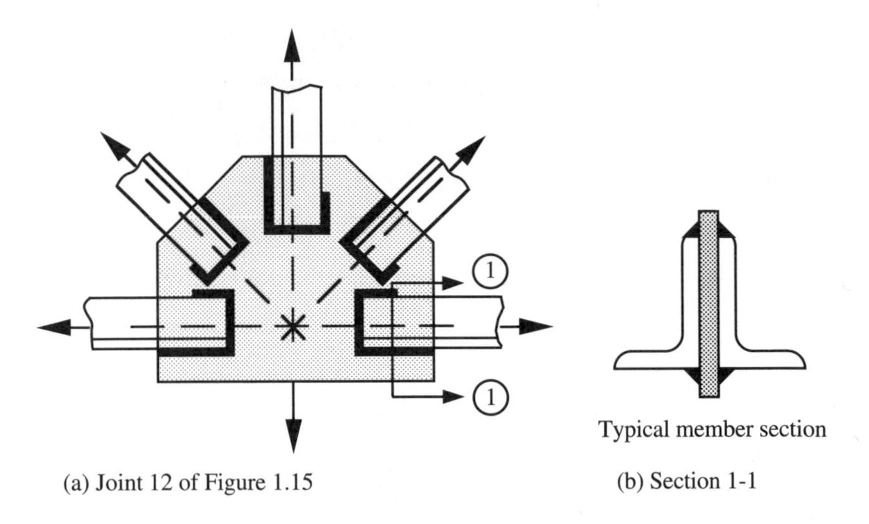

(a) Joint 12 of Figure 1.15 (b) Section 1-1

FIGURE 2.9 Welded truss joint details for double-angle members.

Solution

Assume that *yielding on* A_g is the governing case; we need:

$$(\phi P_n = 0.90 F_y A_g) \geq (\textit{equivalent}\ P_u = 72.9 \text{ kips})$$

$$A_g \geq \{72.9/[0.90(36)] = 2.25 \text{ in.}^2\}$$

Try a pair of L3 x 2 x 1/4: (A_g = 2.38 in.²) ≥ 2.25; weight = 8.1 lb/ft.
When the final design check is performed in Chapter 6 for members 34 and 43 in Figure 1.15 as a tension-plus-bending member with welded-end connections, we will need the governing ϕP_n which is the least of:

1. For *yielding on* A_g

$$\phi P_n = 0.90 F_y A_g = 0.90(36)(2.38) = 77.1 \text{ kips}$$

2. For *fracture on* $A_e = A_g U$

$$U = \text{smaller of} \begin{cases} 1 - \bar{x}/L_c & = 1 - 0.493/4.50 = 0.890 \\ 0.9 & \end{cases}$$

$$\phi P_n = 0.75 F_u A_e = 0.75(58)(2.38)(0.890) = 92.1 \text{ kips}$$

3. For *BSR* of the member end (see Figure 2.10)

$$A_v = L_v t = 4.50(2)(0.25) = 2.25 \text{ in.}^2$$

$$A_t = L_t t = 3.00(2)(0.25) = 1.50 \text{ in.}^2$$

$$0.60\Gamma_u A_v = 0.6(58)(2.25) = 78.3 \text{ kips}$$

$$F_u A_t = 58(1.50) = 87.0 \text{ kips}$$

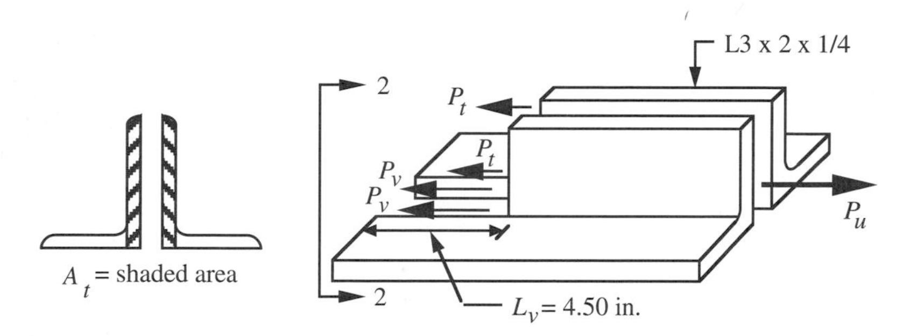

(a) End View 2-2 (b) Block shear rupture of member end

FIGURE 2.10 Block shear rupture of welded member end.

$$\phi P_n = 0.75(F_u A_t + 0.6F_y A_v) = 0.75[87.0 + 0.6(36)(2.25)] = 101.7 \text{ kips}$$

$$(\phi P_n = 102 \text{ kips}) \geq (P_u = 66.3 \text{ kips})$$

The BSR design strength is more than adequate.

Tentatively select a pair of L3 × 2 × 1/4 for members 34 and 43 for which the governing ϕP_n = 77.1 kips is applicable in Chapter 6 when this trial selection is checked as a tension-plus-bending member with welded-end connections.

Example 2.8

See Figure 1.15 and consider the design of a bottom chord member in the truss portion of this structure. As shown in Figure 2.11, the top and bottom chord members of the truss are to be WT sections of A36 steel. Each truss web member is to be a double-angle section with long legs back to back and fillet welded to the WT chord members at each truss joint.

Assume that the same WT section is to be used in Figure 1.15 for members 5 through 14. From Appendix A, for member 10 due to loading 7, we find P_u = 114.2 kips (tension) and M_u = 1.62 ft-kips. In the final design check of these members, P_u and M_u must be accounted for simultaneously as discussed in Chapter 6. In the preliminary design phase of a tension-plus bending member, we account for M_u by using an *equivalent* P_u; in this case, we know (from Chapter 6) that a 20% increase in P_u is adequate. Try *Equivalent* P_u = 1.20(114.2) = 137 kips. Select the lightest available WT7 section of A36 steel for which $\phi P_n \geq 137$ kips.

Solution

Assume that *yielding on* A_g governs ϕP_n; the strength design requirement is

$$(\phi P_n = 0.90F_y A_g) \geq (\textit{equivalent } P_u = 143 \text{ kips})$$

$$A_g \geq \{137/[0.90(36)] = 4.23 \text{ in.}^2\}$$

Fracture on A_e is not applicable.

BSR is not applicable. However, BSR of the web of the chosen WT is applicable in the

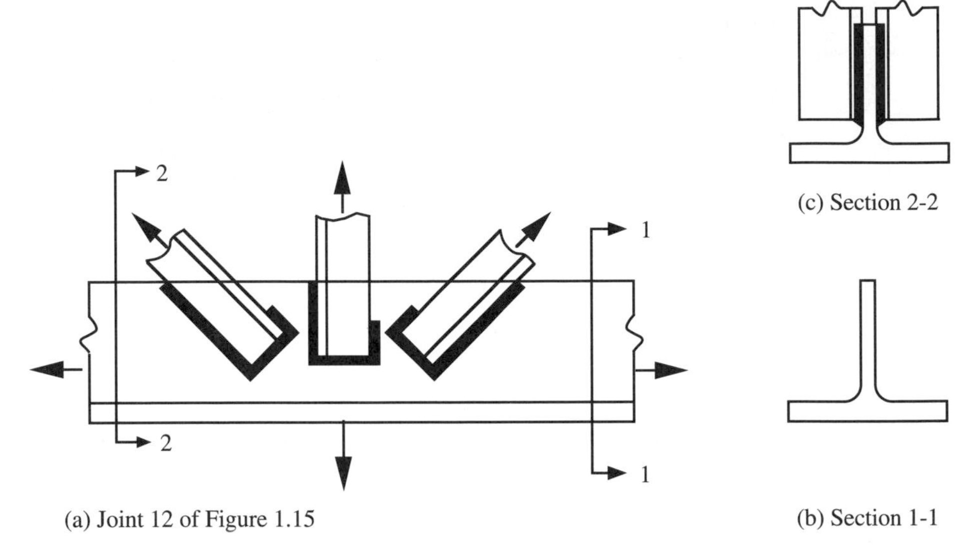

FIGURE 2.11 Truss joint details for WT bottom chord member.

design of members 25 to 43 in Figure 1.15.
Tentatively select a WT7 x 15 for which:

$$\phi P_n = 0.90 F_y A_g = 0.90(36)(4.42) = 143 \text{ kips}$$

ϕP_n = 143 kips is applicable in Chapter 6 when this trial selection is checked as a tension-plus-bending member with welded-end connections.

2.7 SINGLE-ANGLE MEMBERS

On pp. 6-277 to 6-300 of the LRFD Manual, we now find a separate, supplementary specification that deals only with the design of single-angle members. We chose not to discuss single-angle members in this chapter since they usually are subjected to combined bending and axial force, a topic discussed in Chapter 6.

2.8 THREADED RODS

As shown in Figure 1.16, cross braces in roofs and walls may be designed as tension members to resist wind and to provide overall structural stability in a three-dimensional sense for gravity-type loads. If the roof slope in Figure 1.14 had been chosen to be greater than about 15°, sag rods might be designed as tension members to provide lateral support for the weak axis of the purlins. If sag rods were used in Figure 1.14, they would be perpendicular to the purlins and parallel to the roof surface. They would function in a manner similar to the saddle and stirrups for a horseback rider when the rider stands up in the stirrups. Each end of a sag rod would be threaded and passed through holes punched in each purlin web. A nut would be used on each end of a sag rod for anchorage. Adjacent sag rods would be offset in plan view about 6 in. or less to accommodate installation.

If rods are chosen as the tension members for cross bracing, a turnbuckle (see LRFD, p. 8-94) may be used at midlength of the rod in order to take up slack and to pretension the rod. At each end of the rod, clevises (see LRFD, p. 8-92) or welds may be used to fasten the rod to other structural members.

The opening paragraph of LRFD Chapter D (p. 6-44) states that Sec. J3 (p. 6-79) is applicable for threaded rods. LRFD Table J3.2 (p. 6-81) gives the tensile design strength of a threaded rod as $\phi P_n = 0.75(0.75F_uA_g)$.

Example 2.9

Select threaded rods for the cross braces shown in Figure 1.16 using A36 steel.

Solution

Figure 1.14 gives the nominal wind load on the ends of the building as 12-psf pressure on the windward end and 7.5-psf suction on the leeward end. The total factored wind load on the building ends is 1.3(0.012+0.0075)[60(26.75)] = 40.7 kips. At least half of this 40.7-kip wind load should be applied at the roof level, and the remainder of the wind load is applied at the foundation level.

Two pairs of cross braces are shown in Figure 1.16(b). Hence, there are four identical pairs of cross braces that resist wind in the length direction of the building. At any time, only one member in each pair of cross braces is in tension due to wind. The other member in each pair of cross braces is in compression due to wind and buckles at a negligibly small load. When the wind reverses, the other member in each pair of cross braces is in tension and resists the reversed wind. Consequently, for each tension member:

$$L = \sqrt{(25.5)^2 + (30)^2} = 39.37 \text{ ft.}$$

$$P_u = (40.7/2)(339.37/30)/4 = 6.68 \text{ kips}$$

and the strength design requirement is

$$[\phi P_n = 0.75(0.75F_uA_g)] \geq (P_u = 6.68 \text{ kips})$$

$$A_g \geq \left[\frac{P_u}{0.75(0.75F_u)} = \frac{6.68}{0.75(0.75)(58)} = 0.205 \text{ in.}^2 \right]$$

$$\frac{\pi d^2}{4} \geq 0.205 \text{ in.}^2$$

$$d \geq 0.453 \text{ in.}$$

A 1/2 in. diameter rod is acceptable, but no less than a 5/8-in.-diameter rod is preferred for ease of handling during construction. Therefore, use a 5/8-in.-diameter threaded rod A36 steel.

2.9 STIFFNESS CONSIDERATIONS

After the structure in Figure 1.15 is erected, from the factored load combinations in Appendix A, we find that members 34 and 43 are required to resist a maximum axial tension force of 66.4 kips and a maximum axial compression force of 5.14 kips. In the fabrication shop and in shipping the prefabricated truss, a crane is used to lift the prefabricated truss. Unless special lifting procedures are used, members 34 and 43 may be required to resist a larger compressive axial force due to lifting than is required after the structure is erected. The design strength of a compression member is given in Chapter 4. A fundamental parameter in the design strength definition of a pinned-ended compression member is the maximum slenderness ratio L/r of the member, where L is the member length and r is the minimum radius of gyration for the cross section of the member. As shown in Appendix B, the definition of radius of gyration for any cross-sectional axis x is

$$r_x = \sqrt{\frac{I_x}{A}}$$

where

I_x = moment of inertia about the x-axis

A = gross cross-sectional area

I and r are minimum for the minor principal axis, and are needed to obtain the maximum slenderness ratio.

LRFD B7 (p. 6-37) states that $L/r \leq 300$ is preferable for a tension member except for threaded rods. LRFD Commentary B7 (p. 6-177) states that the reason for this advisory upper limit on L/r is to provide adequate bending stiffness for ease of handling during fabrication, shipping, and erection. If the tension member will be exposed to wind or perhaps subjected to mechanically induced vibrations, a smaller upper limit on L/r may be needed to prevent excessive vibrations.

Example 2.10

For the tension member designed in Example 2.7, compute the maximum slenderness ratio.

Solution

From Example 2.7 we find that

1. Member 43 of Figure 1.15 was designed; for this member; L = 90 in.
2. The design choice was a pair of L3 × 2 × 1/4 with the long legs separated by and welded to a gusset plate at each member end. For the purposes of this example, assume that the gusset plate thickness is 3/8 in.

See LRFD, p. 1-98, which shows a sketch of a double-angle section and lists the properties of sections for the X-and Y-axes which are principal axes since the Y-axis is an axis of symmetry. Therefore, for member behavior as a pair of L3 x 2 x 1/4 separated 3/8 in. back to back, the minimum r is the smaller of r_x = 0.957 in. and r_y = 0.891 in.

For behavior as a pair of angles, the *maximum* L/r = (90 in.)/(0.891 in.) = 101, which is much less than the preferred upper limit of 300 for a tension member. However, note that the gusset plate exists only at the member ends. Elsewhere along the member length, there is a 3/8-in. gap between the long legs of the pair of angles. Double-angle behavior is truly ensured only at the points where the pair of angles is tied together (by the gusset plates at the member ends). Therefore, we need to determine the individual behavior of each angle. For one L3 x 2 x 1/4, the *minimum* r is $r_z = 0.435$ in. (from LRFD, p. 1-65) and the *maximum* $L/r = 90/0.435 = 207$, which is less than the preferred upper limit of 300 for a tension member.

If we insert a 3/8-in. plate between the long legs at the midlength of the member and weld the long legs to this plate, we will ensure double-angle behavior at midlength of the member as well as at the member ends. Then, the length for single-angle behavior is $L = 90/2 = 45$ in. and the maximum $L/r = 45/0.435 = 103.5$ for single-angle behavior. Since 103.5 is very nearly equal to 101 (for double-angle behavior), we can conclude that we only need to insert and weld a spacer plate at midlength of the member in order to ensure that double-angle behavior is valid.

PROBLEMS

2.1 A36 steel; 7/8-in. diameter A325N bolts

$s = 2.75$ in. $L_e = 1.5$ in. $g_1 = 3$ in. $g_2 = 3$ in.

The tension member is a pair of L8 x 6 x 1/2 with the long legs bolted to a 10 × 1 gusset plate. For the member, compute the design strength due to:

1. Yielding on A_g
2. Fracture on A_e
3. Block shear rupture

Is the design satisfactory for $P_u = 298$ kips?

2.2 A36 steel; 7/8-in. diameter A325N bolts

$s = 1.5$ in. $L_e = 1.5$ in. $g_1 = 2.25$ in. $g_2 = 2.50$ in.

The tension member is a pair of L6 × 4 × 1/2 with the long legs bolted to a 9 x 1 gusset

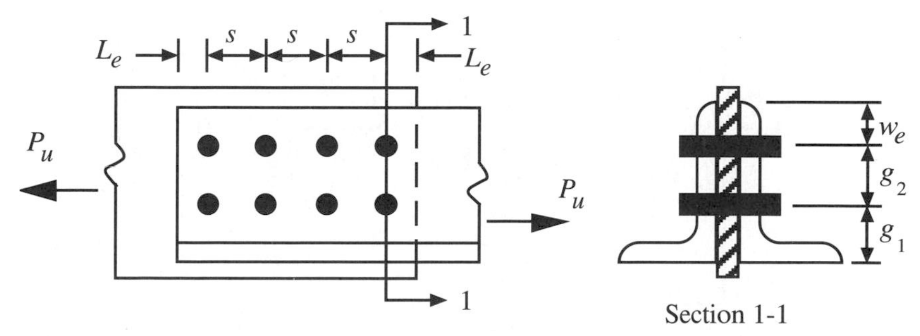

FIGURE P2.1

plate. For the member, compute the design strength due to:

1. Yielding on A_g
2. Fracture on A_e
3. Block shear rupture: Failure mode has a staggered tension path and one shear plane ($L_v = L_e + 6s$).

Is the design satisfactory for $P_u = 270$ kips?

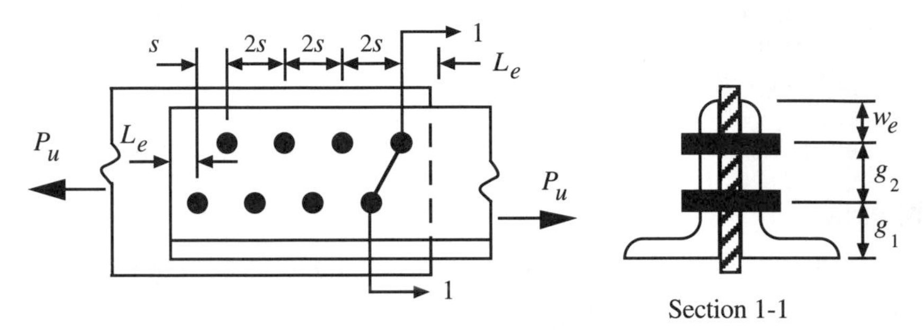

FIGURE P2.2

2.3 A36 steel; 1 in. diameter A325N bolts

$s = 3$ in.; $g = 3$ in.; $L_e = 1.75$ in.

The tension member is a C15×33.9 bolted to a 16×5/8 gusset plate. For the member, compute the design strength due to:

1. Yielding on A_g
2. Fracture on A_e
3. Block shear rupture

Is the design satisfactory for $P_u = 200$ kips?

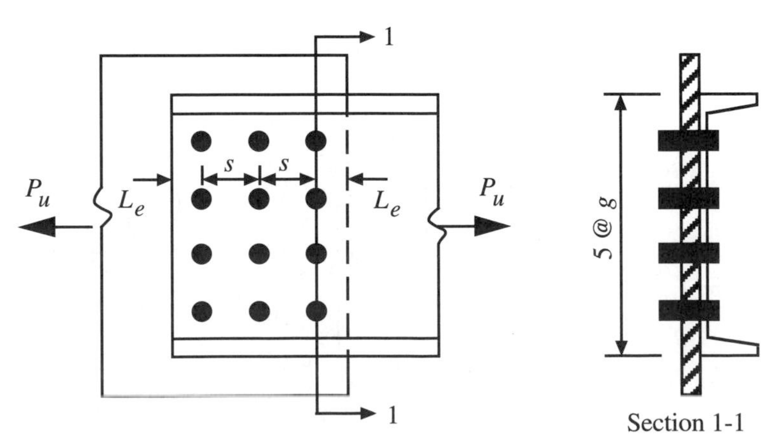

FIGURE P2.3

2.4 A36 steel 1-in.-diameter A325N bolts

$s = 3$ in. $g = 3$ in. $L_e = 1.75$ in.

The tension member is a C15×33.9 bolted to a 16×5/8 gusset plate. For the member, compute the design strength due to:

1. Yielding on A_g
2. Fracture on A_e
3. Block shear rupture

Is the design satisfactory for $P_u = 199$ kips?

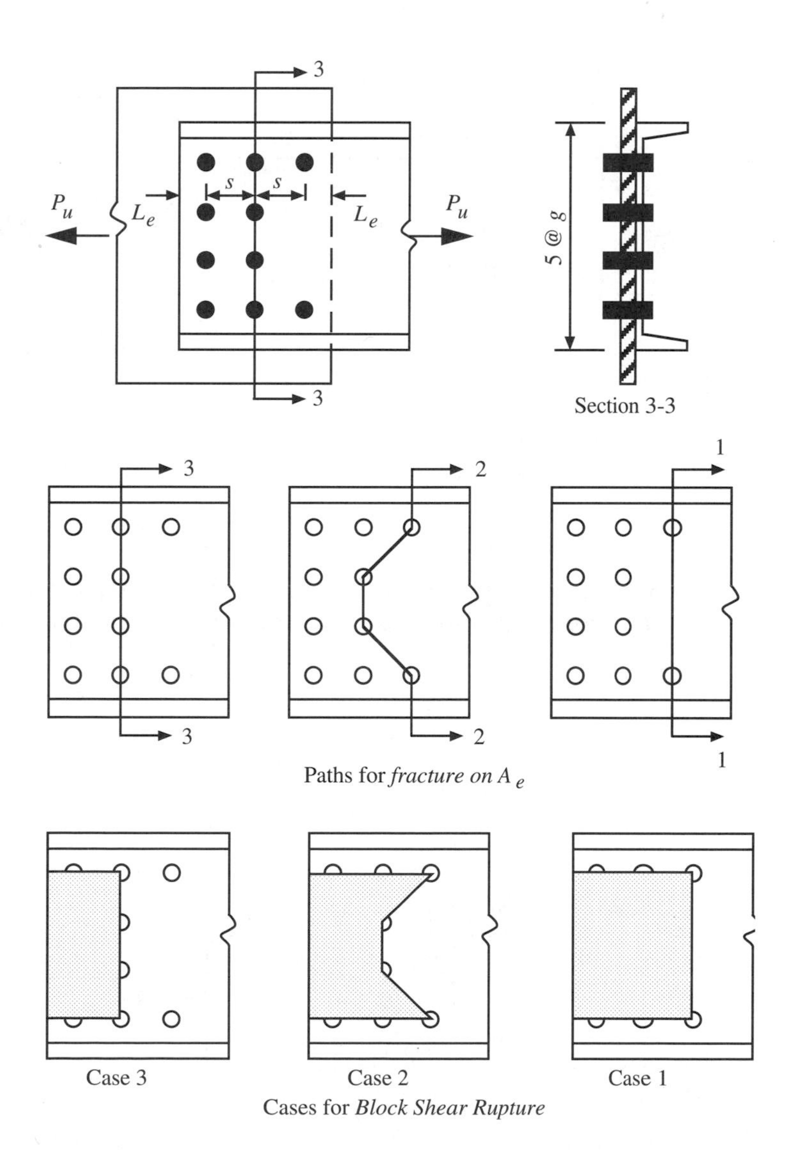

FIGURE P2.4

2.5 A36 steel; 3/4-in.-diameter A325N bolts

$s = 3$ in. $g = 3$ in. $L_e = 1.75$ in.

A pair of $10 \times 3/8$ connector plates is used to butt splice the tension member that is a $10 \times 3/4$ plate. For the member, compute the design strength due to:

1. Yielding on A_g
2. Fracture on A_e
3. Block shear rupture

Is the design satisfactory for $P_u = 240$ kips?

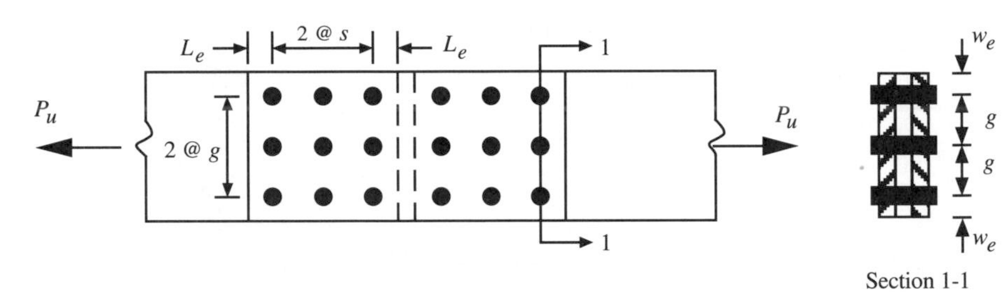

FIGURE P 2.5

2.6 A36 steel; 3/4 in diameter A490X bolts

$s = 3$ in. $g = 3$ in. $L_e = 1.75$ in.

A pair of 10 x 3/8 connector plates is used to butt splice the tension member which is a 10 x 3/4 plate. For the member, compute the design strength due to:

1. Yielding on A_g
2. Fracture on A_e
3. Block shear rupture

Is the design satisfactory for $P_u = 240$ kips?

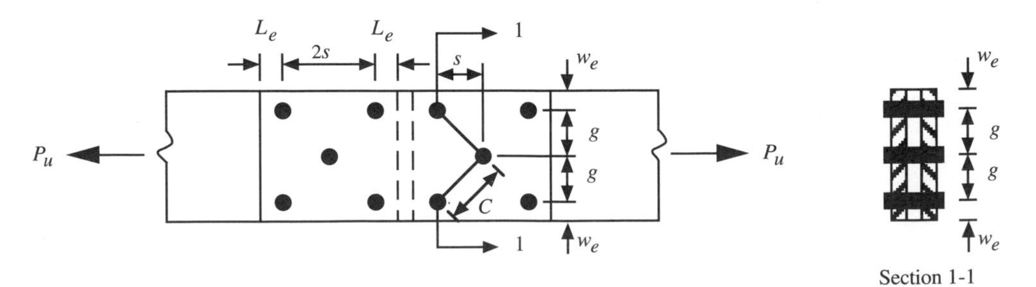

FIGURE P 2.6

2.7 A36 steel; 1-in-diameter A325N bolts

$s = 3$ in. $L_e = 1.75$ in. $g = 5.5$ in.

The member is a W8 x 31 bolted to a pair of 8 x 5/8 gusset plates. For the member, compute the design strength due to:

1. Yielding on A_g
2. Fracture on A_e
3. Block shear rupture

Is the design satisfactory for $P_u = 280$ kips?

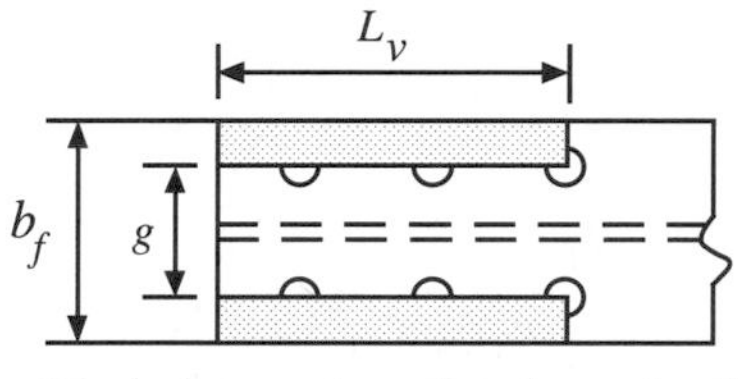

Block shear rupture of each member flange

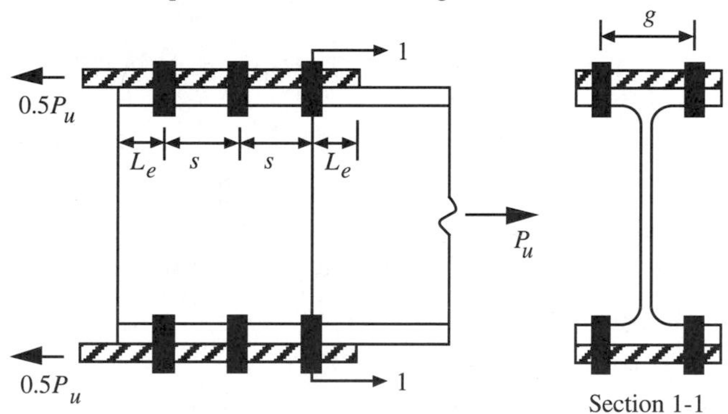

FIGURE P2.7

2.8 A572 Grade 50 steel; 1-in-diameter A490X bolts

$s = 3$ in. $g = 7$ in. $L_e = 1.75$ in.

The tension member is a pair of C15 × 33.9 bolted to a pair of 10 × 5/8 gusset plates. For the member, compute the design strength due to:

1. Yielding on A_g
2. Fracture on A_e
3. Block shear rupture

Is the design satisfactory for $P_u = 495$ kips?

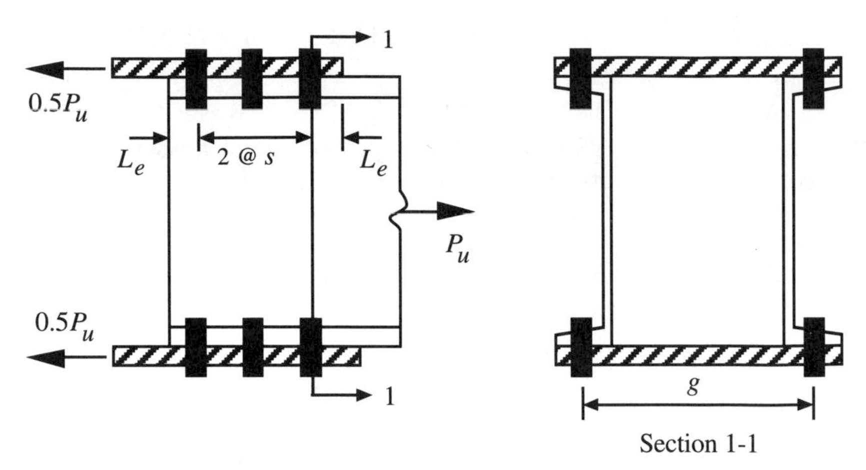

FIGURE P2.8

2.9 A36 steel; 7/8-in.-diameter A325N bolts

$s = 1.50$ in. $L_e = 1.5$ in. $g_1 = 2.25$ in. $g_2 = 2.50$ in. $g = 2.50$ in.

The tension member is a pair of L6×4×1/2 with all legs bolted to 3/4-in.-thick splice plates. The bolt holes are shown on one angle with the short leg flattened down into the plane of the long leg. For the member, compute the design strength due to:

1. Yielding on A_g
2. Fracture on A_e
3. Block shear rupture

Is the design satisfactory for $P_u = 308$ kips?

2.10 A572 Grade 50 steel; E70 electrodes; 1/4-in.-fillet welds

$L_1 = 6$ in. = overlap length of member end on the gusset plate

The tension member is a pair of L5 × 3 × 5/16 with the long legs welded to a 5/8 in. thick gusset plate. For the member, compute the design strength due to:

1. Yielding on A_g
2. Fracture on A_e

Is the design satisfactory for $P_u = 208$ kips?

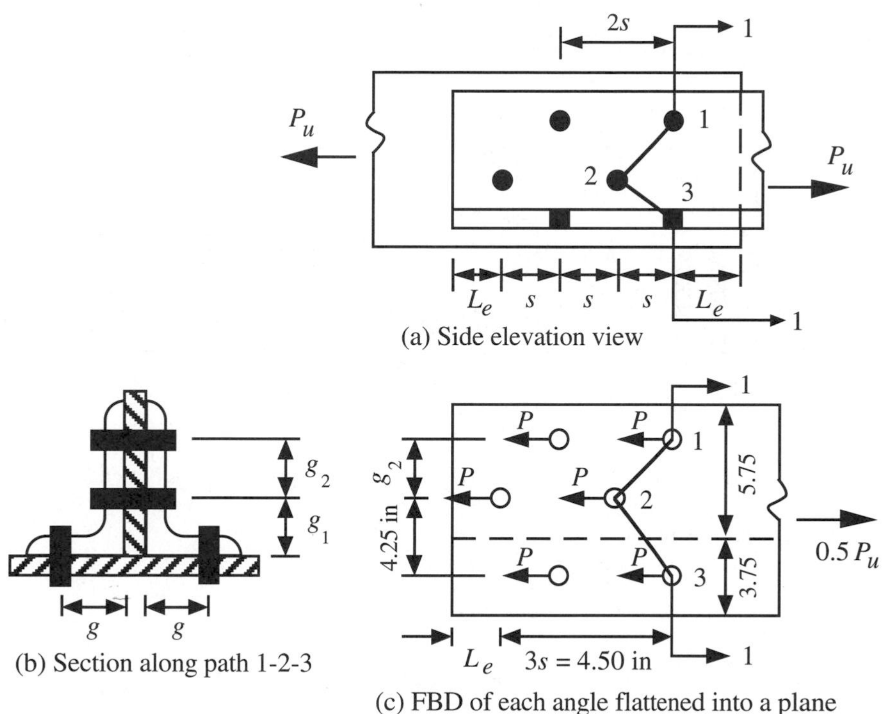

(a) Side elevation view

(b) Section along path 1-2-3

(c) FBD of each angle flattened into a plane

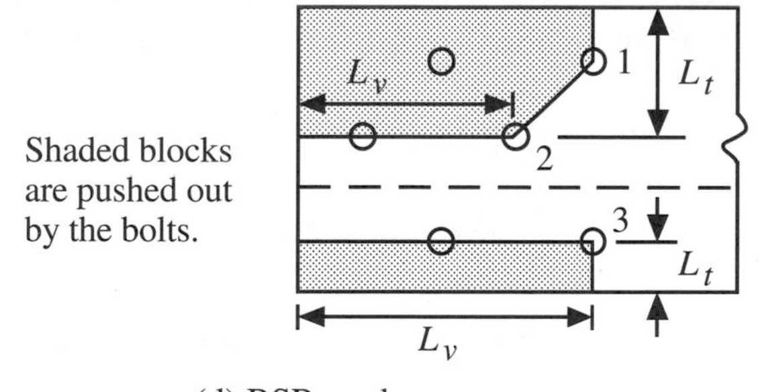

(d) BSR mode

FIGURE P2.9

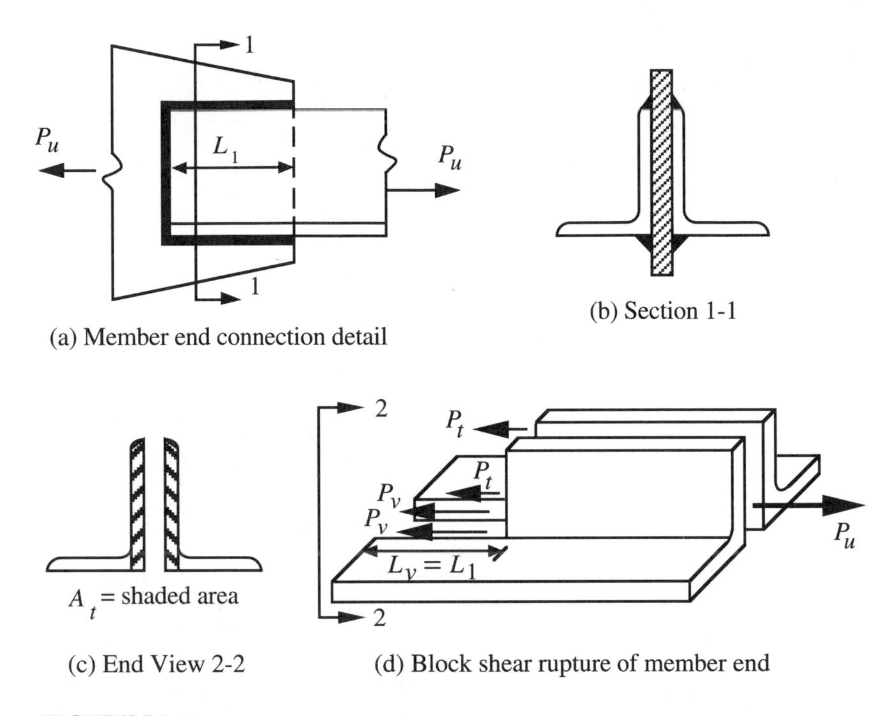

FIGURE P2.10

2.11 A36 steel; E70 electrodes; 1/4-in.-fillet welds

In the truss joint shown, the vertical member is a pair of L5 × 3 × 5/16 with the long legs welded to the web of a WT9 × 23, the horizontal member.
For the vertical member, compute the design strength due to:

1. Yielding on A_g
2. Fracture on A_e

Is the design satisfactory for P_u = 150 kips?

2.12 A36 steel; E70 electrodes; 5/16-in.-fillet welds

L_1 = 18.5 in. = overlap length of each member end on the splice plate

A tension member is a C15 × 33.9 butt spliced with a 1 × 10 × ($2L_1$ + 0.5 in.) plate. For the member, compute the design strength due to:

1. Yielding on A_g
2. Fracture on A_e
3. Block shear rupture

Is the design satisfactory for P_u = 295 kips?

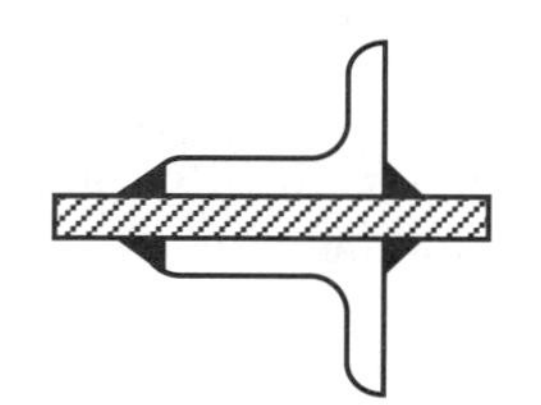

(b) Section 1-1

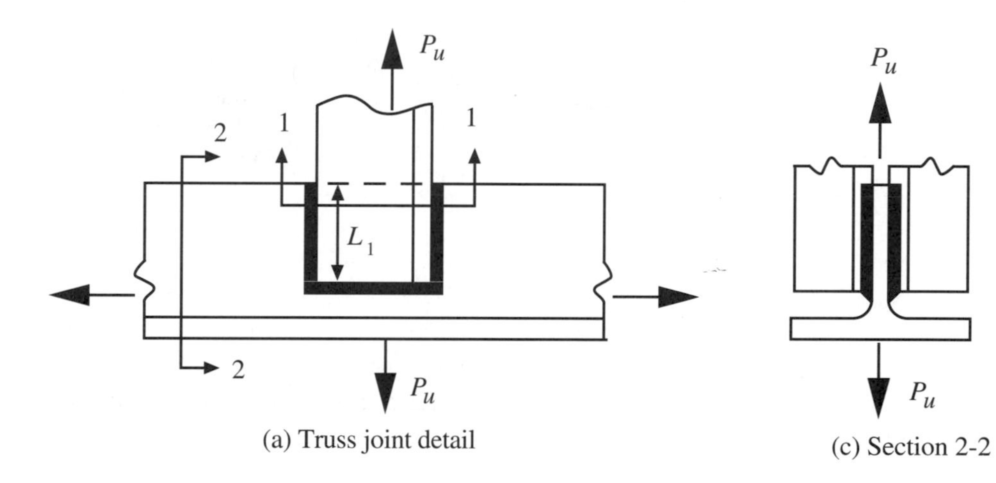

(a) Truss joint detail

(c) Section 2-2

FIGURE P2.11

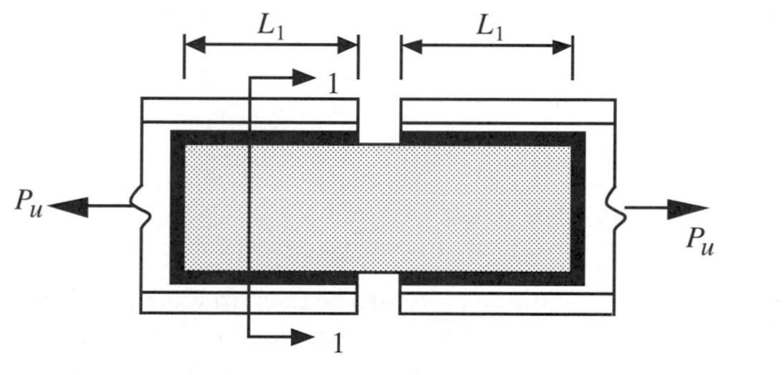

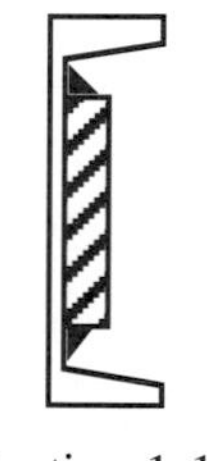

Section 1-1

FIGURE P2.12

2.13 A36 steel; E70 electrodes; 5/16 in. fillet welds;
L_1 = 5.25 in.; P_u = 270 kips

Use the equal-leg and unequal-leg double-angle tables in Part 1 of the LRFD Manual. Select the lightest acceptable pair of angles with a 3/4-in. separation that satisfies the LRFD design requirements for a tension member.

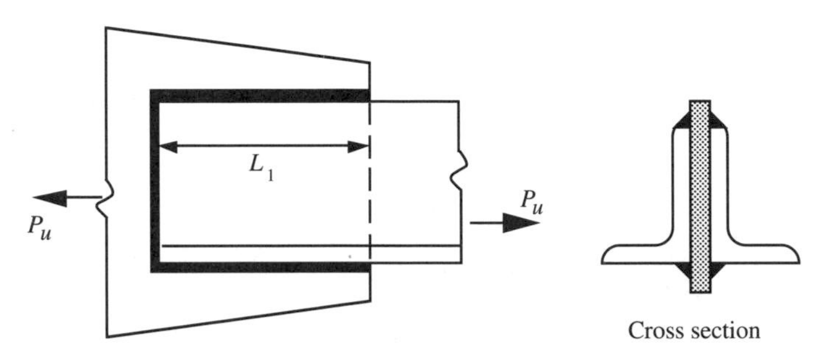

FIGURE P2.13

2.14 A36 steel; E70 electrodes; 5/16 in. fillet welds;
L_1=16in.; P_u = 300 kips

Use the MC (miscellaneous channels) table in Part 1 of the LRFD Manual. Select the lightest acceptable MC section that satisfies the LRFD design requirements for a tension member.

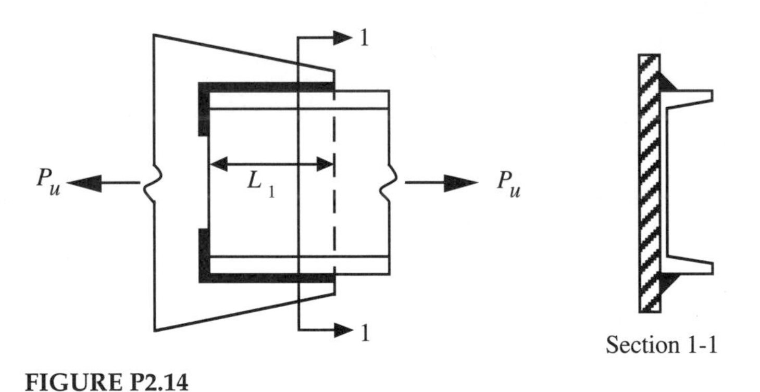

FIGURE P2.14

2.15 A36 steel; E70 electrodes; maximum acceptable fillet welds

Assume that the same double-angle section is to be used in Figure 1.15 for members 5 through 14. From Appendix A, we find that the required axial force in member 10 is 114.2 kips. To account for bending, increase P_u by 25% to 1.25(114.2) = 143 kips. Select the lightest available double-angle section with equal legs that can be used for P_u = 143 kips and A36 steel.

2.16 Solve Problem 2.15 for a double-angle section with long legs back to back.

CHAPTER 3

Connections for Tension Members

3.1 INTRODUCTION

Each end of a tension member is fastened by *connectors* (a group of bolts and/or welds) either to *connecting elements* (plates or rolled sections) or to other members. A *connection* consists of the connecting elements and the connectors.

In the fabrication shop, the connecting elements usually are fastened to the members by welding. At the construction site, the remainder of the connecting process is done either by bolting or by welding. Field bolting requires less skilled labor and can be done under more adverse weather conditions than field welding. If the members are to be connected by field bolting, the bolt holes are punched in the fabrication shop.

A connection may contain a mixture of bolts and welds, only welds, or only bolts. The theoretical analysis techniques for a bolted connection and a welded connection of the same type usually contain some assumed behavioral features that are very similar. Therefore, bolting and welding are discussed in the same chapter.

The connection for a tension member must be designed to develop at least the required force in the tension member. Connections of a structure are designed to have adequate strength to transfer the larger of 10 kips (LRFD J1.7, p. 6-71) per member and the actual member-end forces due to factored loads between the connected member ends.

3.2 CONNECTORS SUBJECTED TO CONCENTRIC SHEAR

Connectors on the ends of a truss member and connectors in the splice of a tension member are examples of connectors subjected to concentric shear. The analysis calculations to determine the governing design strength of the connectors in this type of connection are illustrated in the example problems in this chapter.

LRFD J1.8 (p. 6-72) states that the connectors must be arranged such that the center of gravity of the connector group coincides with the center of gravity of the member, unless provision is made for the eccentricity between the two centers of gravity. However, an exception clause is given in LRFD J1.8 for the end connections of statically loaded single-angle, double-angle, and similar members.

Consider Figures 3.1(a) and (b), which show a tension member fastened, respectively, by welds and bolts to a gusset plate. Along the member at some finite distance from the bolts, all cross-sectional fibers of the member can attain the yield strength when the connectors and gusset plate are stronger than the member. In the region of the member-end connection, the stress distribution due to the applied load is not uniform in the member since some of the cross-sectional elements of the member are not fastened to the gusset plate. Hence, a transition region exists from the connection region to some finite distance from the connection where the stress distribution in the member becomes uniform when yield occurs. Thus, before yielding occurs in the member, the member-end region and the gusset plate usually experience strain-hardening, and fracture can possibly occur either in the gusset plate or in the member-end region.

The behavior of connections is more complex than the behavior of the members joined by the connections. Most connections are highly indeterminate, and plane sections of the connection parts may not remain plane. The length of a connecting element is small compared to the length of the member. Therefore, different simplifying assumptions of behavior are made for the connection than the members joined to the connection. A purely theoretical approach to connections is difficult and very nearly impossible. Finite-element analyses of connections are now possible, but these analyses are dependent on the behavioral assumptions that must be made in the analyses. Therefore, the design of connections is empirical (based on experimental evidence and the structural designer's judgment of how the connections deform). Mathematical models used in the analyses of connections are usually only very rough approximations of the actual behavior of connections.

3.3 BOLTING

In LRFD J3.1 (p. 6-79), we find that:

1. High-strength bolts are to be tightened by the turn-of-nut method, by calibrated torque-wrench, or a direct tension indicator.
2. High-strength bolts in connections not subject to a load that produces tension in the bolts, and where loosening or fatigue is not a design consideration, need only be tightened to the snug-tight condition. The definition of snug-tight is the tightness obtained by a few "rattles" of an air-powered wrench or the full effort of a worker using an ordinary spud wrench in bringing all plies of the connected parts into firm contact.

An abbreviated description of the tightening techniques on LRFD, pp. 6-383 to 386, is:

1. Turn-of-nut method. The nut is rotated to a snug-tight condition and then is further rotated by the amount stipulated in Table 5 on LRFD, p. 6-385.
2. Calibrated torque-wrench method
   (a) Manually operated torque-wrench

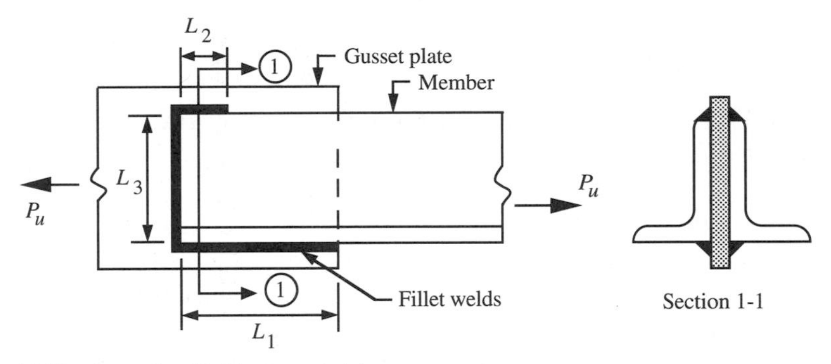

(a) Member end welded to a gusset plate

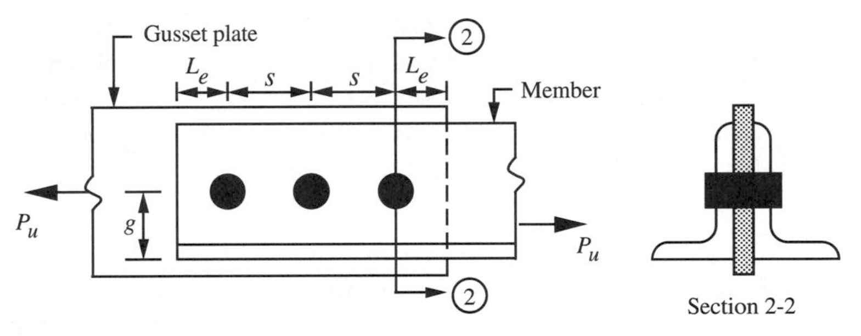

(b) Member end bolted to gusset plate

FIGURE 3.1 Tension member connected to a gusset plate.

For a 0.75-in.-diameter A325 bolt, a 100 pound force applied at the end of a 4 foot long handle is required.

(b) Air-powered wrench
When the air pressure reaches the level that produces the specified torque, a pressure sensor closes the valve at the wrench.

3. Direct tension indicator method

(a) Load indicating washers
A hardened washer with a series of arch-shaped protrusions on one face of the washer. This washer is inserted between the bolt head and the gripped material with the arch-shaped protrusions bearing against the underside of the bolt head. Thus, a gap exists between the bolt head and the surface of the washer. As tightening of the bolt progresses, the arch-shaped protrusions are flattened and the gap decreases. A feeler gage is used to determine when the gap has been reduced sufficiently to produce the required minimum bolt tension.

(b) Load indicating bolts
A bolt with a splined end that extends beyond the threaded portion of the bolt. A specially designed wrench grips the splined end and turns the nut until the splined end shears off. The sheared-off end of the intact bolt is

distinctive and provides an easy means of inspection to ensure that the required minimum bolt tension was achieved.

LRFD J1.8 (p. 6-72) to J1.11 give the applicable bolt grades and the limitations on the usage of bolts.

3.4 TYPES OF CONNECTIONS

There are two types of bolted connections known as *bearing-type* and *slip-critical* (known as *friction-type* prior to 1986). Bolts in a bearing-type connection can be tightened by the turn-of-nut method. In a slip-critical connection, high-strength bolts must be used and must be tightened by using either a direct tension indicator or a calibrated torque-wrench. A field inspection of the bolts in a slip-critical connection must be made to ensure that the required minimum bolt tension was achieved during tightening. Therefore, the costs due to installation and inspection of bolts in a slip-critical connection are greater than in a bearing-type connection. Slip-critical connections are needed for joints subject to fatigue, end connectors in built-up members, bolts in combination with welds, and in other cases where slip is a serviceability concern of the structural engineer.

3.4.1 Slip-Critical Connections

When high-strength bolts in a connection are installed with a specified initial tension by using either the calibrated torque-wrench method or the direct tension indicator method, a *slip-critical* connection is obtained. The required minimum bolt tension is 70% of the minimum tensile strength of the bolt. Due to the specified initial tension in the bolts, the connected pieces are precompressed sufficiently such that the transfer of member-end forces is made by means of the friction developed between the surfaces of the precompressed pieces.

LRFD J3.8 (p. 6-83) requires that (1) a sufficient number of fully tensioned high-strength bolts must be provided to prevent slippage of the member end being connected and (2) bearing at the bolt holes must be checked as required in a bearing-type connection (in case slippage does occur, the connection must function properly as a bearing-type connection).

From LRFD Appendix J3 (p. 6-130), the design requirement for a slip-critical connection on the end of a tension member subjected to factored loads is

$$\phi R_n \geq P_u$$

where

$$P_u = \text{required strength of the tension member}$$

$$R_n = 1.13\mu T_m N_b N_s$$

$$\mu = \text{mean slip coefficient chosen from}$$

(a) $\mu = 0.33$, for unpainted clean mill scale steel surfaces
(b) $\mu = 0.50$, for unpainted blast-cleaned steel surfaces
(c) $\mu = 0.40$, for hot-dip galvanized and roughened surfaces

(d) μ = established from tests (see LRFD, p. 6-389)

$$T_m = 0.7 \quad \text{(nominal bolt strength)}$$

$$N_b = \text{number of bolts}$$

$$N_s = \text{number of slip planes}$$

(a) ϕ = 1.00, for standard holes

(b) ϕ = 0.85, for oversized and short-slotted holes

(c) ϕ = 0.70, for long-slotted holes transverse to load direction

(d) ϕ = 0.60, for long-slotted holes parallel to load direction

When an applied tension force exists parallel to the bolt length direction at the joint and has a required tensile strength of T_u, the design requirement for a slip-critical connection on the end of a tension member subjected to factored loads is

$$\phi R_n [1 - T_u/(1.13 T_m N_b)] \geq P_u$$

3.4.1 BEARING-TYPE CONNECTIONS

Bolts in a bearing-type connection need only be snug-tight and can be either A307 bolts or high-strength bolts (A325, A449, A490). The properties for each of these bolt grades were given in Section 1.1.1 and are not repeated here. The snug-tight condition of tightening bolts removes the slack in the connection, prevents loosening of the nuts, and prevents play at the member ends fastened to the connection. The initial tension in snug-tightened bolts only generates a small clamping force in the connected parts. This clamping force is assumed to be zero in the factored load analysis of bearing-type connections and slippage of the connection occurs. As shown in Figure 3.2, when slippage occurs, the member end moves in one direction and the connecting elements move in the opposite direction until the bolts in the connection stop the slippage. The tension force is transferred from the member end to the connecting elements by bearing on the bolts such that the bolts are subjected to shear.

3.5 BOLTS IN A BEARING-TYPE CONNECTION

As shown in Figures 3.2(b) and 3.2(c), failure of a bolt in a bearing-type bolted connection is due to shear. Each bolt can be subjected either to single shear or to double shear. Only the double-shear strength is used for the case of multiple shear since the simultaneous occurrence of a shear failure on more than two planes is highly improbable.

LRFD J3.6 (p. 6-83) and Table J3.2 (p. 6-81) give the single-shear design strength of one bolt in a bearing-type connection as

$$\phi R_n = \phi F_v A_{gv}$$

where

$$\phi = 0.75$$

$$A_{gv} = \pi d^2/4 = \text{gross single-shear area of a bolt}$$

$$d = \text{unthreaded bolt diameter}$$

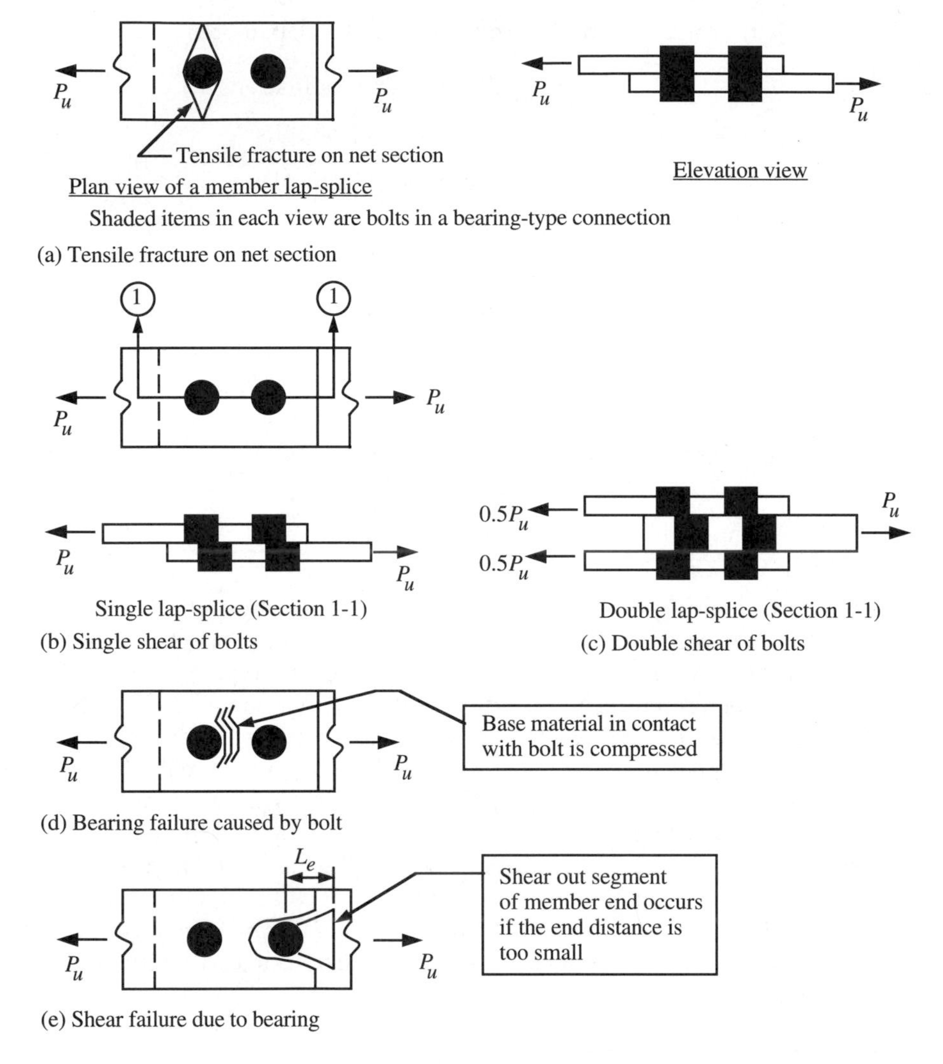

FIGURE 3.2 Bolted lap-splice of a tension member.

$$F_v = \textit{nominal shear strength} \quad \text{(see Table 3.1)}$$

The shear design strength for a bolt subjected to double shear is two times the single-shear strength.

For a bolt group in a bearing-type bolted connection on the end of a tension member, the strength design requirement is

$$\phi R_n \geq P_u$$

ϕR_n = sum of shear design strength of the bolts

P_u = required strength in the tension member

Table 3.1 **Design Shear Strength of Bolts**

Bolt Type	F_v (ksi)
A307	24
A325N	48
A325X	60
A490N	60
A490X	75

Note: N denotes threads included in the shear plane(s), X denotes threads excluded from the shear plane(s).

Adapted from LRFD, Table J3.2.

Example 3.1

In the bearing-type bolted connection of Figure 3.3(a), the gusset plate is 0.50 in. thick, 6.50 in. wide, and has 0.75-in.-diameter A325X bolts in it. Use $L_e = 2$ in. and $s = 3$ in. The member is a pair of L3.5×3×0.25. The gusset plate and member are made of A36 steel. Find the shear design strength of the bolt group.

Solution

From Table 3.1 for A325X bolts, $F_v = 60$ ksi. For double shear of one 3/4-in.-diameter bolt,

$$A_{gv} = 2(\pi d^2/4) = 2[\pi(0.75)^2/4] = 0.8836 \text{ in.}^2$$

$$\phi R_n = \phi F_v A_{gv} = 0.75[60(0.8836)] = 39.76 \text{ kips/bolt.}$$

For the three-bolt group,

$$\phi R_n = 3(39.76) = 119.3 \text{ kips}$$

Example 3.2

In the bearing-type bolted connection of Figure 3.3(a), the gusset plate is 0.50 in. thick, 6.50 in. wide, and has 0.875-in.-diameter A325N bolts in it. Use $L_e = 2$ in. and $s = 3$ in. The member is a pair of L3.5×3×0.25. The gusset plate and member are made of A36 steel. Find the number of bolts required for $P_u = 100$ kips.

Solution

From Table 3.1 for A325N bolts, $F_v = 48$ ksi. For double shear of a 1-in.-diameter bolt:

$$A_{gv} = 2(\pi d^2/4) = 2[\pi(0.875)^2/4] = 1.2026 \text{ in.}^2$$

$$\phi R_n = \phi F_v A_{gv} = 0.75[48(1.2026)] = 43.29 \text{ kips/bolt.}$$

For the bolt group, the design requirement is

$$(\text{Number of bolts})(\phi R_n/\text{bolt}) \geq (P_u = 100 \text{ kips})$$

$$(\text{Number of bolts}) \geq [100 \text{ kips}/(43.29 \text{ kips/bolt}) = 2.30 \text{ bolts}]$$

Use three bolts.

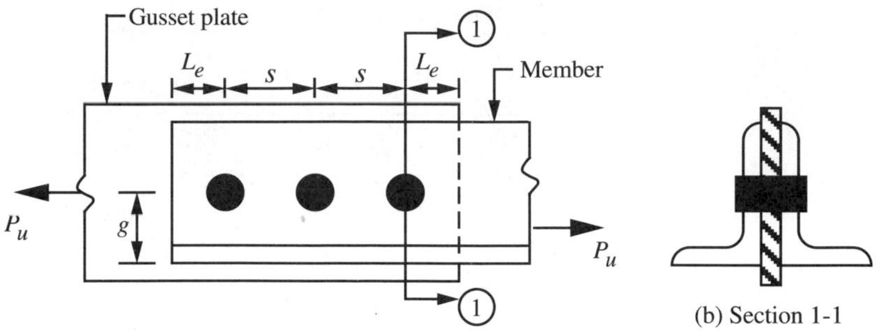

(b) Section 1-1

(a) Member end bolted to gusset plate

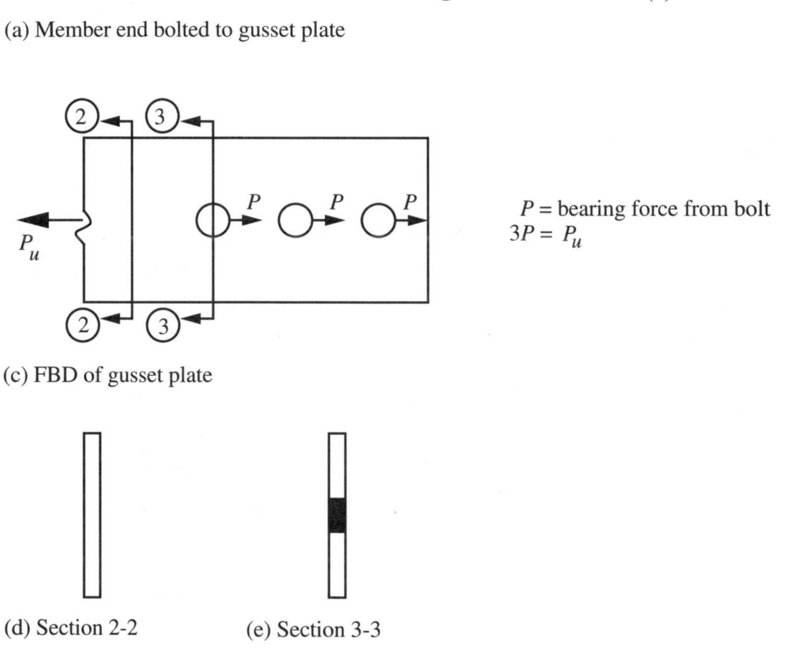

(c) FBD of gusset plate

(d) Section 2-2

(e) Section 3-3

FIGURE 3.3 Tension member bolted to a gusset plate.

3.6 BEARING AT THE BOLT HOLES

As shown in Figures 3.2(d) and (e), bolts in a bearing-type bolted connection can cause a bearing failure at the bolt holes. Bearing failure is analogous to laying a steel bolt across a stick of margarine. The bolt is harder than margarine. The bolt weight bears on and compresses the portion of the margarine stick beneath the bolt. Similarly, since bolts are much harder than the steel in the member and the gusset plate, almost all deformation due to bearing occurs in the member and in the gusset plate.

The definitions of bearing strength at a bolt hole in a bearing-type connection are given in LRFD J3.10 (p. 6-85). The parameters that appear in the bearing strength definitions at a standard bolt hole are

$$\phi = 0.75$$

$$R_n = \text{nominal bearing strength, ksi}$$

L_e = length (in.) parallel to P_u from free end of connected part to center of nearest bolt hole (minimum acceptable values of L_e are given in LRFD Table J3.4, p. 6-82)

s = spacing distance (in.) parallel to P_u between centers of bolt holes

d = diameter of bolt, (in.)

t = thickness of critical connected part, (in.)

F_u = tensile strength of connected part, (ksi)

At the bolt hole nearest the free end of a connected part,

1. When $L_e \geq 1.5d$, and

(a) Deformation of the bolt holes is a design consideration

$$R_n = 2.4dtF_u$$

(b) Deformation of the bolt holes is not a design consideration

$$R_n = L_e tF_u \leq 3.0dtF_u$$

2. When $L_e < 1.5d$,

$$R_n = L_e tF_u \leq 2.4dtF_u$$

At the other bolt holes,

1. When $s \geq 3.0d$, and

(a) Deformation of the bolt holes is a design consideration

$$R_n = 2.4dtF_u$$

(b) Deformation of the bolt holes is not a design consideration

$$R_n = (s - 0.5d)tF_u \leq 3.0dtF_u$$

2. When $s < 3.0d$,

$$R_n = (s - 0.5d)tF_u \leq 2.4dtF_u$$

For bearing at the bolt holes in a bolted connection on the end of a tension member, the strength design requirement is

$$\phi R_n \geq P_u$$

ϕR_n = sum of bearing design strength at the bolt holes

P_u = required strength in the tension member

Example 3.3

In the bearing-type bolted connection of Figure 3.3(a), the gusset plate is 0.50 in. thick, 6.50 in. wide, and has 0.75-in.-diameter A325X bolts in it. Use L_e = 1.5 in. and s = 2.5 in. The member is a pair of L3.5 × 3 × 0.25. The gusset plate and member are made of

A36 steel. Deformation of the bolt holes is a design consideration. Find the bearing design strength of the connection.

Solution

From LRFD Table J3.4, p. 6-82, we find that the minimum acceptable value of $L_e = 1.25$ in. at a sheared edge. ($L_e = 1.5$ in.) ≥ 1.25 in. as required.
For the member, the thickness bearing on the bolt is $t = 2(0.25) = 0.50$ in. For the gusset, the thickness bearing on the bolt is $t = 0.50$ in. The governing thickness is the thinner part bearing on the bolt, which is $t = 0.50$ in.
At the bolt hole nearest the governing free end

$$(L_e = 2 \text{ in.}) \geq [1.5d = 1.5(0.75) = 1.125 \text{ in.}]$$

$$\phi R_n = 0.75(2.4dtF_u) = 0.75(2.4)(0.75)(0.5)(58) = 39.15 \text{ kips}$$

At each of the other bolt holes,

$$(s = 2.5 \text{ in.}) \geq [3d = 3(0.75) = 2.25 \text{ in.}]$$

$$\phi R_n = 0.75(2.4dtF_u) = 39.15 \text{ kips}$$

For the three-bolt group. the bearing design strength is

$$\phi R_n = 3(39.15) = 117.45 \text{ kips}$$

3.7 CONNECTING ELEMENTS IN A BOLTED CONNECTION

In the fabrication shop, connections for building trusses usually are made by welding. If the entire truss is too large to ship, large segments of the truss are fabricated, shipped, and joined together by bolts on the job site to form the complete truss.

Figure 3.2 shows the modes of failure for a bearing-type bolted lap splice of a tension member. These modes of failure and some additional ones are applicable for the connection in Figure 3.3, which shows a tension member fastened by bolts in a bearing-type connection to a gusset plate. Along the member at some finite distance from the bolts, all cross-sectional fibers of the member can attain the yield strength when the bolts and member-end connection are stronger than the member. In the region of the member-end connection, the stress distribution due to the applied load is not uniform in the member since some of the cross-sectional elements of the member are not bolted to the gusset plate. Hence, a transition region exists from the connection region to some finite distance from the connection where the stress distribution in the member becomes uniform when yield occurs. Thus, before yielding occurs in the member, the member-end region and the gusset-end region containing the bolt holes usually experience strain-hardening, and fracture can possibly occur through the bolt holes either in the gusset plate or in the member. If all cross-sectional fibers of a member yield in tension, the member elongates excessively, which can precipitate failure somewhere in the structural system, of which the tension member is a part. Although the gusset plate is much shorter than the member, the same logic of excessive elongation of the gusset plate due to yielding in the gross section also may precipitate failure somewhere in the structural system. Hence, tensile yielding in the gross section of the gusset plate is considered to be a limiting condition of failure. In Figure 3.3, fracture of the gusset plate can occur on Section 3-3, which has a bolt hole in it. Also, block shear rupture of the gusset plate can occur through all bolt holes on a section parallel to the member length (see Figure 3.4).

At a truss joint, more than one member is usually fastened to a gusset plate. In the following discussion, we are only concerned with the design strength of the gusset plate at the end of each tension member attached by bolts in a bearing-type connection. The simplest possible case is one tension member attached to a gusset plate. Equally simple is a lap splice for a tension member [see Figure 3.2(c)]. A lap-splice plate and the simplest case of a gusset plate are very short tension members. Therefore, except for the definition of fracture on the critical net section, the tensile design strength definitions of a lap splice and the simplest case of a gusset plate are identical to the definitions of the design strength for a tension member.

For connecting elements subjected to tension in a bearing-type bolted connection, the strength design requirement is

$$\phi R_n \geq P_u$$

$$\phi R_n = \text{design strength of the connecting elements}$$

$$P_u = \text{required strength in the tension member}$$

The design strength definitions of the connecting elements are:

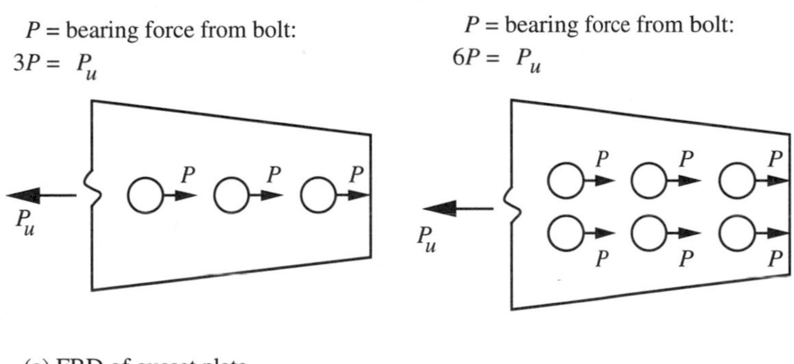

(a) FBD of gusset plate

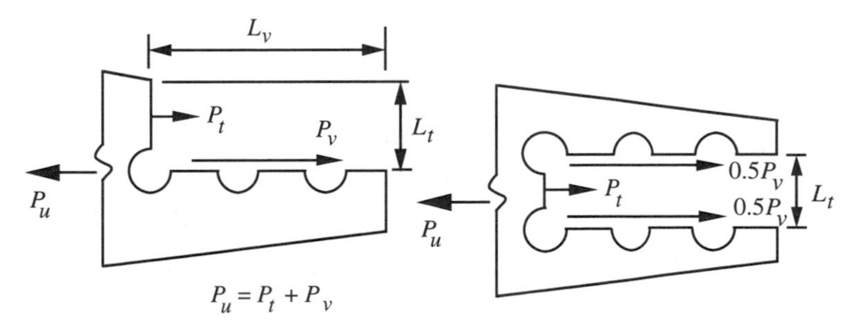

(b) Gusset plate block shear rupture FBD

FIGURE 3.4 Examples of block shear rupture.

1. *Yielding on* A_g (LRFD J5.2, p. 6-88) [see Figures 3.3(c) and (d) for an example]

$$\phi R_n = 0.90 F_y A_g$$

where

$$F_y = \text{yield strength}$$

$$A_g = \text{gross area of the plate} = t_p b_p$$

$$t_p = \text{thickness of the plate}$$

$$b_p = \text{width of the plate perpendicular to } P_u$$

2. *Fracture on* A_n (LRFD J5.2, p. 6-88, and B2, p. 6-34) [see Figures 3.3(c) and 3.3(e) for an example] LRFD Commentary J5.2 (p. 6-229) states that the $0.85A_g$ limitation on A_n of connecting elements was chosen to agree with test results and provide a reserve capacity. This limitation recognizes the limited inelastic deformation capacity in a connecting-element length that is very short compared to a tension-member length.

$$\phi R_n = 0.75 F_u A_n$$

where

$$F_u = \text{tensile strength}$$

$$A_n = \text{smaller of} \begin{cases} A_g - A_{\text{holes}} \\ 0.85 A_g \end{cases}$$

$$A_g = \text{gross area of the plate} = t_p b_p$$

$$A_{\text{holes}} = \Sigma d_h t_h = \text{sum of bolt hole areas in the cross section}$$

$$d_h = \text{actual diameter of hole} + 1/16 \text{ in.}$$

$$t_h = \text{thickness of hole}$$

For a staggered path across the plate,

$$A_g = t_p b_p + \sum_{i=1}^{ns} \left(\frac{s^2 t}{4g} \right)$$

where

$$ns = \text{number of staggers on the path}$$

$$s = \text{pitch} = \text{stagger component parallel to } P_u$$

$$g = \text{gage} = \text{stagger component perpendicular to } P_u$$

$$t = \text{thickness of plate}$$

3. *Block shear rupture*, abbreviated BSR (LRFD J4.3, pp. 6-87 and p. 6-228) (see Figure 3.4 for some examples)

When $F_uA_{nt} \geq 0.6F_uA_{nv}$,

$$\phi R_n = 0.75(F_uA_{nt} + 0.6F_yA_{gv})$$

When $0.6F_uA_{nv} \geq F_uA_{nt}$,

$$\phi R_n = 0.75(0.6F_uA_{nv} + F_yA_{gt})$$

where

A_{gv} = gross area on BSR shear plane(s), (in.^{2})

A_{gt} = gross area on BSR tension plane, (in.^{2})

$A_{nv} = A_{gv}$ - [A_{holes} on BSR shear plane(s)]

$A_{nt} = A_{gt}$ - [A_{holes} on BSR tension plane]

All bolt holes are standard holes except where otherwise specified. In the fabrication shop, standard bolt holes are punched in the member unless the material thickness exceeds the hole diameter. The nominal diameter of a standard bolt hole (see LRFD Table J3.3, p. 6-82) is the bolt diameter plus 1/16 in. For tensile strength calculations, the hole diameter is defined in LRFD B2 as the nominal diameter of the hole plus 1/16 in.

Example 3.4

In the bearing-type bolted connection of Figure 3.3(a), the gusset plate is 0.50 in. thick, 6.50 in. wide, and has 0.75-in.-diameter A325N bolts in it. Use L_e = 2 in. and s = 3 in. The member is a pair of L3.5 × 3 × 0.25. The gusset plate and member are made of A36 steel. Find the governing design strength of the gusset plate.

Solution

The governing design strength is the least ϕR_n value obtained from:

1. *Yielding on* $A_g = b_pt_p = 6(0.5) = 3.00$ in.^{2}

$$\phi R_n = 0.90F_yA_g = 0.9(36)(3.00) = 97.2 \text{ kips}$$

2. *Fracture on* A_n [see Figures 3.3(c) and (e)]

d_h = actual diameter of hole + 1/16 in. = 3/4 + 1/16 = 7/8 in.

$$A_{\text{hole}} = d_ht_h = 0.875(0.5) = 0.4375 \text{ in.}^2$$

$$A_g - A_{\text{holes}} = 3.00 - 0.4375 = 2.5625 \text{ in.}^2$$

$$0.85A_g = 0.85(3.00) = 2.55 \text{ in.}^2 \quad \text{(governs } A_n\text{)}$$

$$\phi R_n = 0.75F_uA_n = 0.75(58)(2.55) = 110.9 \text{ kips}$$

3. *Block shear rupture* [see Figure 3.4(b), one row of bolts case] For one-bolt hole,

$$A_{\text{hole}} = 0.4375 \text{ in.}^2 \quad \text{(from preceding item)}$$

$$A_{gv} = [2 + 2(3)](0.5) = 4.00 \text{ in.}^2$$
$$A_{nv} = A_{gv} - [A_{\text{holes}} \text{ on BSR shear plane(s)}]$$
$$A_{nv} = 4.00 - 2.5(0.4375) = 2.906 \text{ in.}^2$$
$$A_{gt} = (2.50)(0.5) = 1.25 \text{ in.}^2$$
$$A_{nt} = A_{gt} - [A_{\text{holes}} \text{ on BSR tension plane}]$$
$$A_{nt} = 1.25 - 0.5(0.4375) = 1.03 \text{ in.}^2$$
$$F_u A_{nt} = 58(1.03) = 59.8 \text{ kips}$$
$$0.6F_u A_{nv} = 0.6(58)(2.906) = 101.1 \text{ kips}$$
$$\phi R_n = 0.75(0.6F_u A_{nv} + F_y A_{gt})$$
$$\phi R_n = 0.75[101.1 + 36(1.03)] = 103.6 \text{ kips}$$

For the design shown in Figure 3.3(a), a summary of our computed design strengths is:

1. For the gusset plate (from Example 3.4),
   (a) $\phi R_n = 97.2$ kips due to *yielding on* A_g
   (b) $\phi R_n = 103.6$ kips due to BSR
2. For bearing at the bolt holes (from Example 3.3, $\phi R_n = 117.4$ kips)
3. For shear of bolt group (from Example 3.1, $\phi R_n = 119.3$ kips)
4. For the member (from Example 2.1),
   (a) $\phi P_n = 101.4$ kips due to *yielding on* A_g
   (b) $\phi P_n = 96.1$ kips due to BSR of the member end

The governing design strength is 96.1 kips, which is the least of the design strengths, and the LRFD strength design requirement is ($\phi P_n = 96.1$ kips) $\geq P_u$.

3.8 WELDING

Welding of steel is the process of joining two pieces of steel by melting a similar metal into the joint between the two pieces of steel. After the molten metal solidifies and cools to the temperature of the surrounding air, the weld is stronger than an identically sized portion of steel from the base material (the two pieces of steel being joined).

There are numerous welding processes, but the most important one for structural engineers is electric arc welding. For automated shop welds, the *submerged* (hidden) *arc welding* (SAW) process and the *electroslag* process (for groove welds primarily; however, this process can be used for fillet welds) are used. For field welds, the *shielded metal arc welding* (SMAW) process is used. In the SMAW process, a coated wire called a weld *electrode* is placed in an electrode holder connected to a variable source of electric power. When the free end of the electrode is located close enough to the base material, an arc forms between the electrode tip and the base material. The temperature of the arc is about 10,000°F. As the weld electrode melts, the coating on the electrode forms a gaseous shield to protect the molten metal from the air (prevents oxidation), stabilizes the arc, and makes more effective use of the arc energy. The temperature of the molten metal is about 6500°F. The temperature in the base material beneath the deposited weld can be as high as 3500°F, which causes some melting of the base material and intermixing of the two molten metals. Residual stresses form in the base material when the temperature exceeds 1400°F [29].

The types of welds used in civil engineering structures are *fillet welds, groove welds, slot welds, plug welds,* and *puddle welds.*

Fillet welds are discussed in Section 3.8 and used in some example problems. About 80% of structural welds are fillet welds since they are the easiest and the least expensive ones to make.

As the name implies, *groove welds* are made in a groove whose edges usually must be specially prepared. Groove welds are the most efficient welds, but they are also the most expensive ones to make. About 15% of structural welds are groove welds.

A *plug weld* is made in a circular hole. A *slot weld* is made in a slot (elongated hole with a circular end or ends). For example, suppose that the length available for fillet welds at the end of a plate is not adequate to provide the required design strength. A slot can be removed from the center of the plate width to increase the length available for welding, and weld material is deposited in the slot. For strength purposes, a slot weld can only be used to resist shear. Plug welds and slot welds are also used in built-up members to stitch a cover plate to the underlying material in order to prevent buckling of the cover plate.

Puddle welds are used to stitch sheets of metal decking to the underlying top flanges of beams that serve as supports for the metal decking. An automatic puddle weld device has been developed to allow the welder to stand upright and make the puddle welds. This device is held somewhat like one holds a walking cane. The electric arc on the bottom tip of the device melts through the deck, forms a puddle of molten metal, and fastens the metal decking to the underlying top flanges of the beams on which the metal decking is supported.

3.9 FILLET WELDS

Figure 3.5(a) and (b) show possible cross sections of a fillet weld. The sides of the largest isosceles right triangle that can be inscribed within the cross section of the weld are called legs. The leg dimension is the fillet weld size S_w.

For the SMAW (shielded metal arc welding) process, the *effective throat thickness* t_e is the shortest distance from the root of the weld to the hypotenuse of the isosceles right triangle; that is, $t_e = 0.707S_w$.

For the SAW (submerged arc welding) process the effective throat thickness is:

1. $t_e = S_w$ when $S_w \leq 3/8$ in.
2. $t_e = 0.707S_w + 0.11$ in. when $S_w > 3/8$ in.

3.9.1 Strength of Fillet Welds

In LRFD Table J2.5, p. 6-78, for fillet welds, we find that weld metal with a strength equal to or less than matching weld metal is permitted to be used. LRFD uses yield strength to denote the grade of a structural steel. The American Welding Society (AWS) uses tensile strength to denote the grade of an electrode. The electrode grades for SMAW are E60, E70, E80, E90, E100, and E110 where E denotes electrode and the numerals are the tensile strength F_u (ksi) of the electrode. From AWS D1.1 Table 4.1 [30], we find the following SMAW matching weld electrodes:

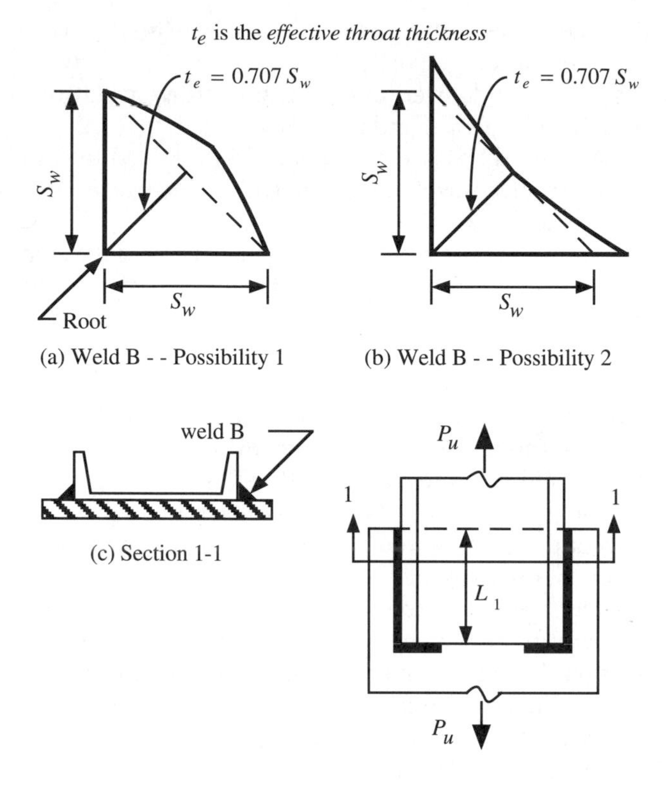

FIGURE 3.5 Possible cross sections of a fillet weld.

1. E70(F_y = 60 ksi; $F_u \geq 72$ ksi) for A36 steel(F_y = 36 ksi; $F_u \geq 58$ ksi)
2. E70(F_y = 60 ksi; $F_u \geq 72$ ksi) for A572 Grade 50 steel(F_y = 50 ksi; F_u = 65 ksi)
3. E80 for A572 Grades 60 and 65 steel
4. E100 for A514 steel when $t > 2.5$ in.
5. E110 for A514 steel when $t \leq 2.5$ in.

When electrodes whose strength is not greater than the strength permitted in LRFD Table J2.5 are used in fillet-welded connections on the ends of tension members, the fillet welds do not separate from the base material. The weld failure mode is fracture of the welds or, as stipulated in LRFD Table J2.5, a shear rupture failure alongside the welds in the base material.

Figures 3.6 to 3.10 show fillet-welded connections on the end of a tension member. As shown in Figure 3.6, the failure mode of longitudinal welds is shear fracture on each weld throat plane. As shown in Figure 3.7, the failure mode of transverse welds is combined tension and shear fracture on each weld throat plane. Figures 3.8 to 3.10 depict the more common situation of longitudinal and transverse welds being used in the same connection. Tests by Butler et al [22] have shown:

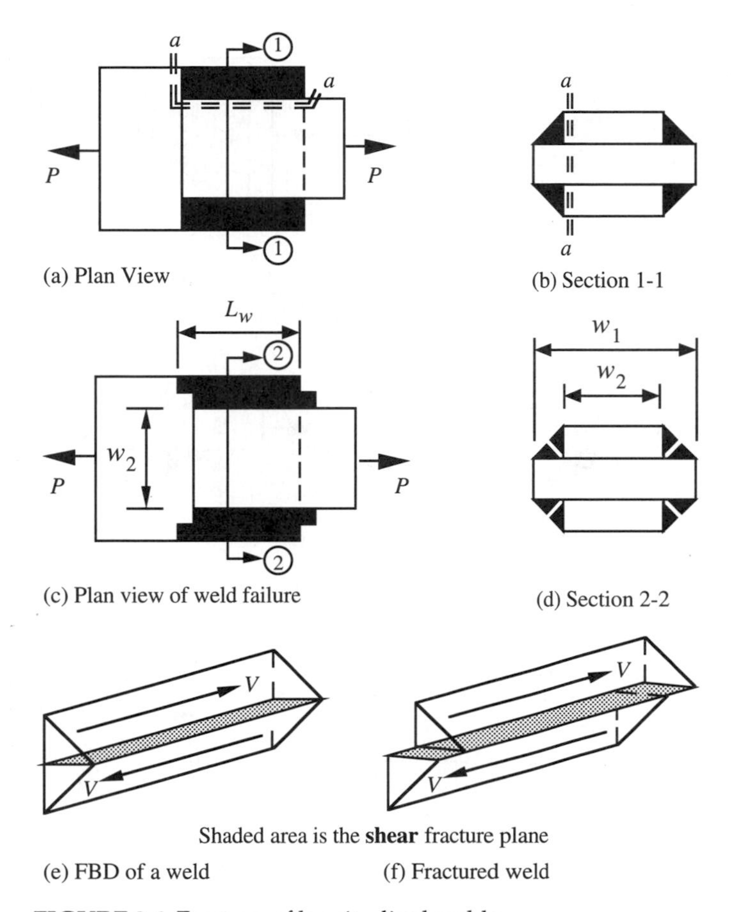

FIGURE 3.6 Fracture of longitudinal welds.

1. a transverse weld is 50% stronger than a longitudinal weld.
2. a longitudinal weld is much more ductile than a transverse weld.

LRFD Appendix J2.4 (p. 6-129) provides an alternative design strength for fillet welds that accounts for a transverse weld being 50% stronger than a longitudinal weld.

For a fillet-welded connection on the end of a tension member, the strength design requirement is:

$$\phi R_n \geq P_u$$

ϕR_n = design strength of the weld group

P_u = required strength in the tension member

We choose to use the alternative design strength definition for fracture on the throat plane of fillet welds (LRFD Appendix J2.4, p. 6-129). For each weld in a fillet-weld group, the design strength is

$$\phi R_n = 0.75(0.60F_{EXX})(0.707S_w)(1.0 + 0.50 \sin^{1.5}\theta)L_w$$

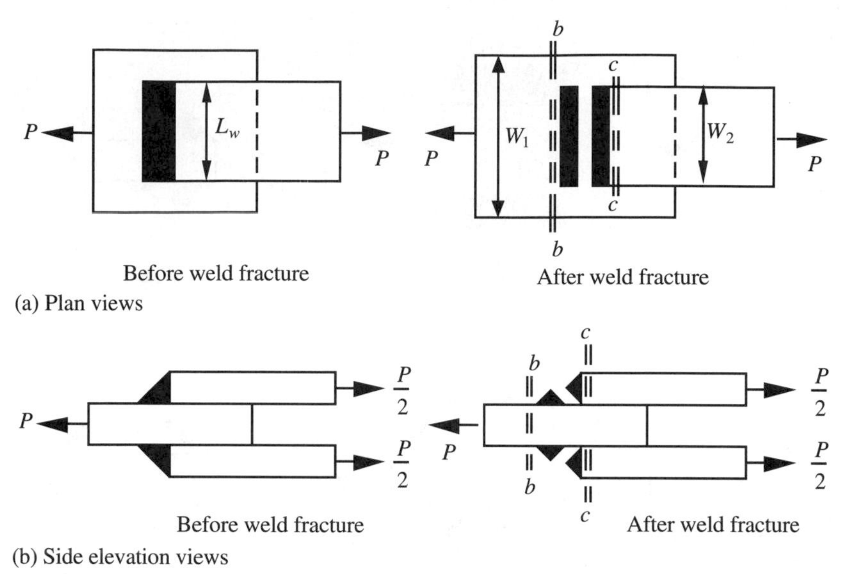

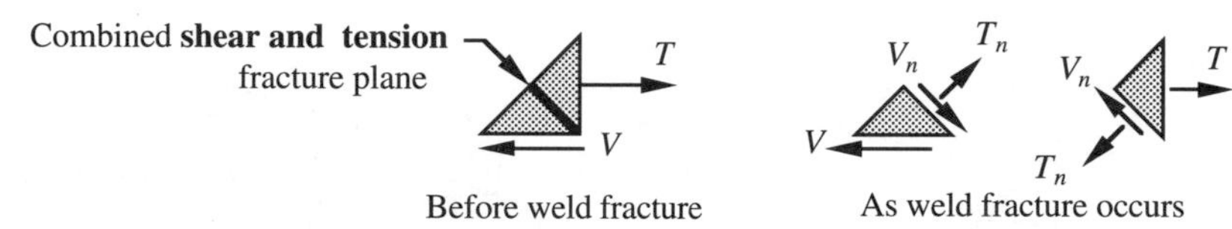

(c) FBD of a transverse weld

FIGURE 3.7 Fracture of transverse welds.

where

$$F_{\text{EXX}} = \text{electrode tensile strength}$$

$$S_w = \text{size of fillet weld}$$

$$L_w = \text{length of a weld}$$

θ = angle (degrees) between P_u and the length-direction axis of a weld

For a longitudinal weld in a tension member connection,

$$\theta = 0$$

$$\phi R_n = 0.75(0.60F_{\text{EXX}})(0.707S_w)L_w$$

For a transverse weld in a tension member connection,

$$\theta = 90°$$

$$\phi R_n = 0.75(0.60F_{\text{EXX}})(0.707S_w)(1.50L_w)$$

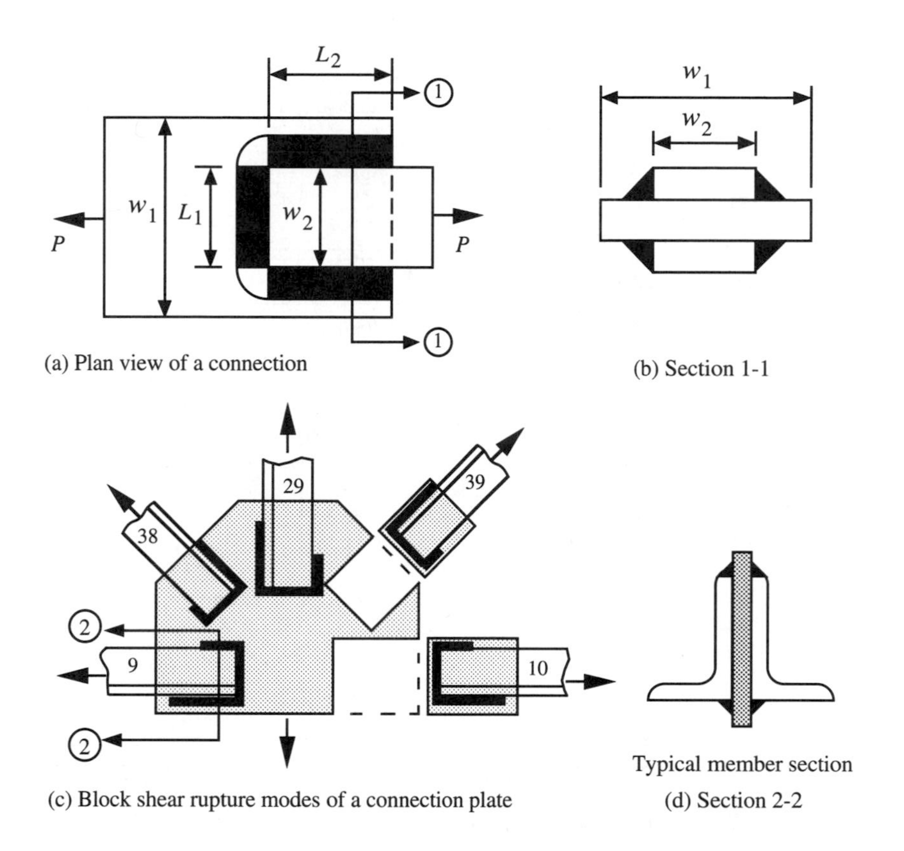

FIGURE 3.8 Weld lengths and block shear rupture of a gusset plate.

Example 3.5

In Figure 3.9(a), E60 electrodes are used for the 1/4 in. fillet welds; the member is a pair of L3×2×5/16; and, the 5/8-in.-thick gusset plate is 5 in. wide. A36 steel is used for the member and gusset plate. Find the design strength of the fillet welds for L_1 = 4.50 in., L_2 = 1.50 in., and L_3 = 3.00 in.

Solution

For a longitudinal weld,

1. Weld fracture:

$$\phi R_n\,/\text{in.} = 0.75(0.60F_{EXX})(0.707S_w)$$

$$\phi R_n\,/\text{in.} = 0.75(0.60)(60)(0.707)(0.25) = 4.77 \text{ kips/in.}$$

2. Shear rupture of the member alongside a weld:

$$\phi R_n\,/\text{in.} = 0.75(0.60F_u)t = 0.75(0.60)(58)(0.3125) = 8.16 \text{ kips/in.}$$

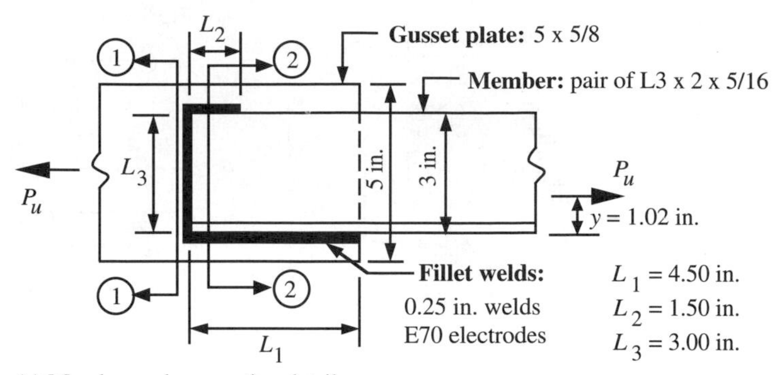

(a) Member end connection detail

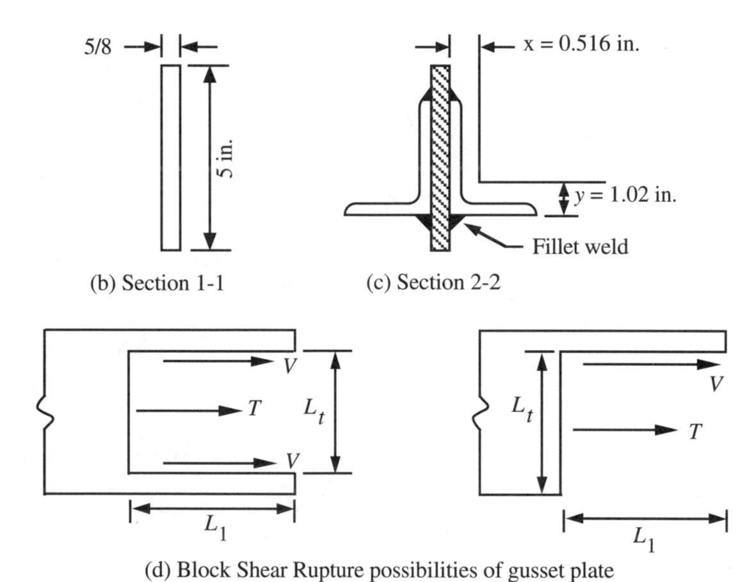

(b) Section 1-1

(c) Section 2-2

(d) Block Shear Rupture possibilities of gusset plate

FIGURE 3.9 Tension member fillet welded to a gusset plate.

Since 8.16 > 4.77, weld fracture governs.

3. Shear rupture of the gusset alongside two welds:

$$\phi R_n\,/\text{in.} = 0.75(0.60F_u)t = 0.75(0.60)(58)(0.625) = 16.3 \text{ kips/in.}$$

Since 16.3 > [2(4.77) = 9.54], weld fracture governs.

For a transverse weld,

1. Weld fracture:

$$\phi R_n\,/\text{in.} = (4.77 \text{ kips/in.})(1.5) = 7.16 \text{ kips/in.}$$

2. Tension rupture of the gusset alongside two welds:

$$\phi R_n\,/\text{in.} = 0.75\Gamma_u t = 0.75(58)(0.625) = 27.2 \text{ kips/in.}$$

Since 27.2 > [2(7.16) = 14.3], weld fracture governs.

For the fillet weld group, the design strength is

$$\Sigma[(\phi R_n/\text{in.})L_w] = 2[(4.77)(4.50 + 1.50) + (7.16)(3.00)] = 100.2 \text{ kips}$$

3.9.2 Design of Fillet Welds

The maximum fillet weld size that can be deposited in one pass is $S_w = 5/16$ in. Most of the heat given off in the welding process is absorbed by the parts being joined. Heat in the deposited weld is absorbed faster by the thicker part being joined. Ductility is adversely affected when the weld metal cools too rapidly. Also, the thicker base material is stiffer and provides more restraint to shrinkage of the deposited weld during cooling. To ensure that enough heat is available in the deposited weld for proper fusion of the weld to the base material, a minimum fillet weld size is specified in LRFD Table J2.4 (p. 6-75) as a function of the thicker part being joined, and $L_w \geq 4S_w$ is stipulated in LRFD J2.2b (p. 6-75). $L_w > 70S_w$ is not permitted for a longitudinal weld. End returns [shown in Figure 3.10(d) and numerically illustrated in Example 3.7] are not required, but they increase the ductility of the connection. When an end return is not provided, the distance from the weld termination point to the end of the connected part must be at least S_w. Other limitations and reasons for them are given, respectively, in LRFD J2.2b (p. 6-75) and C-J2.2b (p. 6-220).

Only the design of SMAW fillet welds will be illustrated. Figures 3.10 (c–e) show the possible weld arrangements on the end of a pair of L2.5 × 2 × 5/16 used as a tension member. In Examples 3.6 to 3.8, we will design SMAW fillet welds for each of the weld arrangements to resist $P_u = 84.9$ kips. We will use E70 electrodes and A36 steel for the member and the gusset plate ($t = 3/8$ in.). Information applicable in each of the examples is:

1. For thicker part joined, $t = 3/8$ in. (gusset plate), minimum $S_w = 3/16$ in.
2. $t = 5/16$ in. of the angle section governs maximum $S_w = 5/16 - 1/16 = 1/4$ in.
3. We choose to use $S_w = 1/4$ in.
4. For a longitudinal weld:

   (a) Weld fracture:

   $$\phi R_n/\text{in.} = 0.75(0.60F_{\text{EXX}})(0.707S_w)$$

   $$= 0.75(0.60)(70)(0.707)(0.25) = 5.57 \text{ kips/in.}$$

   (b) Shear rupture of the member alongside a weld:

   $$\phi R_n/\text{in.} = 0.75(0.60F_u)t = 0.75(0.60)(58)(0.3125) = 8.16 \text{ kips/in.}$$

   Since 8.16 > 5.57, weld fracture mode governs.

   (c) Shear rupture of the gusset alongside two welds:

   $$\phi R_n/\text{in.} = 0.75(0.60F_u)t = 0.75(0.60)(58)(0.375) = 9.79 \text{ kips/in.}$$

   Since 9.79 < [2(5.57) = 11.14], shear rupture mode governs.

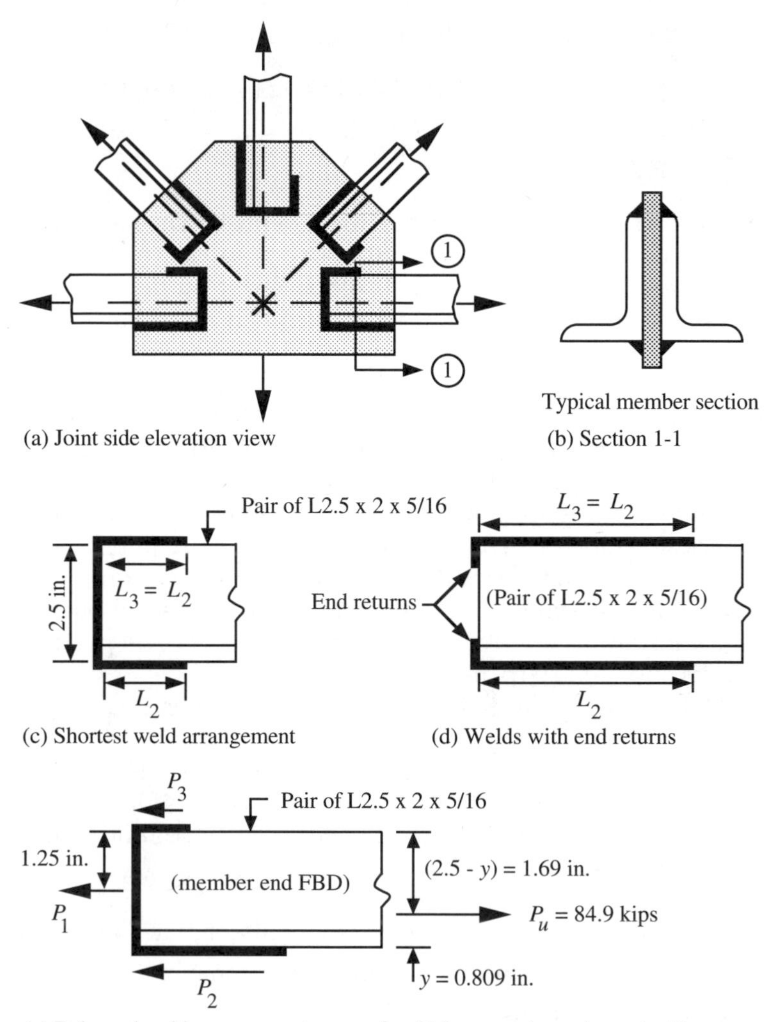

FIGURE 3.10 Welded truss joint details for double-angle members.

5. For a transverse weld,

   (a) Weld fracture:

   $$\phi R_n/\text{in.} = (5.57\ \text{kips/in.})(1.5) = 8.35\ \text{kips/in.}$$

   (b) Tension rupture of the gusset alongside two welds:

   $$\phi R_n/\text{in.} = 0.75F_ut = 0.75(58)(0.375) = 16.3\ \text{kips/in.}$$

   Since 16.3 < [2(8.35) = 16.7], tension rupture mode governs.

Note: This failure mode is not listed in LRFD Table J2.5, but it can govern as shown when Appendix J2.4 is used to obtain the design weld strengths.

6. Other design strength values for the design in Figure 3.10(a) are:

   (a) For *yielding on* A_g of the member, $\phi P_n = 84.9$ kips.
   (b) For *fracture on* A_e of the member, $L_c \geq 2.19$ in. is required for $\phi P_n \geq 84.9$ kips.
   (c) For BSR of the gusset, $L_v \geq 3.00$ in. is required for $\phi P_n \geq 84.9$ kips.

Example 3.6

For the weld arrangement shown in Figure 3.10(c), note that $L_1 = 2.50$ in. is given and $L_2 = L_3$ is required. Design the welds for $P_u = 84.9$ kips using the applicable information in the last paragraph preceding this example.

Solution

From items 5(b) and 4(c) of the list preceding this example, the design strength of the fillet weld group is

$$\Sigma[(\phi R_n\,/\text{in.})L_w] = (16.3)(2.50) + (9.79)(L_2 + L_3)$$

$$= 40.75 \text{ kips} + 19.58L_2$$

The strength design requirement is

$$\Sigma[(\phi R_n\,/\text{in.})L_w] \geq P_u$$

$$(40.75 \text{ kips} + 19.58L_2) \geq 84.9 \text{ kips}$$

$$L_2 \geq 2.25 \text{ in.}$$

$$L_2 \geq (L_v \geq 3.00 \text{ in.}) \text{ for the required BSR strength [see item 6c]}$$

Use $L_3 = L_2 = 3.00$ in. for each L2.5 x 2 x 5/16.

Example 3.7

For the arrangement shown in Figure 3.10(d), note that there are end returns and $L_2 = L_3$. Using the minimum permissible length for each end return and the applicable information preceding Example 3.6, design the welds for $P_u = 84.9$ kips.

Solution

Let L_{er} = length of an *end return*. From LRFD J2.2b, at the free end of a tension member:

1. $L_{er} \geq [2S_w = 2(0.25) = 0.5$ in.] is required.
2. If we want the end return to be fully effective in our weld strength calculations, $L_{er} \geq [4S_w = 4(0.25) = 1.00$ in.] is required.
3. When $L_{er} < 4S_w$, then $S_w = L_{er}/4$ (or, when rupture of the base material governs, one-fourth of the design strength for rupture of the base material) must be used in the weld strength calculations.

For each end return, choose $L_{er} = 0.5$ in. Since the end returns are transverse welds, $(L_{er} = 0.5 \text{ in.}) < [4S_w = 4(0.25) = 1.00 \text{ in.}]$, and item (5b) in the itemized list preceding Example 3.6 is applicable, the total usable design strength for the end returns is

$$\phi R_n/\text{in.} = 2(16.3 \text{ k/in.})(0.5 \text{ in.})/4 = 4.08 \text{ kips}$$

Also, item (4c) in the itemized list preceding Example 3.6 is applicable. The design strength of the fillet weld group is

$$\Sigma[(\phi R_n/\text{in.})L_w] = 4.08 \text{ kips} + (9.79)(L_2 + L_3)$$

$$= 4.08 \text{ kips} + 19.58L_2$$

The strength design requirement is

$$\Sigma[(\phi R_n/\text{in.})L_w] \geq P_u$$

$$(4.08 \text{ kips} + 19.58L_2) \geq 84.9 \text{ kips}$$

$$L_2 \geq 4.13 \text{ in.}$$

$$L_2 \geq (L_v \geq 3.00 \text{ in.}) \text{ for the required BSR strength} \quad [\text{see item 6c}]$$

From LRFD J2.2b, we find that $L_2 \leq [70S_w = 70(0.25) = 17.5 \text{ in.}]$ is required. Use $L_3 = L_2 = 4.25$ in. with 0.5-in. end returns for each L2.5 × 2 × 5/16.

Example 3.8

The weld lengths L_2 and L_3 in a balanced-weld arrangement are chosen such that the centroid of the weld forces coincides with P_u which is the axial force in the member. For the balanced-weld arrangement shown in Figure 3.10(e), note that $L_2 > L_3$, and use the applicable information preceding Example 3.6 to design the welds for $P_u = 84.9$ kips.

Solution

From items 5(b) and 4(c) of the list preceding Example 3.6, the design strength of the fillet weld group is

$$P_1 = (16.3)(2.50) = 40.75 \text{ kips}$$

$$P_2 = 9.79L_2$$

$$P_3 = 9.79L_3$$

The strength design requirement is

$$(P_1 + P_2 + P_3) \geq (P_u = 84.9 \text{ kips})$$

$$[40.75 + 9.79(L_2 + L_3)] \geq 84.9$$

$$(L_2 + L_3)] \geq 4.51 \text{ in.}$$

At P_3 in Figure 3.10(e), $\Sigma M = 0$ gives

$$2.5P_2 + 1.25P_1 = 1.69(P_u)$$

$$2.5(9.79L_2) + 1.25(40.75) = 1.69(84.9)$$

$$L_2 = 3.78 \text{ in.}$$

$L_2 \geq (L_v \geq 3.00$ in.) for the required BSR strength [see item 6(c)]

$L_3 \geq (4.51 - 3.78 = 0.73$ in.) is required. $L_3 \geq [4S_w = 4(0.25) = 1.00$ in.] also is required. Use $L_2 = 3.75$ in. and $L_3 = 1.00$ in. on each L2.5 × 2 × 5/16.

3.10 CONNECTING ELEMENTS IN A WELDED CONNECTION

A *gusset plate* (a member-end connector plate) may have an irregular shape [see Figure 3.10(a)] to accommodate the fastening of several member ends at a joint. The gusset plate shape in Figure 3.9(a) is the simplest form of a gusset plate. More than one connector plate or more than one connecting element may be used.

Figure 3.8(c) shows examples of block shear rupture of the connecting element at the end of members 10 and 39. Block shear rupture is a tearing failure mode that can occur along the perimeter of welds. As shown in Figures 3.8(c) and (g), a block of material can be torn out of the connection. As shown in Figure 3.11, if fracture starts to occur on the tension plane of the block, yielding simultaneously occurs on the shear plane(s) of the block. Fracture occurs on the block plane(s) that provides the larger possible fracture force.

In Figure 3.8(a), the connecting elements can fail by yielding on their gross sections. For the middle plate, block shear rupture can occur as depicted for members 10 and 39 in Figure 3.8(c).

In Figure 3.6, the connecting elements can fail by yielding on their gross sections or by shear rupture between the pair of dotted lines along path a-a. If width w_1 of the middle plate is larger than implied, block shear rupture can occur for the middle plate as depicted for members 10 and 39 in Figure 3.8(c).

In Figure 3.7, the connecting elements can fail by yielding on their gross sections b-b and c-c. If the width w_1 of the middle plate is large, block shear rupture can occur for the middle plate as depicted for members 10 and 39 in Figure 3.8(c).

The LRFD design strength definitions for failure of the connecting element(s) in a welded connection are:

1. *Yielding on* A_g of the connecting element(s) [LRFD J5.2(a), p. 6-88] (see Figure 3.9 Section 3-3 for an example)

$$\phi R_n = 0.90F_yA_g$$

where

$$F_y = \text{yield strength}$$

$$A_g = \text{gross area of the plate} = t_pb_p$$

$$t_p = \text{thickness of the plate}$$

$$b_p = \text{width of the plate perpendicular to } P_u$$

2. *Block shear rupture* (LRFD J4.3, p. 6-87)

(see Figure 3.8(c) for examples) For welded connections, there are no holes; therefore, let

$$A_v = \text{gross area on BSR shear plane(s)}$$

$$A_t = \text{gross area on BSR tension plane}$$

When $F_u A_t \geq 0.6 F_u A_v$,

$$\phi R_n = 0.75(F_u A_t + 0.6 F_y A_v)$$

When $0.6\ F_u A_v > F_u A_t$

$$\phi R_n = 0.75(0.6 F_u A_v + F_y A_t)$$

3. *Shear rupture strength* (LRFD J4.1, p. 6-87).

   [see path a-a on the plates in Figures 3.6(a) and (b) for an example]

$$\phi R_n = 0.75(0.6 F_u A_v)$$

and the parameters are as previously defined for block shear rupture.

Example 3.9

Find the design strengths of the $5 \times 5/8$ gusset plate of A36 steel in Figure 3.9(a). See Figure 3.11 for the BSR failure mode of this gusset plate.

Solution

1. *Yielding on* $A_g = 5(0.625) = 3.125\ \text{in.}^2$

$$\phi R_n = 0.9 F_y A_g = 0.9(36)(3.125) = 101.25\ \text{kips}$$

2. *Block shear rupture*

$$F_u A_t = (58\ \text{ksi})(4.25\ \text{in.})(0.625\ \text{in.}) = 154.06\ \text{kips}$$

$$0.60 F_y A_v = 0.60(58\ \text{ksi})(6.75\ \text{in.})(0.625\ \text{in.}) = 146.81\ \text{kips}$$

$$\phi R_n = 0.75[F_u A_t + 0.6 F_y A_v]$$

$$\phi R_n = 0.75[154.06 + 0.6(36)(6.75)(0.625)] = 183.9\ \text{kips}$$

For the design shown in Figure 3.9(a), a summary of our computed design strengths is:

1. For the gusset plate (from Example 3.9)
   (a) $\phi R_n = 101.25$ kips due to yielding on A_g
   (b) $\phi R_n = 183.9$ kips due to BSR
2. For the weld group (from Example 3.5, $\phi R_n = 100.2$ kips)
3. For the member (from Example 2.6)

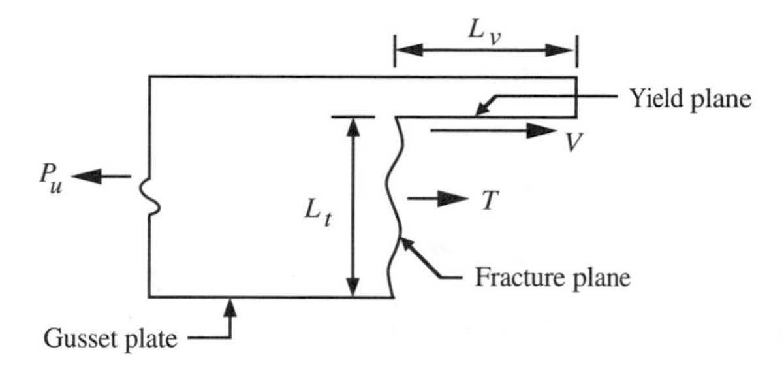

(a) Tensile fracture and shear yielding possibility

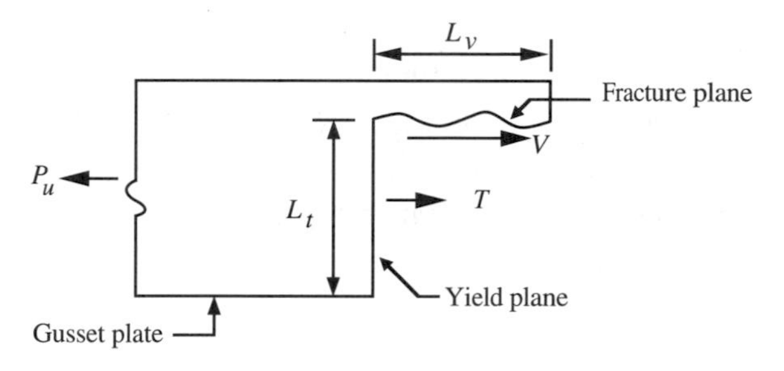

(b) Tensile yielding and shear fracture possibility

For the 5/16 by 5 in. gusset plate shown in Figure 3.9(a),

L_t = 3.50 + (5 - 3.50)/2 = 4.25 in.

L_V = 6.75 in.

Figure 3.11 Block shear rupture possibilities for a gusset plate.

(a) $\phi P_n = 94.9$ kips due to yielding on A_g
(b) $\phi P_n = 114$ kips due to fracture on A_e
(c) $\phi P_n = 127$ kips due to BSR

The governing design strength is 94.9 kips, which is the least of the design strengths, and the LRFD strength design requirement is ($\phi P_n = 94.9$ kips) $\geq P_u$.

PROBLEMS

3.1 A36 steel; 7/8-in.-diameter A325N bolts

$s = 2.75$ in. $L_e = 1.5$ in. $g_1 = 3$ in. $g_2 = 3$ in. $w_e = 3$ in.

The tension member is a pair of L8 × 6 × 1/2 with the long legs bolted to a 10 × 1 gusset plate. Compute the design strength for:

1. Bolt shear
2. Bearing at the bolt holes
3. The gusset plate (yielding on A_g; fracture on A_n)

Is the design satisfactory for $P_u = 298$ kips?

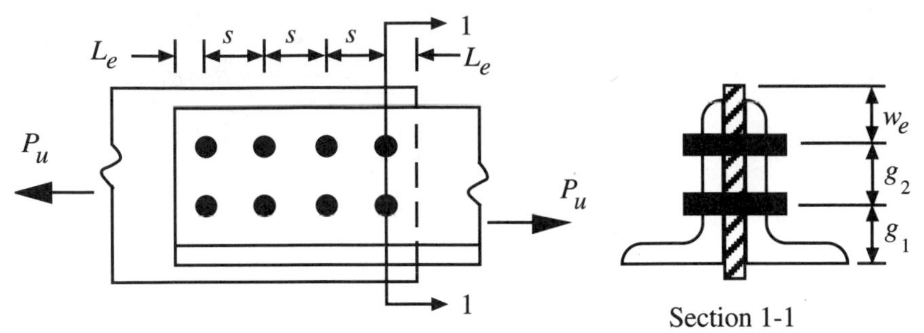

FIGURE P3.1

3.2 A36 steel; 7/8-in.-diameter A325N bolts

$s = 1.5$ in.; $L_e = 1.5$ in.; $g_1 = 2.25$ in.; $g_2 = 2.50$ in.; $w_e = 2.25$ in.

The tension member is a pair of L6 × 4 × 1/2 with the long legs bolted to a 9 × 1 gusset plate. Compute the design strength for:

1. Bolt shear
2. Bearing at the bolt holes
3. The gusset plate (yielding on A_g; fracture on A_n for each path with full P_u on it)

Is the design satisfactory for $P_u = 270$ kips?

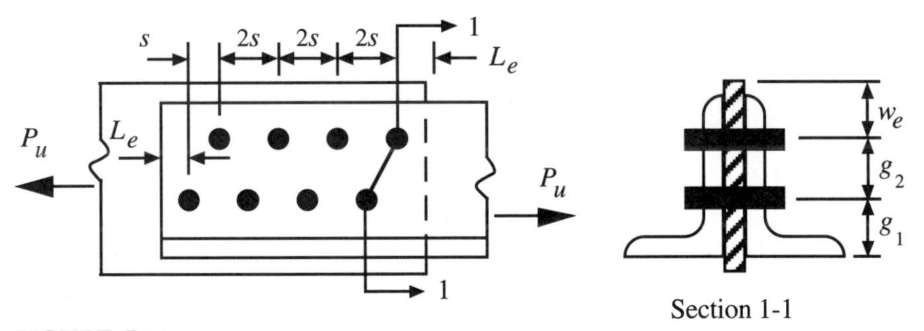

FIGURE P3.2

3.3 A36 steel; 1-in.-diameter A325N bolts

$s = 3$ in.; $g = 3$ in.; $L_e = 1.75$ in.; $w_e = 3.5$ in.

The tension member is a C15 × 33.9 bolted to a 16 × 5/8 gusset plate. Compute the design strength for:

1. Bolt shear
2. Bearing at the bolt holes
3. The gusset plate (yielding on A_g; fracture on A_n)

Is the design satisfactory for $P_u = 200$ kips?

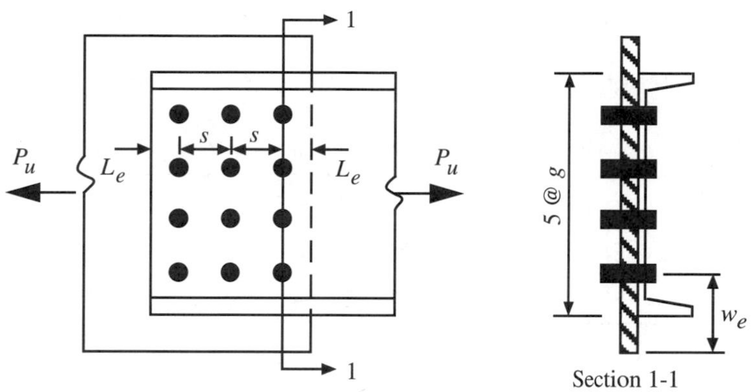

FIGURE P3.3

3.4 A36 steel; 1-in.-diameter A325N bolts

$$s = 3 \text{ in.} \qquad g = 3 \text{ in.} \qquad L_e = 1.75 \text{ in.} \quad w_e = 3.5 \text{ in.}$$

The tension member is a C15 × 33.9 bolted to a 16 × 5/8 gusset plate. Compute the design strength for:

1. Bolt shear
2. Bearing at the bolt holes
3. The gusset plate (yielding on A_g; fracture on A_n)

Is the design satisfactory for $P_u = 200$ kips?

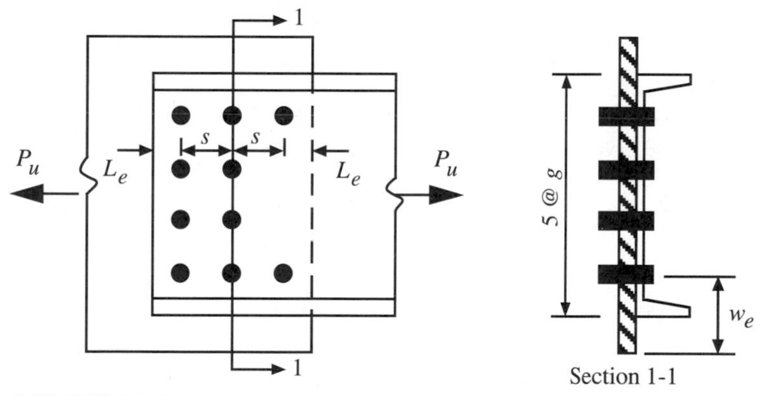

FIGURE P3.4

3.5 A36 steel; 3/4 in diameter A325N bolts

$$s = 3 \text{ in.} \qquad g = 3 \text{ in.} \qquad L_e = 1.75 \text{ in.}$$

A pair of 10 × 3/8 connector plates is used to butt splice the tension member which is a 10 × 3/4 plate. Compute the design strength for:

1. Bolt shear
2. Bearing at the bolt holes

3. The splice plates (yielding on A_g; fracture on A_n)

Is the design satisfactory for $P_u = 240$ kips?

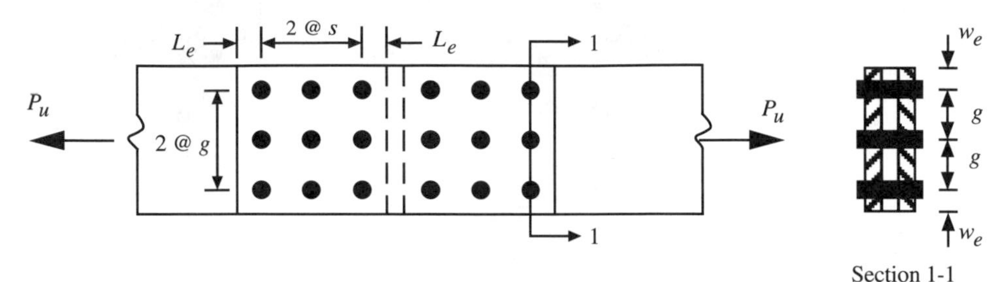

FIGURE P3.5

3.6 A36 steel; 3/4-in.-diameter A490X bolts

$$s = 3 \text{ in.} \qquad g = 3 \text{ in.} \qquad L_e = 1.75 \text{ in.}$$

A pair of 10 × 3/8 connector plates is used to butt splice the tension member which is a 10 × 3/4 plate. Compute the design strength for:

1. Bolt shear
2. Bearing at the bolt holes
3. The splice plates (yielding on A_g; fracture on A_n for each path with full P_u on it)

Is the design satisfactory for $P_u = 240$ kips?

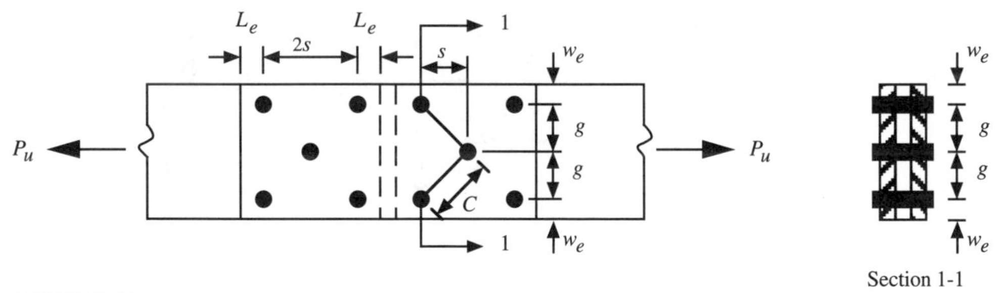

FIGURE P3.6

3.7 A36 steel; 1-in. diameter A325N bolts

$$s = 3 \text{ in.} \qquad L_e = 1.75 \text{ in.}; \qquad g = 5.5 \text{ in}$$

Each flange of the member(W8 × 31) is bolted to an 8 x 5/8 gusset plate at the member end. Compute the design strength for:

1. Bolt shear
2. Bearing at the bolt holes
3. The gusset plate (yielding on A_g; fracture on A_n)

Is the design satisfactory for $P_u = 280$ kips?

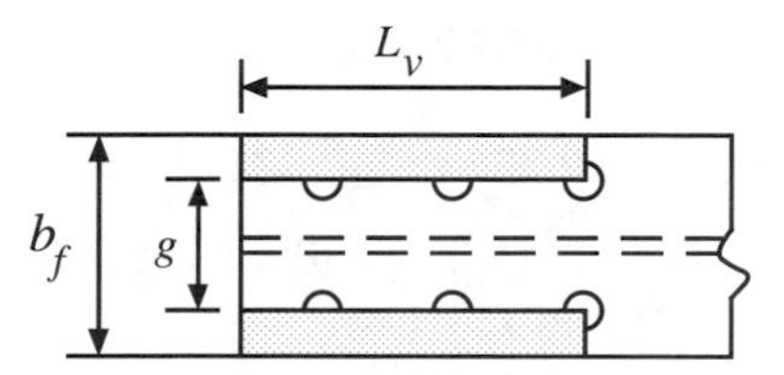

Block shear rupture of each member flange

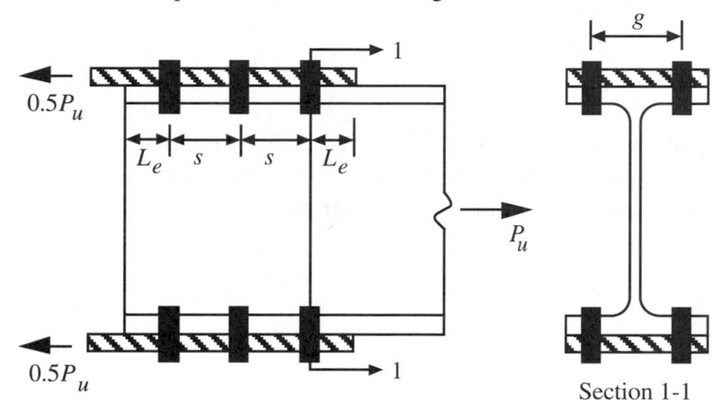

FIGURE P3.7

3.8 A572 Grade 50 steel; 1-in. diameter A490X bolts

$$s = 3 \text{ in.}; \quad g = 7 \text{ in.}; \quad L_e = 1.75 \text{ in.}$$

The tension member is a pair of C15 x 33.9 bolted to a pair of 10 x 5/8 gusset plates. Compute the design strength for:

1. Bolt shear
2. Bearing at the bolt holes
3. The gusset plates (yielding on A_g; fracture on A_n; block shear rupture)

Is the design satisfactory for $P_u = 495$ kips?

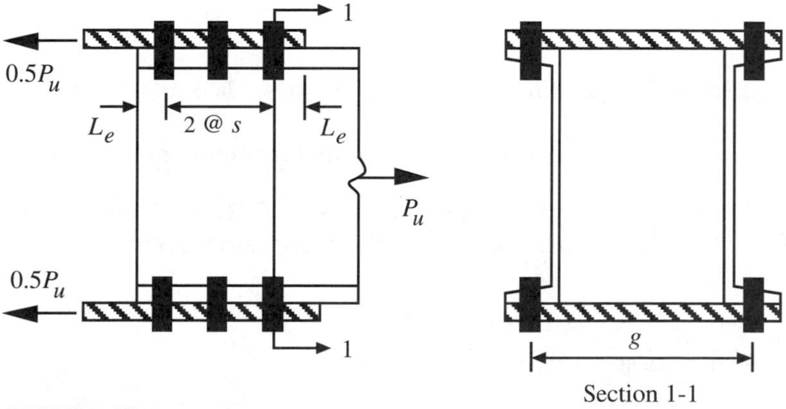

FIGURE P3.8

3.9 A36 steel; 7/8-in.-diameter A325N bolts

$s = 1.50$ in.; $L_e = 1.5$ in.; $g_1 = 2.25$ in.; $g_2 = 2.50$ in.; $g = 2.50$ in.

The tension member is a pair of L6 × 4 × 1/2 with all legs bolted to 3/4 in. thick splice plates. Compute the design strength for:

1. Bolt shear
2. Bearing at the bolt holes

Is the design satisfactory for $P_u = 308$ kips? Design the splice plates. That is, specify the minimum acceptable combination of widths for the vertical and horizontal plates such that the pair of plates is satisfactory for $P_u = 308$ kips due to yielding on A_g; fracture on A_n ; and block shear rupture.

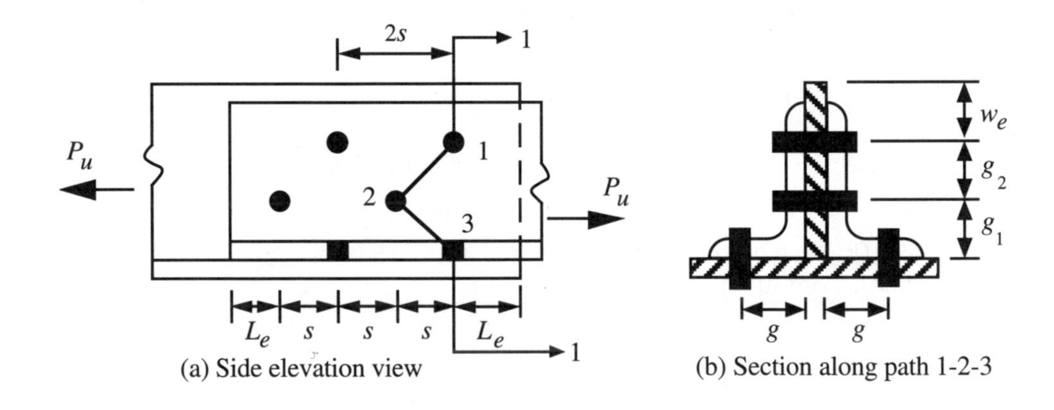

FIGURE P3.9

3.10 A572 Grade 50 steel; E70 electrodes; 1/4-in. fillet welds; $w_e = 1.5$ in.

$L_1 = 6$ in. = overlap length of member end on the gusset plate

The tension member is a pair of L5 x 3 x 5/16 with the long legs welded to a 5/8 in. thick gusset plate. Compute the design strength for:

1) fracture of the fillet welds
2) the gusset plate (yielding on A_g)

Is the design satisfactory for $P_u = 208$ kips?

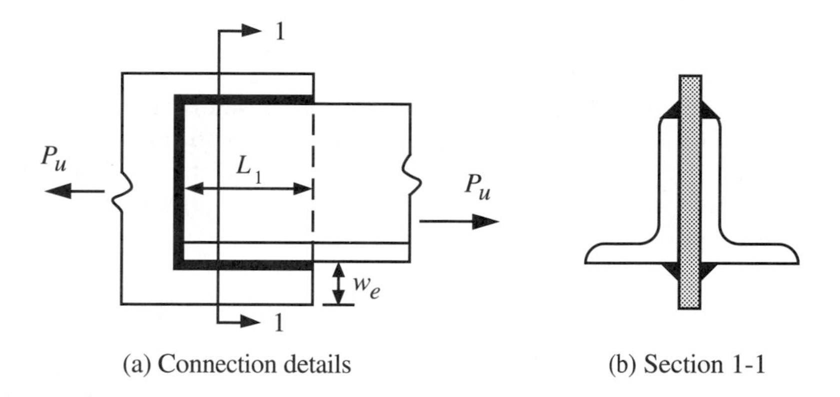

(a) Connection details (b) Section 1-1

FIGURE P3.10

3.11 A36 steel; E70 electrodes; 1/4-in. fillet welds

In the truss joint shown, the vertical member is a pair of L5 x 3 x 5/16 with the long legs welded to the web of a WT9 x 23 which is the horizontal member. $L_1 = (d - k - S_w)$ = overlapping length of the L5 x 3 x 5/16 member end on the web of the WT9 x 23. S_w = size of the fillet weld; d and k are properties of the WT9 x 23. Compute the design strength for:

1) the welds
2) block shear rupture of the connector plate (web of WT9 x 23).

Is the design satisfactory for P_u = 150 kips?

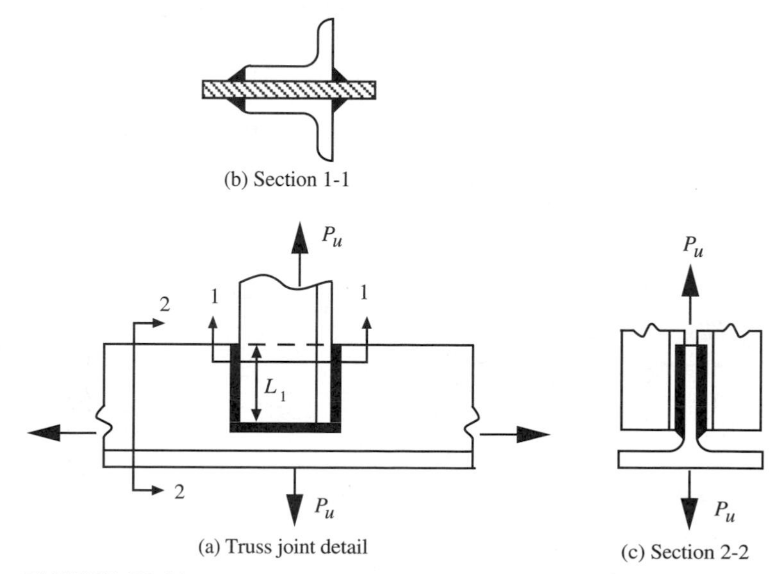

(b) Section 1-1

(a) Truss joint detail (c) Section 2-2

FIGURE P3.11

3.12 A36 steel; E70 electrodes; 5/16 inch fillet welds; $P_u = 300$ kips

A tension member (C15 × 33.9) is welded to a $t \times 16.5$ in. gusset plate. Compute the minimum acceptable thickness for the gusset plate due to yielding on A_g. Use the preferred increment for thickness on LRFD p1-133 and choose the actual minimum acceptable thickness that can be used. Each end return of the weld group is $0.25d$, where d = depth of C15 × 33.9 section. Find the minimum value of L_1 that satisfies the design requirement for the welds.

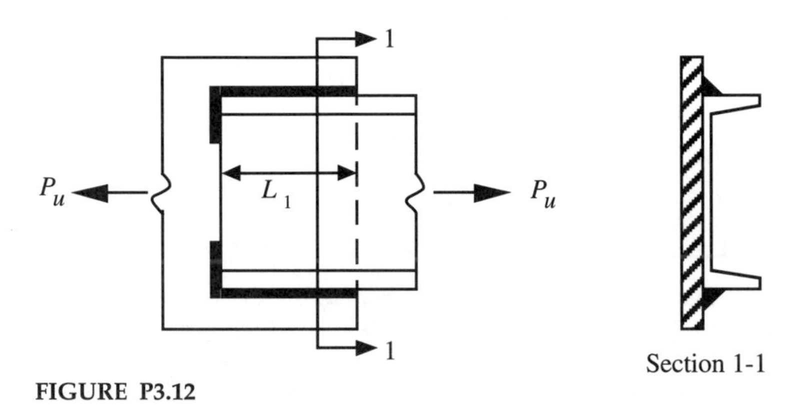

FIGURE P3.12

3.13 A36 steel; E70 electrodes; 3/8-in. fillet welds; $L_1 = 5.25$; $P_u = 270$ kips

A pair of L6 × 4 × 7/16 with long legs welded to a $w \times 3/4$ in. gusset plate serves as a tension member. Compute the minimum acceptable value of w for the gusset plate due to yielding on A_g. Assume that the preferred increment for width is 1/8 in. and choose the actual minimum acceptable w that can be used. Compute the design strength for:

1. Fracture of the fillet welds
2. The gusset plate (yielding on A_g; block shear rupture)

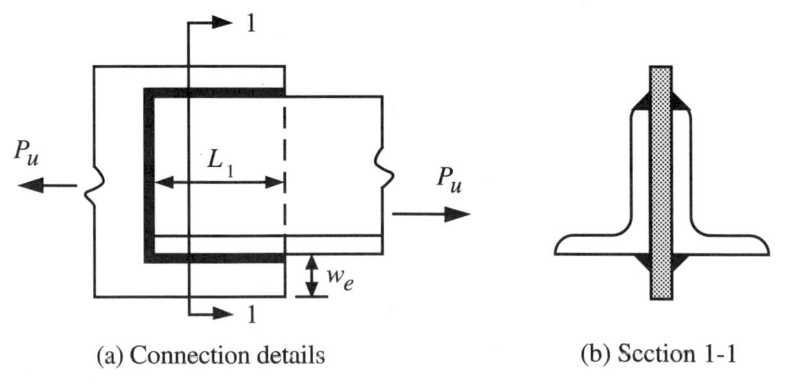

FIGURE P3.13

3.14 A36 steel; E70 electrodes; use the maximum acceptable fillet weld size

A tension member (C15 × 33.9) is to be butt-spliced for P_u = 323 kips with a $t \times 14 \times (2L_1 + 0.5$ in.) splice plate. Choose the minimum acceptable splice plate thickness (see LRFD, p.1-133). Fillet welds are located as shown in the cross section and on the back side of the web of the C15 × 33.9 at the ends of the splice plate. Find the minimum acceptable value of L_1. Ignore block shear rupture.

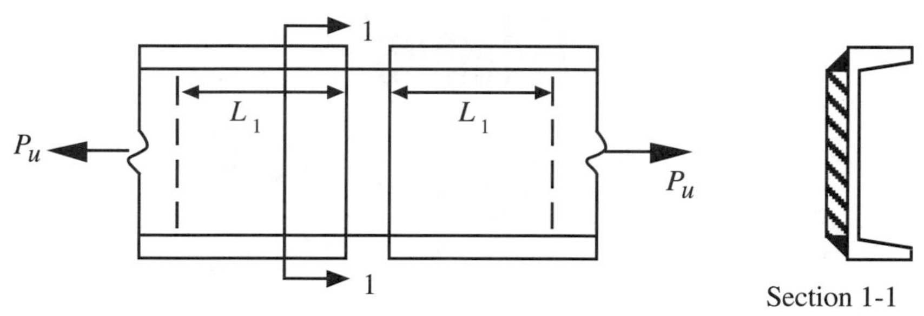

FIGURE P3.14

CHAPTER 4

Columns

4.1 INTRODUCTION

A member subjected only to an axial compressive force is called a *column*. Practically speaking, it is impossible for a member to be subjected only to an axial compressive force. When a lab test of such a member is conducted, locating the centroid of the member's cross section at the member ends in order to apply the axial force concentrically cannot be done perfectly. Also, the member is not perfectly straight. Consequently, the initial crookedness increases as the axial compression force is applied. Hence, the member is subjected to a combination of bending and axial compression; such a member is called a beam-column. However, as we will see in Chapter 6, the design strength definition of a *beam-column* is an interaction equation that contains a term due only to column action and another term due only to bending action. This chapter deals with column action only. Chapter 5 deals with bending action only.

In a truss analysis, only an axial tension or an axial compression force is assumed to exist in each member. Also, there are other situations where the designer deems that bending is negligible and considers only *column action* in the design of a compression member. Consequently, for practical reasons as well as for the subsequent treatment of beam-columns, we need to consider column action as a separate topic.

Figure 4.1 shows a compression member bolted to a gusset plate. Examination of the FBD in Figure 4.1(c) shows that the maximum compressive force in the net section for a bearing-type bolted connection is always less than the axial compressive force in the gross section. Therefore, if the only holes in a compression member are at the member ends for a bolted connection, the net section is not involved in the design strength definition of a column.

As shown in Figure 4.2, when the axial compressive force in a pinned-ended member reaches a certain value called the *critical load*, or *buckling load*, the member buckles. If the column is a W section, the *buckled shape* is a half sine wave in the *xz*-

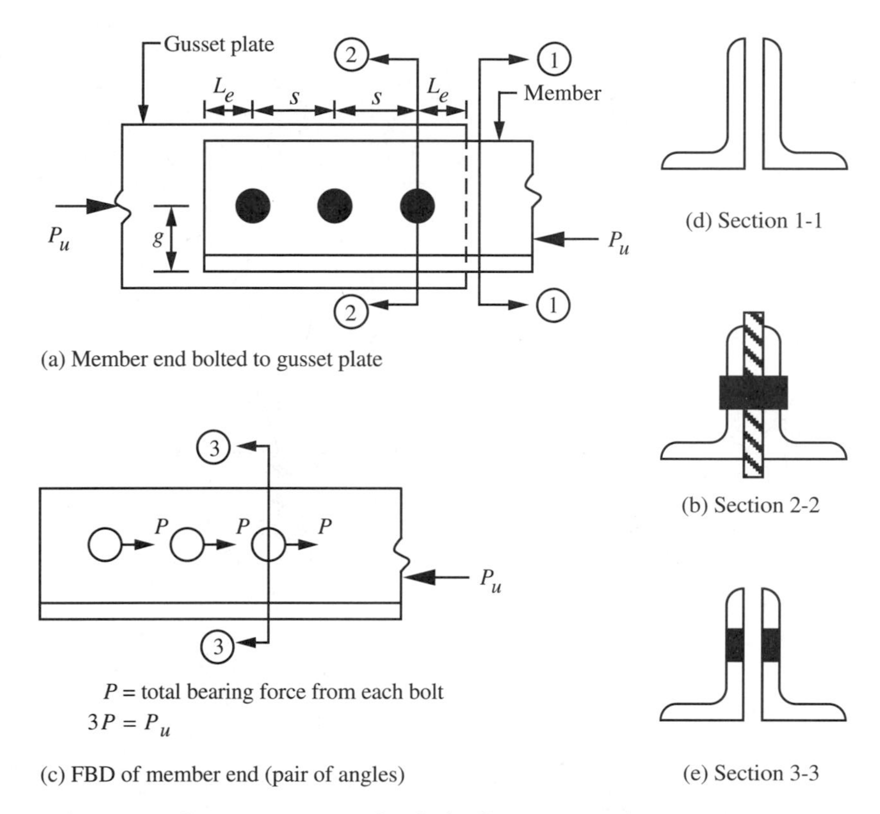

FIGURE 4.1 Compression member bolted to a gusset plate.

plane. This type of buckling is called *column buckling*. If the flange and web elements of the W section are not properly configured, *local buckling* of these cross-sectional elements can occur before column buckling occurs. Therefore, both types of buckling must be discussed in this chapter. For the W sections usually chosen to serve as a column, the flange and web elements have been configured such that local buckling does not occur before column buckling occurs. The simplest type of column buckling is *flexural buckling,* which denotes that the member bends about one of the principal axes when the column buckles. No twisting of the cross section occurs for flexural buckling. If the cross section twists when column buckling occurs, this is called *flexural-torsional buckling*.

4.2 ELASTIC EULER BUCKLING OF COLUMNS

As shown in Figure 4.3, the boundary conditions of a column significantly influence the buckled shape, which can be used to compute the buckling load. The chord length between the points of inflection on the buckled shape is called the *effective length KL*, where L is the member length. For an isolated column, values of K are shown in Figure 4.3 for various boundary conditions. In Section 4.5, we will discuss how to determine K for a column that is integrally connected to other members in a structure. For simplicity, we choose to use a pinned-ended column [case (d) of Figure 4.3] in our initial discussion. For a pinned-ended column, note that $K = 1$ and $KL = L$.

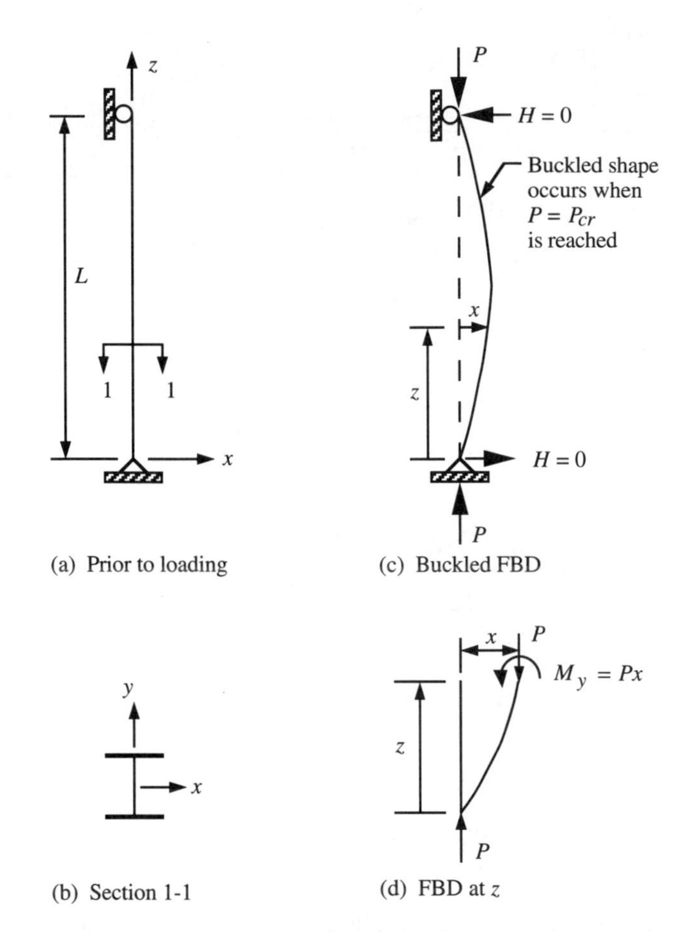

FIGURE 4.2 Elastic pinned-ended column.

The first theoretical buckling load solution was published in 1744 by Leonhard Euler [4] for a flagpole column [case (a) of Figure 4.3]. In 1759, Euler [5] published the solution for a pinned-ended column. In his solutions, Euler assumed that the member was elastic, prismatic, and perfectly straight before the axial compressive load was applied. As the axial load was slowly applied, he assumed that the member remained elastic and perfectly straight until the value of the axial load reached the *critical load*, or *buckling load*. Then, he reasoned that the member had reached a state of critical equilibrium and buckled into an assumed shape that was dependent on the boundary conditions at the member ends. Consequently, as Euler chose to do, we refer to the value of the axial compressive load at which the member buckles as the *critical load*.

The prismatic, pinned-ended member in Figure 4.2(a) is assumed to be perfectly straight before the axial compressive load, P, is slowly applied. Also, the member is assumed to be weightless and elastic. When P reaches the critical load value P_{cr} the member bows into a bent configuration, and the cross section does not twist when bowing occurs. Bowing or bending occurs about the cross-sectional axis of least resistance, which is the minor principal axis y. Note that the member deflects in the

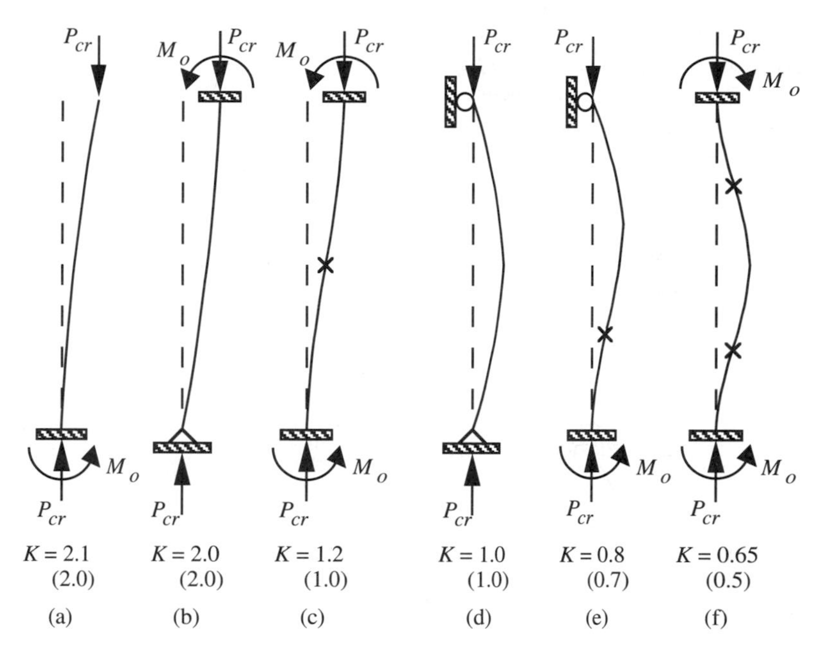

FIGURE 4.3 Effective length of isolated columns.

direction of the major principal axis x, which is perpendicular to the bending axis. At an arbitrary distance z from the origin of the member, the bending moment in the buckled member is

$$M_y = Px \tag{4.1}$$

which can be substituted into the moment curvature relation

$$-\left(\frac{d^2x}{dz^2}\right) = \left(\frac{M_y}{EI_y} = \frac{Px}{EI_y}\right) \tag{4.2}$$

to obtain the governing differential equation

$$\frac{d^2x}{dz^2} + \left(\frac{P}{EI_y}\right)x = 0 \tag{4.3}$$

For mathematical convenience, let $c^2 = P/(EI_y)$; then Eq. (4.3) becomes

$$\frac{d^2x}{dz^2} + c^2x = 0 \tag{4.4}$$

The solution of Eq. (4.4) is

$$x = A \sin cz + B \cos cz \tag{4.5}$$

At $z = 0$, the boundary condition is $x = 0$, which requires $B = 0$. At $z = L$, the boundary condition is $x = 0 = A \sin(cL)$, which has a non-trivial solution only when $cL = n\pi$, where n is an integer number, and

$$c^2 = \frac{P}{EI_y} = \left(\frac{n\pi}{L}\right)^2 \tag{4.6}$$

For a column pinned on both ends as shown in Figure 4.2, $n = 1$ is applicable and gives the least buckling load, which is

$$P_{cr} = \frac{\pi^2 EI_y}{L^2} \tag{4.7}$$

For mathematical convenience, let

$$I_y = A_g r_y^2$$

where

A_g = gross area of the cross section

r_y = radius of gyration about y axis

By substituting $I_y = A_g r_y^2$ into Eq. (4.7) and then dividing both sides of the resulting equation by A_g, we obtain the *critical stress*:

$$F_{cr} = \frac{P_{cr}}{A_g} = \frac{\pi^2 E}{(L/r_y)^2} \tag{4.8}$$

Note that we could not find a value for the coefficient A in Eq. (4.5). We only determined that A must be greater than zero. Therefore, the buckled shape in Figure 4.2 is a half sine wave with an indeterminate amplitude, $A > 0$.

4.3 EFFECT OF INITIAL CROOKEDNESS ON COLUMN BUCKLING

Perfectly straight members cannot be manufactured. As shown in Figure 4.4, each rolled steel section has an initial curvature upon arrival at the fabrication shop. Note that e is the maximum deviation from a straight line connecting the member ends and L is the member length. If $e > L/1000$, some of the crookedness must be removed since $e \le L/1000$ is required for the member to be acceptable. For a rolled steel section, the *average value of* $e = L/1500$ (see LRFD Commentary E2, p. 6-192).

The behavior of an initially crooked member subjected to axial compression is shown in Figure 4.5, where the maximum out-of-straightness was assumed to occur at midheight of the member. L/r is a fundamental parameter in Eq. (4.8), which is the critical stress definition for an initially straight, pinned-ended member. L/r is called the slenderness ratio, in which L is the member length and r is the radius of gyration for the axis about which bending would occur for Euler buckling (initially perfectly straight member). The effect of initial crookedness on the critical load is greatest in the range $50 < L/r < 135$ for a column of A36 steel. Based on experimental and theoretical investigations in which the cross section did not twist when buckling

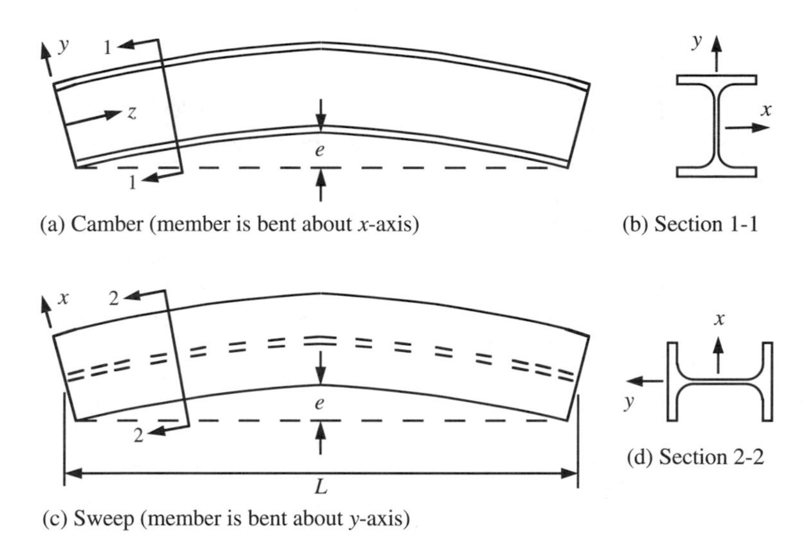

(a) Camber (member is bent about x-axis)

(b) Section 1-1

(c) Sweep (member is bent about y-axis)

(d) Section 2-2

FIGURE 4.4 Initial crookedness: camber and sweep.

occurred, the LRFD Specification writers chose the following definitions for an elastic, prismatic, pinned-ended column with an out-of-straightness of $e = L/1500$:

$$P_{cr} = \frac{0.877\,\pi^2 EI_y}{L^2} \tag{4.9}$$

$$F_{cr} = \frac{P_{cr}}{A_g} = \frac{0.877\,\pi^2 E}{(L/r)_y^2} \tag{4.10}$$

If the column is not pinned-ended, then

$$P_{cr} = \frac{0.877\,\pi^2 EI}{(KL)^2} \tag{4.11}$$

$$F_{cr} = \frac{P_{cr}}{A_g} = \frac{0.877\,\pi^2 E}{(KL/r)^2} \tag{4.12}$$

which can be written as

$$F_{cr} = \left(\frac{0.877}{\lambda_c^2}\right) F_y \tag{4.13}$$

where

$$\lambda_c = \frac{KL/r}{\pi}\sqrt{\frac{F_y}{E}} \tag{4.14}$$

λ_c is computed for the principal axis having the larger slenderness ratio, KL/r, and L

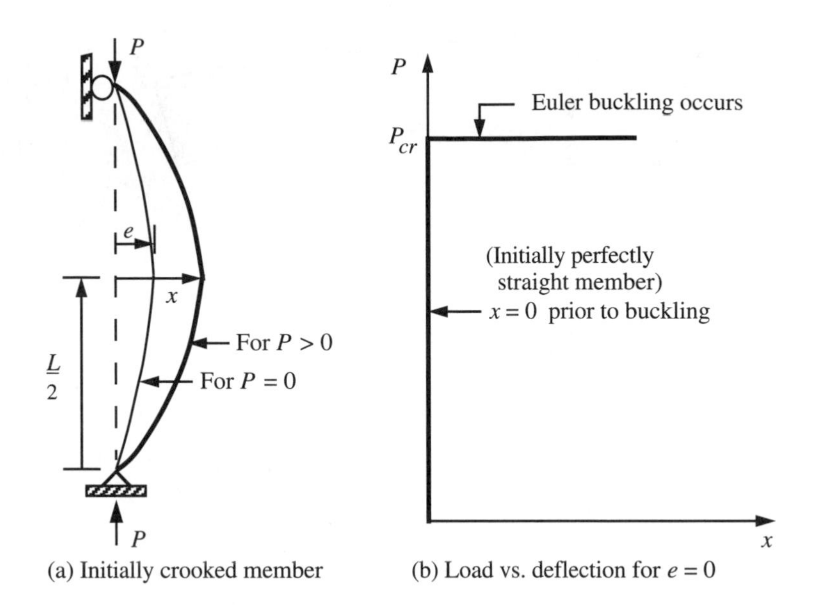

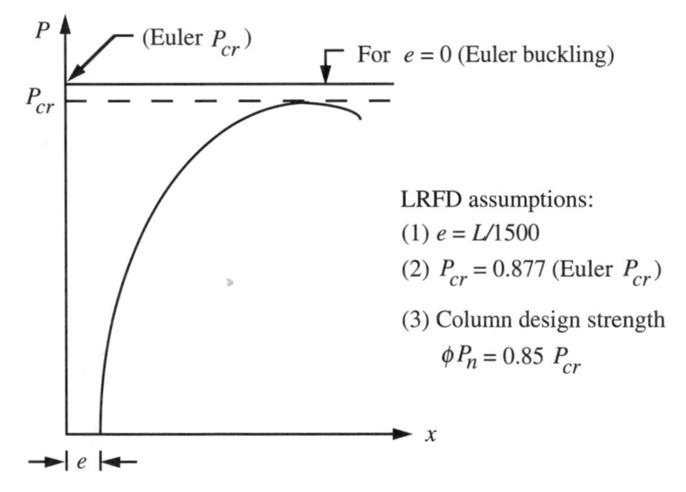

FIGURE 4.5 Initially crooked, elastic, pinned-ended column.

is the distance between braced points for each principal axis. When $\lambda_c \geq 1.5$, Eq. (4.13) is applicable.

For a column that is elastic when buckling occurs, the LRFD design requirement is

$$\phi_c P_n = P_u \tag{4.15}$$

where

$\phi_c P_n$ = column design strength

P_n = nominal column strength

$\phi_c = 0.85$ = column strength reduction factor

$$\phi_c P_n = 0.85 A_g F_{cr} = 0.85 A_g \left(\frac{0.877}{\lambda_c^2} \right) F_y \tag{4.16}$$

P_u = required column strength

and P_u is obtained from a structural analysis for factored loads.

4.4 INELASTIC BUCKLING OF COLUMNS

As explained in Section 2.4, the residual stress pattern shown in Figure 4.6(a) is representative of some W sections. The *maximum compressive residual stress* f_{rc} occurs at the flange tips and at midheight of the web, and the *maximum tensile residual stress* f_{rt} occurs at the junction of the flanges and the web.

Consider an axial compression laboratory test of a pinned-ended W section member of A36 steel. If we assume that the member is perfectly straight and twisting does not occur during buckling, elastic buckling occurs when [see Figures 4.4(a) and (b)] the maximum compressive residual stress plus the applied stress is less than F_y. For A36 steel, when $f_{rc} + P_u/A_g < 36$ ksi, none of the compression fibers are yielding and elastic buckling occurs [see Figure 4.6(c)]. If we choose L/r small enough to prevent elastic buckling from occurring, the compressive stress–strain curve [see Figure 4.6(c)] for a W section containing residual stresses becomes nonlinear after the flange tips begin to yield and inelastic buckling occurs. As shown in Figure 4.6(c), the slope of the compressive stress–strain curve for inelastic buckling is called the *tangent modulus of elasticity* E_t.

If twisting does not occur during buckling of a prismatic, pinned-ended column that was originally perfectly straight, the inelastic critical load is

$$P_{cr} = \frac{\pi^2 E_t I_y}{L^2} \tag{4.17}$$

If the column has an out-of-straightness of $e = L/1500$, the LRFD inelastic critical load definition can be written as

$$P_{cr} = \frac{0.877 \pi^2 E_t I_y}{L^2} \tag{4.18}$$

and the corresponding inelastic critical stress can be written as

$$F_{cr} = \frac{P_{cr}}{A_g} = \frac{0.877 \pi^2 E_t}{(L/r)_y^2} \tag{4.19}$$

and E_t is derived below [see Eq. (4.23)].

For a column that is not pinned-ended, but has an out-of-straightness of $e = L/1500$, the inelastic LRFD definitions can be written as

$$P_{cr} = \frac{0.877 \pi^2 E_t I_y}{(KL)_y^2} \tag{4.20}$$

$$F_{cr} = \frac{P_{cr}}{A_g} = 0.877 \left[\frac{\pi^2 E_t}{(KL/r)^2} \right] \tag{4.21}$$

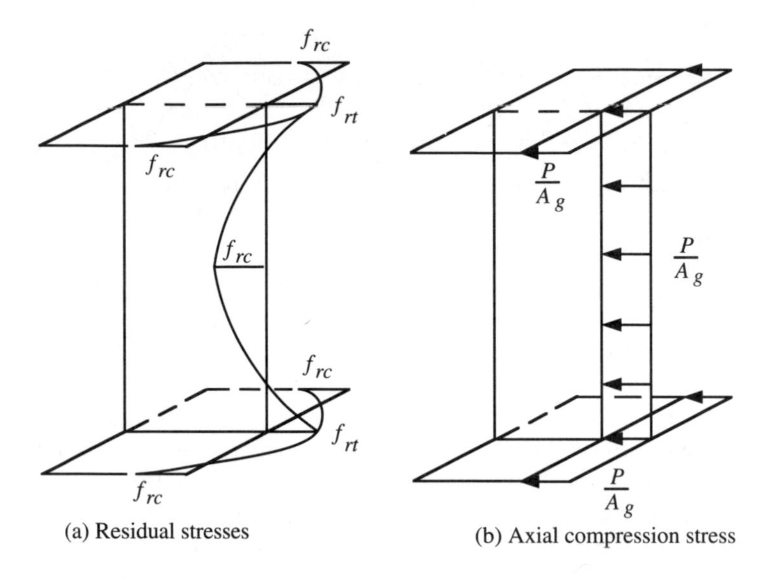

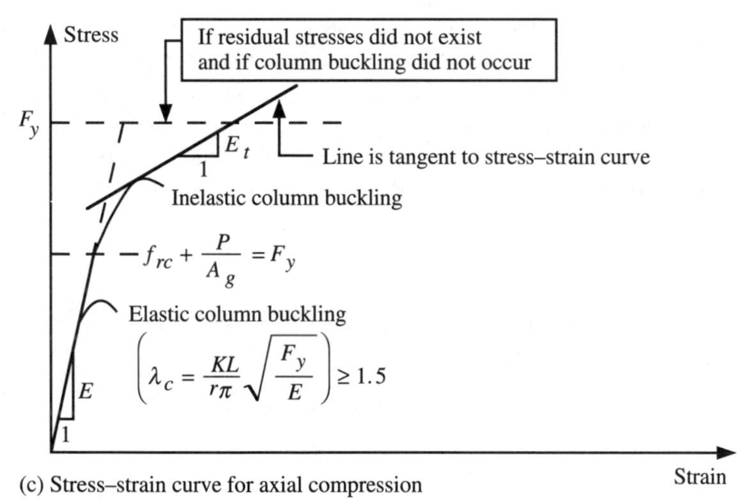

FIGURE 4.6 Perfectly straight W section member.

As in LRFD E2 (p. 6-47), Eqn (4.21) can be written as:

$$F_{cr} = (0.658)^{\lambda_c^2} F_y \tag{4.22}$$

Equation (4.22) is applicable when $\lambda_c \leq 1.5$, where λ_c is Eq. (4.14) for the principal axis having the larger slenderness ratio, *KL/r*.

We can equate Eq. (4.19) and (4.22) and solve for

$$\tau = \frac{E_t}{E} = \frac{\lambda_c^2 (0.658)^{\lambda_c^2}}{0.877} \tag{4.23}$$

which is useful for inelastic buckling problems.

For a column that is inelastic when buckling occurs, the LRFD design requirement is

$$\phi_c P_n = P_u \tag{4.24}$$

where

$\phi_c P_n$ = column design strength

P_n = nominal column strength

$\phi_c = 0.85$ = column strength reduction factor

$$\phi_c P_n = 0.85 A_g F_{cr} = 0.85 A_g (0.658)^{\lambda_c^2} F_y \tag{4.25}$$

P_u = required column strength

and P_u is obtained from a structural analysis for factored loads. Equation (4.25) is applicable when $\lambda_c \leq 1.5$, where λ_c is Eq. (4.14) for the principal axis having the larger slenderness ratio, KL/r, and L is the distance along the member between braced points for each principal axis.

See Figure 4.7 in which we summarized the LRFD definitions of the *nominal column strength* (critical load) for a prismatic, axially loaded compression member that:

1. Does not twist when column buckling occurs by bending about the principal axis having the *larger* KL/r ratio.
2. Contains a realistic, representative, residual stress pattern.
3. Has an out-of-straightness of $e = L/1500$.
4. Is composed of compression elements for which local buckling does not occur before column flexural buckling occurs.

Local buckling does not occur before column flexural buckling occurs if the width–thickness ratio of each compression element does not exceed the applicable λ_r in LRFD Table B5.1 (p. 6-38). Example 4.1 illustrates the definitions of the width–thickness ratios and the applicable λ_r expressions for a W section.

4.5 EFFECTIVE LENGTH

The *effective length* (equivalent pinned-ended length) KL of a column is the chord length between the points of inflection ($M = 0$ points) on the buckled column shape and L is the actual length of the column between braced points for each principal axis. See Figure 4.3 for some examples of buckled shapes and K values for isolated, individual columns. KL must be determined for each principal axis of the cross section. For example, an individual W section column may be fixed at the base and free at the top [see case (a) of Figure 4.3] for bending about the major principal axis, but may be fixed at the base and hinged at the top [see case (e) of Figure 4.3] for bending about the minor principal axis.

For a column in an unbraced frame, the unbraced frame (sidesway uninhibited) nomograph on LRFD p. 6-186 can be used to obtain an approximate value of $K > 1$ for in-plane, elastic column buckling. For a column in a braced frame, the braced frame (sidesway inhibited) nomograph on LRFD p. 6-186 can be used to obtain $K \leq$

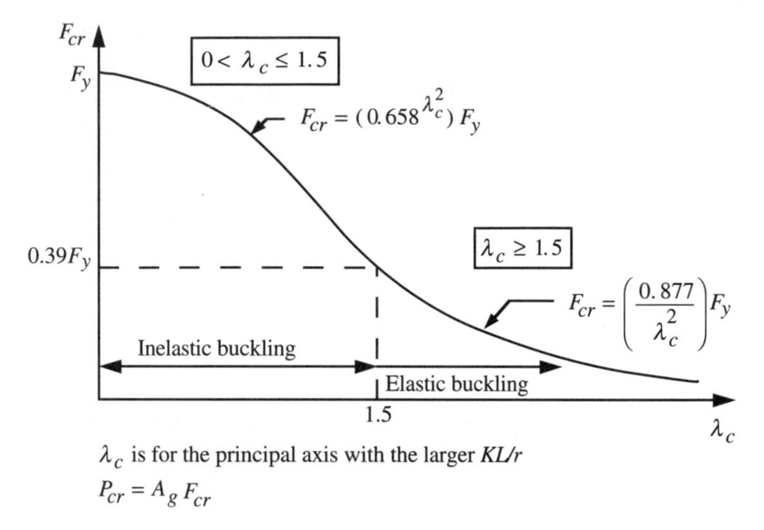

FIGURE 4.7 LRFD column design strength.

1, or we can conservatively use $K = 1$. Julian and Lawrence [31] made the assumptions stated on LRFD p. 6-186 in deriving the equations for the effective length factor K for a column in a plane frame. They used one of the derived equations in preparing the *sidesway uninhibited* nomograph on LRFD p. 6-186 for an unbraced frame. The other derived equation was used in preparing the *sidesway inhibited* nomograph on LRFD p. 186 for a braced frame. For discussions on the derivation of these equations, see Kavanagh [11], Galambos [8], and Chen and Lui [34]. Instead of using the nomographs on LRFD p. 6-186 to determine K, we prefer to use the following approximate version of the formulas used to generate the nomographs. These approximate formulas have appeared in the French design rules since 1966 [32]. Consider column m, which spans between joints i and j in a plane frame. For column m in a braced frame, the effective length factor K_m for a principal axis of bending is

$$K_m = \frac{3G_iG_j + 1.4(G_i + G_j) + 0.64}{3G_iG_j + 2.0(G_i + G_j) + 1.28} \tag{4.26}$$

For column m in an unbraced frame,

$$K_m = \sqrt{\frac{1.6G_iG_j + 4.0(G_i + G_j) + 7.5}{G_i + G_j + 7.5}} \tag{4.27}$$

The relative stiffness at a typical joint is

$$G = \frac{\sum\left(\frac{I}{L}\right)_c}{\sum\left(\frac{I}{L}\right)_g} \tag{4.28}$$

Example 4.1

An A36 steel W14 × 90 section is used as an axially loaded compression member and the effective length is $KL = 20$ ft for both principal axes. Use LRFD Table B5.1 (p. 6-38) to show that local buckling does not occur before column buckling occurs. Then, use the LRFD definitions summarized in Figure 4.7 to find the column design strength.

Solution

From LRFD p. 1-30, for a W14 × 90:

$$A = 26.5 \text{ in.}^2 \qquad d = 14.02 \text{ in.} \qquad k = 1.375 \text{ in.}$$

$$\frac{0.5b_f}{t_f} = 10.2 \qquad \frac{h}{t_w} = 25.9 \qquad I_y = A_g r_y{}^2$$

An axis of symmetry is a principal axis; therefore, the x- and y-axes are the principal axes. Since the x-axis has the larger moment of inertia value, the x-axis is the major principal axis and the y-axis is the minor principal axis. Properties for these axes are

$$I_x = 999 \text{ in.}^4 \qquad r_x = 6.14 \text{ in.}$$

$$I_y = 362 \text{ in.}^4 \qquad r_y = 3.70 \text{ in.}$$

Check the b/t requirements of the compression elements (LRFD Table B5.1, p.6-38):

1. For the flange (*unstiffened element*),

$$\left(\frac{b}{t} = \frac{0.5b_f}{t_f} = 10.2\right) \le \left(\lambda_r = \frac{95}{\sqrt{F_y}} = \frac{95}{\sqrt{36}} = 15.8\right)$$

   Flange local buckling does not govern ϕP_n.

2. For the web (*stiffened element*),

$$\left(\frac{b}{t} = \frac{h}{t_w} = 25.9\right) \le \left(\lambda_r = \frac{253}{\sqrt{F_y}} = \frac{253}{\sqrt{36}} = 42.2\right)$$

   Web local buckling does not govern ϕP_n.

The formulas in Figure 4.7 are valid since local buckling does not govern ϕP_n. Since $(r_x = 6.14 \text{ in.}) > (r_y = 3.70 \text{ in.})$ and $(KL)_x = (KL)_y = 20 \text{ ft} = 240$ in., we know that $(KL/r)_y > (KL/r)_x$ and the cross section bends about the y-axis when column buckling occurs. Therefore,

$$\left[\lambda_c = \frac{(KL/r)_y}{\pi}\sqrt{\frac{F_y}{E}} = \frac{(240/3.40)}{\pi}\sqrt{\frac{36}{29{,}000}} = 0.7275\right] < 1.5$$

$$\lambda_c^2 = 0.529$$

$$F_{cr} = \left(0.658^{\lambda_c^2}\right) F_y = \left(0.658^{0.529}\right)(36 \text{ ksi}) = 28.85 \text{ ksi}$$

$$\phi_c P_n = 0.85 A_g F_{cr} = 0.85(26.5)(28.85) = 650 \text{ kips}$$

Example 4.2

An A36 steel W14×90 section is used as an axially loaded compression member and the effective length is KL = 20 ft for both principal axes. Use the LRFD column design table on p. 3-20 to find the column design strength.

Solution

Enter LRFD p. 3-20 at $(KL)_y = 20$ ft for a W14×90 and $F_y = 36$ ksi. Find $\phi P_{ny} = 650$ kips. In Example 4.1 we found $\phi P_n = \phi P_{ny} = 650$ kips, which agrees with the value in the LRFD column design table. Therefore, Example 4.1 shows how an entry in the LRFD column design table for W sections was obtained.

Example 4.3

An A36 steel W14×90 section is used as an axially loaded compression member with $(KL)_x = 20$ ft and $(KL)_y = 10$ ft. For this section, we showed in Example 4.1 that local buckling does not govern ϕP_n. Therefore, find the column design strength.

Solution

From LRFD p. 1-30, for a W14 × 90:

$$A = 26.5 \text{ in.}^2 \qquad d = 14.02 \text{ in.} \qquad k = 1.375 \text{ in.} \qquad \frac{0.5 b_f}{t_f} = 10.2 \qquad \frac{h}{t_w} = 25.9$$

$$\left(\frac{KL}{r}\right)_x = \frac{20(12)}{6.14} = 39.09$$

$$\left(\frac{KL}{r}\right)_y = \frac{10(12)}{3.17} = 32.43$$

Since $(KL/r)_x > (KL/r)_y$, the cross section bends about the x axis when column buckling occurs.

$$\left[\lambda_c = \frac{(KL/r)_x}{\pi}\sqrt{\frac{F_y}{E}} = \frac{20(12)}{6.14\pi}\sqrt{\frac{36}{29{,}000}} = 0.438\right] < 1.5 \qquad \lambda_{cy}^2 = 0.192$$

$$F_{cr} = (0.658)^{\lambda_c^2} F_y = (0.658)^{0.192}(36) = 33.2 \text{ ksi}$$

$$\phi P_n = \phi P_{cr} = 0.85 A_g F_{cr}$$

$$\phi P_n = 0.85(26.5)(33.2) = 748 \text{ kips}$$

Example 4.4

An A36 steel W14×90 section is used as an axially loaded compression member with $(KL)_x = 20$ ft and $(KL)_y = 10$ ft. Use the LRFD column table to find the column design strength, which is governed by the larger of $(KL)_y$ and $(KL)_x/(r_x/r_y)$.

Solution

Enter LRFD p. 3-20 for a W14×90 and $F_y = 36$ ksi. At $(KL)_y = 10$ ft, find $\phi P_{ny} = 767$ kips. Column design strength values for $(KL)_x$ are not given, but they can be easily obtained from the given information. At the bottom of the LRFD column table, find $r_x/r_y = 1.66$ for a W14 x 90. Enter the table at $(KL)_x/(r_x/r_y) = 20/1.66 = 12.05$ ft; use linear interpolation to find

$$\phi P_{nx} = 749 - 0.05\,(749\text{-}738) = 748.45 \text{ kips}$$

$$\phi P_n = 748 \text{ kips} \qquad (\text{smaller of } \phi P_{nx} \text{ and } \phi P_{ny})$$

which agrees with the solution obtained in Example 4.3.

Example 4.5

The following are given $P_u = 300$ kips; $(KL)_x = 20$ ft $(KL)_y = 10$ ft and A36 steel. For each nominal depth listed, find the lightest W section that satisfies the LRFD specifications for axial compression:

1. W14
2. W12
3. W10
4. W8

Solution

The design requirement is $\phi P_n \geq (P_u = 300 \text{ kips})$.

We start by selecting a section that satisfies the design requirement for the y-axis. For this selected section we use its r_x/r_y ratio listed at the bottom of the column table and compute $(KL)_x/(r_x/r_y)$. If $(KL)_x/(r_x/r_y) > (KL)_y$, column buckling occurs with the section bending about the x axis, and we must enter the column table with an assumed value for $(KL)_x/(r_x/r_y)$ in order to choose a section that satisfies the design requirement. Assume *that* r_x/r_y will be the same as it was for the section selected for the y-axis. If r_x/r_y for the selected section differs significantly from the assumed value, compute a revised value of $(KL)_x/(r_x/r_y)$ to use in making the next selection.

1. Select the lightest W14 that satisfies the design requirement. For $F_y = 36$ ksi, enter LRFD p. 3-21 at $(KL)_y = 10$ ft and find the least value that exceeds 300 kips. If y-axis bending governs the column design strength, we find that a W14 × 43 $[(\phi P_{ny} = 312) \geq 300]$ is the lightest choice.

For a W14 × 43, $r_x/r_y = 3.08$ is found at the bottom of the table and is used to compute the following assumed value for entering the table to make the selection that satisfies the design requirement for the x axis:

$$(KL)_x/(r_x/r_y) = 20/3.08 = 6.49 \text{ ft}$$

$$\text{W14} \times 43 \qquad [\phi P_n = (\phi P_{nx} = 350)] > (\phi P_{ny} = 312) \geq 300$$

This is the lightest W14 that satisfies the design requirement. Note that the assumed value of $r_x/r_y = 3.08$ used in entering the table at 6.49 ft was the same as the r_x/r_y value for the section selected for the x-axis. That is, the assumption made was correct.

2. Select the lightest W12 that satisfies the design requirement. For $F_y = 36$ ksi, enter LRFD p. 3-25 at $(KL)_y = 10$ ft and find

$$\text{W12} \times 45 \qquad (\phi P_{ny} = 330) \geq 300$$

Enter at $(KL)_x/(r_x/r_y) = 20/2.65 = 7.55$ ft and find

$$\text{W12} \times 45 \qquad (\phi P_{nx} = 360) > (\phi P_{ny} = 330) \geq 300$$

Note that when the LRFD column table is being used to determine the column design strength, ϕP_n is governed by the larger of $(KL)_x/(r_x/r_y)$ and $(KL)_y$.

3. Select the lightest W10 that satisfies the design requirement. For $F_y = 36$ ksi, enter LRFD p. 3-27 at $(KL)_y = 10$ ft and find

$$\text{W10} \times 45 \qquad (\phi P_{ny} = 337) \geq 300$$

Enter at $(KL)_x/(r_x/r_y) = 20/2.15 = 9.30$ ft and find

$$\text{W10} \times 45 \qquad (\phi P_{nx} = 346) > (\phi P_{ny} = 337) \geq 300$$

4. Select the lightest W8 that satisfies the design requirement. For $F_y = 36$ ksi, enter LRFD p. 3-28 at $(KL)_y = 10$ ft and find

$$\text{W8} \times 48 \qquad (\phi P_{ny} = 362) \geq 300$$

Enter at $(KL)_x/(r_x/r_y) = 20/1.74 = 11.5$ ft and find

$$\text{W8} \times 48, \qquad (\phi P_{nx} = 342) \geq 300$$

Note: For W8 × 40, $(\phi P_{ny} = 298) \approx 300$, but $(\phi P_{nx} = 280) < 300$. W8 × 40 does not satisfy the design requirement. The lightest W8 choice is

$$\text{W8} \times 48, \qquad (\phi P_{nx} = 362) > (\phi P_{ny} = 342) \geq (P_u = 300 \text{ kips})$$

Example 4.6

Figure 4.8 is an unbraced plane frame. Due to sidesway buckling, all members bend about their x-axis. At the support joints, use the recommended G values given in the last paragraph on LRFD p. 6-186. Use Eqs. (4.27) and (4.28) to find $(KL)_x$ for columns 1 to 5. Also find ϕP_{nx} for members 1 to 5.

Solution

W12 x 120 ($A_g = 35.3$ in.^{2}; $I_x = 1070$ in.^{4}; $r_x = 5.51$ in.)
W30 x 173 ($I_x = 8200$ in.^{4})
W30 x 116 ($I_x = 4930$ in.^{4})
From the last paragraph on LRFD p. 6-186,

$$G_1 = 10$$

$$G_3 = 10$$

$$G_6 = 1$$

At the interior joints of Figure 4.8,

$$G_2 = \frac{\Sigma\left(\frac{I}{L}\right)_c}{\Sigma\left(\frac{I}{L}\right)_g} = \frac{\frac{1070}{20}}{\frac{4930}{30}} = 0.326$$

$$G_4 = \frac{\frac{1070}{20} + \frac{1070}{15}}{\frac{4930}{30} + \frac{8200}{40}} = 0.338$$

$$G_5 = \frac{1070/15}{8200/40} = 0.348$$

$$G_7 = \frac{(1070/20) + (1070/15)}{2(8200/40)} = 0.380$$

$$G_8 = \frac{1070/15}{2(8200/40)} = 0.174$$

From Eq. (4.27), the effective length factors for columns 1 to 5 are

$$K_1 = \sqrt{\frac{1.6G_1G_2 + 4(G_1 + G_2) + 7.5}{G_1 + G_2 + 7.5}} = 1.74$$

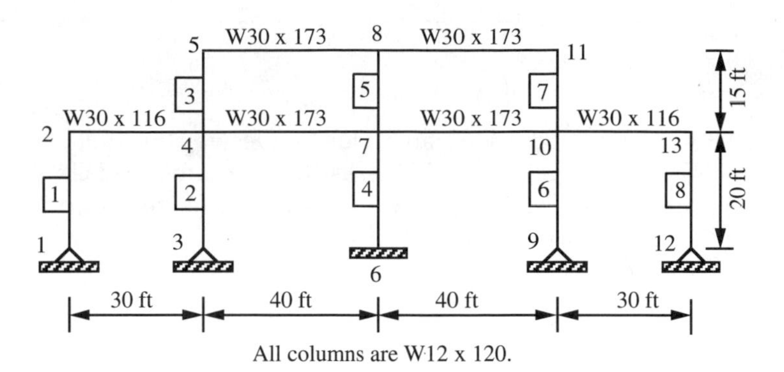

FIGURE 4.8 An unbraced plane frame.

$$K_2 = \sqrt{\frac{1.6G_3G_4 + 4(G_3 + G_4) + 7.5}{G_3 + G_4 + 7.5}} = 1.74$$

$$K_3 = \sqrt{\frac{1.6G_4G_5 + 4(G_4 + G_5) + 7.5}{G_4 + G_5 + 7.5}} = 1.20$$

$$K_4 = \sqrt{\frac{1.6G_6G_7 + 4(G_6 + G_7) + 7.5}{G_6 + G_7 + 7.5}} = 1.24$$

$$K_5 = \sqrt{\frac{1.6G_7G_8 + 4(G_7 + G_8) + 7.5}{G_7 + G_8 + 7.5}} = 1.10$$

On LRFD p. 3-23 for a W12 × 120 and $F_y = 36$ ksi, for each of the columns 1 to 5, enter at $(KL)_x/(r_x/r_y)$ and find ϕP_{nx}:

$$[(KL)_x/(r_x/r_y)]_1 = 1.74(20)/1.76 = 19.77 \text{ ft}$$

$$(\phi P_{nx})_1 = 799 \text{ kips}$$

$$[(KL)_x/(r_x/r_y)]_2 = 1.74(20)/1.76 = 19.77 \text{ ft}$$

$$(\phi P_{nx})_2 = 799 \text{ kips}$$

$$[(KL)_x/(r_x/r_y)]_3 = 1.20(15)/1.76 = 10.23 \text{ ft}$$

$$(\phi P_{nx})_3 = 996 \text{ kips}$$

$$[(KL)_x/(r_x/r_y)]_4 = 1.24(20)/1.76 = 14.09 \text{ ft}$$

$$(\phi P_{nx})_4 = 927 \text{ kips}$$

$$[(KL)_x/(r_x/r_y)]_5 = 1.10(15)/1.76 = 9.375 \text{ ft}$$

$$(\phi P_{nx})_5 = 994 \text{ kips}$$

In Eq. (4.28), all members are assumed to be elastic, and a point of inflection ($M = 0$) is assumed to occur at midspan of each girder in the frame. A girder (restraining member) is a bending member that is attached to one or more column ends at a particular joint in a frame. Each girder end provides a rotational resistance at the column end(s) when column buckling occurs for the frame. At each end of a girder, the girder-end rotational stiffness in Eq. (4.28) is assumed to be $3EI/(L/2) = 6EI/L$. To account for inelastic column buckling and the $M = 0$ point not being at midspan of the elastic girders in an unbraced frame, the definition of the *relative joint stiffness parameter* G is

$$G = \frac{\sum\left(\tau \frac{I}{L}\right)_c}{\sum\left(\gamma \frac{I}{L}\right)_g} \tag{4.29}$$

$$\tau = E_t/E \quad \text{is as defined in Eq.} \quad (4.23)$$

$$\gamma = \frac{\text{actual girder end rotational stiffness}}{(6EI/L)_g} \tag{4.30}$$

For example, when the far end of a girder in an unbraced frame is:

1. fixed, then $\gamma = (4EI/L)/(6EI/L) = 0.667$.
2. hinged, then $\gamma = (3EI/L)/(6EI/L) = 0.5$.

Eqns (4.29) and (4.23) are also valid for braced frames, for which we must use

$$\gamma = \frac{\text{actual girder} - \text{end rotational stiffness}}{(2EI/L)_g} \tag{4.31}$$

For example, when the far end of a girder in a braced frame is:

1. fixed, then $\gamma = (4EI/L)/(2EI/L) = 2$.
2. hinged, then $\gamma = (3EI/L)/(2EI/L) = 1.5$.

Using $\tau = E_t/E = 1$ is conservative, as was done in Eq. (4.28), but the correct γ must be used for each girder.

4.6 LOCAL BUCKLING OF THE CROSS-SECTIONAL ELEMENTS

Suppose an identical amount of material is made into closed shapes (pipes and tubes) and open sections (W, C, and L). When the cross-sectional area is the same for all shapes, the shape with the largest radii of gyration and with compression elements thick enough to prevent local buckling is the most efficient shape for resisting a compression load. Steel pipes (LRFD, p. 3-36) and structural steel tubes (LRFD, p. 3-39) are very good shapes for a column cross section, but attaching other members to these shapes can be difficult and expensive. The open sections are less efficient column sections, but attaching beam and girder sections to them is routine.

Some examples of stiffened and unstiffened compression elements in column cross-sectional shapes are shown in Figure 4.9 to further clarify the definitions for compression elements.

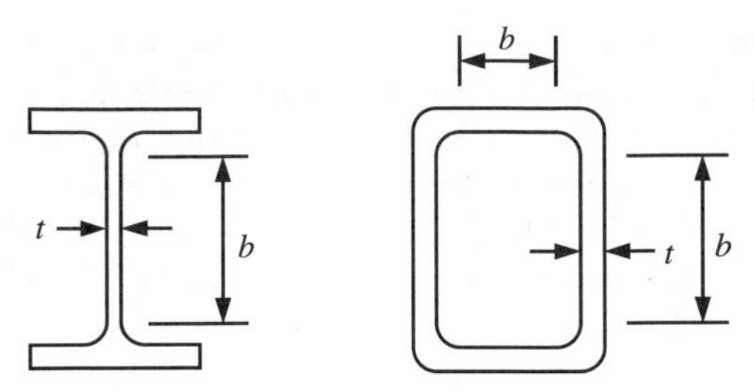

(a) Stiffened compression elements

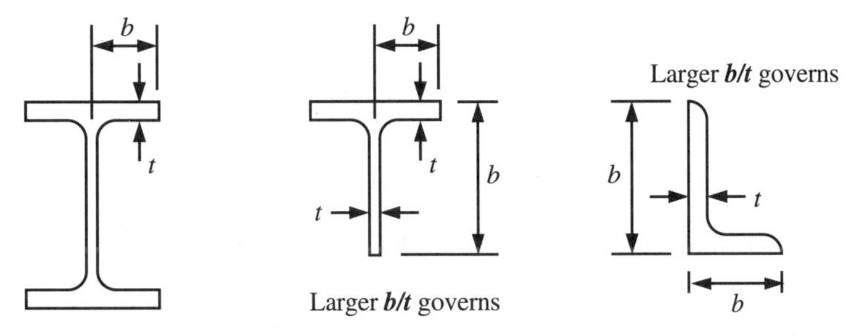

(b) Unstiffened (projecting) compression elements

FIGURE 4.9 Cross-sectional compression elements.

Examples of the *local buckling mode shapes* for the flange and for the web of a W section are shown in Figure 4.10. Also, in Figure 4.10, we assumed that the flange and the web buckled independently of each other.

Theoretical discussions of *elastic buckling of thin-plate elements* are available [6, 7] and give Eq. (4.32) as the *critical stress* for elastic buckling of thin-plate elements subjected to a uniaxial, uniform, compressive stress [see Figure 4.11(a)]:

$$F_{cr} = \frac{k\pi^2 E}{12(1-\nu^2)(b/t)^2} \tag{4.32}$$

where

k = constant depending on a/b and the edge support conditions

E = modulus of elasticity

$\nu = 0.3$ = Poisson's ratio

b/t = width-to-thickness ratio of the plate

a/b = length-to-width ratio of the plate

See Figure 4.11(b) for example values of k. For $a/b \geq 4$, the half wave length of the buckled shape is on the order of the width b.

As shown in Figure 4.10 at any cross section:

1. When *local buckling of a W section flange occurs*, the local buckling mode shape is antisymmetric and the web provides some rotational resistance. Parallel to

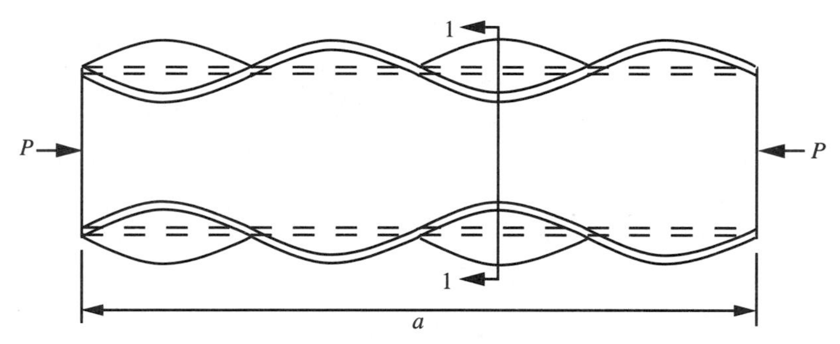

Number of half sine waves is a function of a/b and b/t of flange.

(a) Flange local buckling of a W section column

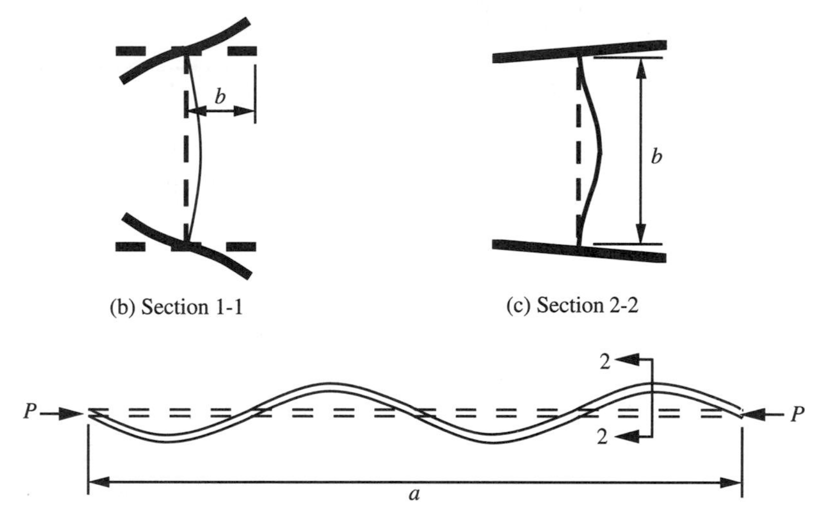

(b) Section 1-1

(c) Section 2-2

Number of half sine waves is a function of a/b and b/t of web.

(d) Web local buckling of a W section column

FIGURE 4.10 Local buckling modes in a W section column.

the applied compressive stress, one flange edge is free and the other edge (at the junction of the flange and web) can be assumed to be such that $k = 0.7$ (about midway between hinged and fixed). *Each half of a W section flange* is an *unstiffened compression element.*

2. When *local buckling of a W section web occurs*, the local buckling mode is symmetric and the flanges provide considerable rotational resistance since for column buckling, the structural designer must prevent twisting of the cross section at the member ends and at any intermediate, weak axis, column-braced points. Therefore, the flanges are restrained at the ends of each unbraced column length, and the torsional resistance of the flanges can be developed in each unbraced column length. At each junction of the web and flanges, the web edge is somewhere between fully fixed and hinged. The web is a stiffened compression element for which we can assume that $k = 5.0$ (longitudinal edges are about one-third fixed).

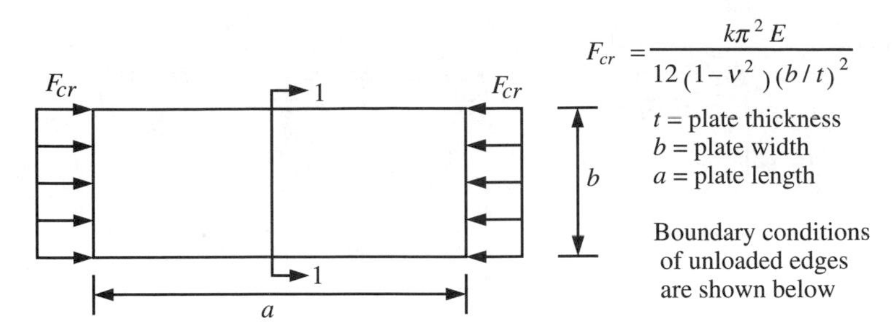

(a) Long plate with loaded edges simply supported

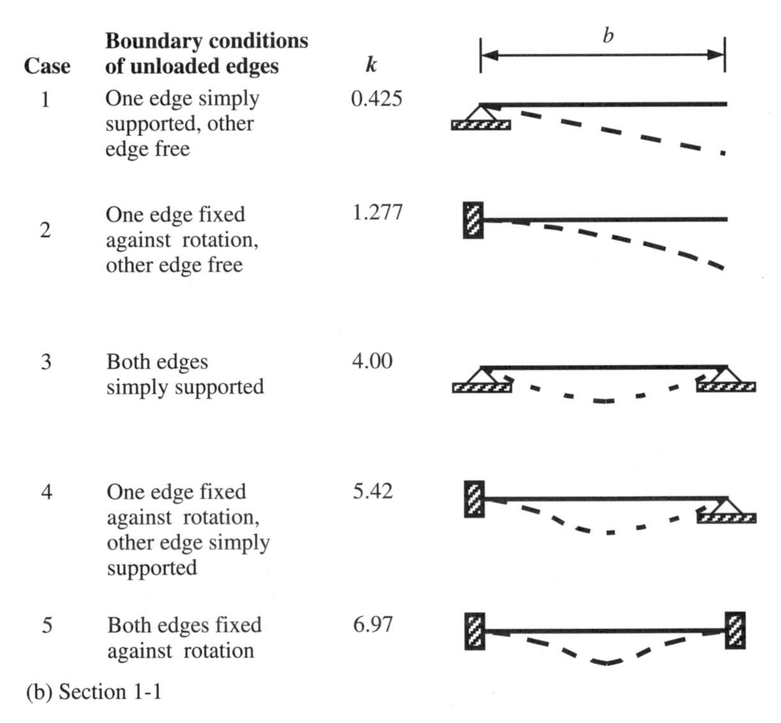

Case	Boundary conditions of unloaded edges	k
1	One edge simply supported, other edge free	0.425
2	One edge fixed against rotation, other edge free	1.277
3	Both edges simply supported	4.00
4	One edge fixed against rotation, other edge simply supported	5.42
5	Both edges fixed against rotation	6.97

(b) Section 1-1

FIGURE 4.11 Coefficients of k for Eq. 4.32 (adapted from [25]).

Note the following:

1. When local buckling of the compression elements in a column cross section occurs, these elements continue to resist some more compressive load until a considerable amplitude of the column buckled shape occurs. However, when column buckling occurs, the member cannot resist any more compressive axial load.
2. Inelastic local buckling of plates can occur when either the b/t ratio or the L/t ratio is small enough.

Fortunately, we seldom have to deal directly with Eq. (4.32). For frequently encountered situations, experts on plate buckling have chosen realistic k values based on currently available theoretical and experimental research, satisfactory performance of existing structures, and engineering judgment to devise definitions of critical stress for local buckling of column cross-sectional elements. For example, buckling experts used $F_{cr}/F_y \approx 0.7$ to account for the presence of residual stresses and imperfections in uniformly compressed elements and made the following choices of k to obtain the indicated λ_r expressions on LRFD p. 6-38:

1. Unstiffened elements:

   Single angles: $k = 0.45$ $\quad \lambda_r = 76/\sqrt{F_y}$

   Flanges: $k = 0.7$ $\quad \lambda_r = 95/\sqrt{F_y}$

   Stems of tees: $k = 1.28$ $\quad \lambda_r = 127/\sqrt{F_y}$

2. Stiffened element

   Web of a W section column: $k = 5.0$ $\quad \lambda_r = 253/\sqrt{F_y}$

When b/t of each compression element in a column cross section is less than λ_r on LRFD p. 6-38, local buckling of a compression element in a column cross section does not occur before column buckling occurs, and the design strength of a column is given by LRFD E2 (p. 6-47).

When local buckling of a compression element in a column cross section occurs and limits the column buckling strength, LRFD B5.3 (p. 6-37) refers the reader to LRFD Appendix B5.3 (p. 6-105) for the reduced design strength definition of a column.

Figure 4.12 provides some explanatory information to aid in coping with LRFD Appendix B5.3 when local buckling of a compression element in a column cross section limits the column design strength. The possible conditions that may be encountered are:

1. The column cross section contains only unstiffened elements [see Figure 4.12(a)]. A *stress reduction factor* Q_s [see LRFD, p. 6-106; Eqs. (A-B5-1 to 6)] must be computed for the unstiffened element having the larger b/t ratio. Q_s must be used in computing the critical stress F_{cr} due to local buckling [LRFD, p. 6-107; either Eq. (A-B5-15) or (A-B5-16)]. Then, the column design strength is computed: $\phi P_n = 0.85 A_g F_{cr}$. If flexural-torsional buckling can occur, Q_s must be used in computing the critical stress F_{cr} due to flexural-torsional buckling [LRFD Eq. (A-E3-1), p. 6-48]. The smaller of F_{cr} due to local buckling and F_{cr} due to flexural-torsional buckling must be used in computing the column design strength: $\phi P_n = 0.85 A_g F_{cr}$. Example 4.13 illustrates the computations involved in computing ϕP_n for a column cross section containing only unstiffened elements.
2. The column cross section contains only stiffened elements [see Figure 4.12b], an *area reduction factor* Q_a must be computed for each stiffened element having $b/t > \lambda_r$. The definition of $Q_a = A_e/A_g$, where $A_e = A_g - \Sigma A_i$ and A_i = the *ineffective area* of a stiffened element. The *ineffective areas* in Figure 4.12(b) are the cross-hatched areas. Q_a must be used in computing

the critical stress F_{cr} [LRFD, p. 6-108; either Eq. (A-B5-15) or (A-B5-16)]. Unfortunately, the ineffective areas are a function of F_{cr}, and an iterative procedure must be used to determine Q_a. Then, the column design strength is computed: $\phi P_n = 0.85 A_g F_{cr}$.

3. The column cross section contains unstiffened and stiffened elements [see Figure 4.12(c)]. $Q = Q_s Q_a$ must be used in computing the *critical stress* F_{cr}. Then, the column design strength is computed: $\phi P_n = 0.85 A_g F_{cr}$. See items 1 and 2, respectively, for the computations of Q_s and Q_a. The governing F_{cr} is the least F_{cr} value computed as described in items 1 and 2.

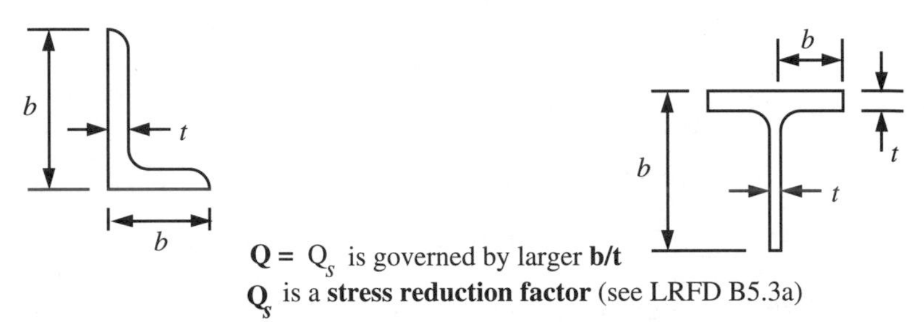

(a) Sections have only unstiffened compression elements

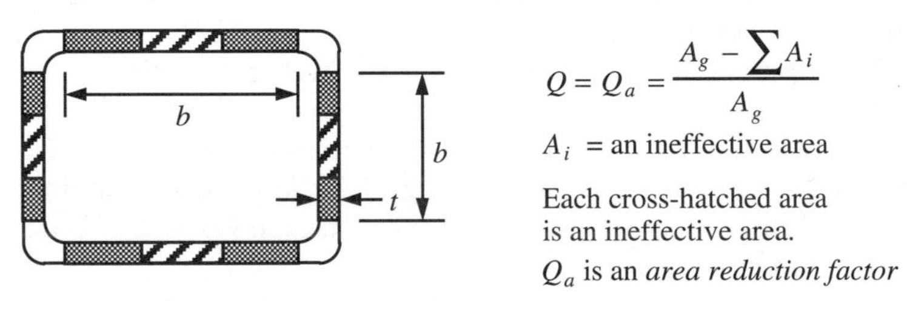

(b) Section has only stiffened compression elements

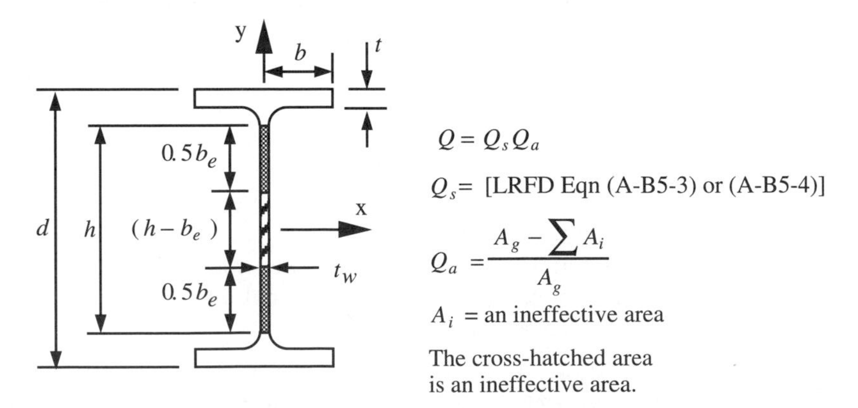

(c) Section has unstiffened and stiffened compression elements

FIGURE 4.12 Slender compression elements.

Example 4.7

The objective of this example is to illustrate how to obtain the column design strength when *local buckling of an unstiffened compression element* occurs and reduces the column buckling strength. For $F_y = 36$ ksi, $(KL)_x = 6$ ft, and a WT8 x 13, find ϕP_{nx}.

Solution

WT8 × 13

$A_g = 3.84$ in.^{2} $\quad d = 7.845$ in. $\quad t_w = 0.250$ in. $\quad r_x = 2.47$ in.
$b_f = 5.50$ in. $\quad t_f = 0.345$ in.

When *b/t* of the flange element and/or stem element exceeds the applicable λ_r from LRFD Table B5.1 (p. 6-38), local buckling may limit ϕP_{nx}:

$$\left(\frac{0.5b_f}{t_f} = \frac{0.5(5.50)}{0.345} = 7.97\right) \leq \left(\lambda_r = \frac{95}{\sqrt{F_y}} = \frac{95}{\sqrt{36}} = 15.83\right)$$

$$\left(\frac{d}{t_w} = \frac{7.845}{0.250} = 31.38\right) > \left(\lambda_r = \frac{127}{\sqrt{F_y}} = \frac{127}{\sqrt{36}} = 21.17\right)$$

FLB does not occur, but stem local buckling may limit ϕP_{nx}. LRFD Appendix B5.3a [item (d) on p. 6-107] and B5.3d must be used to compute ϕP_{nx}:

$$\left(\frac{d}{t_w} = 31.4\right) > \left(\frac{176}{\sqrt{F_y}} = \frac{176}{\sqrt{36}} = 29.3\right)$$

$$Q_s = \frac{20{,}000}{F_y\,(d/t_w)^2} = \frac{20{,}000}{36(31.38)^2} = 0.564$$

$$\lambda_{cx}\sqrt{Q} = \frac{(KL/r)_x}{\pi}\sqrt{\frac{F_y Q}{E}}$$

See LRFD B5.3d: $Q_a = 1.00$ since our section has only unstiffened elements. LRFD Eq. (A-B5-12):

$$Q = Q_s Q_a = Q_s\,(1.00) = Q_s$$

$$\left(\lambda_{cx}\sqrt{Q} = \frac{72}{2.47\,\pi}\sqrt{\frac{36(0.564)}{29{,}000}} = 0.2455\right) < 1.5$$

$$F_{cr} = Q(0.658)^{Q\lambda_{cx}^2} F_y = 0.564(0.658)^{0.062028}(36) = 19.80 \text{ ksi}$$

$$\phi P_{nx} = 0.85 A_g F_{cr} = 0.85(3.84)(19.80) = 64.62 \text{ kips}$$

If we had incorrectly used LRFD E2 to compute ϕP_{nx}, we would have obtained ϕP_{nx} = 112.4 kips. Since 112.4/64.62 = 1.739, if we had not accounted for stem local buckling, we would have overestimated the design strength by 73.9%.

Example 4.8

The objective of this example is to illustrate how to obtain the column design strength ϕP_{ny} for a W21 × 44 and $F_y = 36$ ksi when *local buckling of a stiffened compression element* occurs and reduces the column buckling strength.

Solution

W21 × 44:

$A_g = 13.0$ in. $r_y = 1.26$ in. $t_w = 0.350$ in. $h/t_w = 53.6$ $h = 53.6t_w = 18.76$ in.

From LRFD Table B5.1 (p. 6-38),

$$\left(\frac{0.5b_f}{t_f} = 7.2\right) \le \left(\lambda_r = \frac{95}{\sqrt{36}} = 15.8\right)$$

$$\left(\frac{h}{t_w} = 53.6\right) > \left(\lambda_r = \frac{253}{\sqrt{36}} = 42.17\right)$$

LRFD Appendix B5.3b (item ii) and B5.3c must be used in computing ϕP_{ny} when WLB (web local buckling) may limit the column design strength. Note that FLB does not occur.

The procedure for determining the design strength due to WLB is outlined in this paragraph. First, we must use LRFD Eq. (A-B5-12) to determine a reduced effective width b_e of the web (the stiffened element). See Figure 4.13 where the cross-hatched area is the ineffective area. Then, LRFD Eq. (A-B5-14) is used to determine Q_a, which is an *area reduction factor*. Since $Q_s = 1.00$ [see LRFD B5.3c (item ii)], $Q = Q_s Q_a = 1.00 Q_a = Q_a$. Next, either LRFD Eqn (A-B5-15) or (A-B5-16) is used to compute the critical stress F_{cr}. Finally, $\phi P_{ny} = 0.85 A_g F_{cr}$ gives the column design strength due to WLB.

Unfortunately, f in LRFD Eq. (A-B5-12) is F_{cr}; therefore, the procedure described in the previous paragraph is an iterative procedure. That is, we must assume a value of f in order to determine b_e and at the end of the procedure we find F_{cr}. If the *assumed* $f = F_{cr}$, we compute $\phi P_{ny} = 0.85 A_g F_{cr}$; otherwise, we assume another value of f, determine b_e, and so forth. Fortunately, LRFD Appendix B.5.3c states that A_g and r are for the actual cross section. For this example, we use $A_g = 13.0$ in.^{2} and $r_y = 1.26$ in. as tabulated in the steel manual for the W21 × 44 section.

The $(KL)_y$ value at which WLB ceases to limit ϕP_{ny} is obtained from the condition that $Q_a = 1$, which occurs for the value of f that gives $b_e = h$ in LRFD Eq. (A-B5-12).

$$\left(\frac{h}{t_w} = 53.6\right) = \left(\lambda_r = \frac{253}{\sqrt{f}}\right)$$

$$f = \left(\frac{253}{53.6}\right)^2 = 22.28 \text{ ksi}$$

Then, as shown, we obtain the $(KL)_y$ value at which WLB ceases to govern ϕP_{ny}. Assume that $\lambda_{cy} \le 1.5$:

$$\left[F_{cr} = (0.658)^{\lambda^2_{cx}}\, F_y\right] = (f = 22.28 \text{ ksi})$$

For $F_y = 36$ ksi, we obtain

$$\lambda^2{}_{cy} = \left[\frac{\log(22.28/36)}{\log(0.658)}\right] = 1.1464 \qquad \lambda_{cy} = 1.0707$$

$$\lambda_{cy} = \frac{(KL/r)_y}{\pi}\sqrt{\frac{F_y}{E}} = \frac{(KL)_y}{1.26\pi}\sqrt{\frac{36}{29{,}000}} = 1.0707$$

$$(KL)_y = 120.3 \text{ in.} = 10.02 \text{ ft}$$

For $(KL)_y \ge 10.02$ ft, WLB does not govern ϕP_{ny} and LRFD E2 is applicable for the determination of ϕP_{ny}. However, for $0 \le (KL)_y < 10.02$ ft, LRFD Appendix B5.3b to B5.3d must be used to determine ϕP_{ny} using an iterative procedure:

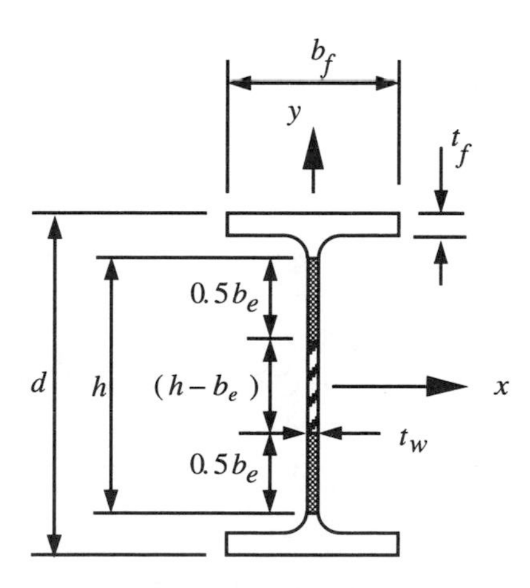

FIGURE 4.13 W section with a slender compression element.

1. At $(KL)_y = 0$, assume that $Q = (Q_a = 0.95)$ and find

$$\lambda_{cy} = \frac{(KL/r)_y}{\pi}\sqrt{\frac{F_y}{E}} = \frac{0}{1.26\pi}\sqrt{\frac{36}{29{,}000}} = 0$$

$$\left(\lambda_{cy}\sqrt{Q} = 0\sqrt{0.95} = 0\right) \le 1.5$$

$$Q\lambda_{cy}^2 = 0.95(0)^2 = 0$$

$$f = F_{cr} = Q(0.658)^{Q\lambda_{cy}^2} F_y = 0.95(0.658)^{0.00}(36) = 34.2 \text{ ksi}$$

2. Using $f = 34.2$ ksi and $h/t_w = 53.6$ in LRFD Eq. (A-B5-12), we find

$$\frac{b_e}{t_w} = \frac{326}{\sqrt{f}}\left[1 - \frac{57.2}{(h/t_w)\sqrt{f}}\right]$$

$$\left[\frac{b_e}{t_w} = \frac{326}{\sqrt{34.2}}\left(1 - \frac{57.2}{53.6\sqrt{34.2}}\right) = 45.57\right] < \left(\frac{h}{t_w} = 53.6\right)$$

$$Q_a = \frac{A - t_w^2(h/t_w - b_e/t_w)}{A} = \frac{13.0 - (0.350)^2(53.6 - 45.57)}{13.0} = 0.924$$

Recall that we had assumed $Q_a = 0.95$; so, perform another iteration cycle.

3. Assume that $Q = Q_a = 0.928$ and find

$$f = F_{cr} = 0.928(0.658)^{0.00}(36) = 33.41 \text{ ksi}$$

4. Using $f = 33.41$ ksi and $h/t_w = 53.6$ in LRFD Eq. (A-B5-12), we find

$$\left[\frac{b_e}{t_w} = \frac{326}{\sqrt{33.41}}\left(1 - \frac{57.2}{53.6\sqrt{33.41}}\right) = 45.99\right] < \left(\frac{h}{t_w} = 53.6\right)$$

$$Q_a = \frac{13.0 - (0.350)^2(53.6 - 45.99)}{13.0} = 0.928$$

Recall that we had assumed $Q_a = 0.928$; therefore, we can proceed to the next step.

5. Use $F_{cr} = 33.41$ ksi to compute the column design strength:

$$\phi P_{ny} = 0.85 A_g F_{cr} = 0.85(13.0)(33.41) = 369 \text{ kips}$$

The iterative procedure illustrated for $(KL)_y = 0$ is applicable for $0 \le (KL)_y < 10.02$ ft and $0.928 \le Q_a < 1$ in this example.

4.7 FLEXURAL-TORSIONAL BUCKLING OF COLUMNS

In the previous discussions in this chapter, we assumed that the buckled shape of a column was due to bending about the principal axis with the *larger KL/r* value and that the cross section did not twist when column buckling occurred. This is called the *flexural mode* of column buckling. However, it is possible that a doubly symmetric cross section (W section, e.g.) only twists when the column buckles; this is called the *torsional mode* of column buckling. Singly symmetric cross sections (T section, channel, equal-leg angle) in which the shear center does not coincide with the centroid (see Figure 4.14) and unsymmetric cross sections (an angle with unequal legs and built-up sections) bend and twist when the column buckles; this is called the *flexural-torsional mode* of column buckling. Theoretical discussions of torsional and flexural-torsional modes of elastic column buckling are available [6–8].

For unsymmetric cross sections, the *critical load* must be determined from the flexural-torsional buckling mode (see LRFD E3, pp. 6-47 and 6-48).

Consider a column whose cross section is a W section. When the member–end supports and any intermediate weak-axis column braces prevent twisting of the cross section at these points, only flexural column buckling can occur, providedthat local buckling is prevented. However, if the intermediate weak-axis column braces are designed to prevent only a translation perpendicular to the weak axis and do not prevent twist of the cross section, the unbraced length for torsion is the member length, whereas the maximum distance between the intermediate braces is the unbraced length for flexure. Consequently, when the intermediate weak-axis column braces do not prevent twist of the cross section, the critical load is the smaller value obtained from the torsional buckling mode [see LRFD, p. 6-110; Eq (A-E3-5)] and the flexural buckling mode.

Two WT7×45 sections are obtained by cutting the web of a W14×90 at middepth along the length direction. Suppose that we use a WT7×45 as a 6-ft-long compression member in a truss. The column design strength for this member can be found from LRFD p. 3-96. In a truss analysis, each member is assumed to have pinned ends. Therefore, using $(KL)_x = (KL)_y = 6$ ft and $F_y = 36$ ksi on LRFD p. 3-96, we find that $\phi P_n = 366$ kips (smaller of $\phi P_{nx} = 366$ kips and $\phi P_{ny} = 389$ kips).

See Figure 4.15a. When $-y$ is the gravity direction, w = member weight/ft causes the member to bend as shown. Since w passes through the shear center, the member bends about the x-axis and deflects in the gravity direction without twisting at any cross section along the member length. Now, if we slowly apply the P forces on the member ends, these P forces cause the amplitude of the deflected shape to increase as we increase P. This enables us to conclude that for bending about the x-axis of a WT section used as a column, the WT section does not twist and *flexural column buckling* is the buckling mode. If b/t of each element in the cross section satisfies LRFD B5.1 (p. 6-32), the applicable definition of ϕP_{nx} is given in LRFD E2 (p. 6-39). If b/t of any element in the cross section exceeds λ_r in LRFD B5.1, then local buckling governs ϕP_{nx} and LRFD B5.3(pp. 6-87 to 89) must be used to compute ϕP_n.

See Figure 4.15c. When x is the gravity direction, w = member weight/ft causes the member to bend as shown. Since w does not pass through the shear center, the member deflects in the gravity direction and twists (θ_z occurs) at each cross section along the member length, except at the member ends where $\theta_z = 0$ is required (see LRFD B6, p. 6-37). Now, if we slowly apply the P forces on the member ends, these

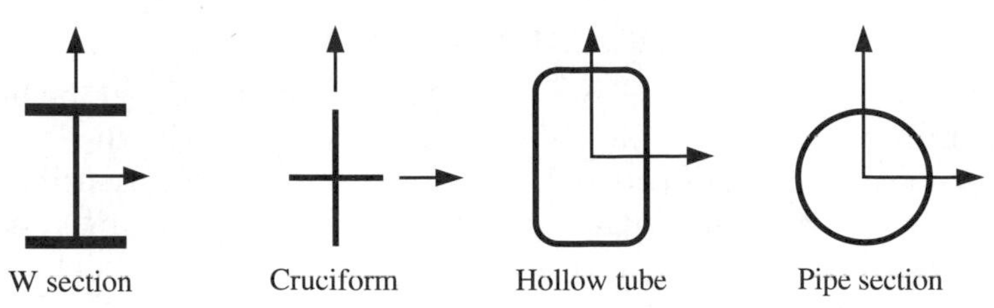

(a) Doubly symmetric sections

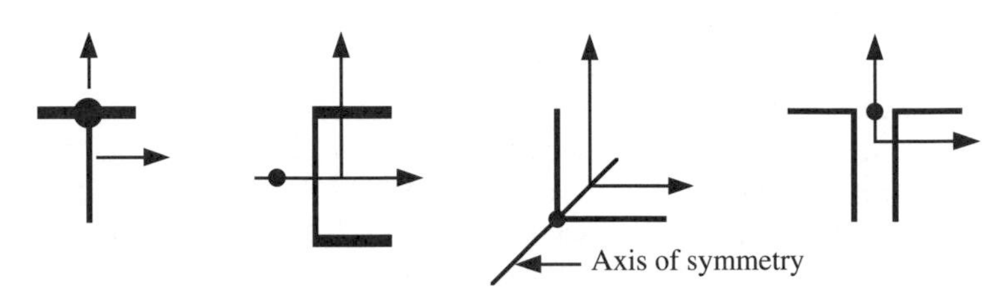

(b) Singly symmetric sections

FIGURE 4.14 Column cross sections.

axial forces P cause the deflected shape to increase as we increase P. This enables us to conclude that for bending about the y-axis of a WT section used as a column, the WT section bends and twists and *flexural-torsional buckling* is the buckling mode. When b/t of each element in the cross section satisfies LRFD B5.1 (p. 6-38), in which case LRFD E3 (p. 6-48) is the applicable definition of ϕP_{ny}. When b/t of any unstiffened element in the cross section exceeds λ_r in LRFD B5.1, then local buckling may govern ϕP_n and LRFD B5.3a (p. 6-106) must be used to compute Q_s, which is a strength reduction parameter in the applicable definition of F_{cr} (LRFD E3d, p. 6-108).

Example 4.9

$$\text{WT7} \times 45 \qquad F_y = 36$$

The objective of this example is to illustrate how $\phi P_{nx} = 366$ kips and $\phi P_{ny} = 389$ kips on LRFD p. 3-96 were computed for WT7 × 45 $F_y = 36$ ksi at $(KL)_x = (KL)_y = 6$ ft.

Solution

WT7 × 45

$$A = 13.2\text{ in}^2 \qquad d = 7.01\text{ in.} \qquad t_w = 0.440\text{ in.} \qquad b_f = 14.52\text{ in.}$$
$$t_f = 0.710\text{ in.} \qquad r_x = 1.66\text{ in.} \qquad r_y = 3.70\text{ in.}$$

On LRFD p. 3-96, $\phi P_{nx} = 366$ kips due to *flexural buckling* was computed as follows:

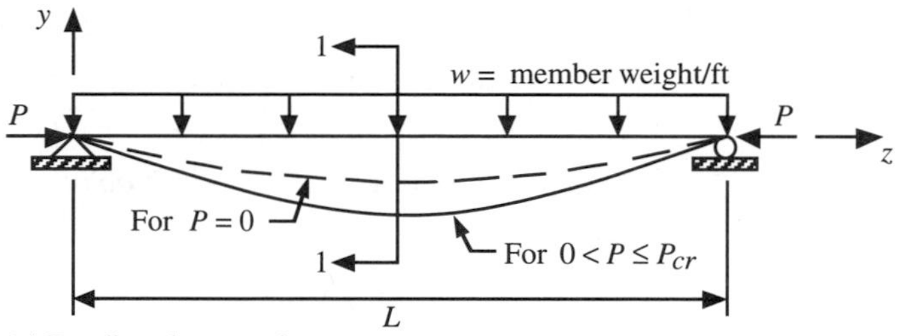

(a) Bending about x axis

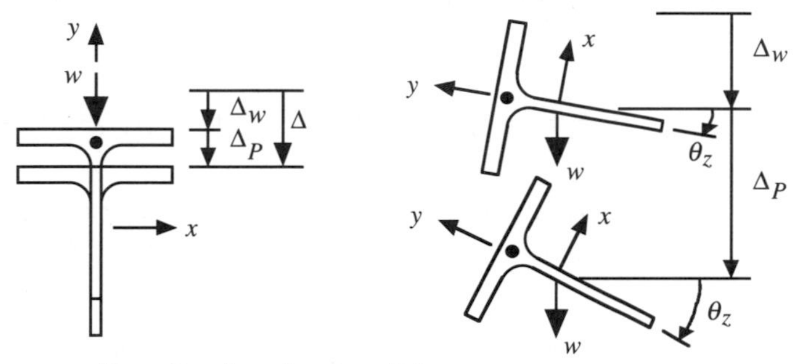

(b) Section 1-1 (c) Section 2-2

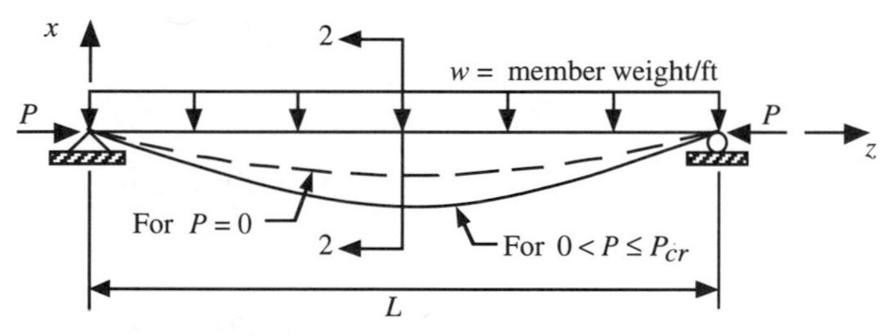

(d) Bending about y axis

FIGURE 4.15 Buckling modes of a WT section column.

$$\left(\frac{0.5b_f}{t_f} = \frac{0.5(14.520)}{0.710} = 10.2\right) \le \left(\lambda_r = \frac{95}{\sqrt{36}} = 15.8\right)$$

$$\left(\frac{d}{t_w} = \frac{7.010}{0.440} = 15.9\right) \le \left(\lambda_r = \frac{127}{\sqrt{36}} = 21.2\right)$$

Local buckling does not govern ϕP_{nx}.

Flexural buckling [LRFD E2 (p. 6-39)] is applicable:

$$\left(\lambda_{cx} = \frac{72/1.66}{\pi}\sqrt{\frac{36}{29,000}} = 0.486437\right) \qquad \lambda_{cx}^2 = 0.2366$$

$$\phi P_{nx} = 0.85(13.2)(36)(0.658^{0.2366}) = 365.8 \text{ kips}$$

On LRFD p.3-96, ϕP_{ny} = 389 kips due to *flexural-torsional buckling* was computed as follows. From LRFD p. 1-166:

$$J = 2.03 \text{ in}^4 \qquad \bar{r}_0 = 4.12 \text{ in.} \qquad H = 0.968$$

From LRFD p.6-19:

$$G = 11{,}200 \text{ ksi.}$$

In the ϕP_{nx} solution, we found that local buckling does not govern ϕP_n. Therefore, we proceed to LRFD p. 6-48:

$$\left(\lambda_{cy} = \frac{72/3.70}{\pi}\sqrt{\frac{36}{29{,}000}} = 0.218239\right) \qquad \lambda_{cy}^2 = 0.04763$$

$$F_{cry} = (0.658)^{\lambda_{cy}^2} F_y = (0.658)^{0.04763}(36) = 35.29 \text{ ksi}$$

$$F_{crz} = \frac{GJ}{A\bar{r}_o^2} = \frac{11{,}200(2.03)}{13.2(4.12)^2} = 101.5 \text{ ksi}$$

$$F_{crft} = \left(\frac{F_{cry} + F_{crz}}{2H}\right)\left[1 - \sqrt{1 - \frac{4F_{cry}F_{crz}H}{(F_{cry} + F_{crz})^2}}\right]$$

$$F_{crft} = \left[\frac{35.29 + 101.5}{2(0.968)}\right]\left[1 - \sqrt{1 - \frac{4(35.29)(101.5)(0.968)}{(35.29 + 101.5)^2}}\right] = 34.71 \text{ ksi}$$

$$\phi P_{ny} = 0.85 A_g F_{crft} = 0.85(13.2)(34.71) = 389.4 \text{ kips}$$

Our computed values of ϕP_{nx} = 366 kips and ϕP_{ny} = 389 kips agree with those tabulated on LRFD p. 3-96 for $(KL)_x = (KL)_y$ = 6 ft and F_y = 36 ksi.

For comparison purposes, compute ϕP_{ny} due to flexure only:

$$\left[\lambda_{cy} = \frac{6(12)/3.70}{\pi}\sqrt{\frac{36}{29{,}000}} = 0.218\right] \qquad \lambda_{cx}^2 = 0.0476$$

$$\phi P_{ny} = 0.85(13.2)(36)(0.658^{0.0476}) = 396 \text{ kips}$$

This is 396/389 = 1.018 times larger than the correct value (*flexural-torsional buckling* solution).

4.8 BUILT-UP COLUMNS

The flexural column buckling behavior of two individual C sections that are not interconnected in any way is shown in Figure 4.16(a) for bending about the y-axis. Note that slippage between the two sections occurs everywhere along the member length, except at midlength of the member. Maximum slippage occurs at each

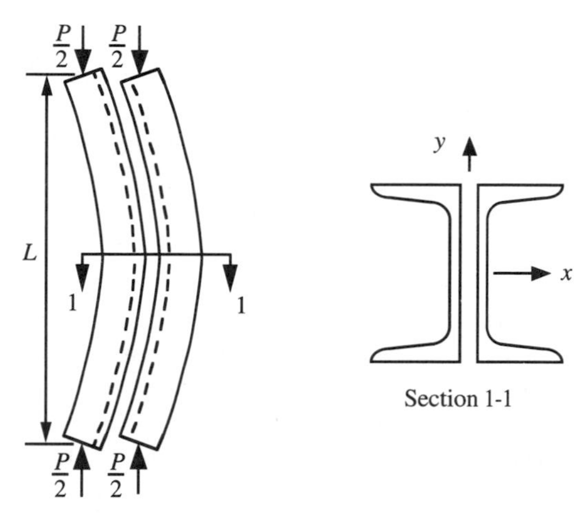

(a) Behavior of two individual C sections

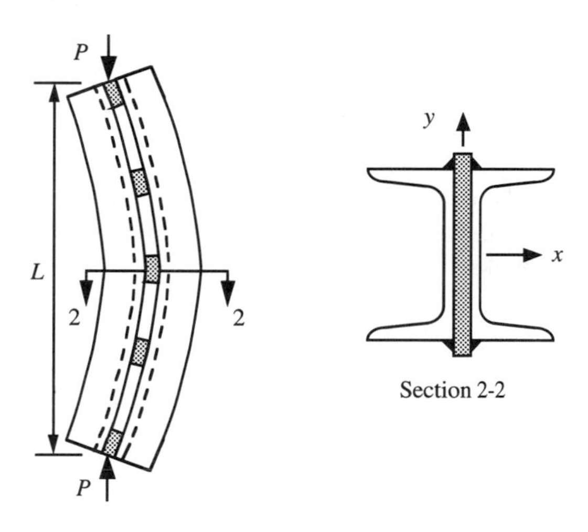

(b) Behavior of two interconnected C sections

FIGURE 4.16 Flexural buckling of a doubly symmetric, built-up column.

member end. As shown in Figure 4.16(b), if a 3/8-in.-thick spacer plate is inserted between the channel webs at the proper locations and welded or fully-tensioned bolted to the C sections, a built-up column is obtained. For simplicity, we refer to the spacer plate and its welds or bolts as a connector. Connectors are uniformly spaced along the member length. The distance between connectors or between a connector and a member end is chosen such that the column design strength of two single channels is not less than the column design strength of the built-up section (a pair of channels). The flexural buckling behavior of the built-up column is as shown in Figure 4.16(b) since the connectors prevent any slippage between the two joined sections. Therefore, the connectors are subjected to shear when the built-up column bends about the y-axis. LRFD E4 (p. 6-48) specifies that the definitions of $(KL/r)_m$ on LRFD p. 6-48 are to be substituted for $(KL/r)_y$ in computing the column flexural strength

(LRFD E2, p. 6-47) for the y-axis of our built-up section (a pair of C sections). When the built-up column bends about the x-axis, the connectors are not subjected to shear. LRFD E2 is applicable for computing the column flexural strength for x-axis bending.

For built-up columns, LRFD E4 (p. 6-48) is applicable. Suppose that the built-up column is a pair of L4 × 3 × 3/8 with long legs back to back separated by and connected to a 3/8-in.-thick gusset plate at each member end. In order to ensure double-angle member behavior, the two angles will be connected to each other by placing a 3/8-in.-thick spacer plate between the long legs at one or more intermediate locations along the member length. Each spacer plate is either fully-tensioned bolted or welded to the two angles and becomes a connector. Let a = connector spacing and r_z = radius of gyration of the minor principal axis of a single angle. The connector locations must be such that Ka/r_z for each single angle is not greater than three-fourths the *maximum* KL/r value for double-angle behavior. L is the member length for double-angle behavior.

Example 4.10

A pair of A36 steel L4 × 3 × 3/8 with long legs back to back is used as a 6-ft-long compression member in a truss. The long leg of each angle is adequately welded at the member ends to a 3/8-in.-thick gusset plate. A 3/8-in.-thick spacer plate is inserted between the long legs of the angles at intervals of 2 ft along the member. Each spacer plate is welded to the long leg of each angle and becomes an intermediate connector for the built-up member.

In a truss analysis, each member is assumed to have pinned ends. Therefore, using $(KL)_x = (KL)_y$ = 6 ft and F_y = 36 ksi on LRFD p. 3-70, we find ϕP_n = 119 kips (smaller of ϕP_{nx} = 128 kips and ϕP_{ny} = 119 kips) and two intermediate connectors are required. The connector spacing is

$$a = \frac{L}{n+1} = \frac{72 \text{ in.}}{2+1} = 24 \text{ in.}$$

The purpose of this example is to illustrate how ϕP_{nx} = 128 kips and ϕP_{ny} = 119 kips were obtained on LRFD p. 3-70 for the problem described when a = 24 in. (two intermediate connectors).

Solution

Double-angle section properties are

$$A = 4.97 \text{ in.}^2 \qquad r_x = 1.26 \text{ in.} \qquad y = 1.28 \text{ in.} \qquad r_y = 1.31 \text{ in.}$$

$$I_x = Ar_x^2 = 4.97(1.26)^2 = 7.89 \text{ in.}^4$$

$$I_y = Ar_y^2 = 4.97(1.31)^2 = 8.53 \text{ in.}^4$$

The shear center is on the y-axis at midthickness of the shorter angle legs. Twist of the cross section is prevented only at the member ends:

$$x_o = 0; \quad x_o^2 = 0$$

$$y_o = y - \frac{t}{2} = 1.28 - \frac{0.375}{2} = 1.0925; \qquad y_o^2 = 1.19$$

$$\bar{r}_o^2 = x_o^2 + y_o^2 + \frac{(I_x + I_y)}{A} = 0 + 1.19 + \frac{7.93 + 8.53}{4.97} = 4.50$$

$$H = 1 - \frac{(x_o^2 + y_o^2)}{\bar{r}_o^2} = 1 - \frac{0 + 1.19}{4.50} = 0.736$$

$$\bar{r}_o = \sqrt{4.50} = 2.12$$

Note: We could have used LRFD p. 1-173 for a pair of L4×3×3/8 with long legs back to back with a 3/8 in. separation to find

$$\bar{r}_o = 2.12 \text{ and } H = 0.735$$

From LRFD Table B5.1 (p. 6-38),

$$\left(\frac{b}{t} = \frac{4}{0.375} = 10.7\right) \le \left(\frac{76}{\sqrt{36}} = 12.7\right)$$

Local buckling does not limit the column design strength.

1. For *x-axis bending* of a double-angle section, LRFD E2 (p. 6-47) is applicable.

$$\left(\frac{KL}{r}\right)_x = \frac{72}{1.26} = 57.14$$

$$\left(\lambda_{cx} = \frac{57.14}{\pi}\sqrt{\frac{36}{29{,}000}} = 0.6409\right) < 1.5; \qquad \lambda_{cx}^2 = 0.4107$$

$$\phi P_{nx} = 0.85(4.97)(0.658)^{0.4107}(36) = 128 \text{ kips}$$

This agrees with $\phi P_{nx} = 128$ kips on LRFD p. 3-70. Also see item 3.

2. For *y-axis bending* of the double-angle section, flexural-torsional buckling occurs. LRFD E4 (p. 6-48) and LRFD E3 are applicable.

$$\left(\frac{KL}{r}\right)_o = \left(\frac{KL}{r}\right)_y = \frac{72}{1.31} = 54.96$$

$$\alpha = \frac{h}{2r_{ib}} = \frac{x + 0.5s}{r_{ysa}} = \frac{0.782 + 0.5(0.375)}{0.879} = 1.103$$

$$\frac{a}{r_{ib}} = \frac{24}{0.879} = 27.30$$

$$\left(\frac{KL}{r}\right)_m = \sqrt{\left(\frac{KL}{r}\right)_o^2 + 0.82\frac{\alpha^2}{(1+\alpha^2)}\left(\frac{a}{r_{ib}}\right)^2}$$

$$\left(\frac{KL}{r}\right)_m = \sqrt{(54.96)^2 + \frac{0.82(1.103)^2}{\left[1+(1.103)^2\right]}(27.30)^2} = 57.93$$

$$\left(\lambda_{cy} = \frac{57.93}{\pi}\sqrt{\frac{36}{29{,}000}} = 0.6497\right) < 1.5; \quad \lambda_{cy}^2 = 0.42210$$

$$F_{cry} = (0.658)^{0.4221}(36) = 30.17 \text{ ksi}$$

For torsional buckling only, the section behaves as two single angles. On LRFD p. 1-159, for a single angle (L4 × 3 × 3/8) we find:

$$J = 0.123 \text{ in.}^4; \quad \bar{r}_o = 1.98 \text{ in.}$$

$$F_{crz} = \frac{GJ}{A\bar{r}_o^2} = \frac{11{,}200(0.123)}{2.48(1.98)^2} = 141.69 \text{ ksi}$$

From LRFD p. 6-48, Eq. (E3-1),

$$F_{crft} = \frac{F_{cry} + F_{crz}}{2H}\left[1 - \sqrt{1 - \frac{4F_{cry}F_{crz}H}{(F_{cry} + F_{crz})^2}}\right]$$

$$F_{crft} = \frac{30.17 + 141.69}{2(0.736)}\left[1 - \sqrt{1 - \frac{4(30.17)(141.69)(0.736)}{(30.17 + 141.69)^2}}\right] = 28.30 \text{ ksi}$$

$$\phi P_{ny} = \phi A_g F_{crft} = 0.85(4.97)(28.30) = 119.55 \text{ kips}$$

This agrees with $\phi P_{ny} = 119$ kips on LRFD p. 3-70. Also see item 3.

3. LRFD E4 (p. 6-48) requires the connector spacing a to be chosen such that

$$\left(\frac{Ka}{r}\right)_i \le 0.75 \text{ times the larger of } \begin{cases}(KL/r)_x \\ (KL/r)_m\end{cases}$$

$$\left[\left(\frac{Ka}{r}\right)_i = \left(\frac{a}{r}\right)_z = \frac{24 \text{ in.}}{0.644 \text{ in.}} = 37.27\right] \le \left[0.75(57.93) = 43.45\right] \text{ as required}$$

Therefore, $a = 24$ in. ($n = 2$ intermediate connectors) is an acceptable choice for this built-up member. Between two adjacent connectors, the built-up column behaves as two single angles. For *z-axis bending* of each L4 × 3 × 3/8, LRFD E2 is applicable:

$A = 2.48 \text{ in.}^2$, $r_z = 0.644$ in., $a = 24$ in. $(KL/r)_z = 37.27$

$$\left(\lambda_{cz} = \frac{37.27}{\pi}\sqrt{\frac{36}{29{,}000}} = 0.418\right) < 1.5; \quad \lambda_{cz}^2 = 0.1747$$

For the two single angles:

$$\phi P_{nz} = 2[0.85(2.48)(0.658)^{0.1747}(36)] = 141 \text{ kips}$$

$$\frac{\phi P_{nx}}{\phi P_{nz}} = \frac{128}{141} = 0.908 \text{ and } \frac{\phi P_{ny}}{\phi P_{nz}} = \frac{119}{141} = 0.844$$

The built-up column design strength is only 9.2% less than that for two single angles with a pinned-ended length *a*.

Example 4.11

The top chord members of the truss in Figure 1.15 are to be selected. The same pair of angles with long legs back to back and welded to 3/8-in.-thick gusset plates is to be used for members 15 to 24. Select the lightest acceptable pair of A36 steel angles. Specify the number of intermediate connectors that are needed.

Solution

From Appendix A for member 20 and loading 7, $P_u = 122.5$ kips and $M_u = 1.91$ ft-kips. In the final design check of these members, P_u and M_u must be accounted for simultaneously, as discussed in Chapter 6. In the preliminary design phase of a compression-plus-bending member, we account for M_u by using an *equivalent* P_u; in this case, we know (from Chapter 6) that a 10% increase in P_u is adequate. Try *equivalent* $P_u = 1.10(122.5) = 135$ kips:

$$L_{20} = \frac{(12 \text{ in./ft})\sqrt{(30)^2 + (2.5)^2}}{5} = 72.25 \text{ in.} = 6.02 \text{ ft}$$

At $(KL)_x = (KL)_y = 6.02$ ft on LRFD p. 3-69 we find

1. $L4 \times 3.5 \times 3/8$ ($\phi P_n = 135$) $\geq$ (*equivalent* $P_u = 135$) and 18.2 lb/ft.
2. Two intermediate welded or fully-tensioned bolted connectors are needed. Therefore, the intermediate connector spacing must not exceed 72.25 in./(2 + 1) = 24.08 in.

Example 4.12

The objective of this example is to illustrate how two entries in the column table were obtained for a double-angle section when local buckling limits the design strength. For $F_y = 50$ ksi, $(KL)_x = (KL)_y = (KL)_z = 10$ ft, and a pair of $L5 \times 3 \times 1/4$ with the long legs back to back and 3/8 in. separators, find ϕP_{nx} and ϕP_{ny}.

Solution

$$\left(\frac{b}{t}=20\right)>\left(\lambda_r=\frac{76}{\sqrt{50}}=10.7\right)$$

Therefore, LRFD Appendix B5.3a and c must be used in computing ϕP_{nx}, which is limited by local buckling of the angle legs with the larger b/t value.

$$\left(\lambda_r=\frac{76}{\sqrt{50}}=10.7\right)<\left(\frac{b}{t}=20\right)<\left(\frac{76}{\sqrt{50}}=21.92\right)$$

$$Q_s=1.340-0.00447(b/t)\sqrt{F_y}=1.340-0.00447(20.0)\sqrt{50}=0.708$$

This agrees with $Q_s=0.708$ given on LRFD p. 1-97. Q_s is a stress reduction factor, Q_sF_{crx} is the local buckling stress, and F_{crx} is defined on LRFD p. 6-108 [either LRFD Eq. (B5-15) or (B5-16) is applicable].

$$\lambda_{cx}=\frac{(KL/r)_x}{\pi}\sqrt{\frac{F_y}{E}}=\frac{120/1.62}{\pi}\sqrt{\frac{50}{29{,}000}}=0.979$$

$$Q=Q_sQ_a=Q_s(1.00)=Q_s=0.708$$

$$\left(\lambda_{cx}\sqrt{Q}=0.979\sqrt{0.708}=0.824\right)<1.5$$

$$Q\lambda_{cx}^2=(0.824)^2=0.679$$

$$\phi P_{nx}=0.85A_gQ\left(0.658^{Q\lambda_{cx}^2}\right)F_y$$

$$\phi P_{nx}=0.85(3.88)(0.708)(0.658)^{0.679}(50)=87.9\text{ kips}$$

This agrees with $\phi P_{nx}=88$ kips given on LRFD p. 3-68 at $(KL)_x=10$ ft for a pair of L5 × 3 × 1/4 and $F_y=50$ ksi.

Solution for ϕP_{ny}

On LRFD p. 3-68 at $(KL)_y=10$ ft for a pair of L5 x 3 x 1/4 and $F_y=50$ ksi, two connectors are required.

$$a=(120\text{ in.})/(2+1)=40\text{ in.}\qquad a/r_i=40/0.663=60.33$$

$$(KL/r)_y=120/1.21=99.17$$

$$a/r_i\le[0.75(KL/r)_y=74.38]\quad\text{as required}$$

$$\alpha=\frac{h}{2r_{ib}}=\frac{x+0.5s}{r_{ysa}}=\frac{0.657+0.5(0.375)}{0.861}=0.981$$

$$\left(\frac{KL}{r}\right)_m=\sqrt{\left(\frac{KL}{r}\right)_o^2+0.82\frac{\alpha^2}{(1+\alpha^2)}\left(\frac{a}{r_{ib}}\right)^2}$$

$$\left(\frac{KL}{r}\right)_m = \sqrt{(99.17)^2 + \frac{0.82(0.981)^2}{\left[1+(0.981)^2\right]}\left(\frac{40}{0.861}\right)^2} = 103.45$$

LRFD Appendix B5.3d, p. 6-108, must be used to compute F_{cry}:

$$\left(\lambda_{cy}\sqrt{Q} = \frac{103.45\sqrt{0.708}}{\pi}\sqrt{\frac{50}{29{,}000}} = 1.1505\right) < 1.5; \quad Q\lambda^2{}_{cy} = 1.3236$$

$$F_{cry} = 0.708(0.658)^{1.3236}(50) = 20.34 \text{ ksi}$$

LRFD E3, p. 6-48, must be used to compute F_{crft}.

From LRFD p. 1-158, for a single $L5 \times 3 \times 1/4$:

$$J = 0.0438 \qquad \bar{r}_o = 2.45$$

$$F_{crz} = \frac{GJ}{A\bar{r}_o^2} = \frac{11{,}200(0.0438)}{1.94(2.45)^2} = 42.13 \text{ ksi}$$

From LRFD p. 1-173, for a pair of $L5 \times 3 \times 1/4$ with $s = 3/8$ in.:

$$H = 0.634 \qquad \bar{r}_o = 2.54$$

$$F_{crft} = \frac{F_{cry} + F_{crz}}{2H}\left[1 - \sqrt{1 - \frac{4F_{cry}F_{crz}H}{(F_{cry} + F_{crz})^2}}\right]$$

$$F_{crft} = \frac{20.34 + 42.13}{2(0.634)}\left[1 - \sqrt{1 - \frac{4(20.34)(42.13)(0.634)}{(20.34 + 42.13)^2}}\right] = 16.47 \text{ ksi}$$

$$\phi P_{ny} = \phi A_g F_{crft} = 0.85(3.88)(16.47) = 54.3 \text{ kips}$$

This agrees with $\phi P_{ny} = 55$ kips on LRFD p. 3-68.

4.9 SINGLE-ANGLE COLUMNS

If a compression member is a single-angle section with only one angle leg fastened at the member ends to a gusset plate, the compression force is applied eccentrically loaded as shown on LRFD p. 3-104. The member is subjected to biaxial bending plus compression and should be treated as a beam-column as illustrated in LRFD Example 3-8 (p. 3-104).

A separate specification and commentary devoted only to single-angle members are given on LRFD pp. 6-277 to 6-300 and must be used to check the design strength requirements for single-angle members. LRFD column tables are given on LRFD pp. 3-107 to 3-116.

4.10 STORY DESIGN STRENGTH

Suppose the plane frame in Figure 4.17(a) is sufficiently braced out of plane such that only in-plane sidesway buckling can occur. Based on the LRFD definitions for column design strength, factored loads applied as shown in Figure 4.17(b) are the maximum acceptable ones for the sidesway buckling mode. The factored loads shown in Figure 4.17(b) were computed as follows:

1. For the exterior columns (W14 × 48) $F_y = 36$ ksi,

$$G_i = G_{bottom} = 10$$

$$G_j = G_{top} = \frac{485/15}{1350/30} = 0.719$$

$$K_{ext} = \sqrt{\frac{1.6G_iG_j + 4(G_i + G_j) + 7.5}{G_i + G_j + 7.5}} = 1.84$$

$$(KL)_x/(r_x/r_y) = 1.84(15)/3.06 = 9.02 \text{ ft}$$

$$\phi P_{nx} = 365 \text{ kips} \quad \text{(from LRFD p. 3-21)}$$

2. For the center column (W14 x 61) $F_y = 36$ ksi,

$$G_i = G_{bottom} = 10$$

$$G_j = G_{top} = \frac{640/15}{2(1350/30)} = 0.474$$

$$K_c = \sqrt{\frac{1.6G_iG_j + 4(G_i + G_j) + 7.5}{G_i + G_j + 7.5}} = 1.78$$

$$(KL)_x/(r_x/r_y) = 1.78(15)/2.44 = 10.94 \text{ ft}$$

$$\phi P_{nx} = 471 \text{ kips} \quad \text{(from LRFD, p. 3-21)}$$

Note that the factored loads in Figure 4.17(b) were obtained from the LRFD design requirement of $\phi P_{nx} \geq P_u$ for each column in the frame. The applied joint loads that produce sidesway buckling for Figure 4.17(b) are $P_{nx} = P_u/\phi = 365/0.85 = 429.4$ kips at the corner joints and $P_{nx} = P_u/\phi = 471/0.85 = 554.1$ kips at the center joint. Also note that Eqs. (4.27) and (4.28) are based on the assumption that all columns in the frame buckle simultaneously. Since the axial force and the axial deformation in each girder are negligibly small when sidesway buckling of the frame occurs, the top end of all columns translates the same amount Δ. When sidesway buckling occurs, the sum of the top-end column moments is $\Sigma P_{nx}\Delta = [2(429.4) + 554.1]\Delta = 1412.9\Delta$. In an acceptable LRFD design, the sum of the top end column moments is $\Sigma P_u\Delta = [2(365) + 471]\Delta = 1201\Delta$. Yura [17] illustrated that the applied load configuration shown in Figure 4.17(b) is not the only acceptable set of P_u values that can be deduced from the sidesway buckling mode. He illustrated that other sets of P_u values are acceptable, as we show in Figure 4.17(c), when $\Sigma P_u \leq$ the story design strength = $\Sigma(\phi P_{nx})$ and when P_u in each column is less than its $\phi P_{n(ns)}$ value, which is obtained from the no-sway buckling mode. To numerically illustrate the preceding statement, suppose the

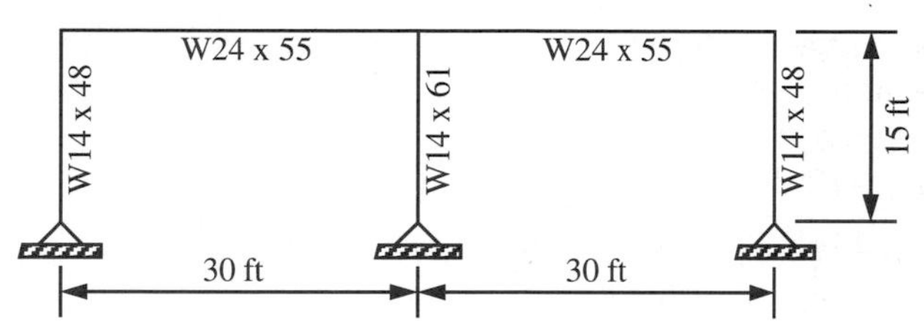

A36 steel.
All members bend about their major principal axis when sidesway buckling occurs.

(a) An isolated plane frame

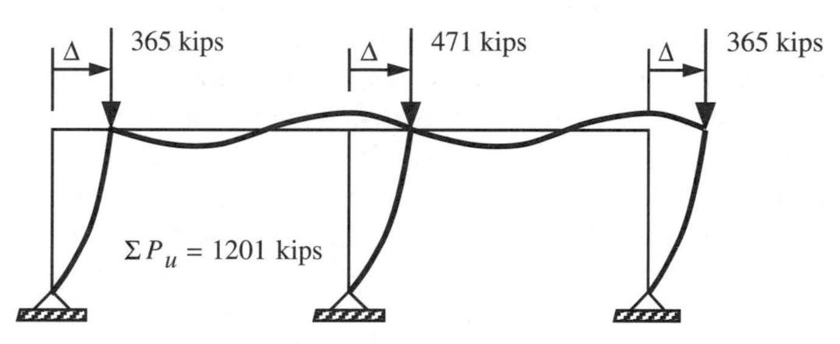

(b) Design loads obtained from LRFD Figure C-C2.2

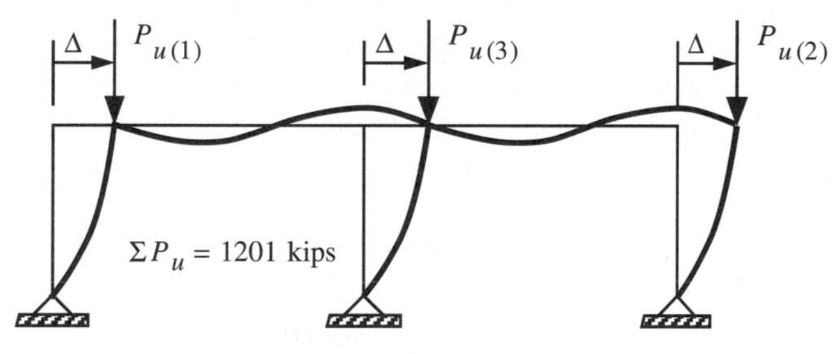

(c) Other possible design load configurations (see text discussion)

FIGURE 4.17 Sidesway buckling of a plane frame.

columns in Figure 4.17(a) are braced out of plane such that $(KL)_y = 7.5$ ft. For the in-plane *no-sway buckling mode*, $K_x < 1$, and we can conservatively use $(KL)_x = 15$ ft for each column. From LRFD p. 3-21, we find for no-sway buckling that:

1. For each W14 × 48,

$$[(KL)_y = 7.5 \text{ ft}] > [(KL)_x/(r_x/r_y) = 15/3.06 = 4.90 \text{ ft}]$$

$$\phi P_{n\,(ns)} = \phi P_{ny} = 384 \text{ kips}$$

2. For the W14 × 61,

$$[(KL)_y = 7.5 \text{ ft}] > [(KL)_x/(r_x/r_y) = 15/2.44 = 6.15 \text{ ft}]$$

$$\phi P_{n\,(ns)} = \phi P_{ny} = 510 \text{ kips}$$

Thus, we can choose any set of P_u values for Figure 4.17(c) that satisfies the following requirements:

1. $\Sigma P_u \leq 1201$ kips
2. $P_{u\,(1)} \leq [\phi P_{n\,(ns)} = 384$ kips]
3. $P_{u\,(2)} \leq [\phi P_{n\,(ns)} = 384$ kips]
4. $P_{u\,(3)} \leq [\phi P_{n\,(ns)} = 510$ kips]

For example, if we choose $[P_{u\,(1)} = P_{u\,(2)} = 346$ kips$] \leq [\phi P_{n\,(ns)} = 384$ kips] and $[P_{u\,(3)} = 509$ kips$] \leq (\phi P_{n\,(ns)} = 510$ kips), then $[\Sigma P_u = 2(346) + 509 = 1201] \leq 1201$ kips. This chosen set of P_u values satisfies the preceding requirements and is an acceptable set of P_u values for Figure 4.17(c). As shown in the following discussion, the story design strength may be applicable for all columns in a story of an entire structure.

The one-story structure in Figure 4.18 is unbraced in each direction. Note that each column cross section in Figure 4.18(a) is rotated 90° with respect to its neighboring cross sections. This was done to provide some major principal-axis column bending strength for resisting sidesway buckling in both directions (x and y). As shown in Figure 4.18(b), the connection of the beam end to the minor axis of the column was chosen for illustration purposes as a hinge. A concrete slab exists on top of the beams shown in Figure 4.18(a) to provide a flat roof surface. This concrete roof slab is stiff in the xy-plane and ensures that the top ends of all columns translate the same amount in the *x-direction* if sidesway buckling occurs in the XZ-plane. Similarly, the top ends of all columns translate the same amount in the *y-direction* if sidesway buckling occurs in the YZ-plane. Therefore, each of the three plane frames in each direction contributes to the story buckling strength (total roof level load that causes sidesway buckling to occur). In the following discussion, we show how to:

1. Determine the story design strength due to sidesway buckling of a single-story structure.
2. Allocate the column axial compressive forces due to a uniformly distributed factored load on the roof.

Only two bays in each direction were chosen in Figure 4.18(a) to simplify the illustrated calculations. The concepts in the following discussion are applicable to any number of bays in each direction, but the calculations for more than two bays in each direction would overshadow what we are trying to convey. Also, rigid connections could have been chosen where hinges are shown.

Figure 4.19 is an enlargement of Figure 4.18(b) and shows the estimated ϕP_{nx} values for each column due to sidesway buckling. These values were estimated as shown in the following discussion.

Column 2 in Figure 4.19(a) is pinned on both ends and does not provide any sidesway buckling resistance. For columns 1 and 3 in Figure 4.19(a),

$$G_i = G_{\text{bottom}} = 10$$

$$G_j = G_{\text{top}} = \frac{485/15}{1350/60} = 1.44$$

$$K_1 = K_3 = \sqrt{\frac{1.6G_iG_j + 4(G_i + G_j) + 7.5}{G_i + G_j + 7.5}} = 2.01$$

$$(KL)_x/(r_x/r_y) = 2.01(15)/3.06 = 9.85 \text{ ft}$$

$$\phi P_{nx} = 355 \text{ kips} \quad \text{(from LRFD p. 3-21)}$$

Columns 4 and 6 in Figure 4.19(b) are pinned on both ends and do not provide any sidesway buckling resistance. For column 5 in Figure 4.19(a), we obtain

$$G_i = G_{\text{bottom}} = 10$$

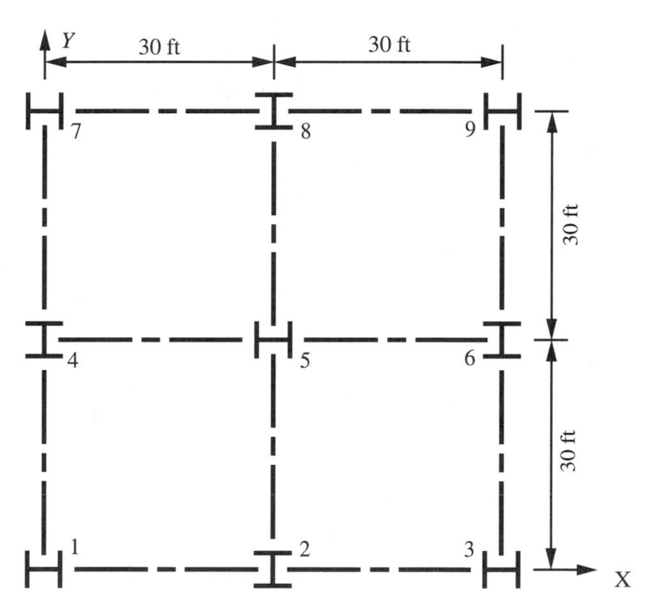

(a) Roof framing plan

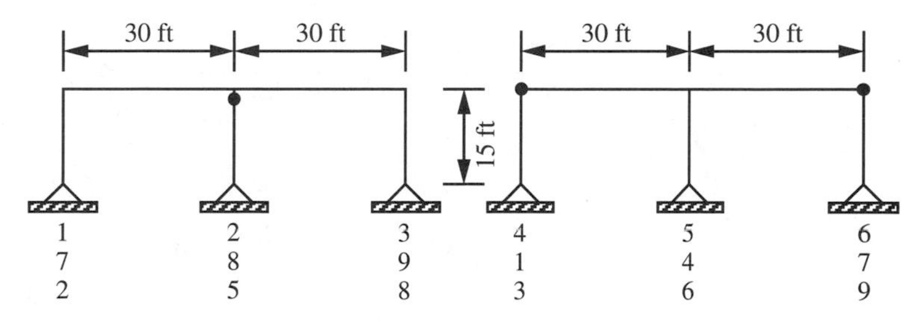

(b) Plane frame views

FIGURE 4.18 One-story unbraced frame.

$$G_j = \frac{\Sigma\left(\frac{I}{L}\right)_c}{\Sigma\left(\gamma\frac{I}{L}\right)_g} = \frac{485/15}{2[0.5(1350/30)]} = 0.719$$

$$K_5 = \sqrt{\frac{1.6G_iG_j + 4(G_i + G_j) + 7.5}{G_i + G_j + 7.5}} = 1.84$$

$$(KL)_x/(r_x/r_y) = 1.84(15)/3.06 = 9.02 \text{ ft}$$

$$\phi P_{nx} = 365 \text{ kips (from LRFD p. 3-21)}$$

For sidesway buckling in the x-direction of Figure 4.18(a), the lower bound estimate of the story design strength = $\Sigma(\phi P_{nx}) = 2(2)(355) + 365 = 1785$ kips. As mentioned, the concrete roof slab ensures that Δ in Figures 4.18(a) and (b) is identical, and for the buckled configuration we obtain $P\Delta = 1785\Delta$, which is the secondary moment produced by the factored loads due to a sidesway deflection of Δ.

For sidesway buckling in the y-direction of Figure 4.18(a), the lower bound

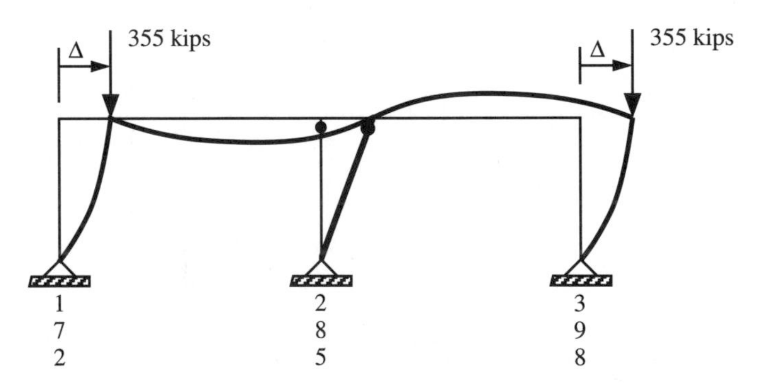

(a) One pinned-ended column case

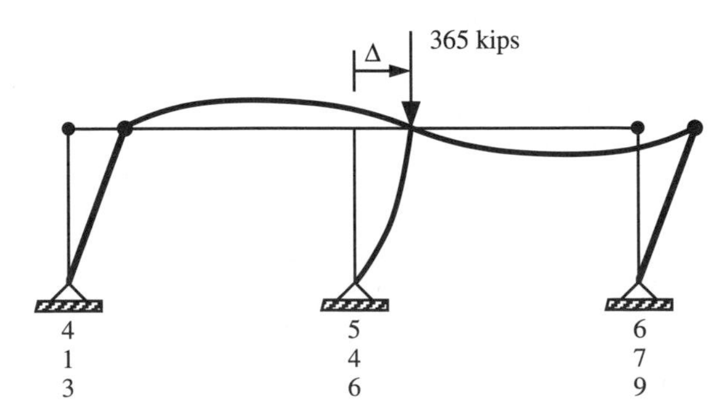

(b) Two pinned-ended columns case

FIGURE 4.19 Design loads for sidesway buckling of a plane frame.

estimate of the story design strength = $\Sigma(\phi P_{nx}) = 2(365) + 2(355) = 1440$ kips, which is less than 1785 kips obtained for the x-direction. Therefore, the lower bound estimate of the story design strength = 1440 kips (y-direction governs).

Some of the columns in Figure 4.18 do not provide any resistance to sidesway buckling. However, for a uniformly distributed factored loading on the roof, all columns in Figure 4.18(a) will have an axial compression force in them. For preliminary design purposes, if we compute the axial compression force in each column using tributary loads as shown in Figure 4.20 and ignore the primary bending moment (M_{ux}) at the top ends of the columns with rigid beam-to-column connections, we find

1. P_u in the corner columns (1, 3, 7, and 9),
2. $2P_u$ in the side columns (2, 4, 6, and 8), and
3. $4P_u$ in the center column (5).

Thus, the total factored roof loading = $4(P_u) + 4(2P_u) + 4P_u = 16P_u$.
Note: In Chapter 6, we will discuss how to compute an *equivalent* P_u that accounts

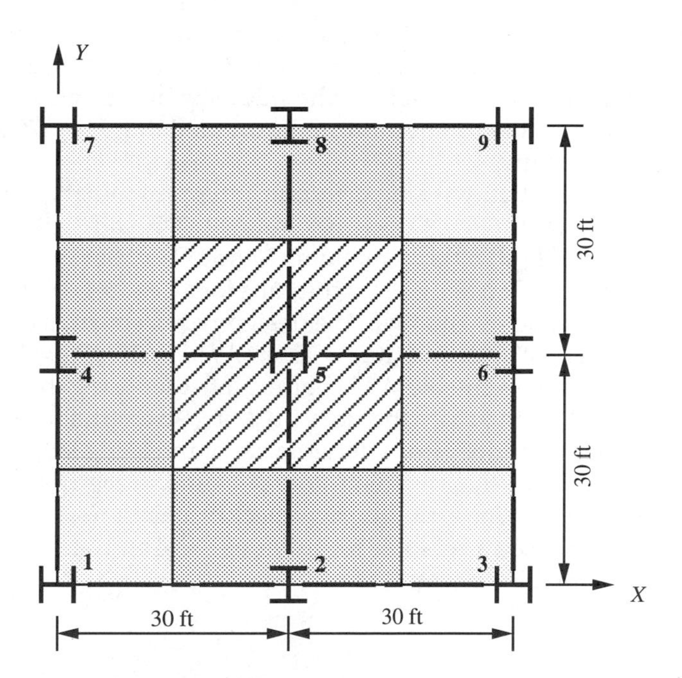

Shaded tributary areas are shown for each column.
For a uniformly distributed load on the roof surface,
column loads from the tributary areas are shown below.
Note: These loads are valid for preliminary design purposes.

P_u on columns 1, 3, 7, and 9
$2P_u$ on columns 2, 4, 6, and 8
$4P_u$ on column 5

FIGURE 4.20 Uniformly loaded structure.

for the effect due to M_{ux}. In the present discussion, we assume that the effect due to M_{ux} is negligible since we do not know how to account for it at the present time.

The lower bound estimate of the story design strength = 1440 kips, and the total factored roof loading cannot exceed 1440 kips for a satisfactory LRFD design. Thus, $16P_u = 1440$ kips and $P_u = 90$ kips, $2P_u = 180$ kips, and $4P_u = 360$ kips.

For no sidesway buckling (braced frame buckling) of Figure 4.18(a), $K = 1$ for all pinned-ended columns and $K < 1$ for the columns rigidly connected at the tops to the girders. If we conservatively use $K = 1$ for the latter columns, $(KL)_x = (KL)_y = 15$ ft and $\phi P_n = (\phi P_{ny} = 270$ kips) for each column. Since $[\Sigma(\phi P_{ny}) = 9(270) = 2430$ kips$] > 1440$ kips, the governing story design strength is 1440 kips due to sidesway buckling in the *y-direction*.

Column 5 of Figure 4.20 is the most heavily loaded column. For the sidesway buckling mode, this column has an axial compression force of $(4P_u = 360$ kips$) > (\phi P_n = 270$ kips for no sway buckling), and a stronger column needs to be chosen. Since $(2P_u = 180$ kips$) < (\phi P_n = 270$ kips for no-sway buckling) and $(P_u = 90$ kips$) < (\phi P_n = 270$ kips for no-sway buckling), sidesway buckling in the *y-direction* governs the design strength of the other columns in Figure 4.18(a).

If the factored loading on the roof is not uniformly distributed as assumed in Figure 4.20, the most heavily loaded column can have $P_u < 270$ kips in the sidesway buckling mode and $P_u = 270$ kips in the no sidesway buckling mode. In the sidesway buckling mode, $\Sigma P_u \leq 1440$ kips with $P_u < 270$ kips in any of the columns are the limiting conditions. That is, in the sidesway buckling mode the distribution of the column P_u values can be arbitrary, provided that $P_u < 270$ kips in any column and $\Sigma P_u \leq 1440$ kips.

Suppose that we are performing the preliminary design of the structure in Figure 4.18(a) for $P_u = 90$ kips in columns 1, 3, 7, and 9; $P_u = 180$ kips in columns 2, 4, 6, and 8; and $P_u = 360$ kips in column 5 and $\Sigma P_u = 1440$ kips. As shown, we can choose to use W14 × 48 and $F_y = 36$ ksi columns for all except column 5. For the no sidesway buckling mode of column 5, the design requirement is $\phi P_{ny} \geq (P_u = 360$ kips). For column 5, $F_y = 36$ ksi and $(KL)_y = 15$ ft; a W14 × 61, $(\phi P_{ny} = 412$ kips$) > (P_u = 360$ kips), is the lightest acceptable W14 choice. Note that using the W14 × 61 for column 5 in Figure 4.18(a) increases the story buckling strength for sidesway deflection in the *x-direction*, but does not affect the story buckling strength for sidesway in the *y-direction*. Since the governing story buckling strength is due to sidesway deflection in the *y-direction*, no further design check calculations are necessary. For preliminary design purposes, a W14 × 48 for columns 1 to 4 and 6 to 9 and a W14 × 61 for column 5 are acceptable choices.

PROBLEMS

4.1 Find ϕP_n for W14 × 90, $F_y = 65$ ksi; $(KL)_x = (KL)_y = 20$ ft.

4.2 Find ϕP_n for W14 × 90, $F_y = 65$ ksi; $(KL)_x = 20$ ft; $(KL)_y = 10$ ft.

4.3 Find ϕP_n for W14 × 90, $F_y = 65$ ksi; $(KL)_x = 12$ ft; $(KL)_y = 6$ ft.

4.4 Use LRFD p. 3-20 and find ϕP_n for W14 × 90, $F_y = 50$ ksi; $(KL)_x = (KL)_y = 20$ ft.

4.5 Use LRFD p. 3-20 and find ϕP_n for W14 × 90, $F_y = 50$ ksi; $(KL)_x = 20$ ft; $(KL)_y = 10$ ft.

4.6 Given $P_u = 550$ kips; $(KL)_x = 20$ ft; $(KL)_y = 10$ ft; $F_y = 50$ ksi, find the lightest acceptable

(a) W14
(b) W12
(c) W10
(d) W8

4.7 $F_y = 36$ ksi. Girders are as shown in Figure P4.7. All members bend about their major axis for in-plane frame buckling. All columns are W12 × 65 and $(KL)_y = L$. Use Eq. (4.27) to find K_x. Find the column design strength for members 1 to 3.

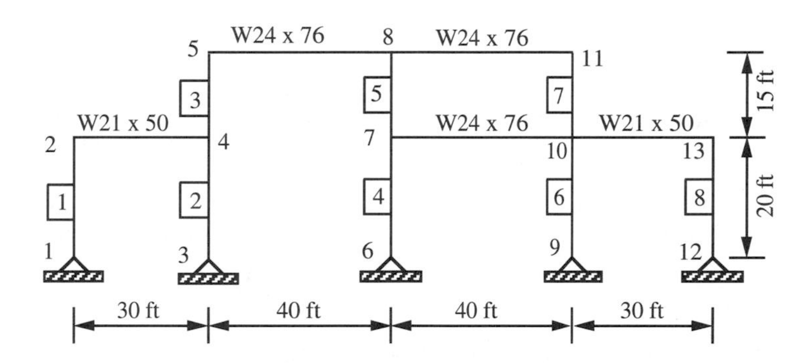

FIGURE P4.7

4.8 Solve Problem 4.7 for $F_y = 50$ ksi.

4.9 $F_y = 36$ ksi. Girders are as shown in Figure P4.7. All members bend about their major axis for in-plane frame buckling. All columns are W12 × 65 and $(KL)_y = L$. Use Eq. (4.27) to find K_x. Find the column design strength for members 6 to 8.

4.10 Solve Problem 4.9 for $F_y = 50$ ksi.

4.11 $F_y = 36$ ksi. Girders are as shown in Figure P4.7. All members bend about their major axis for in-plane frame buckling. All columns are W12 × 65 and $(KL)_y = L$. Use Eq. (4.27) to find K_x. Find the column design strength for members 4 and 5.

4.12 Solve Problem 4.11 for $F_y = 50$ ksi.

4.13 A pair of L6 x 4 x 1/2 with long legs back-to-back and 3/4-in.-thick separators spaced at intervals of 80 in. along the member length is used as a compression member in a truss. $(KL)_x = (KL)_y = 20$ ft. Find ϕP_n for $F_y = 36$ ksi.

4.14 A pair of L6 x 4 x 3/4 with long legs back to back and 3/4-in.-thick separators spaced at intervals of 80 in. along the member length is used as a compression member in a truss. $(KL)_x = (KL)_y = 20$ ft. Find ϕP_n for $F_y = 65$ ksi.

4.15 A pair of C12 × 30 with tie plates spaced at intervals of 4 ft along the member length is used as a compression member in a truss. See Figure P4.15 for the cross-section dimensions. $(KL)_x = (KL)_y = 20$ ft. Find ϕP_n for $F_y = 36$ ksi.

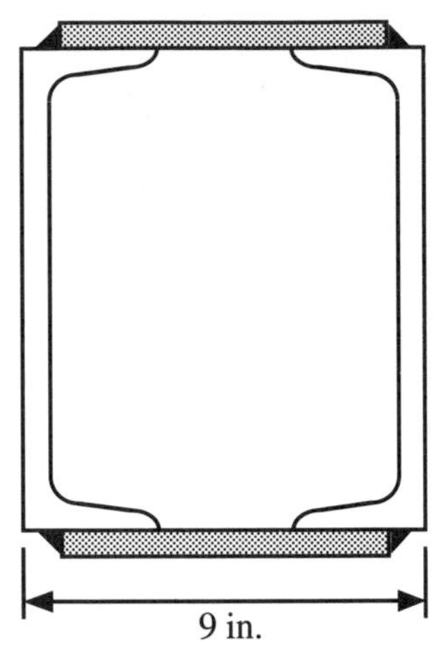

FIGURE P4.15

4.16 $P_u = 360$ kips; $F_y = 36$ ksi; $(KL)_x = (KL)_y = 20$ ft. Find the lightest pair of angles with 3/8-in.-thick separators to serve as a compression member in a truss. Specify the minimum acceptable number of intermediate connectors and the maximum connector spacing.

4.17 Solve Problem 4.16 for $F_y = 50$ ksi.

4.18 $P_u = 240$ kips; $F_y = 36$ ksi; $(KL)_x = (KL)_y = 16$ ft. Find the lightest pair of angles with 3/8-in.-thick separators to serve as a compression member in a truss. Specify the minimum acceptable number of intermediate connectors and the maximum connector spacing.

4.19 Solve Problem 4.18 for $F_y = 50$ ksi.

4.20 $P_u = 240$ kips; $F_y = 36$ ksi; $(KL)_x = 8$ ft; $(KL)_y = 16$ ft. Find the lightest pair of angles with 3/8-in.-thick separators to serve as a compression member in a truss. Specify the minimum acceptable number of intermediate connectors and the maximum connector spacing.

4.21 Solve Problem 4.20 for $F_y = 50$ ksi.

4.22 $P_u = 200$ kips; $F_y = 36$ ksi; $(KL)_x = (KL)_y = 10$ ft. Find the lightest acceptable WT section to serve as a compression member in a truss.

4.23 Solve Problem 4.22 for $F_y = 50$ ksi.

4.24 $P_u = 200$ kips; $F_y = 36$ ksi; $(KL)_x = 5$ ft; $(KL)_y = 10$ ft. Find the lightest acceptable WT section to serve as a compression member in a truss.

4.25 Solve Problem 4.24 for $F_y = 50$ ksi.

4.26 $P_u = 130$ kips; $F_y = 36$ ksi; $(KL)_x = 5$ ft; $(KL)_y = 10$ ft. Find the lightest acceptable WT section to serve as a compression member in a truss.

4.27 Solve Problem 4.26 for $F_y = 50$ ksi.

4.28 A structural tube ST20×12×5/16 (LRFD, p. 1-126) of $F_y = 46$ ksi steel is used as a column for $(KL)_x = 20$ ft and $(KL)_y = 10$ ft. Using LRFD Eq. (A-B5-11) on p. 6-107 and Eqs. (A-B5-15) to (17) on p. 6-108, find the column design strength. The width of each stiffened compression element in the structural tube is the applicable outside dimension minus $2r$. Assume that $r = 2t$ (see last paragraph on LRFD, p. 1-120). Use $F_y = 16.5$ ksi.

4.29 Solve Problem 4.28 for a ST20 × 12 × 3/8.

4.30 Using LRFD Eq. (A-B5-9) on p. 6-107 for $F_y = 36$ ksi, verify that $Q_s = 0.563$ as shown on LRFD p. 1-77 for a WT8 × 13 used in a truss as a compression member. Using LRFD Eqs. (A-B5-15) to (17) on LRFD p. 6-108, compute ϕP_{nx} and ϕP_{ny} for $(KL)_x = (KL)_y = 10$ ft. Using LRFD Eqns (A-E3-1) to (4) and (6) on pages 6-109 and 110, compute ϕP_{ny}. Compare the computed ϕP_{nx} and ϕP_{ny} values to those on LRFD p. 3-95 for a WT8 x 13 and $(KL)_x = (KL)_y = 10$ ft.

4.31 Solve Problem 4.30 for a WT7 x 11.

4.32 Solve Problem 4.30 for a WT6 x 20 and $F_y = 50$ ksi.

4.33 Using LRFD Eq. (A-B5-3) on p. 6-106 for $F_y = 36$ ksi, verify that $Q_s = 0.804$ as shown on LRFD p. 1-101 for a pair of L5 x 3 x 1/4 used in a truss as a compression member, with the short legs separated at the member ends by a 3/8-in.-thick gusset plate. Using LRFD E2 to E4 and Appendix E, compute ϕP_{nx} and ϕP_{ny} for $(KL)_x = (KL)_y = 8$ ft and $(KL)_z = a = 32$ in. Compare your computed design strength values to those on LRFD p. 3-77 for a pair of L5 × 3 × 1/4 and $(KL)_x = (KL)_y = 8$ ft.

4.34 Solve Problem 4.33 for $F_y = 50$ ksi.

4.35 All members of the plane frame in Figure P4.35 are A36 steel and bend about their strong axis when sidesway buckling occurs. All columns are W14 x 74 and all girders are W21 x 44. The exterior columns are rigidly connected to the girder. The interior columns are pinned at each end, but the girders are continuous (contain no hinges). Weak-axis column bracing is provided only at the column ends. Find the story buckling strength. Find the maximum percentage of the story buckling load that can be allocated to each interior column.

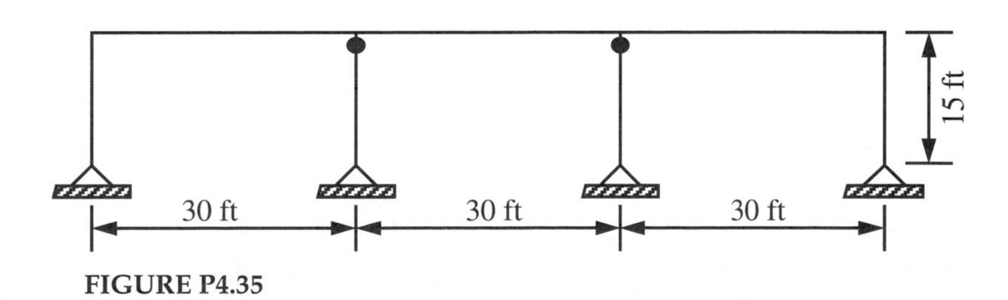

FIGURE P4.35

4.36 Solve Problem 4.35 for $F_y = 50$ ksi steel.

4.37 All members of the plane frame in Figure P4.37 are A36 steel and bend about their strong axis when sidesway buckling occurs. All columns are W14 x 74 and all girders are W21 x 44. The interior columns are rigidly connected to the girder. The exterior columns are pinned at each end. Weak-axis column bracing is provided only at the column ends. Find the story buckling strength. Find the maximum percentage of the story buckling load that can be allocated to each exterior column.

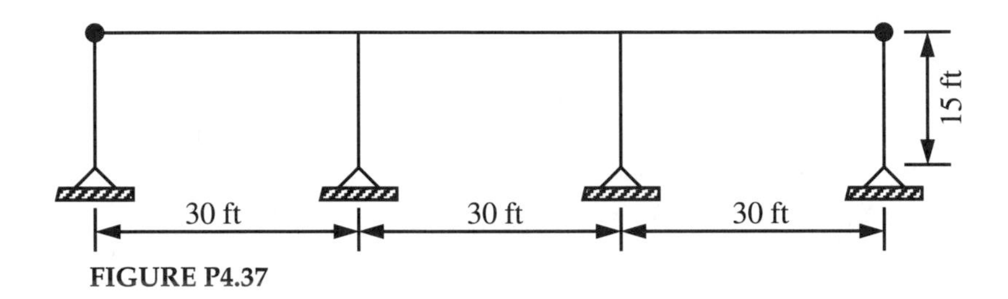

FIGURE P4.37

4.38 Solve Problem 4.37 for $F_y = 50$ ksi steel.

4.39 All members of the plane frame in Figure P4.39 are A36 steel and bend about their strong axis when sidesway buckling occurs. The girder is a W27 x 94. The left exterior column is rigidly connected to the girder, but the right exterior column is pinned at each end. Weak-axis column bracing is provided only at the column ends. For the factored loads shown, choose the lightest acceptable W14 for each column.

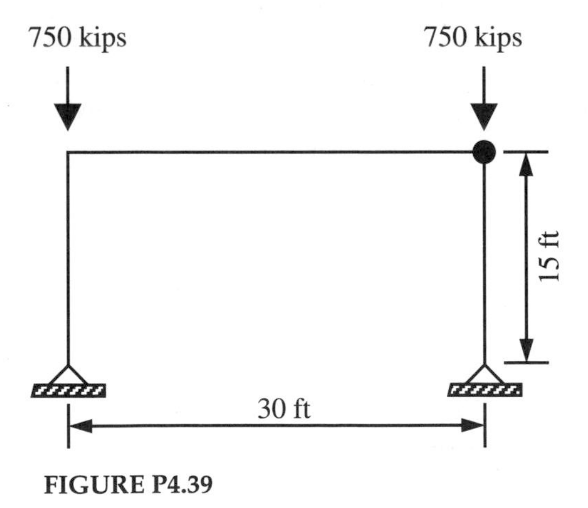

FIGURE P4.39

4.40 All members of the plane frame in Figure P4.40 are A36 steel and bend about their strong axis when sidesway buckling occurs. All girders are W27 x 94. The exterior columns are rigidly connected to the girder. The interior columns are pinned at each end, but the girders are continuous (contain no hinges). Weak-axis column bracing is provided only at the column ends. For the factored loads shown, choose the lightest acceptable W14 for each column.

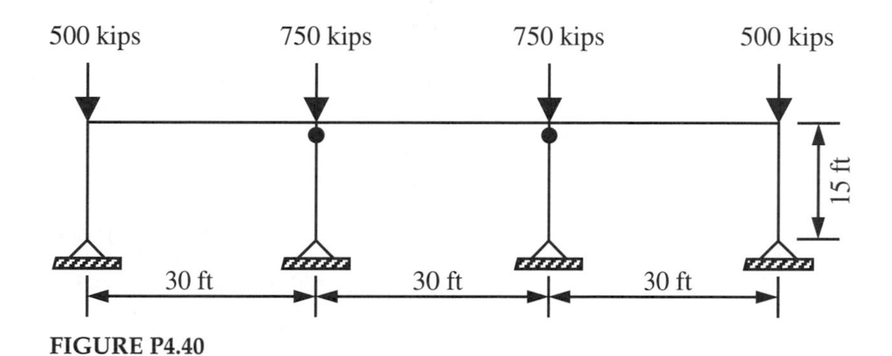

FIGURE P4.40

4.41 All members of the plane frame in Figure P4.41 are A36 steel and bend about their strong axis when sidesway buckling occurs. All girders are W27 x 94. The interior columns are rigidly connected to the girder. The exterior columns are pinned at each end. Weak-axis column bracing is provided only at the column ends. For the factored loads shown, choose the lightest acceptable W14 for each column.

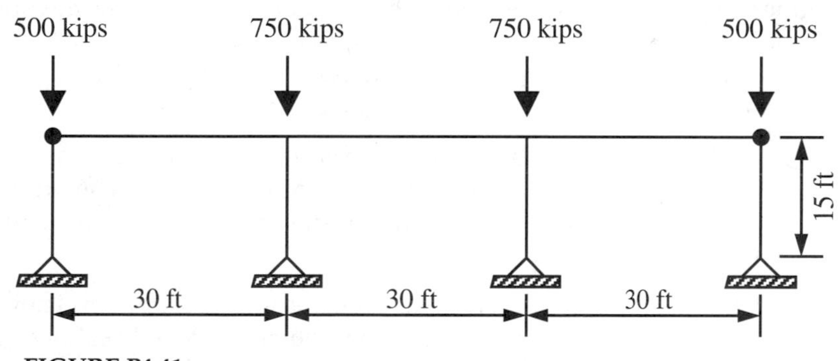

FIGURE P4.41

4.42 Girders are as shown in Figure P4.7. All members bend about the x-axis for in-plane frame buckling. $F_y = 36$ ksi. All columns are W12 x 65. $(KL)_y = L$ for all columns. Use Eq. (4.27) to find K_x. Find the story design strength for members 3, 5, and 7.

4.43 Solve Problem 4.42 for $F_y = 50$ ksi.

4.44 Girders are as shown in Figure P4.7. All members bend about the x-axis for in-plane frame buckling. $F_y = 36$ ksi. All columns are W12 x 65. $(KL)_y = L$ for all columns. Use Eq. (4.27) to find K_x. Find the story design strength for members 1, 2, 4, 6, and 8.

4.45 Solve Problem 4.44 for $F_y = 50$ ksi.

CHAPTER

5

Beams

5.1 INTRODUCTION

A *beam* is defined as any structural member that bends and/or twists due to the applied loads, which do not cause an internal axial force to occur in the member. Therefore, an applied load on a beam cannot have any component parallel to the member length. Concentrated and distributed loads between the member ends, and member-end moments, are examples of applied loads on a beam.

For notational convenience on structural drawings and in structural design calculations, beams are sometimes categorized as follows:

1. *Girders* are the beams spaced at the largest interval in a floor or roof system. They support the most load in a floor or roof system. The primary loads on girders are the reactions of other beams and possibly some columns.
2. *Floor beams* support joists.
3. *Joists* are the most closely spaced beams in a floor system. They support the concrete floor slab. Steel joists may be either rolled sections or fabricated open-web joists (small trusses).
4. *Roof beams* support purlins.
5. *Purlins* are the most closely spaced beams in a roof system. Purlins support the roof surface material and may be open-web joists, hot-rolled sections, or cold-formed sections.
6. *Spandrel beams* support the outside edges of a floor deck and the exterior walls of a building up to next floor level.
7. *Lintels* span over window and door openings in a wall. A lintel supports the wall portion above a window or door opening.
8. *Girts* are exterior wall beams attached to the exterior columns in an industrial-type building. They support the exterior wall and provide bending resistance due to wind.

9. *Stringers* are beams parallel to the traffic direction in a bridge floor system supported at panel points of trusses located on each side of the bridge deck.
10. *Diaphragms* are beams that span between the girders in a bridge floor system and provide some wheel load distribution in the direction perpendicular to traffic.

5.2 DEFLECTIONS

In the second paragraph of LRFD Chapter L (p. 6-98), we find: "Limiting values of structural behavior to ensure serviceability (for maximum deflections, accelerations, etc.) shall be chosen with regard to the intended function of the structure."

LRFD L3.1 states: "Deformations in structural members and structural systems due to service loads shall not impair the serviceability of the structure."

LRFD L3.3 states: "Lateral deflection or drift of structures due to code-specified wind or seismic loads shall not cause collision with adjacent structures nor exceed the limiting values of such drifts which may be specified or appropriate."

The LRFD Specification does not provide any guidelines on the limiting values for beam deflections and for drift (system deflections due to and in the direction of wind, e.g.). Therefore, the structural designer must decide what the appropriate limiting deflection values for each structure are based on experience, judgment, the satisfactory performance of a similar structure, and the owner's intended use of the structure. Table 1.2 gives some suggested limiting values for beam and drift deflections.

Deflections must be considered in the design of almost every structure. In the interest of minimizing the dead weight of high-rise structures, high-strength steel members are used wherever they are economically and structurally feasible. Consider a 20-ft.-long W section member, for example. The member weight is a function of only one variable, the cross-sectional area. As the member weight decreases, the cross-sectional area and the moments of inertia decrease, and the member becomes more flexible and permits larger deflections to occur. Consequently, controlling deflections becomes more of a problem when the dead weight of the steel members is minimized. If a high-rise structure sways too much or too rapidly, the occupants become nauseated or frightened although no structural damage may occur. Similarly, the public becomes alarmed if a floor system of a building or a bridge is too flexible and noticeably sags more than a tolerable amount. Also there are situations, such as a beam over a plate glass window or a water pipe, where excessive deflections can cause considerable damage if they are not controlled by the structural designer. If the roof beams in a flat roof sag too much, water ponds on the roof, causing additional sagging, more ponding, which can rupture the roof surface, and extensive water damage can occur to the contents of the structure. Consequently, it is not unusual for deflections to be the controlling factor in the design of a structure or the design of a member in a structure.

Construction can only be done within tolerable limits. For example, columns cannot be perfectly plumbed and foundations cannot be placed perfectly in plan view nor in elevation view. Deflections that occur during construction due to these imperfect erection conditions, wind, temperature changes, and construction loads must be controlled by the steel erection contractor to ensure the safety of the structure, the construction workers, and the public. Total collapses of steel structures have occurred during construction because the erection contractors did not provide adequate drift control bracing and/or adequate shoring to limit gravity direction deflections during construction.

The deflected shape of a structure due to service conditions sometimes can be economically controlled by precambering the structure. Precambering (see LRFD L1, p. 6-98) is achieved by erecting the structure with built-in deformations such that the structure deflects to or slightly below its theoretical no-load shape when the maximum service loads occur on the structure. For example, suppose that the bottom chords of a simply supported plane truss are deliberately fabricated too short. When the truss is assembled, the interior truss joints displace upwardly. When the service loads occur, each interior truss joint displaces downward to or slightly below its no-load, camberless position.

Current structural design practice is to design the structural members to have adequate strength to resist bending moment, shear, and axial force (if applicable). Then, gravity direction and drift deflections are checked to determine if they are adequately controlled to ensure the desired level of serviceability. Some of the common serviceability problems were stated in Section 1.1.4.

5.3 SHEAR

The design requirement for shear is

$$\phi_v V_n \geq V_u$$

where

$\phi_v = 0.90$

V_u = *required shear strength*

V_n = *nominal shear strength,* which is defined later

The design approach for a beam whose cross section is a W section, for example, is to select the lightest W section that satisfies the design requirement for bending and to check the selected section for all of the other design requirements [shear and serviceability (deflection and vibration control), for example]. If any of the other design requirements are not satisfied for the selected section, the structural designer must either choose another section that satisfies all of the design requirements or appropriately modify the selected section to satisfy all of the violated design requirements.

For W section and C section beams, satisfying the design requirement for shear usually is not a problem, except for the following cases:

1. A beam end is coped (see Figure 5.12(e); LRFD Fig. C-J5.1, p. 6-228; LRFD Fig. 8-59, p. 8-226). If the beam end has a bolted connection (see LRFD Fig. C-J5.1, p. 6-228), the *design shear rupture strength* (see LRFD J4, p. 6-87) must be determined. If the beam end is coped and has a welded connection, the coped web depth must be used in the nominal shear strength definition.
2. Holes are made in the beam web for electrical, heating, and air-conditioning ducts, for example. The net web depth must be used in the nominal shear strength definition. Also, web stiffeners may be needed around the holes to strengthen the web.

3. The beam is subjected to a large concentrated load located near a support. For W and C sections bending about their strong axis, the web resists all of the shear. Most likely, *web stiffeners* (see Sections 5.14 and 5.15; LRFD K1.8 and K1.9, p. 6-96) will be needed at the support and at the concentrated load points, and we would satisfy LRFD Appendix F2 (p. 6-113) in the region between the web stiffeners. For W and C sections bending about their weak axis, each flange resists half of the shear.

If a beam does not have any holes in the web nor any coped ends, the following formulas are applicable for W and C sections subjected to shear in the plane of the web.

1. When $(h/t_w) \leq (418/\sqrt{F_y})$

   shear yielding of the web is the mode of failure, and the nominal shear strength definition is (see LRFD F2.2, p. 6-56)

$$V_n = 0.6F_y t_w d$$

2. When $(418/\sqrt{F_y}) < (h/t_w) \leq (523/\sqrt{F_y})$

   inelastic shear buckling of the web is the mode of failure, and the nominal shear strength definition is

$$V_n = \frac{0.6F_y t_w d(418/\sqrt{F_y})}{(h/t_w)}$$

3. When $(523/\sqrt{F_y}) < (h/t_w) \leq 260$

   elastic shear buckling of the web is the mode of failure, and the nominal shear strength definition is

$$V_n = 132{,}100\left[\frac{t_w d}{(h/t_w)^2}\right]$$

For unsymmetric sections and for weak-axis bending of singly-symmetric or doubly-symmetric sections, the shear design strength definition is given in LRFD H2 (p. 6-60).

For the design shear strength definition of other sections, see LRFD Appendix F2 (p. 6-113) and LRFD Appendix G3 (p. 6-124).

Also, see LRFD Figure C-K1.2 (p. 6-234) for a sketch pertaining to LRFD K1.7 (p. 6-95), which gives the design shear strength definition for a column web panel subjected to high shear.

Since shear usually is not a problem for rolled sections used as a beam, we chose not to have any examples dealing only with shear.

5.4 BENDING BEHAVIOR OF BEAMS

The hot-rolled steel section most commonly used as a beam is the W section [see Figure 5.1(a)], which is doubly symmetric and I-shaped. A W section is configured for economy to provide much more bending resistance about the major principal axis than about the minor principal axis. Therefore, the following discussion begins with the behavior of a W section beam subjected to an applied moment that causes the member to bend about the major principal axis of the cross section.

Figure 5.1(b) shows a W section beam of infinitesimal length subjected only to bending about the major principal axis x of the cross section. A cross section in the xy plane prior to bending is assumed to remain a plane section in the rotated position after bending occurs. Therefore, as shown in Figure 5.1(c), the strain diagram is linear; the *maximum compressive strain* is denoted as ε_c and the *maximum tensile strain* is denoted as ε_t. The rate of change of $\theta_x = d\theta_x/dz$ in Figure 5.1(b) is called the *curvature* ϕ_x, which can be computed as shown in Figure 5.1(c) for small slope theory.

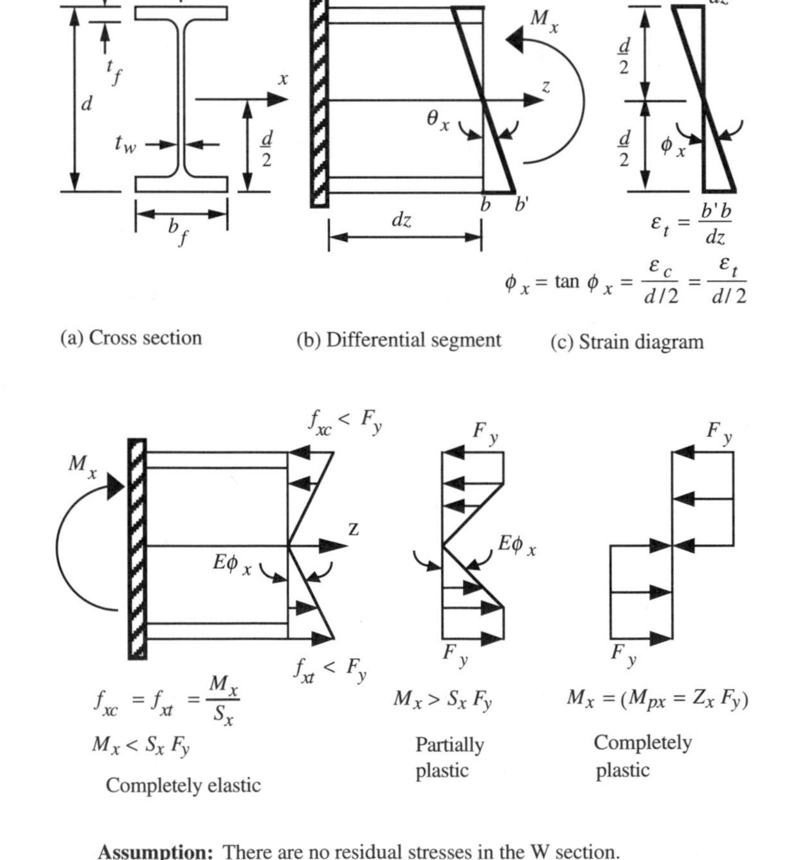

FIGURE 5.1 Uniform bending of infinitesimal-length W section.

To simplify the discussion, we temporarily assume there are not any *residual stresses* in the member. For a particular steel grade, the *stress–strain relation* for each cross-sectional fiber is as shown in Figures 2.1 and 2.2. If the *extreme fiber strains*, ε_c and ε_t, do not exceed the *yield strain*, ε_y, on the appropriate stress–strain curve, the stress diagram is E times the strain diagram (see the completely elastic case in Figure 5.1(d) or, alternatively, the bending stresses can be computed by using the *flexure formula*:

$$f = \frac{-M_x y}{I_x}$$

where the minus sign accounts for the chosen sign convention: a tensile stress is positive and a compressive stress is negative. In the cross-sectional region where y is negative in Figure 5.1(a), only tensile stresses exist in Figure 5.1(d) due to the applied moment in Figure 5.1(b), and tensile stresses are positive. Therefore, the minus sign in the flexure formula is needed for the bending stress to have the correct sign.

In Figure 5.1(d), note that we chose to use arrows instead of signs to denote compressive and tensile stresses acting on the right end of the beam segment. For the completely elastic case in Figure 5.1(d), the extreme fiber stresses due to M_x are denoted as f_{bxc} (maximum compression stress due to bending about the x-axis) and f_{bxt} (maximum tension stress due to bending about x-axis).

Eventually we will have to be prepared to deal with cross sections that are not symmetric about the bending axis. Therefore, we choose to make the following definitions for a cross section that is not symmetric about the x-axis:

1. Let y_c and y_t, respectively, be the absolute value of y in the flexure formula when the stress in the *extreme compression fiber* and in the *extreme tension fiber*, respectively, are computed.
2. Let

$$S_{xc} = \frac{I_x}{y_c}$$

$$S_{xt} = \frac{I_x}{y_t}$$

Then, the maximum compression stress due to *x-axis* bending is

$$f_{xc} = \frac{M_x}{S_{xc}}$$

and the maximum tension stress due to *x-axis* bending is

$$f_{xt} = \frac{M_x}{S_{xt}}$$

If the cross section is symmetric about the *x-axis*, let

$$S_{xc} = S_{xt} = \left(S_x = \frac{I_x}{0.5d} \right)$$

$$f_{xc} = f_{xt} = \frac{M_x}{S_x}$$

The value of S_x, the *elastic section modulus* for x-axis bending, is given in Part 1 of the LRFD Manual for each W section.

When M_x in Figure 5.1(b) produces extreme fiber strains (see Figure 5.1(c)) that exceed the yield strain but are less than the strain-hardening strain, the *bending stresses* are as shown in the partially plastic case (see Figure 5.1(d)) and cannot be computed by the flexure formula. These bending stresses must be obtained from the stress–strain relations (see Figure 2.2). The completely plastic case in Figure 5.1(d) is theoretically impossible since the strain at the bending axis is always zero. However, in laboratory tests the *plastic bending moment*, $M_{px} = Z_xF_y$, where Z_x is the *plastic section modulus* for x-axis bending, can be developed and exceeded as shown in Figure 5.2 on the moment–curvature diagram. When the curvature in Figure 5.2 is only two times the curvature at first yield, $M_x = 0.97M_{px}$. For A36 steel, the curvature at which strain hardening begins to occur is about 12 times the yield curvature; that is, $\phi_{sh} \approx 12\phi_y$.

As shown in Figure 5.3, due to uneven cooling there are residual stresses in a hot-rolled W section. The extreme fiber at each flange tip has the maximum residual compressive stress, F_{rc}, which is assumed to be 10 ksi (see LRFD F1.2a, p. 6-54). Therefore, based on the assumption stated in the note of Figure 5.3, the top corner fibers in Figure 5.3(a) begin to yield when $M_x/S_x + F_{rc} = F_y$. Figure 5.4 shows the effect of residual stresses on the moment–curvature relation for x-axis bending of a W

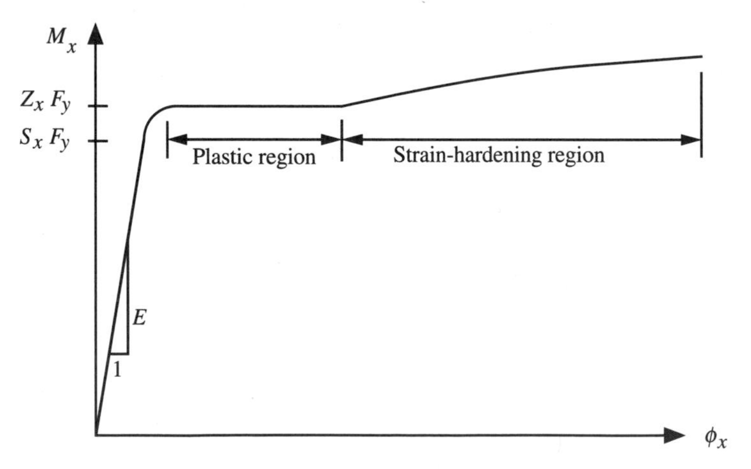

Assumption: There are no residual stresses in the W section.

FIGURE 5.2 Moment-curvature for X-axis bending of a W section.

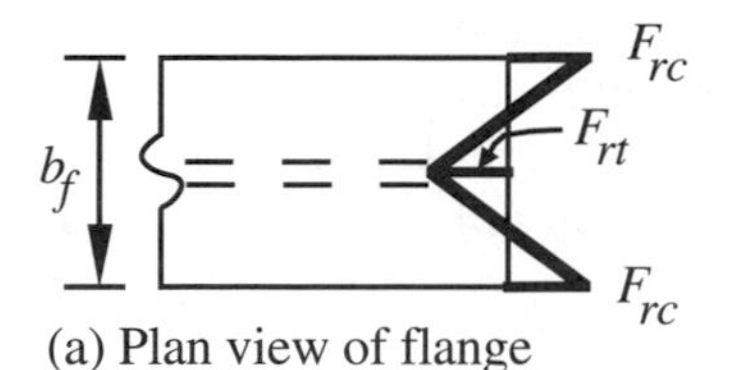

(a) Plan view of flange

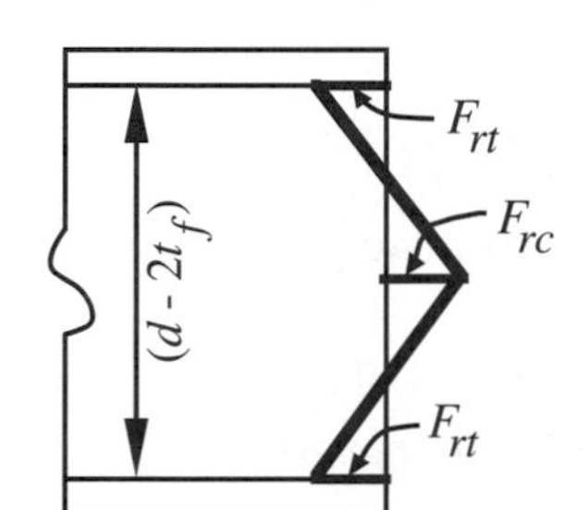

(b) Side elevation view of web

Residual stresses in each flange:
F_{rc} = maximum compressive residual stress
F_{rt} = maximum tensile residual stress

Residual stresses in the web:
F_{rc} and F_{rt} same as noted for flanges

Note: For a linear variation of residual stresses across the flange width and along the depth direction of the web, $F_{rc} = F_{rt} = F_r$

LRFD page 6-18 gives
F_r = 10 ksi for rolled sections

FIGURE 5.3 Residual stresses in a W section.

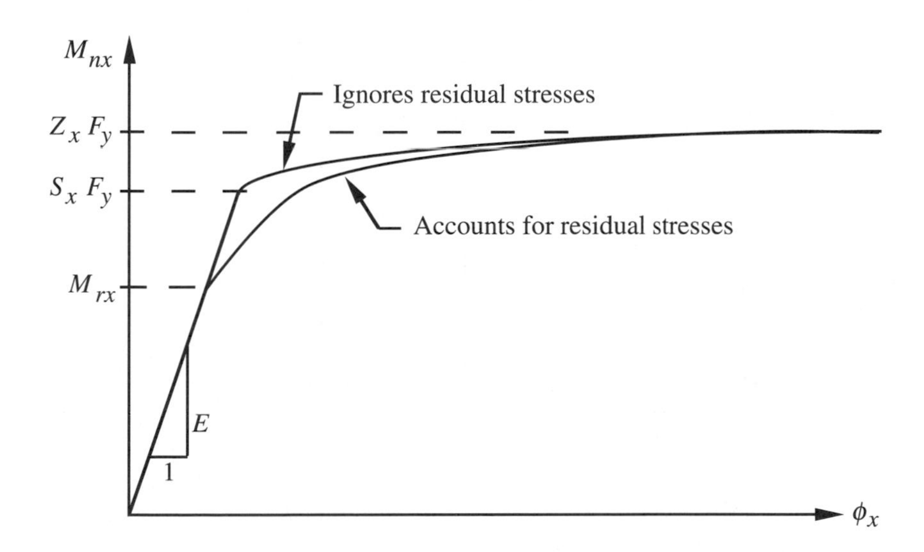

FIGURE 5.4 Effects of residual stresses on moment-curvature.

section. When residual stresses are accounted for, the elastic limit for M_x is $M_{rx} = S_x (F_y - F_{rc})$. Also, the fibers that yielded first will be the first fibers to strain harden. However, the *maximum nominal bending strength* defined in the LRFD Specifications is the internal bending moment for the completely plastic case of Figure 5.1(d).

5.5 PLASTIC BENDING

The purpose herein is to illustrate how

1. to locate the *plastic neutral axis* (PNA)
2. to compute the *plastic moment* for x-axis bending M_{px} of a singly-symmetric, I-shaped section for which:
   a. all elements are made of the same grade of steel
   b. the flange elements are made of a higher grade of steel than the web element.

For illustration purposes, let all of the elements in Figure 5.5 have the same yield strength, $F_y = 36$ ksi. We can treat M_{px} as a $C = T$ couple separated by a lever arm a. Let:

A_c = section area subjected to compression yield stress

A_t = section area subjected to tension yield stress

A = total cross sectional area.

Since $(C = A_cF_y) = (T = A_tF_y)$, this requires that $A_c = A_t = A/2$, which enables us to

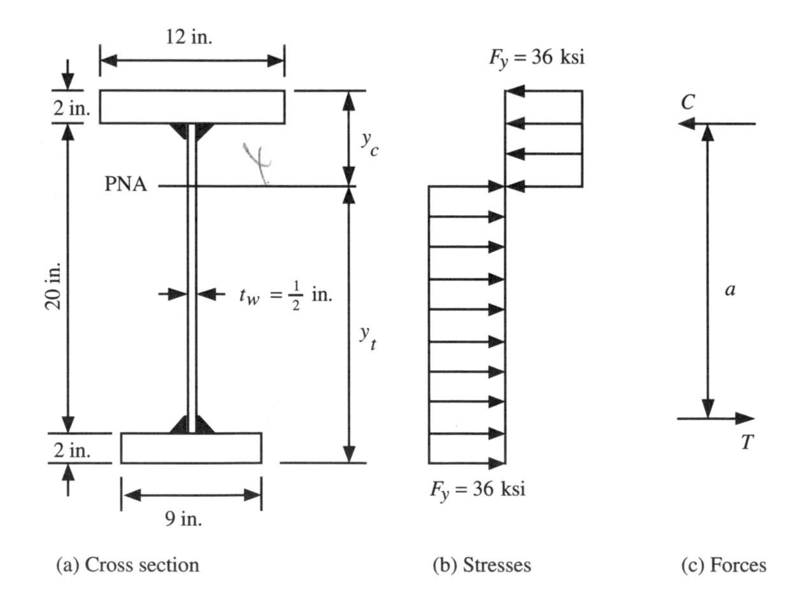

FIGURE 5.5 PNA of a singly-symmetric section.

locate the PNA.

$$A = 2(12) + 20(0.5) + 2(9) = 52 \text{ in.}^2$$

$$\frac{A}{2} = \frac{52}{2} = 26 \text{ in.}^2$$

Since the area of the top flange = 2(12) = 24 in.^{2} is less than $A/2$ = 26 in.^{2}, the PNA is located at

$$y_c = 2 + \frac{(26-24)}{t_w} = 2 + \frac{22}{0.5} = 6.00 \text{ in.}$$

$$y_t = d - y_c = (2 + 20 + 2) - 6.00 = 18.00 \text{ in.}$$

The procedure for computing M_{px} is to decompose the compression and tension regions into shapes for which we can easily find their area and centroid. The internal force acting at each of these centroids is a stress volume = $F_y A_i$. To obtain M_{px}, we use the definition of internal forces and their distances from the PNA as shown in Figure 5.6, and we sum moments at the PNA location.

$$M_{px} = \sum(d_{ci} C_i) + \sum(d_{tj} T_j)$$

$$C_1 = F_y A_{c1} = 36(2)(12) = 864 \text{ kips}$$

$$C_2 = F_y A_{c2} = 36(4)(0.5) = 72 \text{ kips}$$

$$T_1 = F_y A_{t1} = 36(2)(9) = 648 \text{ kips}$$

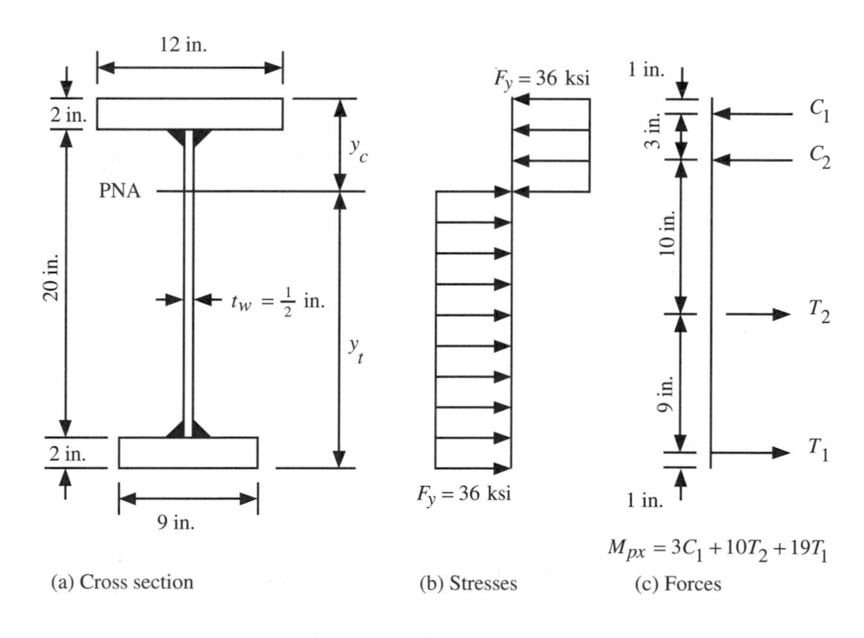

FIGURE 5.6 Plastic moment of a singly-symmetric section.

$$T_2 = F_y A_{t2} = 36(16)(0.5) = 288 \text{ kips}$$

$$d_{c1} = y_c - y_{c1} = 6 - 1 = 5 \text{ in.}$$

$$d_{c2} = y_c - y_{c2} = 6 - (1 + 3) = 2 \text{ in.}$$

$$d_{t1} = y_t - y_{t1} = 18 - 1 = 17 \text{ in.}$$

$$d_{t2} = y_t - y_{t2} = 18 - (1 + 9) = 8 \text{ in.}$$

$$M_{px} = 5(864) + 2(72) + 17(648) + 8(288)$$

$$= 17{,}784 \text{ in.-kips} = 1482 \text{ ft-kips}$$

To ensure elastic behavior due to service condition loads, LRFD F1.1 (p. 6-52) restricts M_{px} to $1.5F_yS_{xt}$ for homogeneous sections. Therefore, we need to compute $1.5F_yS_{xt}$. From the bottom or tension surface of the section, the *elastic neutral axis* (ENA) for this section is located at

$$y_t = \frac{(9)(2)(1)+(20)(0.5)(2+10)+(12)(2)(23)}{18+10+24} = \frac{690}{52} = 13.27 \text{ in.}$$

$$I_x = \frac{12(2)^3}{12} + 24(9.73)^2 + \frac{9(2)^3}{12} + 18(12.27)^2 + \frac{0.5(20)^3}{12} + 10(1.27)^2 = 5360 \text{ in.}^4$$

$$S_{xt} = \frac{I_x}{y_t} = \frac{5360}{13.27} = 403.9 \text{ in.}^3$$

$$1.5F_yS_{xt} = 1.5(36)(403.9) = 21{,}810 \text{ in.-kips} = 1817 \text{ ft-kips}$$

$$M_{px} = \text{smaller of } \begin{cases} (\text{computed } M_{px}) = 1482 \text{ ft-kips} \\ 1.5F_yS_{xt} = 1817 \text{ ft-kips} \end{cases}$$

$$M_{px} = 1482 \text{ ft-kips}$$

For the hybrid, singly-symmetric, I-shaped section in Figure 5.7, let $F_{yw} = 36$ ksi and $F_{yf} = 50$ ksi. The PNA is the axis that subdivides the axial yield force P_y into two equal parts.

$$P_y = F_{yf}(\Sigma A_f) + F_{yw}A_w$$

$$= 50(24 + 18) + 36(10) = 2460 \text{ kips}$$

$$\frac{P_y}{2} = 12.30 \text{ kips}$$

For the top flange,

$$\left[F_{yf}A_f = 50(24) = 1200 \text{ kips}\right] < \left(\frac{P_y}{2} = 1230 \text{ kips}\right)$$

Therefore, the PNA is located at

$$y_c = 2 + \frac{(1230 - 1200)}{36(0.5)} = 3.67 \text{ in.}$$

$$y_t = d - y_c = (2 + 20 + 2) - 3.67 = 20.33 \text{ in.}$$

To obtain M_{px}, we use the definition of internal forces and their distances from the PNA as shown in Figure 5.8, and we sum moments at the PNA location.

$$C_1 = F_{yf}A_{1c} = 50(2)(12) = 1200 \text{ kips}$$

$$C_2 = F_{yw}A_{2c} = 36(1.67)(0.5) = 30.06 \text{ kips}$$

$$T_1 = F_{yf}A_{1t} = 50(2)(9) = 900 \text{ kips}$$

$$T_2 = F_{yw}A_{2t} = 36(18.33)(0.50) = 329.94 \text{ kips}$$

$$d_{1c} = y_c - y_{1c} = 3.67 - 1 = 2.67 \text{ in.}$$

$$d_{2c} = y_c - y_{2c} = 3.67 - \left(1 + \frac{1.67}{2}\right) = 1.835 \text{ in.}$$

$$d_{1t} = y_t - y_{1t} = 20.33 - 1 = 19.33 \text{ in.}$$

$$d_{2t} = y_c - y_{2t} = 20.33 - \left(1 + \frac{18.33}{2}\right) = 10.165 \text{ in.}$$

$$M_{px} = 2.67(1200) + 1.835(30.06) + 19.33(900) + 10.165(329.94)$$

$$= 23{,}649 \text{ in.-kips} = 1972 \text{ ft-kips}$$

To ensure elastic behavior due to service condition loads, LRFD F1.1 (p. 6-52) restricts M_{px} to $1.5F_{yf}S_{xt}$ for hybrid sections. Therefore, we need to compute $1.5F_{yf}S_{xt}$. From the

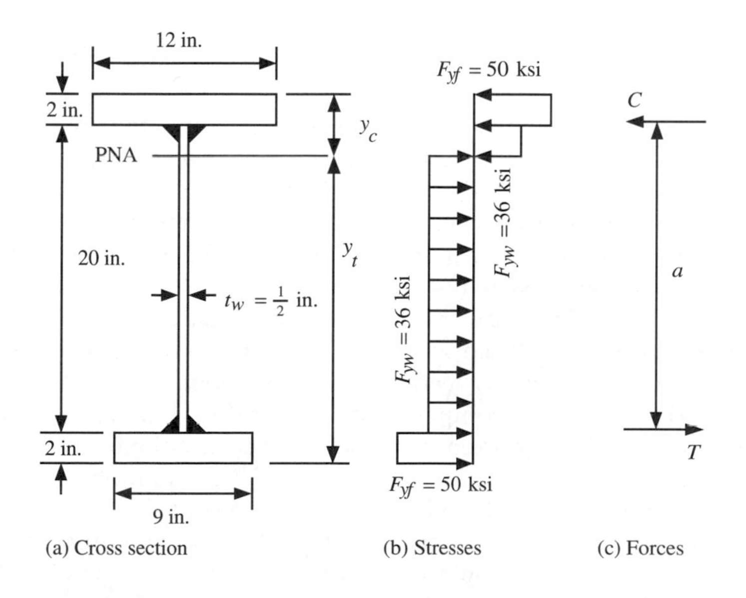

FIGURE 5.7 PNA of a hybrid section.

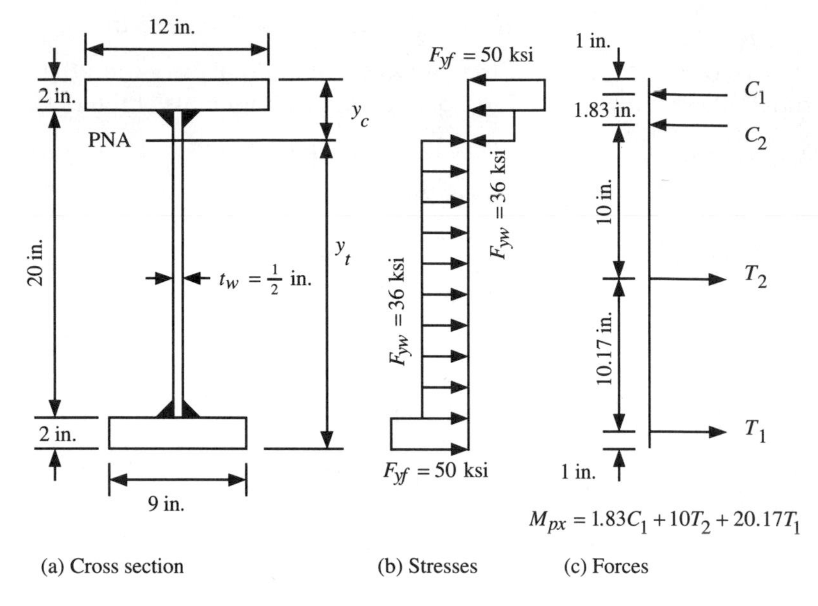

FIGURE 5.8 Plastic moment of a hybrid section.

previous example, the elastic properties for this section are: $y_t = 13.27$ in.; $I_x = 5360$ in.^{4}; $S_{xt} = I_x / y_t = 403.9$ in.^{3}

$$1.5F_{yf}S_{xt} = 1.5(50)403.9 = 30{,}293 \text{ in.-kips} = 2524 \text{ ft-kips}$$

$$M_{px} = \text{smaller of} \begin{cases} (\text{computed } M_{px}) = 1972 \text{ ft-kips} \\ 1.5F_{yf}S_{xt} = 2524 \text{ ft-kips} \end{cases}$$

$$M_{px} = 1972 \text{ ft-kips}$$

Figure 5.9(a) shows a W section beam subjected to applied member end moments that cause bending to occur about the x-axis of the cross section. When the bending moment due to the self-weight of the member is assumed to be negligible, the bending moment is constant [see Figure 5.9(b)] along the member length. As shown in Figure 4.4, the member has some initial crookedness defined as camber (x-axis bending) and sweep (y-axis bending).

As shown in Figure 5.5, the member end moments in Figure 5.9(a) can conceptually be replaced with a couple. Due to the tension force of the couple, below the x-axis of the W section the fibers elongate and their sweep crookedness decreases. Due to the compression force of the couple, above the x-axis of the W section the fibers shorten and their sweep crookedness increases. Thus, the compression flange plus a small portion of the adjoining web can be imagined to be a column that will buckle about the y-axis of the W section when the axial column force C reaches the critical value. Note that column buckling about the x-axis for the imagined column cannot occur. The bottom half of the member is in tension. Any tendency of the imagined column to move some more in the

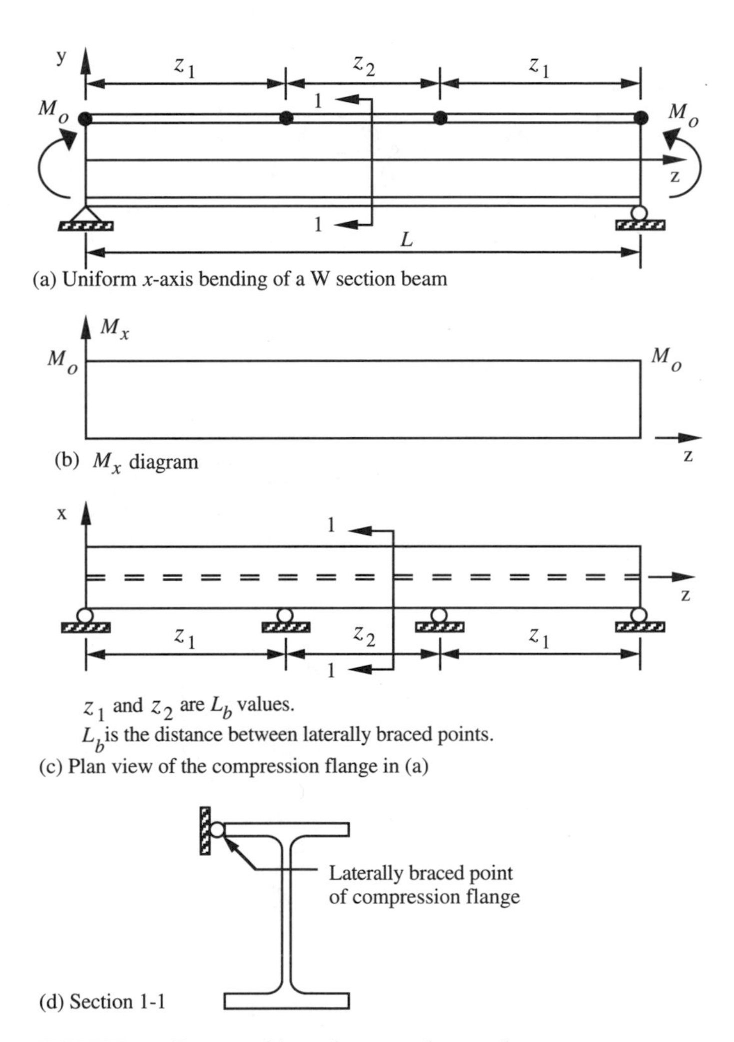

FIGURE 5.9 Beam subjected to member-end moments.

deflected beam direction (in the negative y direction) would be transferred through the web of the W section and resisted by the tension force in the bottom half of the member.

If the intermediate lateral braces for the compression flange in Figure 5.9 are removed, we get Figure 5.10, which is the fundamental case in the LRFD definition of *lateral-torsional buckling* (LTB) of a W section subjected to x-axis bending.

For $L_b > L_r$(defined later in Eq. 5.9) in Figure 5.10, the failure mode is *elastic* LTB and the critical value of M_{ox} is (see [6, p. 253] or [7, p. 160]):

$$M_{cr} = \frac{\pi}{L_b}\sqrt{EI_y GJ + \left(\frac{\pi E}{L_b}\right)^2 I_y C_w} \tag{5.1}$$

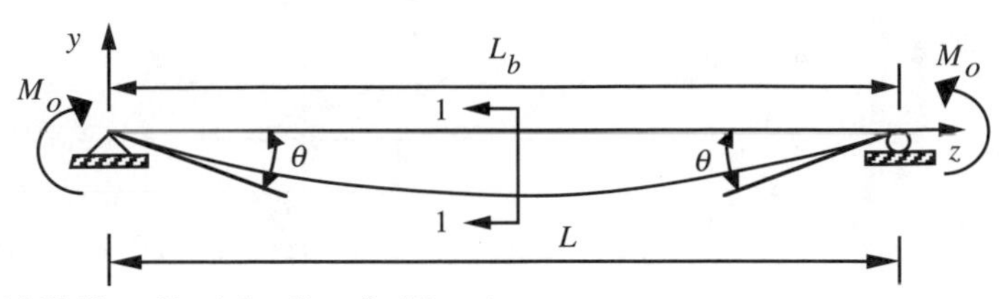

(a) Uniform X-axis bending of a W section

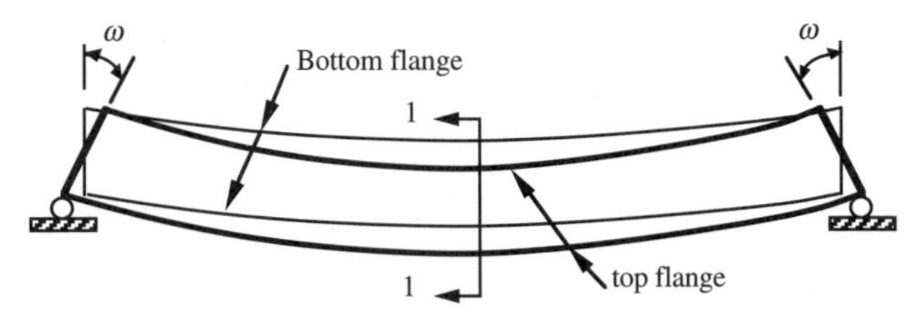

ω is the *warping angle* between the end planes of top and bottom flanges.

(b) Plan view of (a)

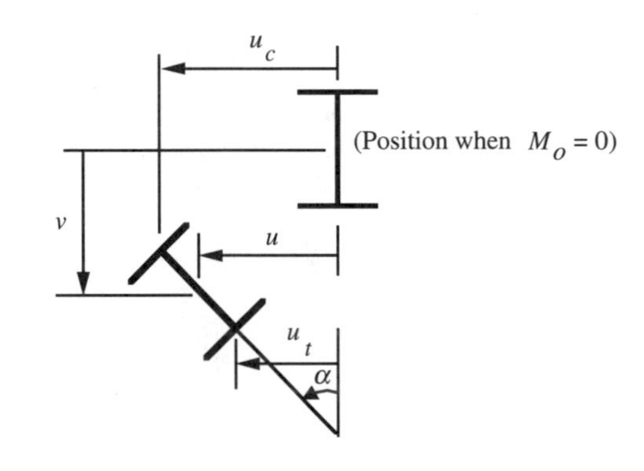

(c) Section 1-1

FIGURE 5.10 Lateral-torsional buckling of a W section beam.

which is LRFD Eq. (F1-13) (p. 6-55) without the C_b parameter. J is the *torsional constant* of a cross section and C_w is the *warping constant* of a cross section. Values of J and C_w are given on LRFD pp. 1-146 to 1-174 for hot-rolled sections. It should be noted that $M_{ox} \leq M_{rx}$ (see Figure 5.4); that is, the effect of residual stresses is accounted for in defining the condition of elastic LTB.

For $L_b \leq L_p$(defined later in Eq. 5.2) in Figure 5.10, the failure mode is *plastic bending moment* for x-axis bending, M_{px}, is reached before LTB occurs and

$$M_{nx} = \left(M_{px} = \text{smaller of} \begin{cases} F_y Z_x \\ 1.5 F_y S_x \end{cases} \right)$$

For $L_p < L_b \leq L_r$ in Figure 5.10, the failure mode is *inelastic* LTB.

A summary of the described behavior for the member in Figure 5.10 is given in Figure 5.11, which also contains some information that we will discuss in the next two sections of this chapter. In Figure 5.11, the small turned-down arcs with numbers

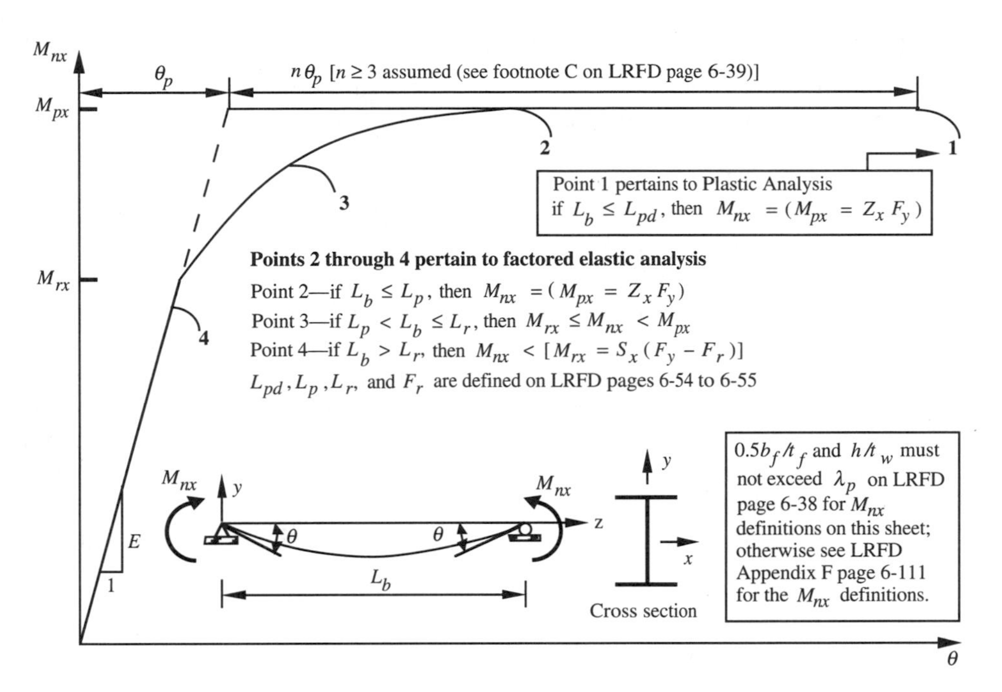

FIGURE 5.11 M_{nx} vs. θ at ends of W section beam subjected to uniform bending.

beneath them denote that either LTB has occurred and/or local buckling (of the compression flange or the compression zone of the web) has occurred.

The behavior of a W section beam subjected to applied member end moments that cause uniform bending to occur only about the y-axis also needs to be discussed. For this type of W section bending, LTB cannot occur. However, *local buckling* of each half flange that is in compression due to bending about the y-axis can occur if the *width–thickness ratio* exceeds a certain limit λ_p (see Section 5.5). When local buckling is prevented, as the applied member-end moment increases from zero up to $M_{ry} = S_y(F_y - F_r)$, where $F_r = 10$ ksi is the maximum residual stress for a rolled section, the stresses are completely elastic and can be computed by using the flexure formula, $f = M_y x/I_y$, and the extreme fiber stresses are $M_y/S_y \pm F_r$ where S_y is the elastic section modulus for y-axis bending, and the minus sign applies for the flange tips, which are in tension due to M_y. When local buckling is prevented, the plastic moment for y-axis bending M_{py} can be reached. To ensure elastic behavior due to service condition loads, LRFD F1.1 (p. 6-52) restricts M_{py} to $1.5F_yS_y$ for homogeneous sections, which means that

$$M_{ny} = \left(M_{py} = \text{smaller of } \begin{cases} F_y Z_y \\ 1.5 F_y S_y \end{cases} \right)$$

The behavior of other cross sections (channels, tees, double angles) subjected to uniform bending about the major principal axis is discussed later. Also, the behavior of a W section beam subjected to unequal applied member-end moments, which cause bending to occur about the x-axis, is discussed later.

5.6 LIMITING WIDTH-THICKNESS RATIOS FOR COMPRESSION ELEMENTS

When a W section beam is subjected to bending about the x-axis, one flange and half of the web are in compression (see Figure 5.1). Therefore, if the W section elements are not properly configured, local buckling of these compression elements can occur and be the phenomenon that controls the design bending strength. As shown in Figure 5.10, LTB of the beam can also occur and be the phenomenon that controls the design bending strength.

Local buckling and LTB are not always independent phenomena. For discussion purposes, suppose that we choose a W10 x 26 section of A36 steel for some beam tests to be conducted in a laboratory. In each beam test described here, local buckling of the web cannot occur (the reader can verify this after we have completed the discussion on the limiting width–thickness ratios for compression elements). In each beam test, an identical W10 x 26 section of length L is subjected to uniform x-axis bending and the only variable is L_b (the distance between the laterally braced points of the compression flange).

In our first beam test, the compression flange is laterally braced at $L_b = L_p$, where

$$L_p = \frac{300 r_y}{\sqrt{F_y}} \tag{5.2}$$

where L_p is the largest possible value of L_b for which M_{px} can be reached. The compression flange will begin to deflect in the lateral direction when M_{px} is reached at about $\theta_x = 2\theta_p$ in Figure 5.11, but the capacity to resist moment will not be reduced until local buckling in the compression flange occurs at $\theta_x \geq 3\theta_p$.

In our second beam test, $L_b = 0.5L_p$ whereas the beam length L is the same as in the first beam test. M_{px} will be reached as before, but the capacity to resist moment will not be reduced until one of the following conditions occurs at $\theta_x \geq 3\theta_p$:

1. Local buckling of the compression flange will occur and either immediately or after a very short time lapse LTB will occur.
2. Lateral-torsional buckling will occur and either immediately or after a very short time lapse local buckling of the compression flange will occur.

In our third beam test, $L_b = 0$ (the compression flange is continuously laterally braced along the beam length L, which is the same as in the previous tests). M_{px} will be reached and local buckling of the compression flange will occur at about $\theta_x = 9\theta_p$, but LTB cannot occur.

The preceding discussion indicates that local buckling and LTB are not necessarily independent phenomena. For simplicity, wherever it is possible to do so, they are treated independently in the research literature and theoretically oriented textbooks [6, 7]. At first glance of LRFD B5 (p. 6-36), local buckling appears to be treated independently. However, on looking more carefully we find that LRFD B5.2 and B5.3 are listed under the topic entitled "Local Buckling." The most complicated and less frequently needed LRFD Specifications are located in the LRFD Appendices. Therefore, an in-depth look at LRFD Appendix F1 (p. 6-111), which is cited in LRFD B5.3, is not appropriate at this point in our discussion. Consequently, the reader will have to accept our statement that local buckling and LTB are not treated as two completely independent topics in the LRFD Specifications. Furthermore, we believe that students will gain a better understanding of the material that needs to be presented now if we show how LTB is related to local buckling in the AISC Specifications.

For a W section or a channel subjected to bending about the x-axis to reach M_{px} (see point 2 on Figure 5.11) and to develop an inelastic rotation of at least $3\theta_p$ before either local buckling or LTB occurs, the following requirements must be satisfied:

1. For the flanges (see first item in LRFD Table B5.1, p. 6-38),

$$\frac{b}{t} \leq \frac{65}{\sqrt{F_y}} \tag{5.3}$$

For a W section [see LRFD B5.1 (first item a)—an unstiffened element]: $b/t = 0.5b_f/t_f$. Note: $0.5b_f/t_f$ is tabulated in the LRFD Manual Part 1 for each W section. For a C section [see LRFD B.1 (first item b)—an unstiffened element]: $b/t = b_f/(\text{average } t_f)$.

2. For the web [see LRFD Table B5.1 (webs in flexural compression)],

$$\frac{h}{t_w} \leq \frac{640}{\sqrt{F_y}} \tag{5.4}$$

See LRFD B5.1 (second item a)—a stiffened element—for the definition of h. See LRFD Manual Part 1 where h/t_w is tabulated for each W section. For a C section, $h = T$.

3. For the compression flange (see LRFD F1.2a, p. 6-53), each $L_b \leq L_p$, where L_b is the *distance between two adjacent laterally braced points.*

The width–thickness *ratio* of each compression element in a beam cross section and the unbraced length of the beam compression flange are parameters we must use in the determination of the *nominal bending strength.* Therefore, the requirements of these parameters are stated in Section 5.8 for each nominal bending strength definition.

5.7 LATERAL SUPPORT

Figures 5.12 and 5.13 show examples of how the compression flange of a W section beam can be laterally braced. For each W section in Figure 5.12, the top flange is assumed to be the compression flange in the following discussion. It must be noted that twisting of the beam cross sections at the beam ends (at the beam supports for gravity direction loads) must be prevented. When an interior lateral brace does not prevent twisting of the cross section, LRFD K1.5 (p. 6-93) must be satisfied.

Continuous lateral support ($L_b = 0$) is provided for the top beam flange in Figure 5.12(a) if the concrete slab is attached to supports that prevent translation of the slab in the direction perpendicular to the web of the W section beam. See Figure 10.3, which shows shear studs welded to the top flange of a W section at specified intervals along the length direction of the W section. When the specified interval between the shear studs is small enough, continuous lateral support is provided for the top beam flange. Also, the shear studs prevent the concrete slab from slipping along the length

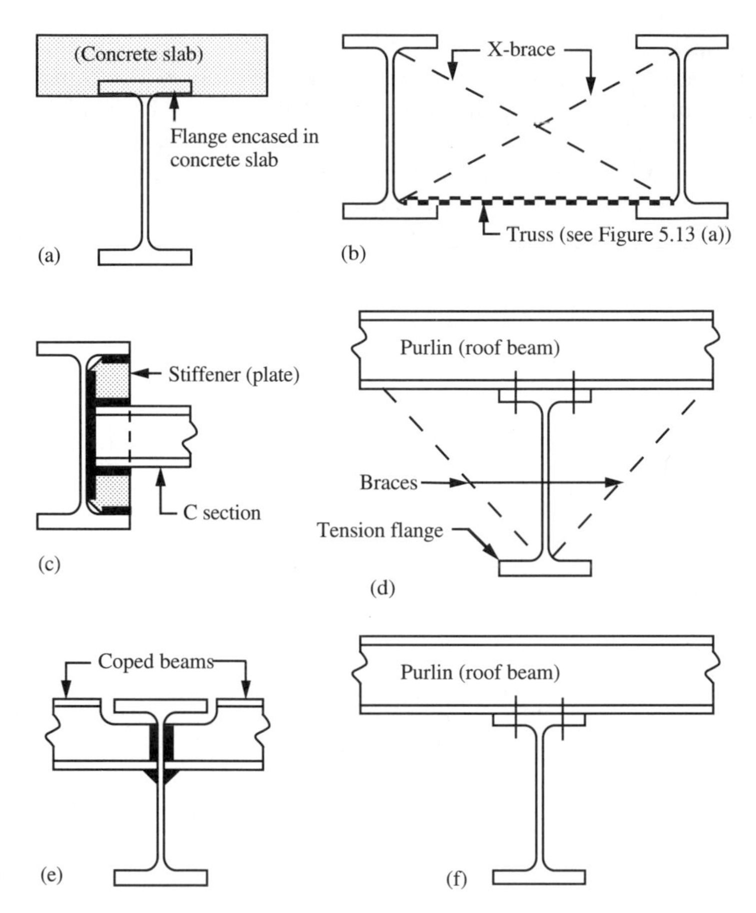

W section in each part is the member being laterally braced.
LRFD K1.5 (p. 6-93) must be satisfied for (a), (e), and (f).
Note: Also see Figure 5.13.

FIGURE 5.12 Examples of lateral support for a W section beam.

direction of the beam, and we have a composite section. A portion of the concrete slab is the top of the composite section, and the W section is the remainder of the composite section whose behavior we will discuss in Chapter 10. When the shear studs in Figure 10.3 are omitted and if the metal decking is adequately welded to the top beam flange at sufficiently small intervals along the beam length, calculations may show that continuous lateral support is provided for the top flange of the beam.

As shown in Figure 5.12(b), X braces can be used to provide lateral bracing at intervals along the beam length. The X braces can be single angles, for example. Alternatively, as shown in Figures 5.12(c) to 5.12(f), cross beams can be used to provide lateral bracing at intervals along the beam length. In Figure 5.13(a), a truss is provided to prevent translation of the cross beams and the X braces in the direction perpendicular to the web of the W section being laterally braced.

Figures 5.12 and 5.13 show some ways that lateral bracing can be provided to ensure that the design bending strength requirement ($\phi M_{nx} \geq M_{ux}$) is satisfied before

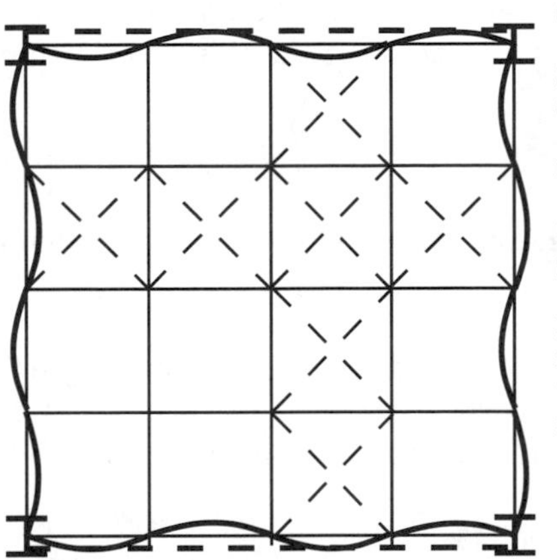

Comments for (a) and (b):
Floor framing plans of a 30 by 30-ft bay between the W section columns.
The main members span between the columns.
Curved lines show the lateral-torsional buckling pattern of the main members.
In (a) the in-plane X-braces and interior members form plane trusses which laterally brace the main members at three locations.

(a) Main members are laterally braced at five locations

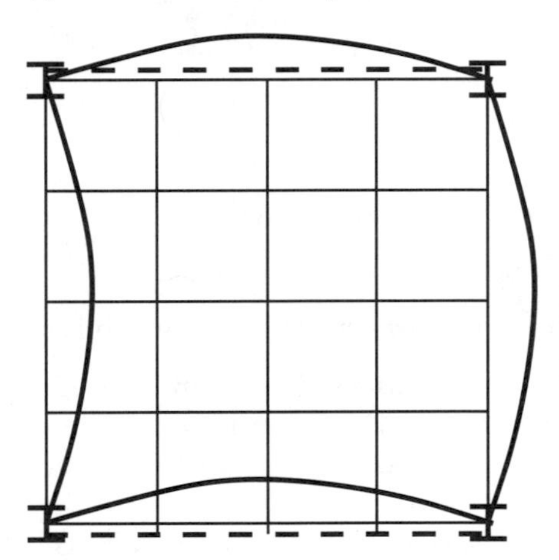

Comments continued:
In (b) there is no in-plane X-bracing and no interior member is capable of providing lateral bracing for any main member.
Dotted lines in (a) and (b) show the location of X-braces in the vertical plane. X-braces, columns, and main members form a plane truss to resist wind in the horizontal direction.

(b) Main members are laterally braced only at the column locations

FIGURE 5.13 Lateral bracing of main beams in a floor system.

LTB occurs. Lateral braces must prevent both twist and lateral deflection of the member's cross section at laterally braced points. Spacing (L_b), stiffness, and strength of the lateral braces must be adequately chosen to prevent LTB before the design bending strength requirement ($\phi M_{nx} \geq M_{ux}$) is satisfied. The requirements for braces in a braced frame, weak-axis column bracing, and lateral bracing of beams are discussed in Chapter 7. The LRFD Specification does not give any guidelines for the bracing requirements of frames, columns, and beams.

5.8 HOLES IN BEAM FLANGES

LRFD B10 (p. 6-37) gives the criteria that must be used to determine when we must account for the holes in a flange of a beam. We are to treat the flange as an isolated tension member and compute its strengths due to fracture on A_e and yielding on A_g. When the fracture-on-A_e strength exceeds the yielding-on-A_g strength, we can ignore the holes and use the flexural properties of the gross section. When the yielding-on-

A_g strength exceeds the fracture-on-A_e strength, the flexural properties must be computed in accordance with the flange having an effective area of

$$A_{fe} = \frac{5}{6}\frac{F_u}{F_y} A_{fn} \tag{5.5}$$

where A_{fn} = net area of flange treated as a tension member. Consequently,

$$A_{\text{holes}} = A_{gf} - A_{fe} \tag{5.6}$$

where A_{gf} = gross area of flange, and a deduction for A_{holes} in the flange must be made in the cross-sectional properties involved in checking the serviceability requirements and in computing the design bending strength.

5.9 DESIGN BENDING STRENGTH

The design requirement for bending strength is

$$\phi_b M_n \geq M_u$$

where

M_u = *required bending strength*

$\phi_b M_n$ = *design bending strength*

ϕ_b = 0.9

M_n = *nominal bending strength,* which is defined later

To begin our discussion, we consider only a beam whose cross section is either a W or C section. For y-axis bending, $\phi_b M_n \geq M_u$ must be satisfied at the point on the beam where the moment due to factored loads is maximum. For x-axis bending, $\phi_b M_{nx} \geq M_{ux}$ must be satisfied in each L_b region.

When the width-thickness ratio of each compression element in either a singly or doubly symmetric cross section of a beam does not exceed the applicable λ_p given in LRFD Table B5.1 (p. 6-38), the LRFD definition of the x-axis flexural design strength is $\phi_b M_{nx}$, where ϕ_b = 0.9 and M_{nx} is the x-axis nominal bending strength.

See LRFD F1.2 (p. 6-53) for the preceding definition of the flexural design strength. We prefer to say this is the definition of the x-axis design bending strength to be consistent with previously introduced LRFD terminologies: design tensile strength (for tension members) and design compressive strength (for columns).

For a beam whose cross section is either a W or C section bending about the x-axis, the nominal bending strength is

1. When $\dfrac{0.5 b_f}{t_f} \leq \dfrac{65}{\sqrt{F_y}}$; $\dfrac{h}{t_w} \leq \dfrac{640}{\sqrt{F_y}}$; $L_b \leq L_p$

$$M_{nx} = (M_{px} = Z_x F_y) \tag{5.7}$$

2. When $\dfrac{0.5 b_f}{t_f} \leq \dfrac{65}{\sqrt{F_y}}$; $\dfrac{h}{t_w} \leq \dfrac{640}{\sqrt{F_y}}$; $L_p < L_b \leq L_r$

$$M_{nx} = C_b \left[M_{px} - \left(M_{px} - M_{rx} \right) \left(\frac{L_b - L_p}{L_r - L_p} \right) \right] \leq M_{px} \tag{5.8}$$

$$L_r = \frac{r_y X_1}{F_y - F_r} \sqrt{1 + \sqrt{X_2 \left(F_y - F_r \right)^2}} \tag{5.9}$$

$$F_r = 10 \text{ ksi}$$

$$X_1 = \frac{\pi}{S_x} \sqrt{\frac{EAGJ}{2}}$$

$$X_2 = \frac{4C_w}{I_y} \left(\frac{S_x}{GJ} \right)^2$$

$$M_{rx} = S_x \left(F_y - F_r \right)$$

$$C_b = \frac{12.5 M_{max}}{2.5 M_{max} + 3M_A + 4M_B + 3M_C} \tag{5.10}$$

This new C_b formula is applicable when the M_x diagram is not constant in an L_b region. The parameters in the C_b formula are:

M_{max} = absolute value of maximum M in an L_b region

M_A = absolute value of M at $0.25L_b$

M_B = absolute value of M at $0.5L_b$

M_C = absolute value of M at $0.75L_b$

Exception:

$C_b = 1.0$ is applicable for cantilevers and overhangs when the free end is not braced.

3. When $\frac{0.5b_f}{t_f} \leq \frac{65}{\sqrt{F_y}}$; $\frac{h}{t_w} \leq \frac{640}{\sqrt{F_y}}$; $L_b > L_r$

$$M_{nx} = (M_{cr} \leq M_{px}) \tag{5.11}$$

$$M_{cr} = \frac{C_b \pi}{L_b} \sqrt{EI_y GJ + \left(\frac{\pi E}{L_b} \right)^2 I_y C_w}$$

Note: M_{nx} vs. L_b defined in items 1, 2 and 3 is conceptually illustrated in Figure 5.14 for $C_b = 1$ in items 2 and 3. See Figure 5.15 for the effect of $C_b > 1$ on the nominal bending strength.

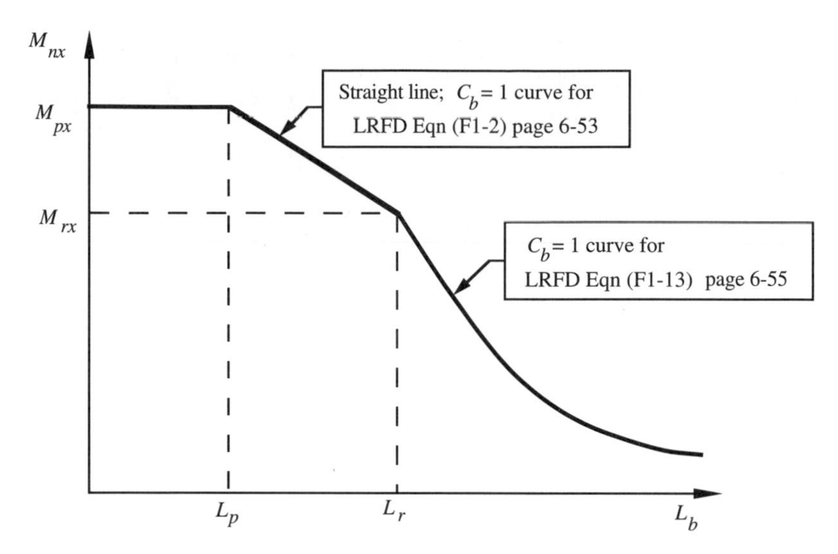

Note : Bending strength information is valid provided
$0.5\ b_f/t_f$ and h/t_w do not exceed λ_p
given on LRFD page 6–38.

FIGURE 5.14 M_{nx} vs L_b of a W section beam subjected to uniform bending.

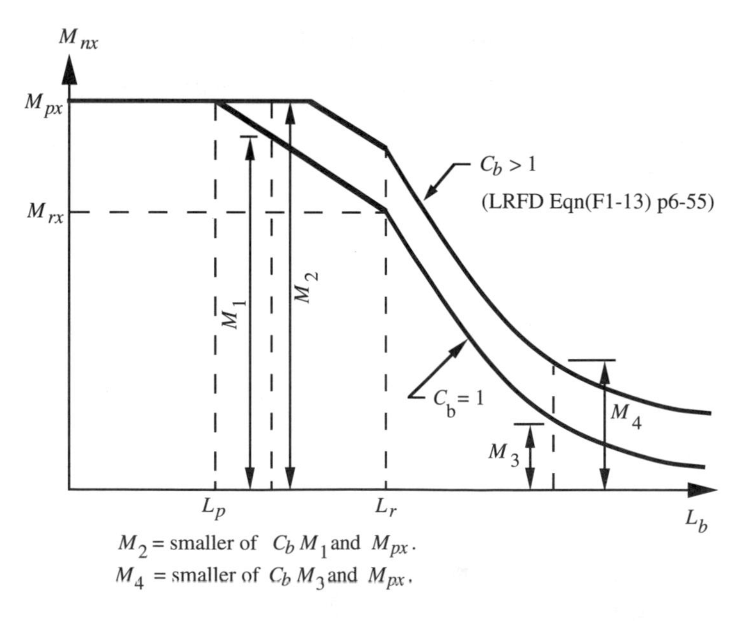

M_2 = smaller of $C_b M_1$ and M_{px}.
M_4 = smaller of $C_b M_3$ and M_{px}.

FIGURE 5.15 M_{nx} vs L_b of a W section beam subjected to non-uniform bending.

When either $0.5b_f/t_f$ or h/t_w of a W section or a channel subjected to x-axis bending exceeds the value of λ_p (LRFD Table B5.1) but does not exceed the value of λ_r (LRFD Table B5.1), LRFD Appendix F (p. 6-111) must be used to determine the nominal bending strength due to inelastic local buckling.

When either $0.5b_f/t_f$ or h/t_w of a W section or a channel subjected to x-axis bending exceeds the value of λ_r (LRFD Table B5.1), LRFD Appendix F must be used to determine the nominal bending strength due to elastic local buckling.

For a beam whose cross section is either a W or C section, when

$$\frac{0.5b_f}{t_f} \le \frac{65}{\sqrt{F_y}}$$

the *y-axis* nominal bending strength is

$$M_{ny} = (M_{py} = Z_y F_y) \tag{5.12}$$

Otherwise, see LRFD Appendix F1.7.

Equation (5.5) also applies for any bending axis of solid circular and square sections as well as for bending about the minor principal axis of any non-built-up section (see LRFD F1.7). For minor axis bending, if the width–thickness ratio of any compression element exceeds the appropriate λ_p (LRFD Table B5.1), LRFD Appendix F must be used to determine the nominal bending strength.

For a beam whose cross section is either a T section or a double-angle section with zero separation between the back-to-back legs, when

$$\frac{0.5b_f}{t_f} \le \frac{95}{\sqrt{F_y}}; \quad \frac{d}{t_w} \le \frac{127}{\sqrt{F_y}}$$

the x-axis nominal bending strength is

$$M_{nx} = \text{smaller of } M_{cr} \text{ and } \begin{cases} 1.5 S_x F_y & \text{when stem in tension} \\ S_x F_y & \text{when stem in compression} \end{cases} \tag{5.13}$$

$$M_{cr} = \frac{\pi}{L_b}\sqrt{EI_y GJ}\left(B + \sqrt{1+B^2}\right) \tag{5.14}$$

$$B = \pm \frac{2.3d}{L_b}\sqrt{\frac{I_y}{J}}$$

where the plus sign for B applies when, due to x-axis bending, the stem (the vertical element in a T-shaped section) is in tension and the minus sign for B applies when the stem is in compression.

The x-axis nominal bending strength for a beam whose cross section is a double-angle section fastened to connector plates between the back-to-back legs at intervals along the member length is as defined in Eqs. 5.13 and 5.14, when

$$\frac{b}{t} \le \frac{76}{\sqrt{F_y}}$$

is true for each leg in each angle.

In Part 4 of the LRFD Manual, we find beam design aids. The information given on the pages prior to each design aid should be read before trying to use it to perform a design.

In the Z_x table on LRFD pp. 4-16 to 4-21, we find that W40 x 174, W14 x 99, W14 x 90, W12 x 65, W10 x 12, W8 x 10, and W6 x 15 have a superscript *b* on them. The W6 x 15 also has a superscript *c* on it. These superscripts *b* (for $F_y = 50$ ksi) and *c* (for $F_y = 36$ ksi) are footnote symbols to warn the reader that, for the indicated value of F_y:

1. *Flange local buckling* (FLB) governed ϕM_{nx} for these sections.
2. The ϕM_{px} value listed for these sections is really ϕM_{nx} due to FLB and the listed L_p value is really the L_b at which FLB ceases to govern ϕM_{nx}.

In this table, except when $F_y = 36$ ksi and 50 ksi, we need the Z_x values to compute $\phi M_{px} = Z_x F_y$, which is valid when

$$\left(\frac{0.5b_f}{t_f}\right) \leq \left(\frac{65}{\sqrt{F_y}}\right) \text{ and } \left(\frac{h}{t_w}\right) \leq \left(\frac{640}{\sqrt{F_y}}\right)$$

Otherwise, local buckling governs ϕM_{nx} and we must use LRFD Appendix F to compute the maximum possible value of ϕM_{nx}.

Starting on LRFD p. 4-113 for $F_y = 36$ ksi and LRFD p. 4-139 for $F_y = 50$ ksi, we find a plot of ϕM_{nx} for $C_b = 1$ vs. L_b for each W section. Historically, these plots have been called beam design charts and they correctly account for those cases for which flange local buckling governs ϕM_{nx}. For each W section in the beam design charts, the shape of the plotted information is similar to Figure 5.14, but the information continues across several LRFD pages.

Example 5.1

A simply supported beam whose span $L = 30$ ft. is subjected only to the following uniformly distributed loads:

$$\text{Dead} = 0.8 \text{ kips/ft (includes estimated beam weight)}$$

$$\text{Live} = 1.4 \text{ kips/ft}$$

There is no design limitation on deflection and the compression flange can be laterally braced such that $L_b \leq L_p$. Select the lightest W section of A36 steel that is acceptable to serve for the described beam.

Solution

From LRFD A4.1 (p. 6-30), the governing factored loading is

$$q_u = 1.2D + 1.6L = 1.2(0.8) + 1.6(1.4) = 3.2 \text{ kips/ft}$$

See Figure 5.16 for the shear and moment diagrams due to this factored loading. The applicable design requirements are:

1. $\phi M_{nx} \geq M_{ux}$ in each $L_b \leq L_p$ region.
2. $\phi V_n \geq V_u$.

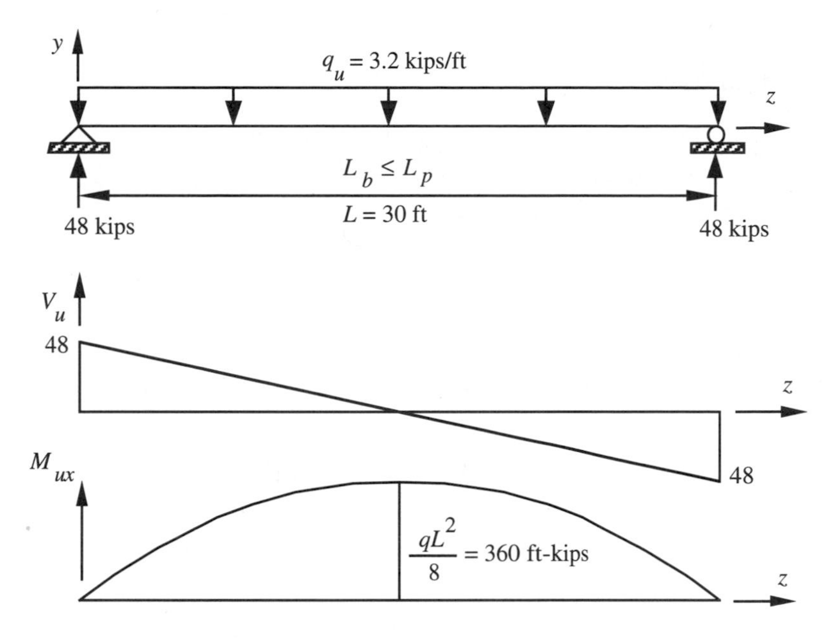

FIGURE 5.16 Load, shear, and moment diagrams for Examples 5.1 to 5.3.

Assume that shear does not govern the selection. Since $L_b \leq L_p$ everywhere along the beam, the selection criterion for the $L_b \leq L_p$ region where the maximum moment occurs is

$$(\phi M_{nx} = \phi M_{px}) \geq (M_{ux} = 360 \text{ ft-kips})$$

On LRFD p. 4-19 and in the ϕM_{px} column for $F_y = 36$ ksi, since each boldface section is the lightest within each group, we look at the boldface sections until we find for a W24 x 55 that

$$(\phi M_{px} = 362 \text{ ft-kips}) \geq (M_u = 360); \quad L_p = 5.6 \text{ ft}$$

Since $L/L_p = 30/5.6 = 5.36$, choose 6 @ ($L_b = 5.0$ ft) = ($L = 30$ ft).

Check shear:

We can find $\phi V_n = 181$ kips from the bottom of LRFD p. 4-46, but we prefer to illustrate how that tabular value was computed:

$$\left(\frac{h}{t_w} = 54.6\right) \leq \left(\frac{418}{\sqrt{F_{yw}}} = \frac{418}{\sqrt{36}} = 69.7\right)$$

$$\phi V_n = 0.9(0.6 F_y t_w d)$$

$$[\phi V_n = 0.9(0.6)(36)(0.395)(23.57) = 181 \text{ kips}] > (V_u = 48)$$

Use W24 x 55 ($F_y = 36$ ksi) and ($L_b = 5.0$ ft) $\leq$ ($L_p = 5.6$ ft).

Example 5.2

For the simply supported beam in Figure 5.16, there is no design limitation on deflection and the compression flange can be laterally braced such that $L_b \le L_p$. Select the lightest W section of A572 Grade 65 steel that is acceptable to serve for the described beam.

Solution

See Figure 5.16 for the factored loading, shear diagram, and moment diagram. LRFD p. 4-46 does not list ϕM_{px} for $F_y = 65$ ksi, so we must proceed as follows:

$$M_{px} = \text{smaller of } \begin{cases} Z_x F_y \\ 1.5 S_x F_y \end{cases}$$

Assume that local buckling does not govern ϕM_{nx}. For hot-rolled W sections,

$$\left(\frac{Z_x}{S_x} \approx 1.15\right) < 1.5$$

Therefore, $M_{px} = Z_x F_y$ and we need

$$[\phi M_{nx} = \phi M_{px} = 0.9 Z_x(65 \text{ ksi})] \ge (M_{ux} = 360 \text{ ft-kips} = 4320 \text{ in.-kips})$$

$$Z_x \ge \left[\frac{4320}{0.9(65)} = 73.8 \text{ in.}^3\right]$$

From LRFD p. 4-19,

$$\text{try W18 x 40:} \quad (Z_x = 78.4) \ge 73.8$$

Check flange and web (see LRFD p. 1-33 for the section properties)

$$\left(\frac{0.5 b_f}{t_f} = 5.7\right) \le \left(\lambda_p = \frac{65}{\sqrt{65}} = 8.06\right) \quad \text{FLB is not applicable.}$$

$$\left(\frac{h}{t_w} = 51.0\right) \le \left(\frac{640}{\sqrt{65}} = 79.4\right) \quad \text{WLB is not applicable.}$$

$\phi M_{nx} = \phi M_{px}$ is valid as we assumed, and we need to choose $L_b \le L_p$:

$$L_p = \frac{300 r_y}{\sqrt{F_y}} = \frac{300(1.27)}{\sqrt{65}} = 47.26 \text{ in.} = 3.94 \text{ ft}$$

Check shear

$$\left(\frac{h}{t_w} = 51.0\right) \le \left(\frac{418}{\sqrt{F_{yw}}} = \frac{418}{\sqrt{65}} = 51.9\right)$$

$$\phi V_n = 0.9(0.6F_y t_w d)$$

$$[\phi V_n = 0.9(0.6)(65)(0.315)(17.9) = 198 \text{ kips}] \ge (V_u = 48)$$

Use W18 x 40 (F_y = 65 ksi) and (L_b = 3.75 ft) ≤ (L_p = 3.94 ft)

Example 5.3

For the simply supported beam in Figure 5.16, the compression flange can be laterally braced such that $L_b \le L_p$. The maximum acceptable deflection due to service live load is *Span*/360 = (360 in.)/360 = 1.00 in. Select the lightest W section of A572 Grade 50 steel that is acceptable to serve for the described beam.

Solution

See Figure 5.16 for the factored loading, shear diagram, and moment diagram. From LRFD p. 4-19,

try W18 x 50: (ϕM_{px} = 379 ft-kips) ≥ (M_{ux} = 360); L_p = 5.8 ft.

Check deflection

Service live load, w = 1.4 kips/ft; limiting deflection = 1.00 in.:

$$\Delta_{max.} = \frac{5wL^4}{384EI} = \frac{5(1.4\,\text{k/ft})(30\,\text{ft})^4(12\,\text{in./ft})^3}{384(29{,}000\,\text{ksi})(800\,\text{in.})} = 1.10 \text{ in.}$$

$$(\Delta_{max.} = 1.10\,\text{in.}) > (\text{limiting deflection} = 1.00\,\text{in.}) \qquad (NG)$$

Try W21 x 50: (ϕM_{px} = 413 ft-kips) ≥ (M_{ux} = 360); L_p = 4.6 ft.

$$\left[\Delta_{max.} = 1.10\left(\frac{800}{984}\right) = 0.894 \text{ in.}\right] \le (\text{limiting deflection} = 1.00\,\text{in.}) \qquad (OK)$$

Check shear

$$\left(\frac{h}{t_w} = 49.4\right) \le \left(\frac{418}{\sqrt{F_{yw}}} = \frac{418}{\sqrt{50}} = 59.1\right)$$

$$[\phi V_n = 0.9(0.6)(50)(20.83)(0.380) = 214 \text{ kips}] \ge (V_u = 48)$$

Use W21 x 50 (F_y = 50 ksi) and (L_b = 3.75 ft) ≤ (L_p = 4.6 ft).

Example 5.4

For the simply supported beam in Figure 5.16, there is no design limitation on deflection and the compression flange can be laterally braced only at the supports and at intervals of $L_b = L/3 = 30/3 = 10$ ft (see Figure 5.17). Select the lightest W section of A36 steel that is acceptable to serve for the described beam.

Solution

$$C_b = \frac{12.5 M_{max}}{2.5 M_{max} + 3M_A + 4M_B + 3M_C}$$

$C_b = 1$ when the moment diagram is constant within an L_b region. The middle L_b region on the moment diagram in Figure 5.17 has the largest moment and, since the moment diagram is close to being constant in this region, we know that $C_b \approx 1$:

$$C_b = \frac{12.5(360)}{2.5(360) + 3(350) + 4(360) + 3(350)} = 1.0135$$

For the end L_b regions, $M_{ux} = 320$ ft-kips and

$$C_b = \frac{12.5(320)}{2.5(320) + 3(110) + 4(200) + 3(270)} = 1.59$$

The selected section must satisfy $\phi M_{nx} \geq M_{ux}$ in all L_b regions. When all regions have the same L_b, we choose the W section for the region where M_{ux}/C_b is greatest, which is the middle L_b region where $C_b = 1.01$ and $M_{ux} = 360$ ft-kips. Since $C_b \approx 1$, we can find the lightest acceptable W section by going directly to LRFD p. 4-128 [the page on which we find $\phi M_{nx} \geq (M_{ux} = 360)$ in the beam design charts for $C_b = 1$]. Plot the point whose coordinates are $(L_b, M_{ux}) = (10$ ft, 360 ft-kips$)$. Any curve that passes through this point or any curve that lies above and to the right of the plotted point is a satisfactory solution since $\phi M_{nx} \geq (M_{ux} = 360$ ft-kips$)$. Solid-line curves indicate the lightest choice in each region. Therefore, the solid line on or closest to, but above and to the right of, the plotted point is the lightest section that satisfies $\phi M_{nx} \geq M_{ux}$.

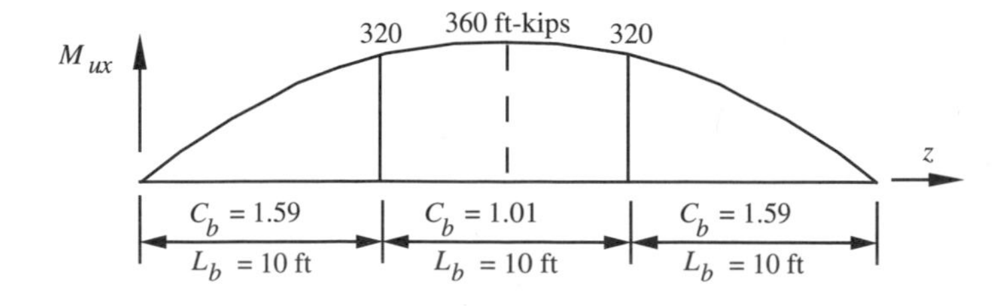

FIGURE 5.17 Moment diagram for Example 5.4.

Try W21 x 62 (ϕM_{nx} = 363 ft-kips at L_b = 10 ft) ≥ (M_{ux} = 360 ft-kips)

Check shear

From LRFD p. 4-48,

$$(\phi V_n = 163 \text{ kips}) \geq (V_u = 48)$$

Use W21 x 62 (F_y = 36 ksi).

Example 5.5

For the simply supported beam in Figure 5.16, there is no design limitation on deflection and the compression flange can be laterally braced only at the supports and at midspan; L_b = 15 ft (see Figure 5.18). Select the lightest W section of A36 steel that is acceptable to serve for the described beam.

Solution (a)

From the last case shown in the figure on LRFD p. 4-9, we find that C_b = 1.30 is applicable for our example. When $C_b > 1$, use the selection procedure given in Figure 5.19, which for our example gives:

1. From LRFD p. 4-19, the lightest section for which $\phi M_{px} \geq M_{ux}$ is

   W24 x 55: (ϕM_{px} = 362 ft-kips) ≥ (M_{ux} = 360 ft-kips)

   L_p = 5.6 ft; L_r = 16.6 ft; ϕM_{rx} = 222 ft-kips; BF = 12.7
2. (L_p = 5.6 ft) < (L_b = 15 ft) ≤ (L_r = 16.6 ft)
3. By Method 1,
   M_{ux}/C_b = 360/1.30 = 277 is found on LRFD p. 4-130. At L_b = 15 ft for W24 x 55, we find that [(ϕM_{nx} for C_b = 1) = 243] < (M_{ux}/C_b = 360/1.30 = 277). A W24 x 55 is not acceptable.
4. In the beam design charts, any section having $\phi M_{px} \geq$ (M_{ux} = 360 ft-kips) whose line passes through or lies above and to the right of (L_b, M_{ux}/C_b) = (15, 277)

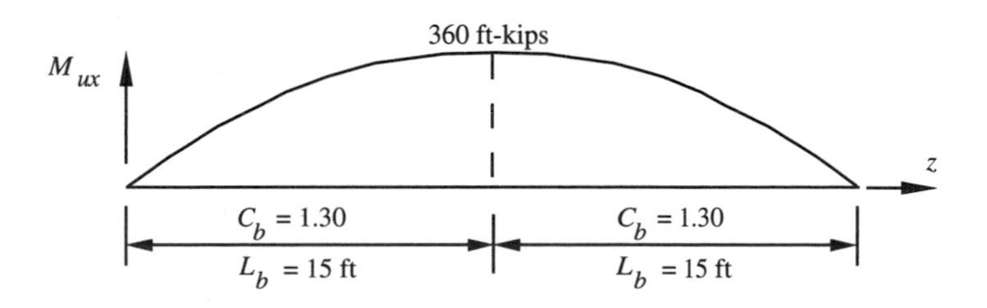

FIGURE 5.18 Moment diagram for Example 5.5.

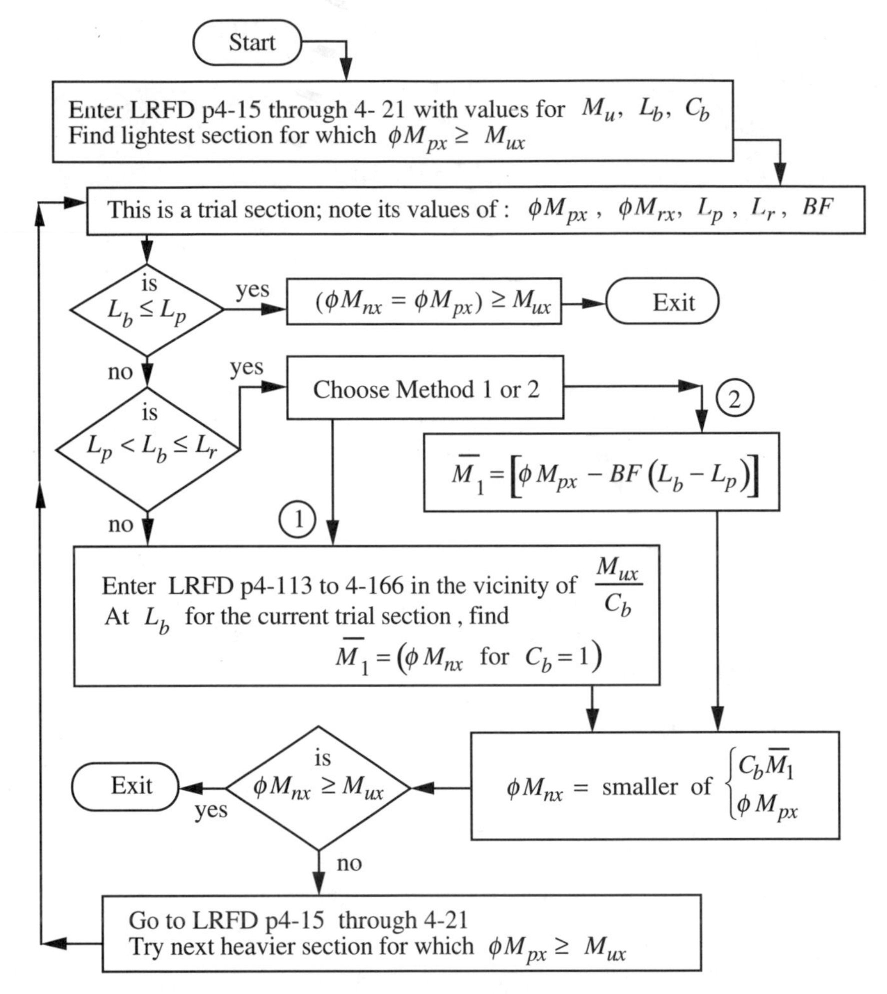

FIGURE 5.19 Beam-design procedure for $F_y = 36$ ksi and $F_y = 50$ ksi.

is acceptable. The first line above the plotted point is W24 x 62. The first solid line above the plotted point is W21 x 62. Both sections are acceptable and have the same weight.

For W21 x 62:

$$\phi M_{nx} = \text{smaller of} \begin{cases} C_b \overline{M}_1 = 1.30(314) = 408 \\ \phi M_{px} = 389 \text{ ft - kips} \end{cases}$$

$$(\phi M_{nx} = 389 \text{ ft-kips}) \geq (M_{ux} = 360) \quad \text{(OK)}$$

$$(\phi V_n = 163 \text{ kips}) \geq (V_u - 48) \quad \text{(OK)}$$

Use W21 x 62 ($F_y = 36$ ksi).

For W24 x 62:

$$\phi M_{nx} = \text{smaller of} \begin{cases} C_b \overline{M}_1 = 1.30(287.5) = 374 \\ \phi M_{px} = 413 \text{ ft-kips} \end{cases}$$

$$(\phi M_{nx} = 374 \text{ ft-kips}) \geq (M_{ux} = 360) \quad \text{(OK)}$$

$$(\phi V_n = 198 \text{ kips}) \geq (V_u = 48) \quad \text{(OK)}$$

Use W24 x 62 (F_y = 36 ksi).

Example 5.6

Lateral braces are provided only at the supports and at midspan. The factored beam weight is accounted for in the concentrated load. There is no limiting deflection criterion to be satisfied. Find the lightest W section of A36 steel (see Figure 5.20).

Solution

The information shown in Figure 5.21 was obtained from Case 13 on LRFD p. 4-194. Find ϕM_{nx}:

$$\text{W18 x 50:} \quad (\phi M_{px} = 273 \text{ ft-kips}) \geq (M_{ux} = 263)$$

$$\text{W21 x 50:} \quad (\phi M_{px} = 297 \text{ ft-kips}) \geq (M_{ux} = 263)$$

Enter the LRFD beam design charts in the vicinity of the larger of

$$M_{ux}/C_b = 263/2.24 = 117 \text{ ft-kips} \quad (\text{for first } L_b \text{ region})$$

$$M_{ux}/C_b = 219/1.67 = 131 \text{ ft-kips} \quad (\text{for second } L_b \text{ region})$$

On LRFD p. 4-162, W18 x 50 and W21 x 50 appear to the right of and above the point whose coordinates are (L_b, M_{ux}/C_b) = (14 ft, 131 ft-kips).
Check W18 x 50.
For the first L_b region.

$$(L_p = 6.9) < (L_b = 14) \leq (L_r = 20.5) \quad \text{and} \quad C_b = 2.24$$

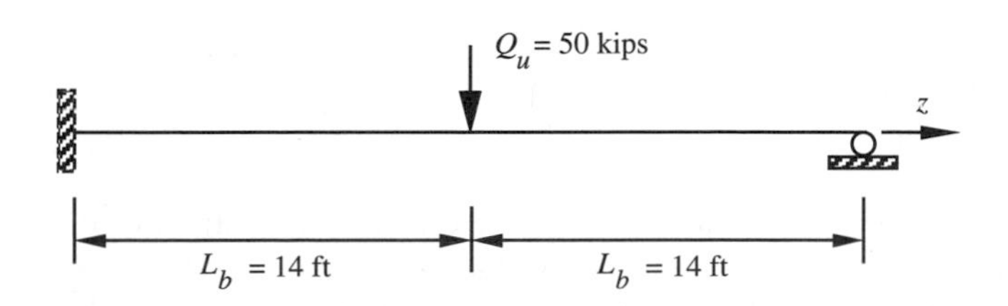

FIGURE 5.20 Sketch for Example 5.6.

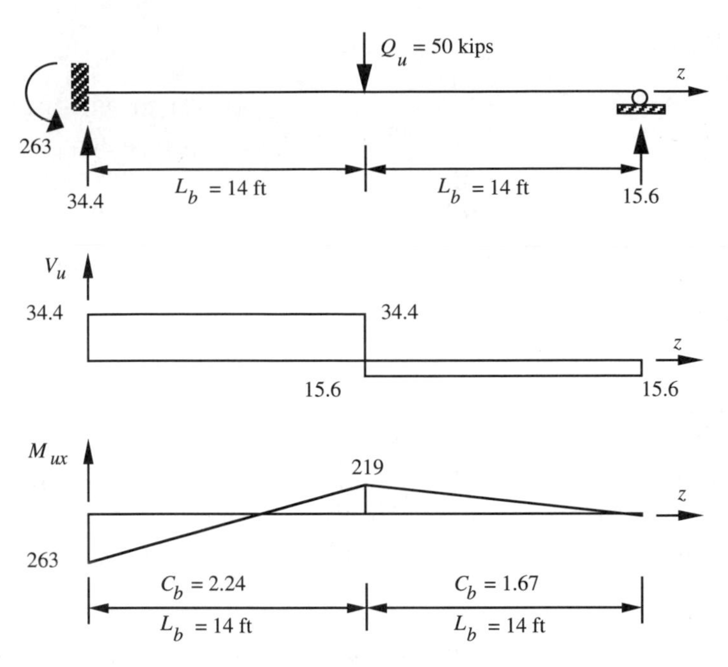

FIGURE 5.21 Load, shear, and moment diagrams for Example 5.6.

$$\phi M_{nx} = \text{smaller of} \begin{cases} C_b \overline{M}_1 = 2.24(221) = 495 \\ \phi M_{px} = 273 \text{ ft-kips} \end{cases}$$

$$(\phi M_{nx} = 273 \text{ ft-kips}) \geq (M_{ux} = 263) \quad \text{(OK)}$$

For the second L_b region:

$$(L_p = 6.9) < (L_b = 14) \leq (L_r = 20.5) \quad \text{and} \quad C_b = 2.24$$

$$\phi M_{nx} = \text{smaller of} \begin{cases} C_b \overline{M}_1 = 1.67(221) = 369 \\ \phi M_{px} = 273 \text{ ft-kips} \end{cases}$$

$$(\phi M_{nx} = 273 \text{ ft-kips}) \geq (M_{ux} = 219) \quad \text{(OK)}$$

Use W18 x 50 (F_y = 36 ksi).

Example 5.7

Repeat Example 5.6 accounting for moment redistribution as permitted by LRFD A5.1 (p. 6–31) for moment diagrams obtained from an elastic factored load analysis. These permitted adjustments in the moment diagram enable the structural designer to obtain a solution that is close to the plastic design solution (discussed in Chapter

11). For W sections used as beams, $M_{px}/M_{(\text{yield})x} = Z_xF_y/S_xF_y$ ranges from 1.10 to 1.18 and the average is 1.14, which is a 14% increase in bending strength for inelastic behavior. The 10% reduction in negative moments at the supports for elastic analyses is close to the redistribution of moment that is obtained in the plastic design solution.

Solution

See LRFD A5.1. The maximum possible reduction in the moment at the left support is 0.1(263) = 26.3 ft-kips. Note that we cannot change the moment at the right support in this example. Try reducing 263 ft-kips by 26 ft-kips and increasing 219 ft-kips by (26 + 0)/2 = 13 ft-kips. The adjusted moment diagram is Figure 5.22.
Try W21 x 44: $(\phi M_{px} = 258 \text{ ft-kips}) \geq (M_{ux} = 237)$
$(L_p = 5.3) < (L_b = 14) \leq (L_r = 15.4)$
Enter the LRFD beam design charts in the vicinity of the larger of

1. $\dfrac{M_{ux}}{C_b} = \dfrac{237}{2.27} = 104 \text{ ft-kips (for first } L_b \text{ region)}$

2. $\dfrac{M_{ux}}{C_b} = \dfrac{232}{1.67} = 139 \text{ ft-kips (for second } L_b \text{ region)}$

For the first L_b region,

$$\phi M_{nx} = \text{smaller of} \begin{cases} C_b\overline{M}_1 = 2.27(173) = 393 \\ \phi M_{px} = 258 \text{ ft-kips} \end{cases}$$

$$(\phi M_{nx} = 258 \text{ ft-kips}) \geq (M_{ux} = 237) \quad \text{(OK)}$$

For the second L_b region,

$$\phi M_{nx} = \text{smaller of} \begin{cases} C_b\overline{M}_1 = 1.67(173) = 289 \\ \phi M_{px} = 258 \text{ ft-kips} \end{cases}$$

$$(\phi M_{nx} = 258 \text{ ft-kips}) \geq (M_{ux} = 232) \quad \text{(OK)}$$

Use W21 x 44.
For Example 5.4, either W18 x 50 or W21 x 50 was needed. Thus, by accounting for LRFD A5.1, we can use W21 x 44, and the savings is 6 lb/ft, which is 100(6/50) = 12% less steel.

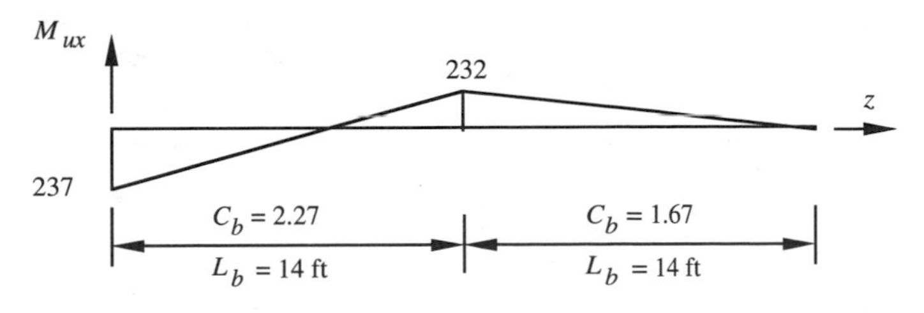

FIGURE 5.22 Moment diagram for Example 5.7.

5.10 When Local Buckling Governs ϕM_{nx}

When $(b/t) > \lambda_p$ (see LRFD B5.1 and Table B5.1, p. 6-38) for any compression element in a beam cross section, local buckling of one or more compression elements occurs before M_{px} can be developed due to bending. Figure 5.23 illustrates local buckling of the compression flange and compression zone of the web due to a bending moment. For any compression element in a beam cross section, when $\lambda_r \geq (b/t) > \lambda_p$, inelastic local buckling occurs and when $(b/t) > \lambda_r$, elastic local buckling occurs. Also, for sections where LTB can occur, LTB may govern ϕM_{nx} when $L_b > L_p$. LRFD Appendix F must be used to determine ϕM_{nx} when inelastic local buckling governs. LRFD Appendices B and F must be used to determine ϕM_{nx} when elastic local buckling governs.

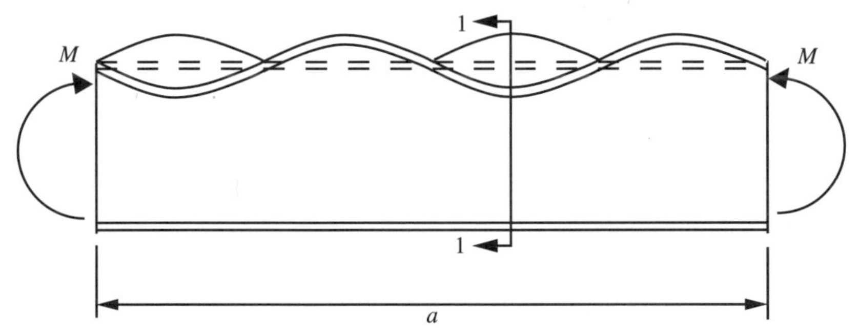

Number of half sine waves is a function of a/b and b/t of flange.
(a) Flange local buckling of a W section beam

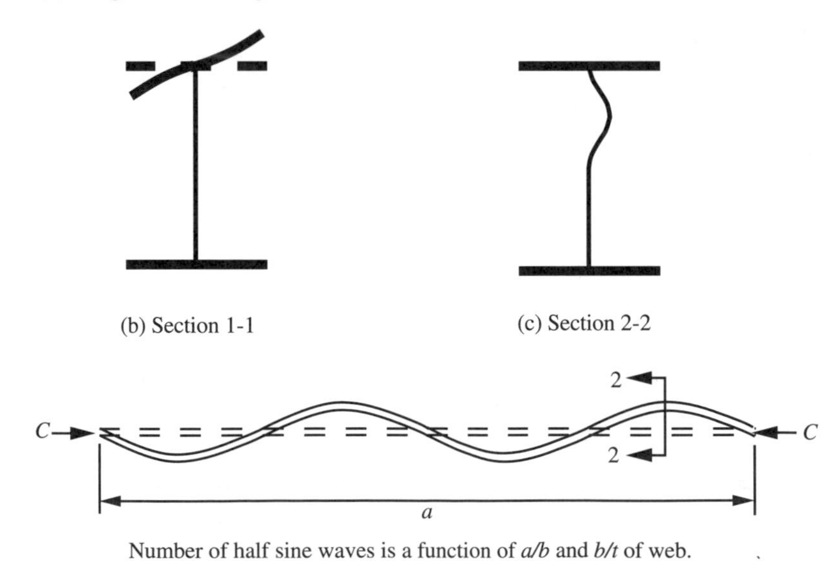

Number of half sine waves is a function of a/b and b/t of web.
(b) Web local buckling of a W section beam

FIGURE 5.23 Local buckling modes of a W section beam subjected to uniform bending.

Example 5.8

This example illustrates some computations of ϕM_{nx} when *flange local buckling* (FLB) and LTB of a W section must be considered.
Find ϕM_{nx} for a W6 x 15 (F_y = 50 ksi) and values of $L_b > L_p$ as specified in the solution.

Solution

W6 x 15:

$$A = 4.43 \text{ in.}^2;\; S_x = 9.72 \text{ in.}^3;\; Z_x = 10.8 \text{ in.}^3;\; I_y = 9.32 \text{ in.}^4;\; r_y = 1.46 \text{ in.}$$

$$0.5b_f/t_f = 11.5;\; h/t_w = 21.6;\; C_w = 76.5 \text{ in.}^6 \text{ and, } J = 0.10 \text{ in.}^4$$

Is LRFD Appendix F applicable?

1. $\left(\dfrac{h}{t_w} = 21.6\right) \le \left(\dfrac{640}{\sqrt{50}} = 90.5\right)$ WLB is not applicable.

2. $\left(\dfrac{0.5b_f}{t_f} = 11.5\right) > \left(\lambda_p = \dfrac{65}{\sqrt{50}} = 9.19\right)$ FLB may govern ϕM_{nx}.

FLB and LTB of the W6 x 15 must be investigated. The definitions given for the first item in LRFD Table A-F1.1 (p. 6-114) are applicable for a W section. As shown in this example, LRFD Eq. (A-F1-3) is applicable for FLB and LRFD Eq. (A-F1-4) is applicable for LTB.
Find ϕM_{nx} for FLB

$$\left(\lambda_r = \frac{141}{\sqrt{50-10}} = 22.3\right) > \left(\frac{0.5b_f}{t_f} = 11.5\right) > \left(\lambda_p = 9.19\right)$$

LRFD Eq. (A-F1-3) is applicable:

$$M_{px} = Z_xF_y = (50)(10.8) = 540 \text{ in.-kips} = 45.0 \text{ ft-kips}$$

$$M_{rx} = (F_y - F_r)S_x = (50 - 10)(9.72) = 388.8 \text{ in.-kips} = 32.4 \text{ ft-kips}$$

$$M'_{nx} = M_{px} - \left(M_{px} - M_{rx}\right)\left(\frac{\lambda - \lambda_p}{\lambda_r - \lambda_p}\right)$$

Note: The M'_{nx} notation was chosen to be compatible with LRFD p. 4-7:

$$M'_{nx} = 45.0 - (45.0 - 32.4)\left(\frac{11.5 - 9.19}{22.3 - 9.19}\right) = 42.8 \text{ ft-kips}$$

$$[\phi M_{nx} = 0.9(42.8) = 38.5 \text{ ft-kips}] < (\phi M_{px} = 40.5 \text{ ft-kips})$$

M'_{nx} is the maximum acceptable design bending strength and is valid for (see last formula on LRFD p. 4-7):

$$L_b \le \left[L'_p = L_p + (L_r - L_p)\left(\frac{M_p - M'_{nx}}{M_p - M_{rx}} \right) \right]$$

$$M'_{nx} = 42.8 \text{ ft-kips} \quad \text{(due to FLB in our example)}$$

$$L_p = \frac{300 r_y}{\sqrt{F_y}} = \frac{300(1.46)}{\sqrt{50}} = 61.94 \text{ in.} = 5.16 \text{ ft}$$

$$X_1 = \frac{\pi}{S_x}\sqrt{\frac{EAGJ}{2}} = \frac{\pi}{9.72}\sqrt{\frac{29{,}000(4.43)(11{,}200)(0.10)}{2}} = 2741.44$$

$$X_2 = \frac{4C_w}{I_y}\left(\frac{S_x}{GJ}\right)^2 = \frac{4(76.5)}{9.32}\left(\frac{9.72}{11{,}200(0.10)}\right)^2 = 0.00247287$$

$$L_r = \frac{r_y X_1}{F_y - F_r}\sqrt{1+\sqrt{1+X_2(F_y - F_r)^2}}$$

$$L_r = \frac{146(2741.44)}{50-10}\sqrt{1+\sqrt{1+0.00247287(50-10)^2}} = 179.73 \text{ in.} = 14.98 \text{ ft}$$

$$L'_p = 5.16 + (14.98 - 5.16)\left(\frac{45.0 - 42.8}{45.0 - 32.4}\right)^2 = 6.89 \text{ ft}$$

Note: See LRFD p. 4-21 for a W6 x 15 ($F_y = 50$ ksi). The preceding calculations show how the $\phi M_{px} = 38.6$ ft-kips and $L_p = 6.8$ ft entries were computed by the person who prepared that page.

Find ϕM_{nx} for LTB

For this example, LTB must be considered when $L_b > (L'_p = 6.89 \text{ ft})$.
Note: In LRFD Appendix F, this condition is shown as $\lambda_p < \lambda \le \lambda_r$ and $\lambda > \lambda_r$.

$$\lambda = \frac{L_b}{r_y}$$

$$\lambda_p = \frac{300}{\sqrt{F_y}} = \frac{300}{\sqrt{50}} = 42.43$$

$$\lambda_r = \frac{X_1}{F_y - F_r}\sqrt{1+\sqrt{1+X_2(F_y - F_r)^2}}$$

$$X_1 = \frac{\pi}{S_x}\sqrt{\frac{EAGJ}{2}} = 2741.44$$

$$X_2 = \frac{4C_w}{I_y}\left(\frac{S_x}{GJ}\right)^2 = 0.00247287$$

$$\lambda_r = \frac{2744.44}{50-10}\sqrt{1+\sqrt{1+0.00247287(50-10)^2}} = 123.105$$

We choose to illustrate the calculations for $L_b = 16$ ft = 192 in.

$$(\lambda = L_b/r_y = 192/1.46 = 131.507) > (\lambda_r = 123.105)$$

From LRFD Eq. (A-F1-4), p. 6-111,

$$M_{nx} = M_{cr} = S_x F_{cr}$$

For the first shape in LRFD Table A-F1.1 (p. 6-114),

$$F_{cr} = \frac{C_b X_1 \sqrt{2}}{\lambda}\sqrt{1+\frac{X_1^2 X_2}{2\lambda^2}}$$

For $C_b = 1$,

$$F_{cr} = \frac{2741.44\sqrt{2}}{131.507}\sqrt{1+\frac{(2741.44)^2(0.00247287)}{2(131.507)^2}} = 36.55 \text{ ksi}$$

$$M_{nx} = M_{cr} = S_x F_{cr} = 9.72(36.55) = 355.3 \text{ in.-kips} = 29.6 \text{ ft-kips}$$

Since (M_{nx} = 29.6 ft-kips) < [M'_{nx} = 42.8 ft-kips (due to FLB)], LTB governs and ϕM_{nx} = 0.9(29.6) = 26.6 ft-kips. On LRFD p. 4-165 for a W6 x 15, the beam chart only goes up to L_r and our $L_b > L_r$. So, we cannot check an entry on the $C_b = 1$ curve due to LTB for $L_b > L_r$. However, we recommend that the point (16 ft, 26.6 ft-kips) be plotted since we have computed it and connect this point with a slightly concave upward line (almost a straight line) to the end of the existing curve.

Example 5.9

For all rolled W sections and $F_y \le 100$ ksi,

$$\left(\frac{h}{t_w}\right) \le \left(\lambda_p = \frac{640}{\sqrt{F_y}}\right)$$

The I-shaped section with the most slender web is an M12 x 11.8 and for $F_y = 100$ ksi (highest available grade of steel),

$$\left(\frac{h}{t_w} = 62.5\right) \leq \left(\lambda_p = \frac{640}{\sqrt{100}} = 64.0\right) \qquad \text{WLB is not applicable.}$$

Therefore, this example illustrates some computations of ϕM_{nx} when WLB and LTB of a built-up W section must be considered. A built-up W section is composed of three plates (two flange plates and a web plate) joined by continuous fillet welds to form an I-shaped section.

For a built-up W57 x 18 x 206 (F_y = 50 ksi) with values of $L_b > L_p$ as specified in the solution, find ϕM_{nx}.

Solution

W57 x 18 x 206 (see LRFD, p. 4-184):
A = 60.5 in.^{2}; d = 58.0 in.; b_f = 18 in.; t_f = 1 in.; h = 56 in.; t_w = 7/16 in.
S_x = 1230 in.^{3}; Z_x = 1370 in.^{3}; I_y = 972.39 in.^{4}; r_y = 4.01 in.; $0.5b_f/t_f = 9$; h/t_w = 56/(7/16) = 128; C_w = 789, 507 in.^{6}; J = 13.56 in.^{4}

We computed the last six of these properties. For a built-up W section composed of three plate elements, the definitions for C_w and J are

$$C_w = \frac{t_f b_f^3 \left(h + t_f\right)^2}{24}$$

$$J = \sum\left(\frac{bt^3}{3}\right)$$

where

b = width of a plate element
t = thickness of a plate element

Is LRFD Appendix F applicable?

1. $$\left(\frac{0.5b_f}{t_f} = 9\right) \leq \left(\lambda_p = \frac{65}{\sqrt{50}} = 9.19\right) \qquad \text{FLB is not applicable.}$$

2. $$\left(\lambda = \frac{h}{t_w} = 128\right) > \left(\lambda_p = \frac{640}{\sqrt{50}} = 90.5\right) \qquad \text{WLB may govern } \phi M_{nx}.$$

WLB and LTB must be considered. The definitions given for the first shape in LRFD Table A-F1.1 (p. 6-94) are applicable. As shown here, for our example LRFD Eq. (A-F1-3) is applicable for WLB and LRFD Eq. (A-F1-4) is applicable for LTB.

Find ϕM_{nx} for WLB

$$\left(\lambda_p = 90.5\right) < \left(\lambda = 128\right) \leq \left(\lambda_r = \frac{970}{\sqrt{50}} = 137.2\right)$$

and LRFD Eq. (A-F1-3) is applicable:

$$M_{px} = Z_x F_y = (1370)(50)/12 = 5708 \text{ ft-kips}$$

$$M_{rx} = S_x F_y = (1230)(50)/12 = 5125 \text{ ft-kips}$$

$$M'_{nx} = M_{px} - \left(M_{px} - M_{rx}\right)\left(\frac{\lambda - \lambda_p}{\lambda_r - \lambda_p}\right)$$

$$= 5708 - (5708 - 5125)\left(\frac{128 - 90.5}{137.2 - 128}\right) = 3332 \text{ ft-kips}$$

$$\phi M'_{nx} = 0.9(3332) = 2998 \text{ ft-kips}$$

M'_{nx} is the maximum acceptable design bending strength when WLB governs and is valid for

$$L_b \le \left[L'_p = L_p + \left(L_r - L_p\right)\left(\frac{M_p - M'_{nx}}{M_p - M_{rx}}\right)\right]$$

$$M'_{nx} = 3332 \text{ ft-kips} \qquad (\text{due to WLB in our example})$$

$$L_p = \frac{300 r_y}{\sqrt{F_y}} = \frac{300(4.01)}{\sqrt{50}} = 170.13 \text{ in.} = 14.186 \text{ ft}$$

$$L_r = \frac{r_y X_1}{F_y - F_r}\sqrt{1 + \sqrt{1 + X_2\left(F_y - F_r\right)^2}}$$

$$X_1 = \frac{\pi}{S_x}\sqrt{\frac{EAGJ}{2}} = \frac{\pi}{1230}\sqrt{\frac{29{,}000(60.5)(11{,}200)(13.56)}{2}} = 932.278$$

$$X_2 = \frac{4C_w}{I_y}\left(\frac{S_x}{GJ}\right)^2 = \left[\frac{4(789.507)}{972.39}\right]\left[\frac{1230}{11{,}200(13.56)}\right]^2 = 0.213025$$

Since $F_r = 16.5$ ksi for welded shapes,

$$L_r = \frac{4.01(932.278)}{50 - 16.5}\sqrt{1 + \sqrt{1 + 0.0213025(50 - 16.5)^2}} = 219.03 \text{ in.} = 18.25 \text{ ft}$$

$$L'_p = 14.18 + (18.25 - 14.18)\left(\frac{5708 - 3332}{5708 - 5125}\right) = 30.77 \text{ ft}$$

Find ϕM_{nx} for LTB

For this example, LTB must be considered when $L_b > (L'_p = 30.77 \text{ ft})$:

$$\lambda = \frac{L_b}{r_y}$$

$$\lambda_p = \frac{300}{\sqrt{F_y}} = \frac{300}{\sqrt{50}} = 42.43$$

$$\lambda_r = \frac{X_1}{F_y - F_r}\sqrt{1+\sqrt{1+X_2\left(F_y - F_r\right)^2}}$$

$$X_1 = \frac{\pi}{S_x}\sqrt{\frac{EAGJ}{2}} = 932.278$$

$$X_2 = \frac{4C_w}{I_y}\left(\frac{S_x}{GJ}\right)^2 = 0.213025$$

$$\lambda_r = \frac{932.278}{50-16.5}\sqrt{1+\sqrt{1+0.213025(50-16.5)^2}} = 54.62$$

For this example, LTB does not govern until

$$\left(F_{cr} = \frac{C_b X_1 \sqrt{2}}{\lambda}\sqrt{1+\frac{X_1^2 X_2}{2\lambda^2}}\right) < \left[\frac{M'_{nx}}{S_x} = \frac{3332(12)}{1230} = 32.5 \text{ ksi}\right]$$

For $C_b = 1$, $F_{cr} = 32.5$ ksi occurs at $\lambda = 114.86$ and $L_b = r_y\lambda = 4.01(114.86) = 460.6$ in. $= 38.38$ ft. We choose to illustrate the calculations for $L_b = 40$ ft $= 480$ in.:

$$\left(\lambda = \frac{L_b}{r_y} = \frac{480}{4.01} = 119.7\right) > [\lambda_r = 54.62]$$

From LRFD Eq. (A-F1-4): $M_{nx} = M_{cr} = S_x F_{cr}$. From the first item in Table A-F1.1 of LRFD Appendix F,

$$F_{cr} = \frac{C_b X_1 \sqrt{2}}{\lambda}\sqrt{1+\frac{X_1^2 X_2}{2\lambda^2}}$$

For $C_b = 1$,

$$F_{cr} = \frac{932.278\sqrt{2}}{119.7}\sqrt{1+\frac{(932.278)^2(0.213025)}{2(119.7)^2}} = 30.09 \text{ ksi}$$

$$M_{nx} = M_{cr} = S_x F_{cr} = (1230)(30.09) = 37{,}006 \text{ in.-kips} = 3083.8 \text{ ft-kips}$$

Since (M_{nx} = 3084 ft-kips) < [M'_{nx} = 3332 ft-kips (due to WLB)], LTB governs and

$$\phi M_{nx} = (0.9)(3084) = 2775 \text{ ft-kips}$$

5.11 BUILT-UP BEAM SECTIONS

A built-up section is composed of two or more rolled shapes. On LRFD pp. 1-105 to 1-119, some properties are provided for frequently used built-up sections called *combination sections*. LRFD p. 1-105 gives an indication of where these built-up sections might be used. Fastening a C section to only the compression flange of a W section (see LRFD, p. 1-106) is frequently done for a crane runway girder and serves two useful purposes: (1) increases L_p and L_r, which allows larger L_b values to be used without LTB governing ϕM_{nx}; and (2) increases the capacity of the compression flange to resist lateral loads (in the x-direction on LRFD, p. 1-105) from the wheels of a crane trolley. A built-up W section (see LRFD, p. 3-183) is composed of three plates joined by welding to form an I-shaped section. In Example 5.9, a built-up W section was used to illustrate the computations when web local buckling governed ϕM_{nx}. The LRFD table of built-up W sections was prepared as an aid in the design of plate girders (discussed in Chapter 9). A plate girder is a flexure member with intermediate web stiffeners or with a slender web $(h/t_w) > 970/\sqrt{F_{yf}}$ and intermediate web stiffeners, where F_{yf} denotes the yield strength of the flange since a hybrid plate girder can be formed. For example, if F_y = 50 ksi is used in the flanges and F_y = 36 ksi is used in the web of a built-up section, this section is classified as a hybrid built-up section. As shown in Section 5.10, bearing stiffeners may be needed at the supports and at any interior concentrated load points to prevent localized failures of the web and/or flanges of a beam or a plate girder. When $(h/t_w) > 418/\sqrt{F_{yf}}$ for a plate girder, transverse stiffeners may also be needed to strengthen the web at other points in addition to the reaction locations and the concentrated load points.

Another type of built-up section is classified as a cover-plated section (see Figures 5.24–27) and is discussed later. A cover plate "covers (or hides)" a flange for some portion of the flange length in the region of the maximum moment. Welding a cover plate on each flange of a W section may be feasible for the following reasons. In the preliminary design stage of a building, the structural engineer estimates the maximum overall beam depth required for each floor level based on a preliminary architectural layout of the building and gives this beam depth to the architect. In the final structural design stage, the loads on some beams may increase due to changes in the architectural layout of the building. However, the agreed on maximum overall beam depth usually cannot be increased, and adding cover plates to a W section to provide the required bending strength may become a feasible solution. When the thickness of each cover plate is governed by the minimum weld size, adding only a cover plate to one flange is generally more economical. Rehabilitation of existing structures is becoming more economical, and strengthening some beams may be feasible by adding only one cover plate to the most accessible flange of each beam. Adding an adequately designed cover plate to the compression flange of a W section increases the LTB strength. Therefore, if only one cover plate can be added to a W section, preferably this plate should be fastened to the compression flange. However, if the thickness of a cover plate added to only one flange exceeds about $1.5t_f$, adding a cover plate to each flange is structurally more efficient.

The structural design procedure for adding a cover plate to each flange of a W section is simpler than for adding only a cover plate to one flange. When lateral braces can only be provided such that $L_b > L_p$, the structural design procedure is

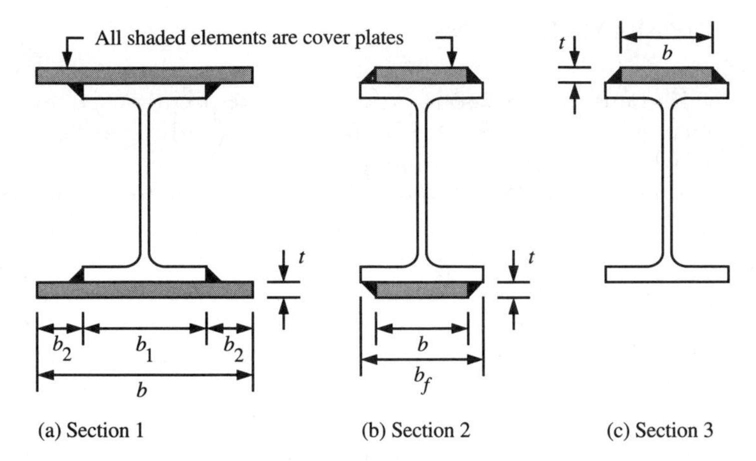

Notation for sketches:
1. b = cover plate width
2. t = cover plate thickness
3. b_f = flange width of W section

From LRFD Table B5.1, limiting width–thickness ratios are:
1. For (a):

For b_1/t: $\lambda_p = 190/\sqrt{F_y}$; $\lambda_r = 238/\sqrt{F_y}$

For b_2/t: λ_p is not applicable ; $\lambda_r = 95/\sqrt{F_y}$

2. For (b) and (c):

For b/t: $\lambda_p = 190/\sqrt{F_y}$; $\lambda_r = 238/\sqrt{F_y}$

FIGURE 5.24 Cover-plated beam sections.

iterative; that is, we must add an estimated plate size either to one flange or to each flange, compute ϕM_{nx}, and if $\phi M_{nx} < M_{ux}$, we must assume a larger plate size and repeat the procedure. Therefore, we illustrate the simplest case first.

Example 5.10

For the member loaded as shown Figure 5.25(a), suppose that we can provide lateral bracing such that $L_b \leq L_p$. For F_y = 36 ksi, the lightest acceptable choice without cover plates is either a W27 x 84 or W24 x 84. Suppose, due to an architectural depth limitation, we desire to use a W21 x 68 (F_y = 36 ksi) section and add a cover plate to each flange as shown in Figures 5.25(c) and 5.25(d).

Solution

As shown in Figures 5.25(b) to 5.25(d), cover plates are required in the region where M_{ux} > 432 ft-kips since ϕM_{nx} = 432 ft-kips for a W21 x 68 without cover plates. From Figures 5.25(b) and 5.26, the design bending strength requirement for the plates is

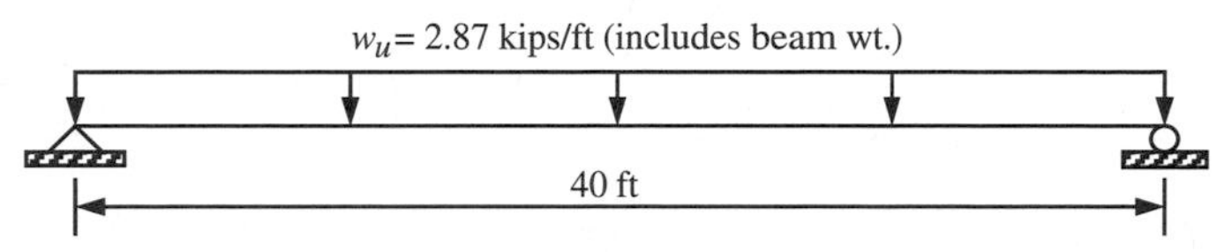

(a) Beam subjected to factored loading

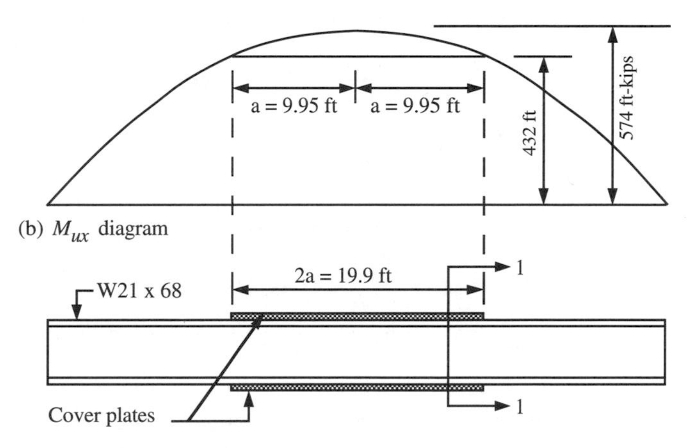

(b) M_{ux} diagram

(c) Beam with cover plates

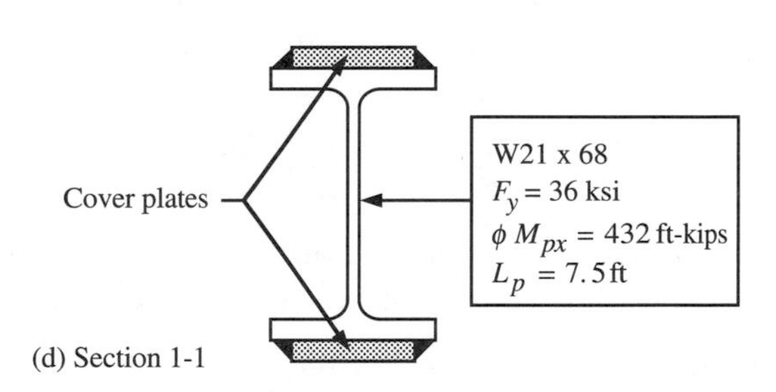

(d) Section 1-1

FIGURE 5.25 Sketches for Example 5.10.

$$\phi(d+t)A_p F_y \geq \left[(574-432) = 142 \text{ ft - kips} = 1704 \text{ in. - kips}\right]$$

For a W21 x 68:

$$d = 21.13 \text{ in.}, b_f = 8.270 \text{ in.}, t_f = 0.685 \text{ in.}, I_y = 64.7 \text{in.}^4, \text{ and } A = 20.0 \text{ in.}^2$$

When t_f is the thicker part joined by fillet welds, from LRFD J2.2b (p. 6-75) the minimum weld size $S_w = 1/4$ in., which requires $t \geq 5/16$ in. and $b \leq [8.270 - 2(0.25) = 7.75$ in.]. Try $t = 3/8$ in. and since $\phi = 0.9$, we need

$$\left(A_p = 0.375b\right) \geq \left[\frac{1704}{0.9(21.13+0.375)(36)} = 2.45 \text{ in.}^2\right]$$

$$(b \geq 6.52 \text{ in.}) \leq 7.77 \text{ in.}$$

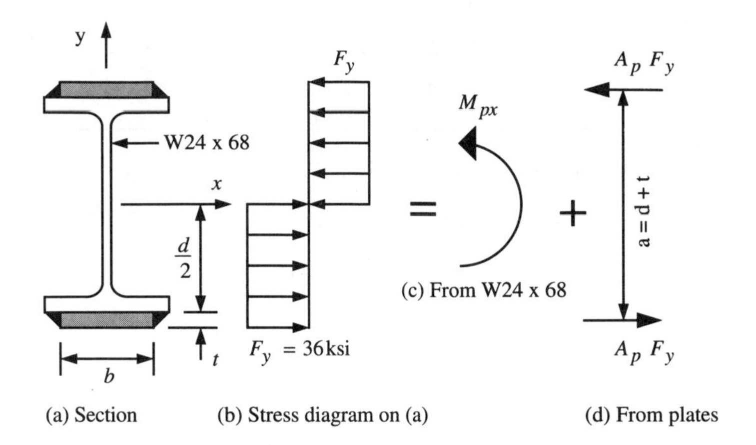

FIGURE 5.26 Plastic moment for a doubly-symmetric cover-plated section.

Check the λ_p requirement for a cover plate choice of b = 6.50 in. and t = 3/8 in.

$$\left(\frac{b}{t} = \frac{6.50}{0.375} = 17.3\right) \leq \left(\frac{190}{\sqrt{36}} = 31.7\right)$$

Use a pair of 6.5 x 3/8 cover plates continuously fillet welded to the W21 x 68 in the region where the cover plates are required. For this built-up section,

$$L_p = \frac{300\,r_y}{\sqrt{F_y}}$$

$$A = 20.01 + 2\,[6.50(0.375)] = 24.875 \text{ in.}^2$$

$$I_y = 64.7 + 2\left[\frac{0.375(6.50)^3}{12}\right] = 81.86 \text{ in.}^4$$

$$r_y = \sqrt{\frac{I_y}{A}} = \sqrt{\frac{81.86}{24.875}} = 1.81 \text{ in.}$$

$$L_p = \frac{300(1.81)}{\sqrt{36}} = 90.71 \text{ in.} = 7.56 \text{ ft}$$

Note: For a W21 x 68 without cover plates, L_p = 7.50 ft. Use (L_b = 80 in.) ≤ (L_p = 7.50 ft).

From Figure 5.25(b), cover plates are required for a theoretical distance a each side of midspan. Since the M_{ux} diagram is a parabola with the vertex at midspan,

$$\left(\frac{a}{20}\right)^2 = \frac{142}{574}$$

$$a = 9.948 \text{ ft} = 119.4 \text{ in.}$$

Is the minimum $S_w = 1/4$ in. adequate? For the built-up section,

$$I_x = 1480 + 2\left[6.5(0.375)\left(\frac{21.13 + 0.375}{2}\right)^2\right] = 2043 \text{ in.}^4$$

At $a = 9.948$ ft from midspan, $V_u = 28.85$ kips. Let P_{uw} denote the required strength per inch for each fillet weld. Since there are two fillet welds per cover plate,

$$P_{uw} = \frac{V_u Q}{2I_x} = \frac{28.85(6.50)(0.375)(21.13 + 0.375)/2}{2(2043)}$$

$$= 0.185 \text{ kips/in.}$$

For our base material of $F_y = 36$ ksi and for E70 electrodes, shear on the throat plane governs the design strength of a fillet weld. The design strength per inch of a longitudinal fillet weld is

$$\phi F_w = 0.75\ [0.6F_{Exx}(0.707S_w)]$$

$$= 0.75(0.6)(70)(0.707S_w) = 22.27S_w$$

The weld design strength requirement is

$$(\phi F_w = 22.27S_w) \geq (P_u = 0.185 \text{ kips/in.})$$

which gives $S_w \geq 0.00831$ in. Therefore, the minimum $S_w = 1/4$ in. is adequate. At midspan the force in each cover plate is

$$P_u = \phi A_p F_y = 0.9(6.50)(0.375)(36) = 79.0 \text{ kips}$$

At $a = 9.948$ ft from midspan, the force in the cover plate is

$$P_u = \left(\frac{432}{574}\right)(79.0) = 59.4 \text{ kips}$$

The total length of 1/4 in. weld required at the end of each cover plate to develop $P_u = 59.4$ kips is

$$[1.5(6.50) + 2L_w] = \frac{P_u}{\phi F_w} = \left[\frac{59.4}{22.27(0.25)}\right] = 10.7 \text{ in.}$$

Each weld group consists of a 6.50-in. transverse weld and two longitudinal welds, each of which has a length of L_w. Since a transverse weld is 1.5 times as strong as a longitudinal weld,

$$1.5(6.50) + 2L_w = 10.7 \text{ in.}$$

$$L_w = 0.475 \text{ in.}$$

Strength requires that the plate be extended by $L_w \geq 0.5$ in. on each end of each cover plate. However, since our (s_w/t = 0.25/0.375 = 0.67) < 0.75, LRFD B10 (p. 6-37) stipulates that the minimum extension for our cover plate is a' = 1.5(6.5) = 9.75 in.

Example 5.11

For the member loaded as shown in Figure 5.25(a), suppose that we can provide lateral bracing such that $L_b \leq L_p$. For F_y = 36 ksi, the lightest acceptable choice without cover plates is either a W27 x 84 or W24 x 84. Suppose that we desire to use a W24 x 68 (F_y = 36 ksi) section and add a cover plate only to the compression flange as shown in Figure 5.27(a).

Solution

As shown in Figures 5.25(b) to 5.25(d), a cover plate is required in the region where M_{ux} > 478 ft-kips since ϕM_{nx} = 478 ft-kips for a W24 x 68 without a cover plate. From Figures 5.27(b) to 5.27(f), the design bending strength requirement for the plate is

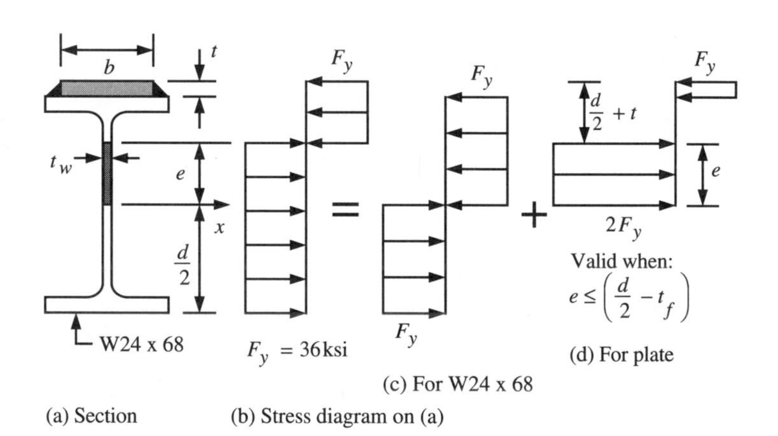

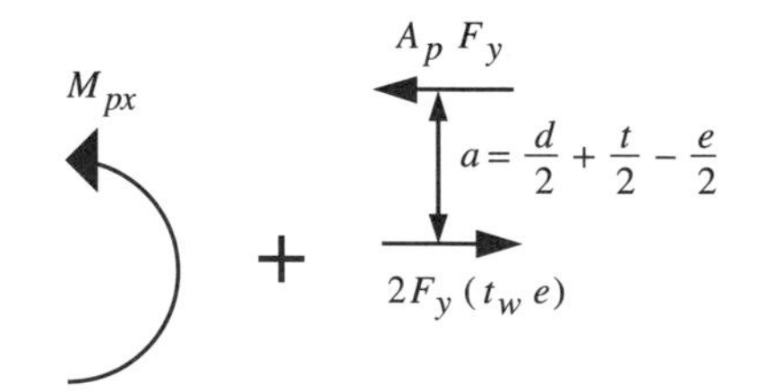

Note: The information in (d) and (f) is valid provided

$$\frac{b}{t} \leq \frac{190}{\sqrt{F_y}} \quad \text{for the cover plate (on the compression flange)}$$

FIGURE 5.27 Plastic moment for a singly-symmetric cover-plated section.

$$\phi\left(\frac{d}{2}+\frac{t}{2}-\frac{e}{2}\right)A_pF_y \geq \left[(574-478)=96 \text{ ft}-\text{kips}=1152 \text{ in.}-\text{kips}\right]$$

From $T = C$ in Figure 5.26(f),

$$e=\frac{A_p}{2t_w}$$

For a W24 x 68:

$$d = 23.73 \text{ in.};\ t_w = 0.415 \text{ in.};\ b_f = 8.965 \text{ in.}$$

$$t_f = 0.585 \text{ in.};\ I_y = 70.4 \text{ in.}^4;\ A = 20.1 \text{ in.}^2$$

When t_f is the thicker part joined by fillet welds, from LRFD J2.2b the minimum $S_w = 1/4$ in., which requires $t \geq 5/16$ in. and $b \leq [8.965 - 2(0.25) = 8.47$ in.]. Try $b = 8$ in. Since $\phi = 0.9$, from the design bending strength requirement of the plate, we need

$$(23.73 + t - 9.64t)t \geq 1152/[0.9(8)(36/2)]$$

$$t \geq 0.447 \text{ in.}$$

Check the λ_p requirement for a cover plate choice of $b = 8$ in. and $t = 1/2$ in.:

$$\left(\frac{b}{t}=\frac{8}{0.5}=16\right)\leq\left(\frac{190}{\sqrt{36}}=31.7\right)$$

Check the restriction on e [see Figure 5.27(d)]:

$$e \leq \left(\frac{d}{2}-t_f\right) \text{is required}$$

$$e=\left[\frac{A_p}{2t_w}=\frac{8(0.5)}{2(0.415)}=4.82 \text{ in.}\right]$$

$$(e=4.82)\leq\left(\frac{23.73}{2-0.585}=11.3\right) \quad (\text{OK})$$

Use an 8 x 1/2 cover plate continuously fillet welded to the W24 x 68 in the region where the cover plates are required. For this built-up section,

$$L_p=\frac{300r_y}{\sqrt{F_y}}$$

$$A = 20.1 + 8(0.5) = 24.1 \text{ in.}^2$$

$$I_y = 70.4+\frac{0.5(8)^3}{12}=91.73 \text{ in.}^4$$

$$r_y = \sqrt{\frac{I_y}{A}} = \sqrt{\frac{91.73}{24.1}} = 1.95 \text{ in.}$$

$$L_b = \frac{300(1.95)}{\sqrt{36}} = 97.5 \text{ in.} = 8.13 \text{ ft}$$

Note: For a W24 x 68 without cover plates, $L_p = 7.8$ ft. Use $(L_b = 80 \text{ in.}) \leq (L_p = 7.8 \text{ ft.})$. We could use $L_b = 8$ ft (see Example 5.12).

From Figure 5.26(b), cover plates are required for a theoretical distance a on each side of midspan. Since the M_{ux} diagram is a parabola with the vertex at midspan,

$$\left(\frac{a}{20}\right)^2 = \frac{96}{574}$$

$$a = 8.18 \text{ ft} = 98.2 \text{ in.}$$

Is the minimum $S_w = 1/4$ in. adequate? With respect to the centroid of the W section, the distance to the centroid of the built-up section is

$$\bar{y} = \frac{4.00(23.73 + 0.5)/2}{4.00 + 20.1} = 2.01 \text{ in.}$$

$$\frac{23.73 + 0.5}{2} - 2.01 = 10.1 \text{ in.}$$

$$I_x = 1830 + 20.1(2.01)^2 = 4.00(10.1)^2 = 2319 \text{ in.}^4$$

At $a = 8.18$ ft $= 98.2$ in. from midspan, $V_u = 23.5$ kips. Let P_{uw} denote the required strength per inch for each fillet weld. Since there are two fillet welds per cover plate,

$$P_{uw} = \frac{V_u Q}{2I_x} = \frac{23.5(4.00)(10.1)}{2(2319)} = 0.205 \text{ kips/in.}$$

The weld design strength requirement is

$$(\phi F_w = 22.27 S_w) \geq (P_{uw} = 0.205 \text{ kips/in.})$$

which gives $S_w \geq 0.00921$ in. Therefore, the minimum $S_w = 1/4$ in. is adequate.
At midspan the force in each cover plate is

$$P_u = \phi A_p F_y = 0.9(8)(0.5)(36) = 130 \text{ kips}$$

At $a = 8.18$ ft from midspan, the force in the cover plate is

$$P_u = \left(\frac{478}{574}\right)(130) = 108 \text{ kips}$$

The total length of 1/4-in. weld required at the end of the cover plate to develop $P_u = 108$ kips is

$$\left[1.5(8)+2L_w\right]=\frac{P_u}{\phi F_w}=\left[\frac{108}{22.27(0.25)}\right]=19.4\text{ in.}$$

Each weld group consists of an 8 in. transverse weld and two longitudinal welds each of which has a length of L_w. Since a transverse weld is 1.5 times as strong as a longitudinal weld,

$$1.5(8) + 2L_w = 19.4\text{ in.}$$

$$L_w = 3.7\text{ in.}$$

Therefore, strength requires that the plate be extended by $L_w \geq 3.75$ in. on each end of each cover plate. However, since our

$$\left(\frac{s_w}{t}=\frac{0.25}{0.5}=0.5\right)<0.75$$

LRFD B10 (p. 6-37) stipulates that the minimum extension for our cover plate is $a' = 1.5(8) = 12$ in.

Use $L_w \geq 3.75$ in. for the longitudinal fillet welds at each end of each cover plate. Note that the actual length required for the cover plate is $2(a + L_w)$.

Example 5.12

For the member loaded as shown Figure 5.25(a), suppose that we can only provide lateral bracing such that ($L_b = 160$ in.) > L_p. Suppose that we desire to use a W24 x 68 ($F_y = 36$ ksi) section and add a cover plate only to the compression flange in the middle $L_b = 160$ in. Note that $C_b = 1$ in this region.

Solution

From Figures 5.27(b) to 5.27(f), the plastic design bending strength requirement for the plate is

$$P_{uw}=\frac{V_uQ}{2I_x}=\frac{23.5(4.00)(10.1)}{2(2319)}=0.205\text{ kips/in.}$$

From $T = C$ in Figure 5.27(f),

$$e=\frac{A_p}{2t_w}$$

However, since $L_b > L_p$ and $C_b = 1$, $\phi M_{nx} < \phi M_{px}$, we recommend the following procedure for estimating the cover plate size when $L_b > L_p$.

For a W24 x 68 ($F_y = 36$ ksi), $\phi M_{px} = 478$ ft-kips, $L_b = 13$ ft 4 in., and $C_b = 1$, from the LRFD beam design charts we find $\phi M_{nx} = 411$ ft-kips. Therefore, assume that we need to increase the plastic design bending strength requirement for the plate by a factor of $478/411 = 1.16$. Also, since we found that $\phi M_{nx} = 411$ ft-kips in the $L_p < L_b \leq L_r$ region and since M_{rx} [see LRFD Eq. (F1-7)] is a function of F_r (10 ksi for W24 x 68 and 16.5 ksi for the cover plated section), we need to increase the plastic design bending strength requirement for the plate by another factor of (36 - 10)/(36 - 16.5) = 1.33. Also, as the single cover plate gets thicker than in Example 5.11, the structural efficiency decreases (*e* increases and *a* decreases as shown in Figure 5.27), and we need to increase the plastic design bending strength requirement for the plate by another factor of 1.05 (our guess for this example). Therefore, the estimated plastic bending strength that will possibly give the required bending strength is 1.16(1.33)(1.05)(1152 in.-kips) = 1866 in.-kips. To maximize r_y, L_p, and L_r, try b = 8 in. The estimated plastic design bending strength requirement for the plate gives

$$(23.73 + t - 9.64t)t \geq \left[\frac{1866}{0.9(8)(36)/2} = 14.4\right]$$

$$t \geq 0.905 \text{ in.}$$

Check the λ_p requirement for a cover plate choice of b = 8 in. and t = 15/16 in.:

$$\left[\frac{b}{t} = \frac{8}{(15/16)} = 8.53\right] \leq \left(\frac{190}{\sqrt{36}} = 31.7\right)$$

Check the restriction on e [see Figure 5.26(d)]:

$$e \leq \left(\frac{d}{2} - t_f\right) \text{ is required}$$

$$e = \frac{A_p}{2t_w} = \frac{8(15/16)}{2(0.415)} = 9.04 \text{ in.}$$

$$(e = 9.04) \leq \frac{23.73}{2 - (15/16)} = 10.9 \quad (\text{OK})$$

For this built-up section,

$$\phi M_{px} = 478 + \phi\left(\frac{d}{2} + \frac{t}{2} - \frac{e}{2}\right)\left(\frac{A_p F_y}{12}\right) = 636 \text{ ft} - \text{kips}$$

$$A = 20.1 + 8(0.9375) = 27.6 \text{ in.}^2$$

$$I_y = 70.4 + \frac{0.9375(8)^3}{12} = 110 \text{ in.}^4$$

$$r_y = \sqrt{\frac{I_y}{A}} = \sqrt{\frac{110}{27.6}} = 2.00 \text{ in.}$$

$$L_p = \frac{300(2.00)}{\sqrt{36}} = 100 \text{ in.} = 8.33 \text{ ft}$$

To find L_r, we need to compute the following:
The ENA is located from the top of the section at,

$$\bar{y} = \frac{7.5(0.9375/2) + 20.1(23.73/2 + 0.9375)}{7.5 + 20.1} = 9.58 \text{ in.}$$

$$I_x = \frac{8(0.9375)^3}{12} + 7.5(9.58 - 0.9375/2)^2$$

$$+1830 + 20.1(23.73/2 + 0.9375 - 9.58)^2 = 2662 \text{ in.}^4$$

$$S_{xc} = \frac{2662}{9.58} = 278 \text{ in.}^3$$

$$S_{xt} = \frac{2662}{23.73 + 0.9375 - 9.58} = 176 \text{ in.}^3$$

$$J = 1.87 + \frac{8(0.9375)^3}{3} = 4.07 \text{ in.}^4$$

$$C_w = \frac{I_c I_t}{(I_c + I_t)} h^2$$

I_c = moment of inertia of the compression flange

I_t = moment of inertia of the tension flange

h = distance between the flange centroids

$h = d + t - t_f = 23.73 + 0.9375 - 0.585 = 24.08$ in.

$$I_t = \frac{0.585(8.9365)^3}{12} = 35.1 \text{ in.}^4$$

$$I_c = I_t + \frac{0.9365(8)^3}{12} = 75.1 \text{ in.}^4$$

$$C_w = 13870 \text{ in.}^6$$

$$X_1 = \frac{\pi}{S_{xc}} \sqrt{\frac{EAGJ}{2}} = 1582$$

$$X_2 = \frac{4C_w}{I_y} \left(\frac{S_{xc}}{GJ} \right)^2 = 0.0163$$

$$L_r = \frac{r_y X_1}{F_y - 16.5} \sqrt{1 + \sqrt{1 + X_2 \left(F_y - 16.5\right)^2}} = 311 \text{ in.} = 25.9 \text{ ft.}$$

$$(L_p = 8.33) < (L_b = 13.33 \text{ ft}) \le (L_r = 25.9)$$

$$\phi M_{rx} = \text{smaller of } \begin{cases} 0.9(F_y - 16.5)S_{xc} \\ 0.9 F_y S_{xt} \end{cases}$$

$$0.9(36 - 16.5)(278) = 4879 \text{ in.-kips} = 407 \text{ ft-kips}$$

$$0.9(36)(176) = 5702 \text{ in.-kips} = 475 \text{ ft-kips}$$

$$\phi M_{rx} = 407 \text{ ft-kips}$$

From LRFD Eq. (F1-3), since $C_b = 1$,

$$\phi M_{nx} = 636 - (636 - 407)\left(\frac{13.33 - 8.33}{25.9 - 8.33}\right) = 571 \text{ ft - kips}$$

$\phi M_{nx} \geq M_{ux}$ is required. Since $\phi M_{nx} = 571 \approx (M_{ux} = 574)$, use a 8 x 15/16 cover plate on the compression flange.

For the W24 x 68,

$$L_b = 13.33 \text{ ft and } \left[C_b \overline{M}_1 = 1.75(411) = 719\right] > \left(\phi M_{px} = 478 \text{ ft - kips}\right)$$

$$(\phi M_{nx} = 478) > \left[M_{ux} = 574 - (574)\left(\frac{13.33/2}{20}\right)^2 = 510\right]$$

and the W24 x 68 is not adequate in the $L_b = 13.33$ ft regions where a cover plate was not provided. Extend the cover plate to $a = 8.18$ ft each side of midspan as in Example 5.10.

The thicker part joined by welding is 15/16 in. Is the minimum S_w = 5/16 in. adequate? With respect to the centroid of the W section, the distance to the centroid of the built-up section is

$$\overline{y} = \frac{7.50(23.73 + 0.9375)/2}{7.50 + 20.1} = 3.35 \text{ in.}$$

$$\frac{23.73 + 0.9375}{2} - 3.35 = 8.98 \text{ in.}$$

$$I_x = 1830 + 20.1(3.35)^2 + 7.50(8.98)^2 = 2660 \text{ in.}^4$$

At a = 8.18 ft = 98.2 in. from midspan, V_u = 23.5 kips. Let P_{uw} denote the required strength per inch for each fillet weld. Since there are two fillet welds per cover plate,

$$P_{uw} = \frac{V_u Q}{2I_x} = \frac{23.5(7.50)(8.98)}{2(2660)} = 0.298 \text{ kips/in.}$$

For our base material of F_y = 36 ksi and for E70 electrodes, shear on the throat plane governs the design strength of a fillet weld. The design strength per inch of a fillet weld is

$$\phi F_w = 0.75\,[0.6F_{\text{Exx}}(0.707S_w)]$$
$$= 0.75\,(0.6)(70)(0.707S_w) = 22.27S_w$$

The weld design strength requirement is

$$(\phi F_w = 22.27S_w) \geq (P_{uw} = 0.298 \text{ kips/in.})$$

which gives $S_w \geq 0.0134$ in. Therefore, the minimum $S_w = 5/16$ in. is adequate. At midspan, the force in each cover plate is

$$P_u = \phi A_p F_y = 0.9(8)(0.9375)(36) = 243 \text{ kips}$$

At $a = 8.18$ ft from midspan, the force in the cover plate is

$$P_u = \left(\frac{478}{636}\right)(243) = 183 \text{ kips}$$

The total length of 5/16 in. weld required at the end of the cover plate to develop $P_u = 183$ kips is

$$\left[1.5(8) + 2L_w\right] = \left[\frac{P_u}{\phi F_w} = \frac{183}{22.27(0.3125)} = 26.3 \text{ in.}\right]$$

Each weld group consists of an 8 in. transverse weld and two longitudinal welds each of which has a length of L_w. Since a transverse weld is 1.5 times as strong as a longitudinal weld,

$$1.5(8) + 2L_w = 26.3 \text{ in.}$$
$$L_w = 7.15 \text{ in.}$$

Therefore, strength requires that the plate be extended by $L_w \geq 7.15$ in. on each end of the cover plate. However, since

$$\left(\frac{S_w}{t} = \frac{0.3125}{0.9375} = 0.333\right) < 0.75$$

LRFD B10 (p. 6-37) stipulates that the minimum cover-plate extension is

$$a' = 1.5(8) = 12 \text{ in.}$$

5.12 BIAXIAL BENDING OF SYMMETRIC SECTIONS

The design requirement for biaxial bending of a W section beam is LRFD Eq. (H1-1b). For $P_u = 0$ and $\phi_b = 0.9$,

$$\frac{M_{ux}}{\phi_b M_{nx}} + \frac{M_{uy}}{\phi_b M_{ny}} \leq 1.00$$

is required.

Example 5.13

In Figure 5.28, a W36 x 170 (F_y = 50 ksi) is used as a simply supported beam with regard to both principal axes. Lateral braces are provided only at the supports. The beam weight is negligible and the only load is 100 kips concentrated at midspan. The load passes through the shear center of the cross section, but the load is not parallel to either principal axis of the cross section. As shown in Figure 5.28, the components of the 100-kip load cause bending moments to occur about both principal axes. Does the described beam satisfy the design requirement for biaxial bending?

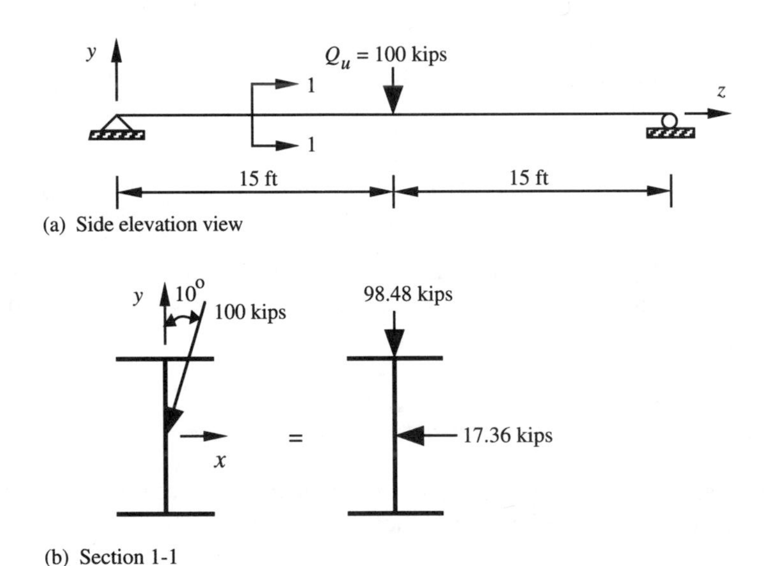

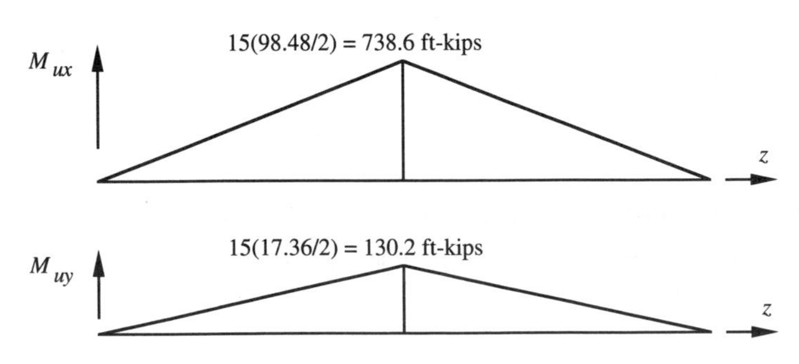

FIGURE 5.28 W section beam subjected to biaxial bending.

Solution

$$M_{ux} = 739 \text{ ft-kips}$$

$$C_b = 1; \quad L_b = 30 \text{ ft}; \quad \phi M_{nx} = 1236 \text{ ft-kips}$$

$$M_{uy} = 130 \text{ ft-kips}$$

For the y-axis design bending strength, see Eq. (5.5) and LRFD F1.7:

$$\phi M_{ny} = 0.9 Z_y F_y = 0.9(83.3)(50) = 3771 \text{ in.-kips} = 314 \text{ ft-kips}$$

$$\left(\frac{739}{1236} + \frac{130}{314} = 0.598 + 0.414 = 1.012 \right) > 1.00 \qquad (\text{NG})$$

Some structural designers might say $1.012 \approx 1.00$ is OK. However, in the process of finding ϕM_{nx}, we noticed that a W27 x 146 ($F_y = 50$ ksi) has slightly more ϕM_{nx} than the W36 x 170. Maybe the W27 x 146 will satisfy the design requirement for biaxial bending.

Try W27 x 146 ($F_y = 50$ ksi):

$$C_b = 1; \quad L_b = 30 \text{ ft}; \quad \phi M_{nx} = 1250 \text{ ft-kips} \quad (\text{from LRFD, p. 4-156})$$

$$\phi M_{ny} = 0.9 Z_y F_y = 0.9(97.5)(50) = 4387.5 \text{ in.-kips} = 366 \text{ ft-kips}$$

$$\left(\frac{739}{1250} + \frac{130}{336} = 0.591 + 0.355 = 0.946 \right) \leq 1.00 \qquad (\text{OK})$$

Use W27 x 146 ($F_y = 50$ ksi).

5.13 BENDING OF UNSYMMETRIC SECTIONS

A single angle with unequal legs and a Z section are examples of an unsymmetric section. The LRFD Specification does not give the design bending strength definition of a Z section, which is usually cold rolled. However, the LRFD Manual contains a separate LRFD Specification and Commentary for Single-Angle Members (pp. 6-277 to 6-300). We choose to illustrate the design checks for an unequal-leg angle used as a beam.

Example 5.14

Is an L6 x 4 x 1/2 section of A36 steel acceptable to use in Figure 5.29 as the simply supported beam? Lateral supports are provided only at the beam ends. The factored loading is $q_u = 0.7$ kips/ft and passes through the shear center.

Solution

In Figure 5.29, W is the major principal axis and Z is the minor principal axis. Note that $q_u = 0.7$ kips/ft is not parallel to either principal axis. The required bending strength and the nominal bending strength for each principal axis must be deter-

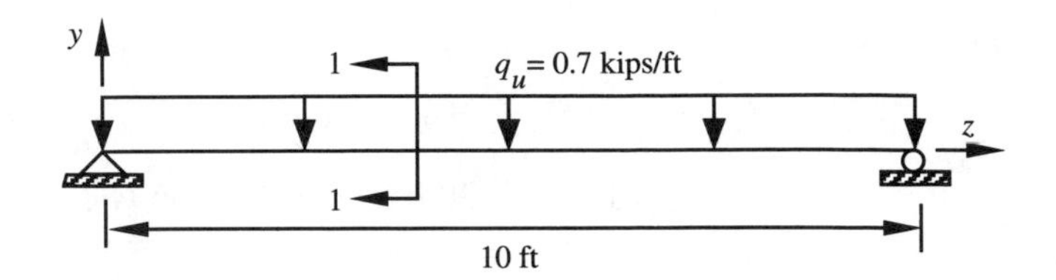

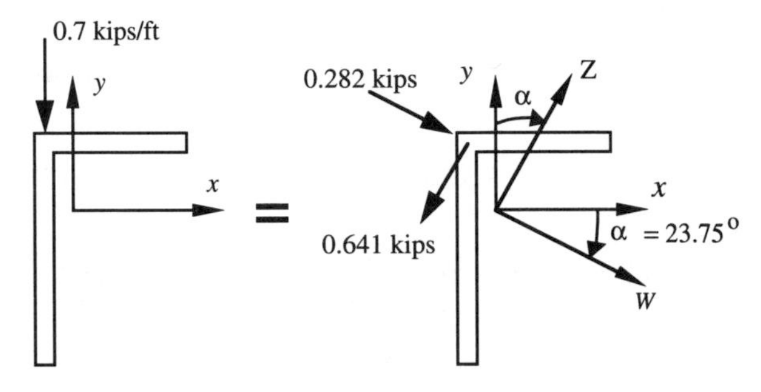

FIGURE 5.29 Bending of an unsymmetric section.

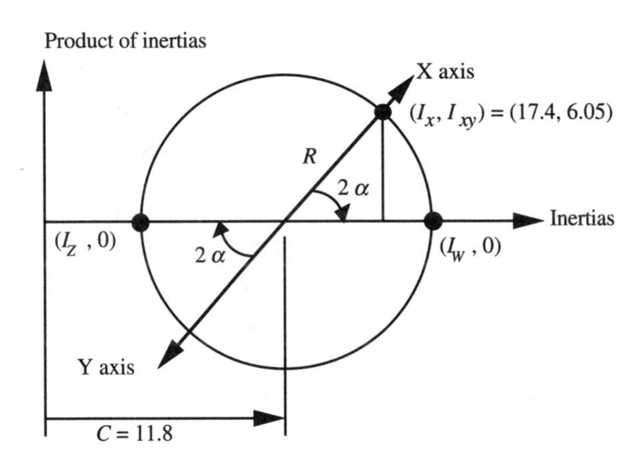

FIGURE 5.30 Mohr's circle of inertia for Example 5.15.

mined. Therefore, we must decompose the load into components parallel to each principal axis in order to find the required bending moment for each principal axis.

Appendix B is used in the calculations shown here to locate the principal axes and to find the section properties for each principal axis of the L6 x 4 x 1/2 (see Figure 5.30). For L6 x 4 x 1/2,

$$A = 4.75; \quad I_x = 17.4; \quad I_y = 6.27; \quad r_Z = 0.870$$

$$\tan\alpha = 0.440; \quad \alpha = 23.75°; \quad 2\alpha = 47.5°$$

$$I_Z = Ar_Z^2 = 4.75(0.870)^2 = 3.60 \text{ in.}^4$$

$$C = \frac{I_x + I_y}{2} = \frac{17.4 + 6.27}{2} = 11.8$$

$$R = C - I_Z = 11.8 - 3.60 = 8.2$$

$$R \sin 2\alpha = (8.2)(0.7372773) = 6.05$$

$$I_W = C + R = 11.8 + 8.2 = 20.0$$

As shown in Figure 5.31, the components of the factored load are

$$q_{uz} = (0.7 \text{ kips/ft}) \cos \alpha = 0.641 \text{ kips/ft}$$

$$q_{uw} = (0.7 \text{ kips/ft}) \sin \alpha = 0.282 \text{ kips/ft}$$

We illustrate how the distances to points A, B, and C in Figure 5.31 were computed:

1. The distance from the *W*-axis to point B is

$$(6 - 1.99)(\cos \alpha - 0.5) \sin \alpha = 3.87 \text{ in.}$$

2. The distance from the *Z*-axis to point B is

$$(6 - 1.99) \sin \alpha - (0.987 - 0.5) \cos \alpha = 1.17 \text{ in.}$$

The required bending strengths for the principal axes are

$$M_{uw} = \frac{(0.641 \text{ kips/ft})(10 \text{ ft})^2}{8} = 8.01 \text{ ft - kips} = 96.1 \text{ in. - kips}$$

$$M_{uz} = \frac{(0.282)(10)^2}{8} = 3.53 \text{ ft - kips} = 42.3 \text{ in. - kips}$$

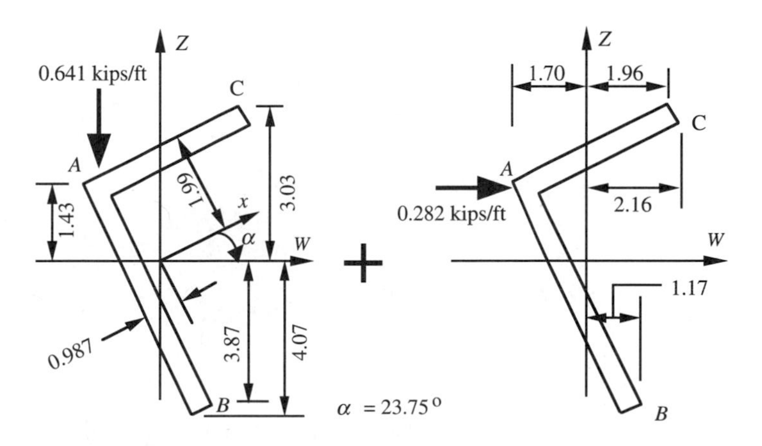

FIGURE 5.31 Loads parallel to principal axes in Example 5.14.

For Z-axis bending, the tip of each angle leg is in tension. From LRFD Eq. (5-2) (p. 6-284), the design bending strength is

$$\phi M_{nz} = [\phi M_{pz} = 0.9(1.25F_y S_{tz})]$$

$$M_{pz} = \text{plastic moment for } Z\text{-axis bending}$$

$$S_{tz} = \text{elastic section modulus for the extreme tension fiber}$$

$$S_{tz} = \frac{I_z}{2.16} = \frac{3.60}{2.16} = 1.67 \text{ in.}^3$$

$$\phi M_{pz} = 0.9(1.25)(36)(1.67) = 67.5 \text{ in.-kips}$$

For W-axis bending, the tip of the 4-in. leg is in compression.

$$\left(\frac{b}{t} = \frac{4}{0.5} = 8\right) \le \left(\frac{65}{\sqrt{F_y}} = \frac{65}{\sqrt{36}} = 10.833\right)$$

Local buckling does not govern ϕM_{nw}. From LRFD Eq. (5-1a) (p. 6-283), the plastic design moment for W-axis bending is

$$\phi M_{pw} = 0.9[1.25M_{(\text{yield})w}]$$

$$M_{(\text{yield})w} = F_y S_{cw}$$

$$S_{cw} = \text{elastic section modulus for the extreme compression fiber}$$

$$S_{cw} = \frac{I_w}{3.03} = \frac{20.0}{3.03} = 6.60 \text{ in.}^3$$

$$M_{yw} = (36)(6.60) = 237.6 \text{ in.-kips}$$

$$\phi M_{pw} = 0.9(1.25)(237.6) = 267.3 \text{ in.-kips}$$

LTB may govern ϕM_{nw}.

The elastic LTB moment from LRFD Eq. (5-6) (p. 6-286) is

$$M_{ob} = \frac{4.9EI_z C_b}{L_b^2}\left[\beta_w + \sqrt{\beta_w^2 + 0.052(L_b t/r_z)}\right]$$

where, from LRFD Table C5.1 (p. 6-298) for an L6 x 4,

$$\beta_w = 3.14$$

$$M_{ob} = \frac{4.9(29{,}000)(3.60)(1.0)}{[10(12)]^2}\left[3.14 + \sqrt{(3.24)^2 + 0.052(120)(0.5)/0.870}\right]$$

$$= 681.26 \text{ in.-kips}$$

$$(M_{ob} = 681.26) > (M_{yw} = 237.6)$$

Due to inelastic LTB, we find from LRFD Eq. (5-3b) (p. 6-284) that

$$\phi M_{nw} = \text{smaller of} \begin{cases} \phi M_{(\text{yield})\,w} \left[1.58 - 0.83\sqrt{M_{(\text{yield})\,w} / M_{ob}}\,\right] = 233 \text{ in.-kips} \\ \phi M_{pw} = 267.3 \text{ in.-kips} \end{cases}$$

Check the design requirement for biaxial bending [LRFD Eq. (61-b) p. 6-287]:

$$\left(\frac{M_{uW}}{\phi M_{nW}} + \frac{M_{uZ}}{\phi M_{nZ}} = \frac{96.1}{233} + \frac{42.3}{67.5} = 1.039\right) > 1.00 \qquad \text{(NG)}$$

An L6 x 4 x 1/2 section of A36 steel is not acceptable to use in Figure 5.29 as the simply supported beam.

5.14 WEB AND FLANGES SUBJECTED TO CONCENTRATED LOADS

In Figure 5.32, both flanges of a W section column are subjected to concentrated tension and compression forces. The applicable LRFD Specifications that must be satisfied are indicated in Figure 5.32. A pair of column web stiffeners will have to be designed for each region in which any of the indicated LRFD Specifications is not satisfied. When more than one of these indicated LRFD Specifications is not satisfied,

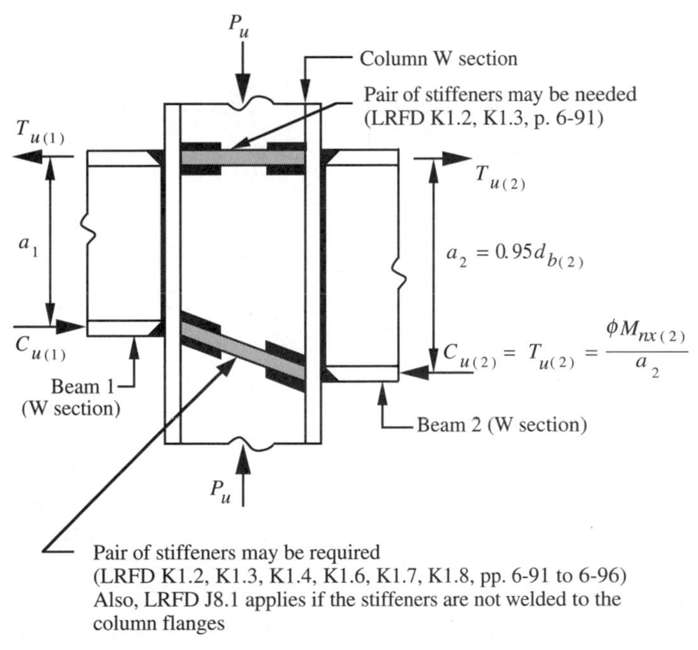

(a) Beam flanges welded to column flanges

See Figure C-K1.3 on LRFD page 6-234 for a set of beam and column forces that exist due to wind plus gravity loads.

(b) Another example of beam flanges welded to column flanges

FIGURE 5.32 Tension and compression forces perpendicular to W section flanges.

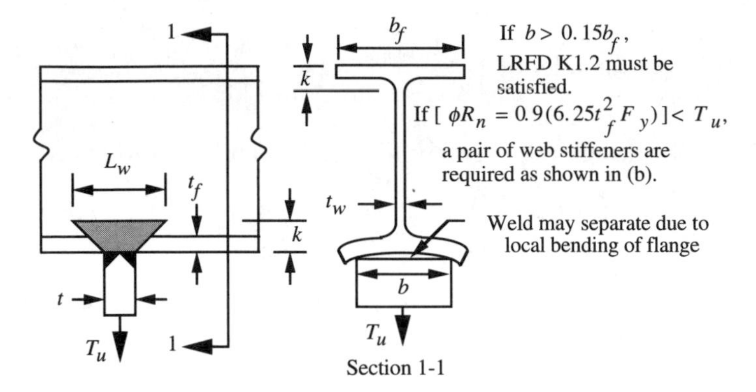

L_w = length along web in which local web yielding may occur

$L_w = 5k + t$ (as defined in LRFD K1.3, p. 6-92)

If [$\phi R_n = 1.0\ (5k + t) t_w F_y$] < T_u,
a pair of web stiffeners is required as shown in (b).

(a) Example of local flange bending (LRFD K1.2, p. 6-91)

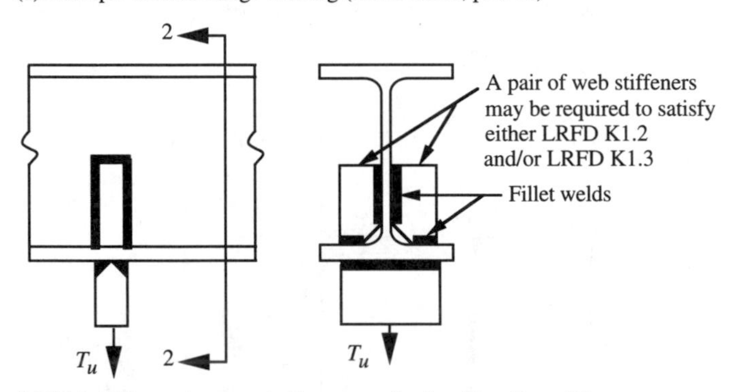

(b) Web stiffeners to prevent either excessive local bending of flange and/or excessive yielding of web

FIGURE 5.33 Tension force perpendicular to a W section flange.

the pair of column web stiffeners will have to be designed as either a tension member [at the top beam flanges in Figure 5.32(a)] or as a compression member [at the bottom beam flanges in Figure 5.32(a), see Section 5.10 for the design procedure] to resist the sum of the excess forces for the violated LRFD Specifications.

In Figure 5.33, a tension force is applied perpendicular to a W section beam flange. For clarity, we chose to show a single plate welded to the flange. However, the single plate could be one of the flanges of another W section whose member end is welded to the bottom flange of the W section shown. In that case, there would be two plates and two T_u forces. The web of the attached W section would also be welded to the bottom flange of the W section shown. The tension force from the attached web would not cause any local bending of the existing W section flange since the webs of both W sections would be in the same plane. However, the tension force from the attached web would have to be accounted for in checking the local web yielding design requirement (LRFD K1.3, p. 6-92). The stress distribution in the web due to T_u is assumed to be uniform on the section at k from the surface of the flange and acting on a *web area* = $t_w L_w$, where $L_w = 5k + t$ and t is the *thickness* of the attached plate. When LRFD K1.2 is not satisfied, a pair of web stiffeners will have to be designed as a tension member to provide the excess amount of the tension force. Also, if LRFD K1.3 is not satisfied, a pair of web stiffeners will have to be designed as a tension member to provide the excess amount of the tension force. When both LRFD K1.2 and K1.3 are not satisfied, a pair of web stiffeners will have to be designed as a tension member to provide the sum of the excess tension force from the indicated specifications.

In Figure 5.34, a W section beam is simply supported at the member ends on a wall. A steel bearing plate must be designed at Section 1-1 to spread the beam reaction uniformly on:

1. The wall beneath the bearing plate (see LRFD J9, p. 6-90). When the bearing plate is not provided,

$$\textit{bearing area} = 2k_1 \text{ (length of beam in contact with the wall)}$$

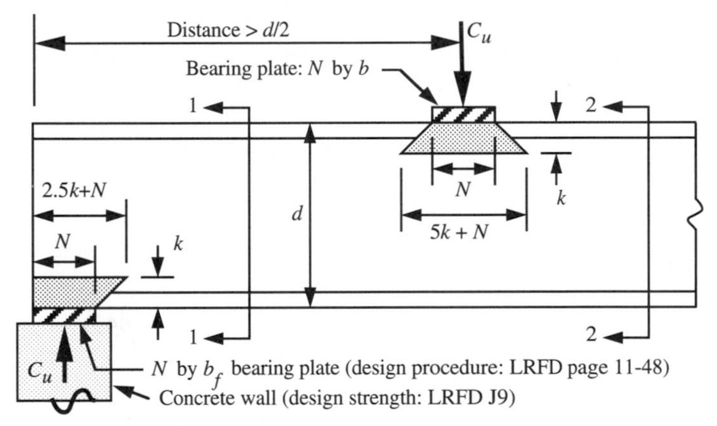

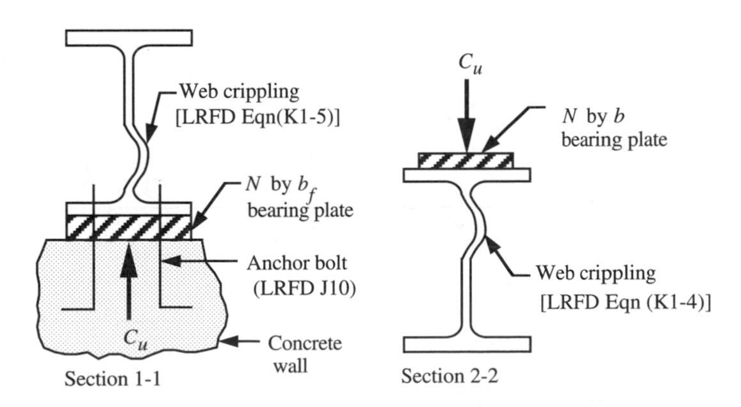

FIGURE 5.34 Compression force perpendicular to a W section flange.

2. The web at the toe of the fillet of the W section beam [see LRFD K1.3 item b, p. 6-92 and K1.4]. See LRFD p. 11-48 for the AISC-recommended procedure. The top flange of the beam must be laterally supported at the member ends (see Section 5.3). When some type of lateral bracing is not otherwise provided, either an end plate must be welded on the W section beam end or web stiffeners must be welded to the web and both flanges to provide lateral bracing for the top flange of the beam at the beam ends.

In Figure 5.34 at Section 2-2, the bearing plate is provided to spread the load uniformly on the web at the toe of the fillet of the W section beam. There is no AISC-recommended procedure for the design of this bearing plate. Alternatively, as shown in Section 5.15, bearing stiffeners [see LRFD K1.8, p. 6-96, and J8, item (a), p. 6-89] can be designed instead of a bearing plate.

Example 5.15

Lateral braces are provided only at the supports and at the concentrated load. There is not any limitation on deflection. The estimated factored beam weight is included in $w_u = 2.80$ kips/ft. Find the lightest W section of A36 steel that satisfies the LRFD Specification for bending strength (see Figure 5.35).

Solution

See Figure 5.36:
W30 x 99 ($F_y = 36$ ksi):

$$(\phi M_{px} = 842 \text{ ft-kips}) \geq (M_{ux} = 840)$$
$$(L_p = 8.8) < (L_b = 20) \leq (L_r = 25.5)$$

From LRFD beam design charts at $L_b = 20$ ft:

$$\phi M_{nx} = \text{smaller of} \begin{cases} C_b \overline{M}_1 = 1.40(628) = 880 \\ \phi M_{px} = 842 \text{ ft-kips} \end{cases}$$

$$(\phi M_{nx} = 842) \geq (M_{ux} = 840) \quad \text{(OK)}$$

Use W30 x 99 ($F_y = 36$ ksi)

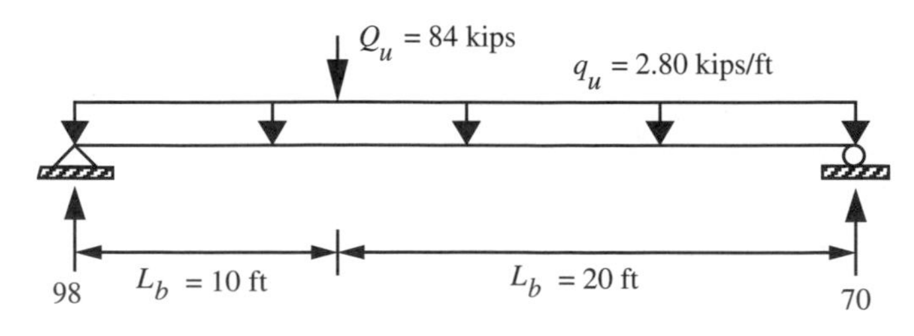

FIGURE 5.35 Load diagram for Example 5.15.

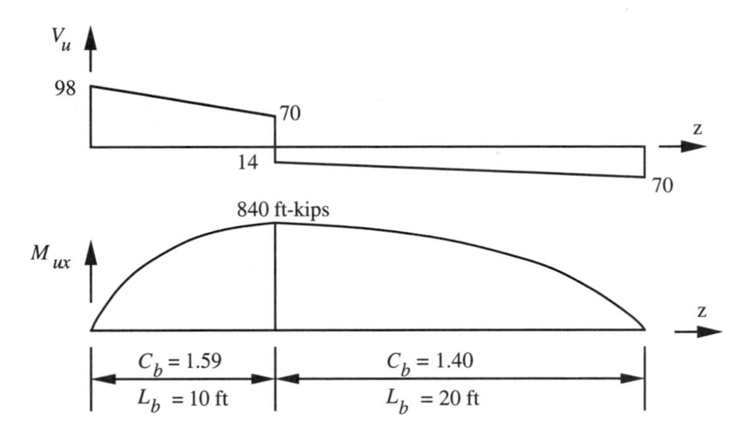

FIGURE 5.36 Shear and moment diagrams for Example 5.15.

Example 5.16

In Example 5.15, suppose that $Q_u = 84$ kips is due to the reaction of a W24 x 84 sitting on the top surface of the W30 x 99 as shown in Figure 5.37. Is a bearing plate needed between the W24 x 84 and W30 x 99? When the answer is yes, specify the value of bearing length N for each W section. Use $F_y = 36$ ksi.

Solution (a):

Check the web of W30 x 99 beneath W24 x 84 without a bearing plate between the flanges of the W sections.
W24 x 84

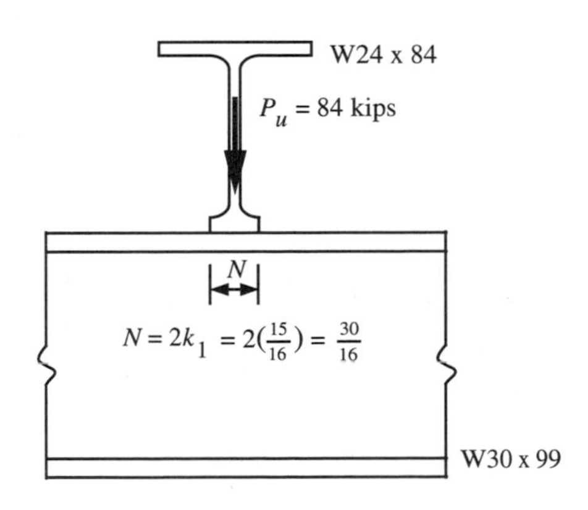

FIGURE 5.37 Sketch for Example 5.16.

$$k_1 = \tfrac{15}{16}; \quad k = 1\tfrac{9}{16}; \quad t_w = 0.470; \quad d = 24.10; \quad t_f = 0.770$$

$$N = 2k_1 = 2(0.9375) = 1.875 \text{ for W24 x 84 on the web of W30 x 99}$$

W30 x 99

$$k_1 = 1; \quad k = 1\tfrac{7}{16}; \quad t_w = 0.520; \quad d = 29.65; \quad t_f = 0.670$$

$$\frac{h}{t_w} = 51.9; \quad b_f = 10.45$$

$$N = 2k_1 = 2(1) = 2.00 \text{ for W30 x 99 on the web of W24 x 84}$$

For *local web yielding* (LRFD K1.3, p. 6-92) of W30 x 99, the design requirement is:

$$[\phi R_n = (5k + N)F_{yw}t_w] \geq (84 \text{ kips})$$

$$[\phi R_n = (5)(1.4375) + (1.875)(36)(0.520) = 170 \text{ kips}] \geq 84 \quad \text{(OK)}$$

For *web crippling* (LRFD K1.4, p. 6-92) of W30 x 99, the design requirement is:

$$\left\{ \phi R_n = 0.75(135)t_w^2 \left[1 + \frac{3N}{d}\left(\frac{t_w}{t_f}\right)^{1.5} \right] \sqrt{\frac{F_{yw}t_f}{t_w}} \right\} \geq 84 \text{ kips}$$

$$\phi R_n = 0.75(135)(0.520)^2 \left[1 + \frac{3(1.875)}{29.65}\left(\frac{0.520}{0.670}\right)^{1.5} \right] \sqrt{\frac{36(0.670)}{0.520}}$$

$$(\phi R_n = 210 \text{ kips}) \geq 84 \quad \text{(OK)}$$

As for *sidesway web buckling* (LRFD K1.5, p. 6-93) of W30 x 99, this specification is applicable when the bottom flange of W30 x 99 is not laterally braced at the $P_u = 84$ kip location. We show the calculations for illustration purposes:

$$\phi R_n = 0.75(135)(0.520)^2 \left[1 + \frac{3(1.875)}{29.65}\left(\frac{0.520}{0.670}\right)^{1.5} \right] \sqrt{\frac{36(0.670)}{0.520}}$$

$$(M_{ux} = 840 \text{ ft-kips}) > \left[S_x F_y = \frac{(269)(36)}{12} = 807 \text{ ft-kips}\right]$$

$$C_r = 480{,}000$$

$$\phi R_n = \frac{0.85 C_r t_w t_f}{(h/t_w)^2} \left[1 + 0.4\left(\frac{h/t_w}{L_b/b_f}\right)^3 \right]$$

$$= \frac{0.85(480{,}000)(0.520)(0.670)}{(51.9)^2} \left[1 + 0.4(2.26)^3 \right] = 296$$

$$(\phi R_n = 296 \text{ kips}) \geq 84 \quad \text{(OK)}$$

A bearing plate is not needed to satisfy any of the applicable LRFD Specifications for the web of the W30 x 99 at the P_u = 84 kip location.

Solution (b):

Check the W24 x 84 web without a bearing plate.
Assume that the P_u = 84 kips reaction is an end reaction (see Figure 5.38).
For *local web yielding* of W24 x 84, the design requirement is:

$$[\phi R_n = (2.5k + N)F_{yw}t_w] \geq (84 \text{ kips})$$

$$[\phi R_n = (2.5)(25/16) + 2)(36)(0.470) = 99.9] \geq 84 \quad \text{(OK)}$$

For *web crippling* of W24 x 84, the design requirement is:

$$\left\{\phi R_n = 0.75(68)t_w^2\left[1 + \frac{3N}{d}\left(\frac{t_w}{t_f}\right)^{1.5}\right]\sqrt{\frac{F_{yw}t_f}{t_w}}\right\} \geq 84 \text{ kips}$$

$$\left(\frac{N}{d} = \frac{2}{24.10} = 0.0830\right) \leq 0.2$$

$$\phi R_n = 0.75(68)(0.470)^2\left[1 + \frac{3(2)}{24.10}\left(\frac{0.470}{0.770}\right)^{1.5}\right]\sqrt{\frac{36(0.770)}{0.470}}$$

$$(\phi R_n = 96.8 \text{ kips}) \geq 84 \quad \text{(OK)}$$

Sidesway web buckling is not applicable since the top flange of the W24 x 84 must be laterally braced at the member end where the W24 x 84 is being checked due to the reaction.

A bearing plate is not needed to satisfy any of the applicable LRFD Specifications for the W24 x 84 web requirements.

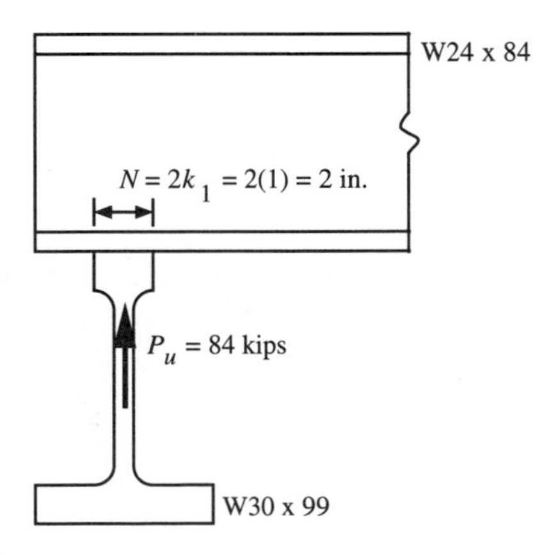

FIGURE 5.38 Sketch near a support in Example 5.16.

Example 5.17

In Example 5.15, suppose that the W30 x 99 is supported at each end on an 8-in. thick concrete wall (see Figure 5.39). Is a bearing plate needed between the W30 x 99 and the concrete walls? When the answer is yes, design the bearing plates. Concrete grade is $f_c' = 3$ ksi. Steel grade is $F_y = 36$ ksi.

Solution

Is a bearing plate needed at the left beam reaction?
Without a bearing plate, assume that the bearing length of the W30 x 99 web on the concrete wall is 6 in. and the bearing width $= 2k_1 = 2(1) = 2$ in. See the note at the end of this example for the bearing width assumption. From LRFD J9 (p6-90),

$$\phi_c P_p = 0.6\left(0.85 f_c' A_1 \sqrt{\frac{A_2}{A_1}}\right)$$

Use maximum allowed $\sqrt{\dfrac{A_2}{A_1}} = 2$

$$A_1 = N(2k_1) = 6(2) = 12 \text{ in.}^2$$

$$[\phi_c P_p = 0.6(0.85 f_c')(2A_1) = 0.6(0.85)(3)(2)(12) = 36.7] < 98$$

A bearing plate must be designed to protect the concrete wall. Use the recommended design procedure on LRFD p. 11-48.

See LRFD Table J3.4 (p. 6-82). Use $N \geq (2)(1.5) = 3$ in., which accommodates up to a 1.125-in.-diameter anchor bolt:

$$N = (\text{bearing length of W30 x 99 web on wall}) - b_1$$

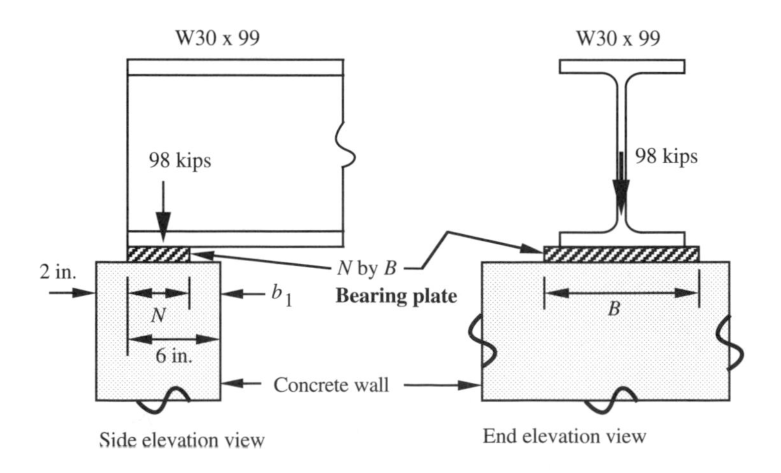

FIGURE 5.39 W section supported by a concrete wall.

$b_1 \geq 1$ in. (our choice); actual b_1 choice is made later

$$N \leq (6 - 1 = 5 \text{ in.})$$

Properties needed for the W30 x 99 from LRFD p. 4-42 are:

$$\phi R_1 = 2.5kF_{yw}t_w = 67.3 \text{ kips}$$

$$\phi R_2 = F_{yw}t_w = 18.7 \text{ kips}$$

$$\phi R_3 = 0.75(0.68)t_w^2 \sqrt{F_{yw}\left(\frac{t_f}{t_w}\right)} = 93.9 \text{ kips}$$

$$\phi R_4 = 0.75(68)t_w^2 \left[1 + \frac{3N}{d}\left(\frac{t_w}{t_f}\right)^{1.5}\right]\sqrt{F_{yw}\left(\frac{t_f}{t_w}\right)} = 6.50 \text{ kips}$$

For *local web yielding*, the design requirement is

$$N \geq \left(\frac{R - \phi R_1}{\phi R_2} = \frac{98 - 67.3}{18.7} = 1.64 \text{ in.}\right)$$

For *web crippling*, the design requirement is

$$N \geq \left(\frac{R - \phi R_3}{\phi R_4} = \frac{98 - 93.9}{6.50} = 0.631 \text{ in.}\right)$$

Use $B \geq (b_f = 10.45$ for W30 x 99) and $6 \geq N \geq 3$.
Try $B = 11$ in.; $A_1 = BN = 11N$. From LRFD J9 (p. 6-90),

$$\phi_c P_p = 0.6\left(0.85 f_c' A_1 \sqrt{\frac{A_2}{A_1}}\right)$$

Use maximum allowed $\sqrt{\dfrac{A_2}{A_1}} = 2$

$$\phi_c P_p = 0.6(0.85f_c)(2A_1) = 0.6(0.85)(3)(2)(11N) = 33.66N$$

$$(\phi_c P_p = 33.66N) \geq 98 \quad \text{is required}$$

$$N \geq 2.91 \text{ in.} \quad \text{is required}$$

LRFD p. 11-50 states that preferably B and N should be in full inches. For $N = 3$ in. and $B = 11$ in.,

$$A_1 = 3(11) = 33 \text{ in.}^2$$

$$n = \left(\frac{B}{2} - k = \frac{11}{2} - 1\tfrac{7}{16} = 4.0625 \text{ in.}\right)$$

$$t \geq \left[n\sqrt{\frac{2.22R}{A_1 F_y}} = 4.0625\sqrt{\frac{2.22(98)}{33(36)}} = 1.74 \text{ in.} \right] \quad \text{is required}$$

From LRFD p. 1-133, we find that when

$$(\text{Width} = 3 \text{ in.}) \leq 6 \quad \text{and} \quad (\text{thickness} = 1.75 \text{ in.}) \geq 0.203$$

the bearing plate is classified as a bar and the preferred thickness increment = 1/8 in. Use an (N = 3) x (B = 11) x (t = 1.75) bearing plate at each reaction to protect the concrete wall. Volume = 3(11)(1.75) = 57.8 in.^{3} Alternatively, we could choose $N = 4$; $B = 11$; $A_1 = 4(11) = 44$ in.^{2} Then,

$$t \geq \left[n\sqrt{\frac{2.22R}{A_1 F_y}} = 4.0625\sqrt{\frac{2.22(98)}{44(36)}} = 1.51 \text{ in.} \right] \quad \text{is required.}$$

A 4 x 11 x 1.5 bearing plate would be acceptable. Volume = 4(11)(1.5) = 66 in.^{3} For the least weight, use the 3 x 11 x 1.75 bearing plate.

Note: Without a bearing plate, we chose to say that the *bearing width* = $2k_1$ = 2 in. The reason is that the W30 x 99 flange is not perfectly flat; the flange either curls up or down at each cross section. If the flange curls down, it will flatten out when the beam reaction occurs. However, if the flange curls up, the nut on the anchor bolts would have to be tightened enough to make the flange flatten out. For a simply supported beam, this is not desirable on one end of the beam since friction between the concrete wall and the bottom flange of the W30 x 99 would prevent the assumed roller end from sliding until friction was overcome. Thus, when the bottom flange curls up, only the $2k_1$ width portion of the W30 x 99 beam is in contact with the wall along the 6-in. bearing length. If we deemed it appropriate to assume that the bearing width $B = b_f$ = 10.5 in., then

$$\phi_c P_p = 0.6(0.85)(3)(6)(10.5) = 96.4 \text{ kips} \approx 98 \quad \text{(OK)}$$

$$\frac{B}{2} - k = \frac{10.5}{2} - 1\tfrac{7}{16} = 3.8125 \text{ in.}$$

$$t_f \geq \left[n\sqrt{\frac{2.22R}{A_1 F_y}} = 3.8125\sqrt{\frac{2.22(98)}{6(10.5)(36)}} = 1.18 \text{ in.} \right] \quad \text{is required}$$

However, (t_f = 0.670 in.) < 1.18 in. and the W30 x 99 flange thickness is not thick enough to serve as the bearing plate.

5.15 BEARING STIFFENERS

As shown in Figure 5.40, a pair of bearing stiffeners may be required for beams and girders at the supports and at concentrated load points. Let C_u denote the concentrated compressive force. When $C_u > (\phi R_n$ for any of the design strength definitions in LRFD K1.3 through K1.5), a pair of bearing stiffeners is required. LRFD K1.8 (p. 6-96) stipulates the design requirements for a pair of web stiffeners. The stiffeners must be welded to the web of the member being stiffened. For a tensile-concentrated load, the stiffeners must be welded to the loaded flange. At unframed ends of beams and girders, the compression flange must be laterally supported. If the tension flange is adequately anchored to a load-bearing wall as shown in Figure 5.34, for example, the pair of stiffeners at the member end in Figure 5.40 can also provide lateral bracing for the top flange. In that case, the pair of stiffeners must either bear on or be welded to the loaded flange and must be welded to the unloaded flange. If the concentrated compressive load at an unframed member end or at an interior location exceeds ϕR_n given in LRFD K1.4 or K1.6, the stiffeners must be designed as an axially loaded column using LRFD E2:

$$KL = 0.75h$$

$$A_g = (2b + A_w)tb = \text{width of each plate in the pair of stiffeners}$$

$$t = \text{thickness of each plate stiffener}$$

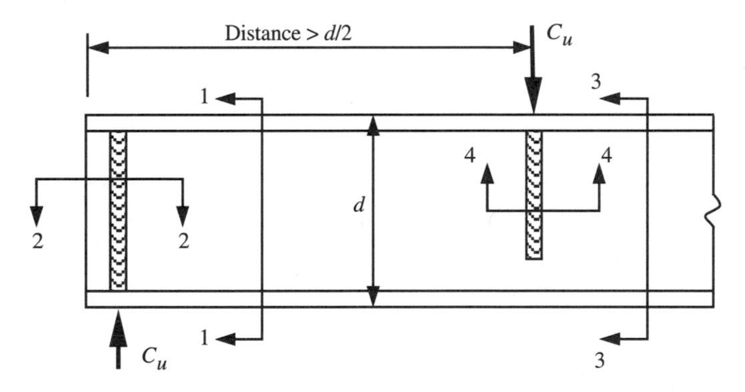

LRFD K1.3 [Eqn (K1-3)], K1.4 apply at the support. If any of the $\phi R_n < C_u$, a pair of web stiffeners is required (Sect. 1-1).

LRFD K1.3 [Eqn (K1-2)], K1.4, K1.5 apply at the interior load. If any of the $\phi R_n < C_u$, a pair of web stiffeners is required (Sect. 3-3).

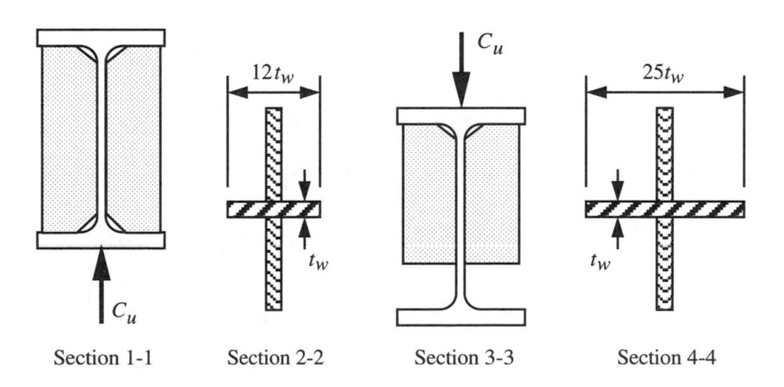

FIGURE 5.40 Bearing stiffeners.

A_w = the area of the beam web that is effective in participating with the pair of stiffeners as a column.

At a member-end stiffener,

$$A_w = 12t_w$$

t_w = thickness of the beam web

At an interior stiffener,

$$A_w = 25t_w$$

In computing the moment of inertia of this column, A_w can be ignored, but A_w should not be ignored in computing the radius of gyration. If the concentrated load at an interior location exceeds ϕR_n given in LRFD K1.2 or K1.3, the stiffeners must extend at least to middepth of the beam. For a compressive load, these stiffeners must be designed as an axially loaded column using LRFD E2 and $A_g = (2b + A_w)t$.

For bearing stiffeners that are not welded to the compression flange, LRFD J8, item (a), p. 6-89, is applicable and requires

$$(\phi R_n = 0.75)(2.0F_y A_{pb}) \geq C_u$$

A_{pb} = projected bearing area

Example 5.18

The beam in Figure 5.40 is a W30 x 99 and C_u = 220 kips is the end reaction of a W24 x 84 sitting on the top surface of the W30 x 99 without any bearing plate between the flanges of the two members. Design a pair of bearing stiffeners at the interior concentrated load point of the W30 x 99. Use F_y = 36 ksi.

Solution

W24 x 84

$$k_1 = \tfrac{15}{16}; \quad k = 1\tfrac{9}{16}; \quad t_w = 0.470; \quad d = 24.10; \quad t_f = 0.770$$

$$\frac{h}{t_w} = 45.9, \; h = 45.9t_w = 21.573 \text{ in.}$$

$$N = 2k_1 = 2(0.9375) = 1.875 \text{ for W24 x 84 on the web of W30 x 99}$$

W30 x 99

$$k_1 = 1; \quad k = 1\tfrac{7}{16}; \quad t_w = 0.520; \quad d = 29.65; \quad t_f = 0.670$$

$$\frac{h}{t_w} = 51.9, \; h = 45.9t_w = 26.99 \text{ in.}$$

$$N = k_1 = 2(1) = 2.00 \text{ for W30 x 99 on the web of W24 x 84}$$

For *local web yielding* (LRFD K1.3, p. 6-92) of W30 x 99, the design requirement is

$$[\phi R_n = (5k + N)F_{yw}t_w] \geq 220 \text{ kips}$$

$$[\phi R_n = (5)(1.4375) + (30/16)(36)(0.520) = 170 \text{ kips}] < 220 \quad \text{(NG)}$$

A pair of bearing stiffeners is required.

For *web crippling* (LRFD K1.4, p. 6-92) of W30 x 99, the design requirement is

$$\left\{ \phi R_n = 0.75(135)t_w^2 \left[1 + \frac{3N}{d}\left(\frac{t_w}{t_f}\right)^{1.5} \right] \sqrt{\frac{F_{yw}t_f}{t_w}} \right\} \geq 220 \text{ kips}$$

$$\phi R_n = 0.75(135)(0.520)^2 \left[1 + \frac{3(1.875)}{29.65}\left(\frac{0.520}{0.670}\right)^{1.5} \right] \sqrt{\frac{36(0.670)}{0.520}}$$

$$(\phi R_n = 210 \text{ kips}) < 220 \quad \text{(NG)}$$

A pair of bearing stiffeners is required. They must be designed such that LRFD E2 is satisfied.

For a pair of plates, the bearing area required by LRFD J8 p. 6-89 is

$$A_{pb} \geq \left(\frac{P_u}{0.75(1.8F_y)} = \frac{220}{0.75(1.8)(36)} = 4.53 \text{ in.}^2 \right)$$

$$\frac{b_f - t_w}{2} = \frac{10.45 - 0.520}{2} = 4.97 \text{ in.}$$

For each plate, use $b = 4.5$ in.

$$\left[A_{pb} = (b - k_1)t = (4.5 - 1.0)t \right] \geq \left(\frac{4.53}{2} = 2.27 \text{ in.}^2 \right)$$

is required. Therefore, $t \geq 0.647$ in. is required. Try $t = 0.625$ in.:

$$\left(\frac{b}{t} = \frac{4.5}{0.625} = 7.2 \right) \leq \left(\frac{95}{\sqrt{36}} = 15.8 \right)$$

$$A_g = (2b + 25t_w)t = [2(4.5) + 25(0.520)](0.625) = 13.75 \text{ in.}^2$$

$$I = \frac{t(2b + t_w)^3}{12} = \frac{0.625[2(4.5) + 0.520]^3}{12} = 44.9 \text{ in.}^4$$

$$r = \sqrt{\frac{I}{A}} = \sqrt{\frac{44.9}{13.75}} = 1.81 \text{ in.}$$

$$KL = 0.75h = 0.75(26.75) = 20.1 \text{ in.}$$

$$\left[\lambda_c = \frac{KL}{r\pi}\sqrt{\frac{F_y}{E}} = \frac{20.1}{1.81\pi}\sqrt{\frac{36}{29{,}000}} = 0.1243 \right] < 1.5$$

$$\lambda_c^2 = 0.1545$$

$$[\phi P_n = 0.85(13.75)(0.658)^{0.1545}(36) = 418 \text{ kips}] \geq (P_u = 220)$$

Use a pair of 4.5 x 0.625 x 28.25 plates for the bearing stiffener. Use the minimum permissible fillet weld sizes.

Example 5.19

The beam in Figure 5.40 is a W30 x 99 and $C_u = 220$ kips is the end reaction of a W24 x 84 sitting on the top surface of the W30 x 99 without any bearing plate between the flanges of the two members. Design a pair of bearing stiffeners at the unframed member end of the W24 x 84. Use $F_y = 36$ ksi.

Solution

For *local web yielding* of W24 x 84, the design requirement is

$$[\phi R_n = (2.5k + N)F_{yw}t_w] \geq 220 \text{ kips}$$

$$\phi R_n = [2.5(1.5625) + 2](36)(0.470) = 99.9$$

$$[\phi R_n = 99.9] < 220 \quad \text{(NG)}$$

A pair of bearing stiffeners is required.
For *web crippling* of W24 x 84, the design requirement is

$$\left\{ \phi R_n = 0.75(68)t_w^2 \left[1 + \frac{3N}{d}\left(\frac{t_w}{t_f}\right)^{1.5} \right] \sqrt{\frac{F_{yw}t_f}{t_w}} \right\} \geq 220 \text{ kips}$$

$$\phi R_n = 0.75(68)(0.470)^2 \left[1 + \frac{3(2)}{24.10}\left(\frac{0.470}{0.770}\right)^{1.5} \right] \sqrt{\frac{36(0.770)}{0.470}}$$

$$(\phi R_n = 96.8 \text{ kips}) < 220 \quad \text{(NG)}$$

A pair of bearing stiffeners is required and must be designed such that LRFD E2 is satisfied.

For a pair of plates, the bearing area required by LRFD J8, item (a), p. 6-89 is

$$A_{pb} \geq \left(\frac{P_u}{0.75(1.8F_y)} = \frac{220}{0.75(1.8)(36)} = 4.53 \text{ in.}^2 \right)$$

$$\frac{b_f - t_w}{2} = \frac{9.45 - 0.470}{2} = 4.28 \text{ in.}$$

For each plate, use $b = 4.25$ in.

$$\left[A_{pb} = (b - k_1)t = (4.5 - 1.0)t \right] \geq \left(\frac{4.53}{2} = 2.27 \text{ in.}^2 \right)$$

is required. Therefore, $t \geq 0.616$ in. is required. Try $t = 0.625$ in.

$$\left(\frac{b}{t} = \frac{4.25}{0.625} = 6.8\right) \leq \left(\frac{95}{\sqrt{36}} = 15.8\right)$$

$$A_g = (2b + 12t_w)t = [2(4.25 + 12(0.470)](0.625) = 8.84 \text{ in.}^2$$

$$I = \frac{t(2b+t_w)^3}{12} = \frac{0.625[2(4.25)+0.470]^3}{12} = 37.6 \text{ in.}^4$$

$$r = \sqrt{\frac{I}{A}} = \sqrt{\frac{37.6}{8.84}} = 2.06 \text{ in.}$$

$$KL = 0.75h = 0.75(21) = 15.8 \text{ in.}$$

$$\left[\lambda_c = \frac{KL/r}{\pi}\sqrt{\frac{F_y}{E}} = \frac{15.8/2.06}{\pi}\sqrt{\frac{36}{29{,}000}} = 0.0860\right] < 1.5$$

$$\lambda_c^2 = 0.00740$$

$$[\phi P_n = 0.85(8.84)(0.658)^{0.00740}(36) = 270 \text{ kips}] \geq (P_u = 220)$$

Use a pair of 4.25 x 0.625 x 22 plates for the bearing stiffener. Use the minimum permissible fillet weld size.

PROBLEMS

Notation

In all problems, q_u includes the factored beam weight. If a value of q_u is not given, ignore the beam weight.

5.1 In Figure P5.1, the built-up section consists of three plates continuously and adequately welded together. The 0.75 in. x 10 in. top flange is the compression flange.

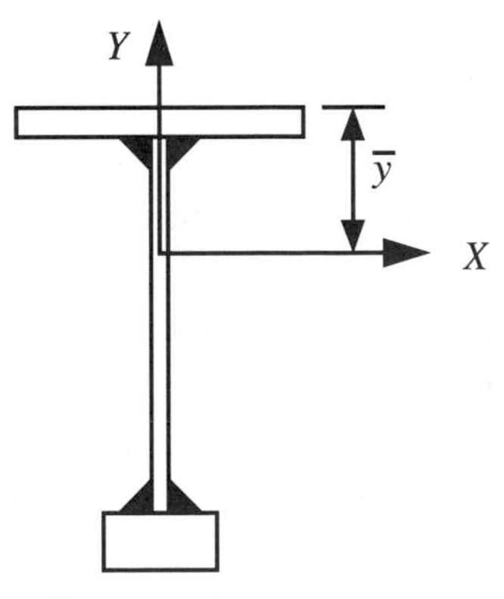

Cross section

FIGURE P5.1

The tension flange is 1.5 in. x 4 in. The web plate is 0.5 in. x 14 in.. Find ϕM_{px} and ϕM_{py} for $F_y = 50$ ksi. <u>*Reminder*</u>: M_p for each principal axis cannot exceed $1.5F_yS$ where S is the applicable elastic section modulus. $S_{xt} = 98.5$ in.^{3} and $S_y = I_y / [0.5(10)]$. (Answer: $\phi M_{px} = 470$ ft-kips.)

5.2 For a W21 x 44(see LRFD, p. 1-32), ignore the fillets and find ϕM_{px} and ϕM_{py} for $F_y = 36$ ksi. Compute $Z_x = M_{px} / F_y$ and $Z_y = M_{py} / F_y$ and compare them to the values listed on LRFD p. 1-33. (Answer: $Z_x = M_{px} / F_y = 92.27$ in.^{3})

5.3 For a WT7 x 45(see LRFD p. 1-78), ignore the fillets and find ϕM_{px} and ϕM_{py} for $F_y = 36$ ksi. For ϕM_{px} , assume that the stem is in tension. Compute Z_x and Z_y and compare them to the values listed on LRFD p. 1-33. (Answer: $Z_x = M_{px} / F_y = 11.37$ in.^{3})

5.4 Simply supported beam. Span = 36 ft; $q_u = 3$ kips/ft; $L_b \le L_p$. No restriction on deflection. Find the lightest acceptable W section that satisfies LRFD F1 for:

(a) $F_y = 36$ ksi

(b) $F_y = 50$ ksi

(c) $F_y = 65$ ksi

For each chosen W section, find L_p.

5.5 Simply supported beam. Span = 48 ft; $q_u = 1.5$ kips/ft; $L_b \le L_p$. No restriction on deflection. Find the lightest acceptable W section that satisfies LRFD F1 for:

(a) $F_y = 36$ ksi

(b) $F_y = 50$ ksi

(c) $F_y = 65$ ksi

For each chosen W section, find L_p.

5.6 Simply supported beam. Span = 36 ft; $q_u = 3$ kips/ft; $L_b \le L_p$. Deflection due to service live load = 1.25 kips/ft is not to exceed span/240. Find the lightest acceptable W section that satisfies LRFD F1 for:

(a) $F_y = 36$ ksi

(b) $F_y = 50$ ksi

(c) $F_y = 65$ ksi

For each chosen W section, find L_p.

5.7 Simply supported beam. Span = 48 ft; $q_u = 1.5$ kips/ft; $L_b \le L_p$. Deflection due to service live load = 0.625 kips/ft is not to exceed span/240. Find the lightest acceptable W section that satisfies LRFD F1 for:

(a) $F_y = 36$ ksi

(b) $F_y = 50$ ksi

(c) $F_y = 65$ ksi

For each chosen W section, find L_p.

5.8 Simply supported beam. Span = 36 ft; q_u = 3 kips/ft; L_b = 12 ft. No restriction on deflection. Find the lightest acceptable W section that satisfies LRFD F1 for:

(a) F_y = 36 ksi

(b) F_y = 50 ksi

5.9 Simply supported beam. Span = 48 ft; q_u = 1.5 kips/ft; L_b = 16 ft. No restriction on deflection. Find the lightest acceptable W section that satisfies LRFD F1 for:

(a) F_y = 36 ksi

(b) F_y = 50 ksi

5.10 Simply supported beam. Span = 36 ft; q_u = 3 kips/ft; L_b = 18 ft. No restriction on deflection. Find the lightest acceptable W section that satisfies LRFD F1 for:

(a) F_y = 36 ksi

(b) F_y = 50 ksi

5.11 Simply supported beam. Span = 48 ft; q_u = 1.5 kips/ft; L_b = 12 ft. No restriction on deflection. Find the lightest acceptable W section that satisfies LRFD F1 for:

(a) F_y = 36 ksi

(b) F_y = 50 ksi

5.12 For Figure P5.12 and F_y = 36 ksi, find the lightest acceptable W section.

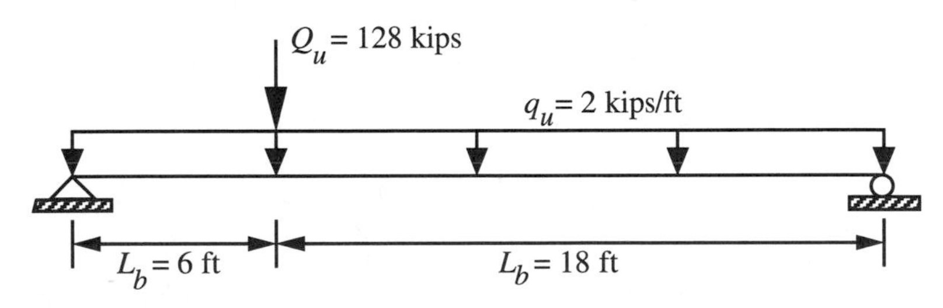

FIGURE P5.12

5.13 For Figure P5.13 and F_y = 36 ksi, find the lightest acceptable W section. Lateral braces are provided only at the supports and at the cantilever tip.

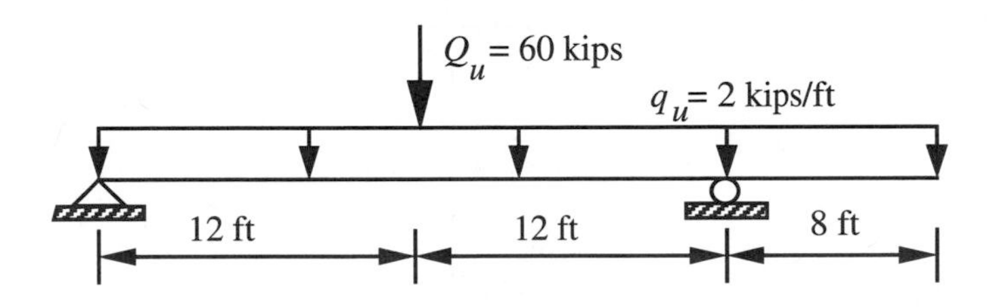

FIGURE P5.13

5.14 Is a W24 x 76 acceptable for the loading and L_b values shown in Figure P5.12 when F_y = 50 ksi?

Q_u is due to the interior reaction of a W27 x 84 sitting on the top flange of the W24 x 76. Is a bearing plate required to protect the web of the W27 x 84? If the answer is yes, find the minimum acceptable value of N. Is a bearing plate required to protect the web of the W24 x 76? If the answer is yes, find the minimum acceptable value of N.

Each support for the W24 x 76 is an 8-in.-thick concrete wall of f_c' = 3 ksi. For N = 6 in., use the recommended design procedure on LRFD p. 3-49 and design the bearing plate for the left reaction.

5.15 The member in Figure P5.15 is a C12 x 20.7 for which F_y = 50 ksi, L_b = 8 ft, and Q_u is applied through the shear center. Ignore the beam weight. Find the maximum acceptable value of Q_u.

5.16 The built-up section properties given in Problem 5.1 for Figure P5.1 are applicable for a beam loaded as shown in Figure P5.16. Additional section properties are as follows: PNA is 10 in. from the bottom of the section and ϕM_{px} = 470 ft-kips for F_y = 50 ksi. ENA is 7.32 in. from the top of the section, r_y = 1.86 in., $S_{xc} = I_x/7.32$, and $S_{xt} = I_x/(16.25 - 7.32)$. The shear center is 2.10 in. from the top of the section, J = 6.50 in.^{4}, and C_w = 1622 in.^{4} The three plates are continuously and adequately fillet welded along the member length. Is the design acceptable for the loading shown in Figure P5.16?

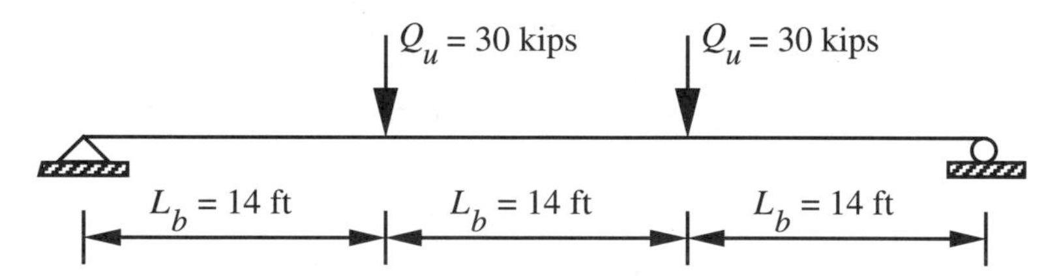

Cross section is shown in Figure P5.1 and defined in Problem 5.1

FIGURE P5.16

5.17 In Figure P5.17, Q_{ux} and Q_{uy} are located at midspan of a simply supported W24 x 131. L_b = span = 20 ft. Is the design acceptable for F_y = 36 ksi?

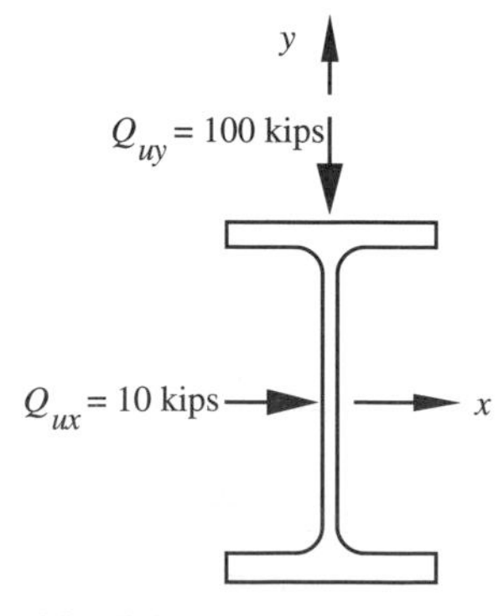

FIGURE P5.17

5.18 Solve Problem 5.18 for W30 x 116 and F_y = 36 ksi.

5.19 Solve Problem 5.19 for W27 x 94 and F_y = 50 ksi.

5.20 The Z section loaded as shown in Figure P5.20 is used as a simply supported beam; L_b = span = 12 ft. Does the Z section satisfy the LRFD bending strength requirements for F_y = 36 ksi?

For Figure P5.20, the following information is applicable:

1. The shear center and plastic centroid coincide with the elastic centroid.
2. For the major principal axis, ϕM_{pw} = 37.1 ft-kips.
3. For the minor principal axis, ϕM_{pz} = 11.2 ft-kips.

Verify the following properties:

A = 6.00 in.^{2}; I_x = 31.75 in.^{4}; I_y = 11.5 in.^{4}; I_{xy} = -14.44 in.^{4}

I_w = 39.3 in.^{4}; I_z = 3.99 in.^{4}; r_z = 0.815 in.

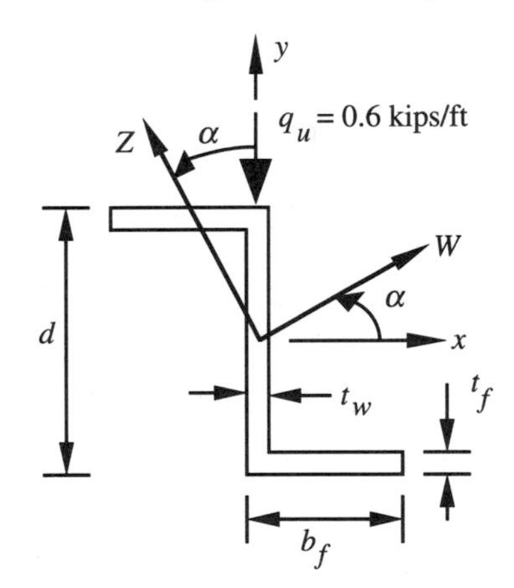

b_f = 3.50 in.; t_f = 0.500 in.
d = 6.00 in.; t_w = 0.500 in.
$h = d - t_f$ = 5.50 in.
$b = b_f - t_w/2$ = 3.25 in.
$t = t_f = t_w$ = 0.500 in.

$$J = t^3(2b + h)/3$$

$$C_w = \frac{b^3 h^2 t}{12}\left(\frac{b + 2h}{2b + h}\right)$$

FIGURE P5.20

5.21 The Z section in Problem 5.20 is simply supported on rafters and loaded as shown in Figure P5.21. The rafters are 18 ft on center; that is, L_b = span = 18 ft for the Z section. Does the Z section satisfy the LRFD bending strength requirements for F_y = 36 ksi?

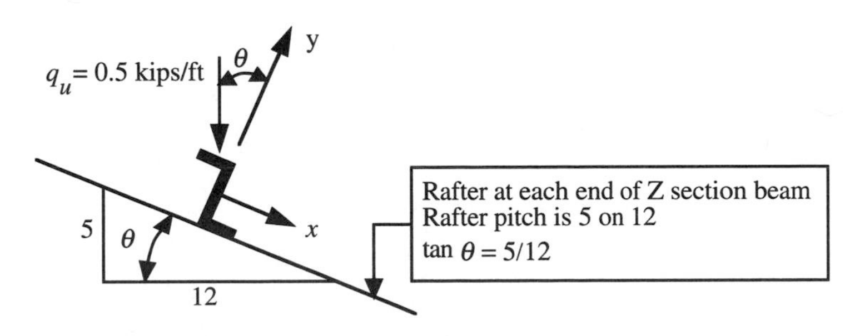

FIGURE P5.21

5.22 Solve Problem 5.21 for 1 on 2 pitch and L_b = span = 20 ft.

5.23 Solve Problem 5.22 for $F_y = 50$ ksi and $q_u = 0.7$ kips/ft.

5.24 For $F_y = 36$ ksi, find ϕM_{nx} due to FLB of a W53 x 18 x 143 (see LRFD, p. 4-184). For this welded, built-up section, we find for the first shape on LRFD p. 6-114 that

$$\lambda_r = \frac{106}{\sqrt{F_{yw} - 16.5}}$$

5.25 Solve Problem 5.24 for $F_y = 50$ ksi.

5.26 For $F_y = 36$ ksi, find ϕM_{nx} due to WLB of a W45 x 16 x 183 (see LRFD, p. 4-184). For this welded, built-up section, we find for the first shape on LRFD p. 6-114 that

$$\lambda_r = \frac{970}{\sqrt{F_{yf}}}$$

5.27 Solve Problem 5.26 for $F_y = 50$ ksi.

5.28 For a W49 x 16 x 129 (see LRFD, p. 4-184) and $F_y = 36$ ksi, find

(a) ϕM_{nx} due to FLB,

(b) ϕM_{nx} due to WLB

(c) The minimum value of L_b for which LTB governs ϕM_{nx} when $C_b = 1$.

Sketch the ϕM_{nx} vs. L_b beam chart for $C_b = 1$; show the values you computed in (a-c) on the sketch.

5.29 Solve Problem 5.28 for $F_y = 50$ ksi.

CHAPTER

6

Members Subject to Bending and Axial Force

6.1 INTRODUCTION

In this chapter, we discuss the behavior and design of members in frames for which LRFD Chapters C (p. 6-41) and H (p. 6-59) are applicable.

Figure 6.1(a) is an example of an industrial building during the early stages of construction. After the steel framework is erected, a flat roof, sidewalls (XY-planes), and a roll-up door are to be installed in each endwall (YZ-planes). The roll-up doors serve two functions: (1) as the endwalls when the doors are closed and (2) as openings to permit vehicles to enter the building. For discussion purposes, assume that the diagonal members are either single angles or threaded rods and all other members are W sections.

In Figure 6.1(b), the weak axis of the columns (vertical members) and the strong axis of the beams (horizontal members) are chosen as the bending axis. Since the weak-axis bending strength of the columns is much less than the strong-axis bending strength of the beams, all members in Figure 6.1(b) are connected such that a negligible moment is transferred between the member ends at each joint. Since all members are pinned-ended, diagonal members are provided to brace (stabilize) the frame against horizontal-in-plane movement and to resist the wind force on the endwalls (roll-up doors). The diagonal members are either single angles or threaded rods, which are strong in tension and weak in compression. When the wind force direction is as shown in Figure 6.1(b), member 1 is in compression and only has a very small buckling strength. Therefore, member 1 is assumed to be inactive when the structural analysis due to wind is performed. When the wind force direction in Figure 6.1(b) is reversed, member 1 is in tension and member 2 is assumed to be inactive in the structural analysis due to wind. This type of frame is classified as a *braced frame*. The deflected structure is not shown, but Δ_x at the top right joint is the horizontal component of the elongation in member 2. Thus, the horizontal movement of a braced frame is limited by the choice of stiffness for the braces.

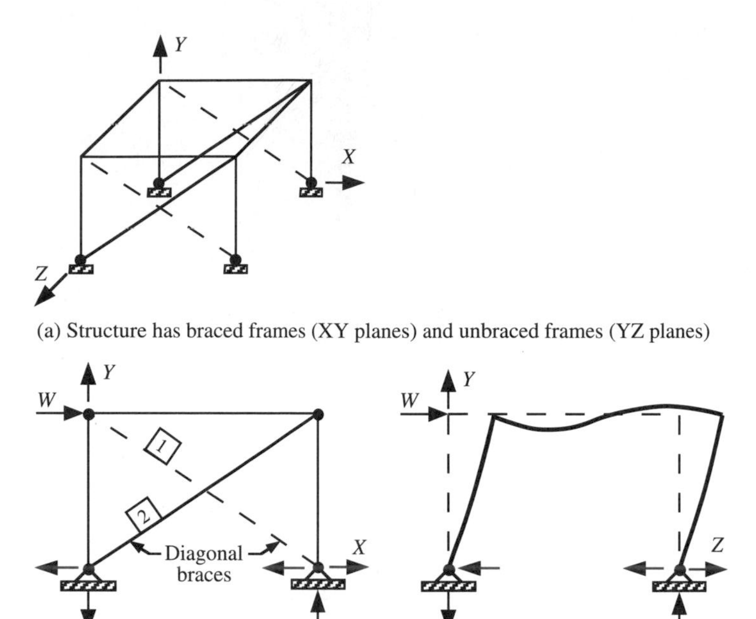

Figure 6.1 Braced and unbraced frames.

Consider Figure 6.1(c). In order to span across the roll-up door openings and to resist lateral forces due to wind on the sidewalls, the strong axis of the beams and the strong axis of the columns are chosen as the bending axis. A connection must be designed to fully transfer bending moment between the connected member ends. This type of frame is classified as an *unbraced frame*. The horizontal movements in an unbraced frame can only be determined from an indeterminate structural analysis that accounts for the bending action of the members in the frame.

Consider the structure represented in Figure 6.1 and the load combinations given in LRFD A4.1, p. 6-30. Due to wind on the finished building and with the roll-up doors closed, there is a pressure on the windward side of the building, a suction on the leeward wall, and a suction on the flat roof. Suction on the roof is an upward load, whereas gravity loads (dead and live) on the roof are downward. All members in the unbraced frame and the roof member in the braced frame are subjected to an axial force and bending due to the load combinations given in LRFD A4.1. For an unbraced frame, the load combinations of most concern to the structural engineer cause an axial compressive force and bending to occur in the members. We refer to a member subjected to bending and an axial compressive force as a *beam-column*.

6.2 MEMBER-SECOND-ORDER (Pδ) EFFECTS

Figure 6.2(a) shows a braced frame for which the exterior walls are attached to girts and the exterior walls are subjected to factored wind loads w_1 and w_2. The length direction of the girts is perpendicular to the plane of Figure 6.1(a). The girts are attached to the exterior flanges of the columns at 4-ft intervals, for example, along the column length. Consider the LRFD (A4-6) load combination assuming that $1.3W$ is greater than $0.9D$. On the roof, the factored wind load is upward and the factored dead load is downward. For the wind direction shown, member 1 would be inactive (would not resist any load). Member 3 is subjected to bending and an axial tensile force. Members 4 and 5 are beam-columns.

Figure 6.2(b) shows a braced frame for which the exterior wall panels are attached to beams. The beams span perpendicularly to the plane of Figure 6.2(b) and are located only at the pinned joint locations. The loading in Figure 6.2(b) is as defined in LRFD load combinations (A4-3), (A4-4), or (A4-6). In each of these cases, the factored wind load on the roof is upward (due to suction), and the other factored loads are downward (due to gravity). Therefore, for the load direction shown on member 4 in Figure 6.2(b), the assumption is that the sum of the factored gravity direction loads exceeds the factored wind load. Member 1 is inactive for the wind direction shown in Figure 6.2(b) and member 4 is a beam-column.

Member 3 in Figure 6.2(a) is subjected to an axial tension force and bending. As shown in Figure 6.3, the deflection and bending moment are decreased due to the

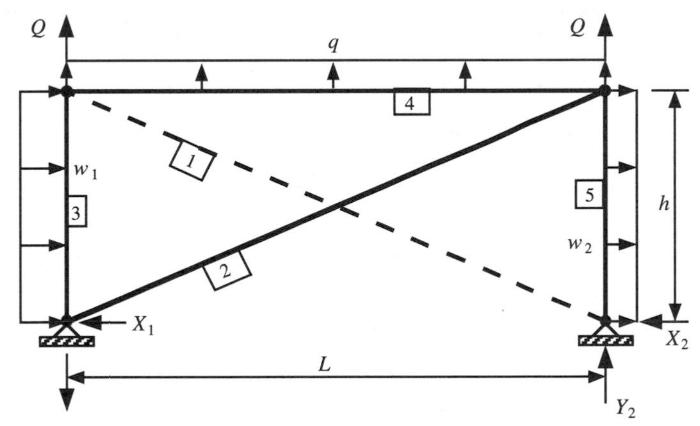

(a) Exterior walls attached to girts on members 3 and 5

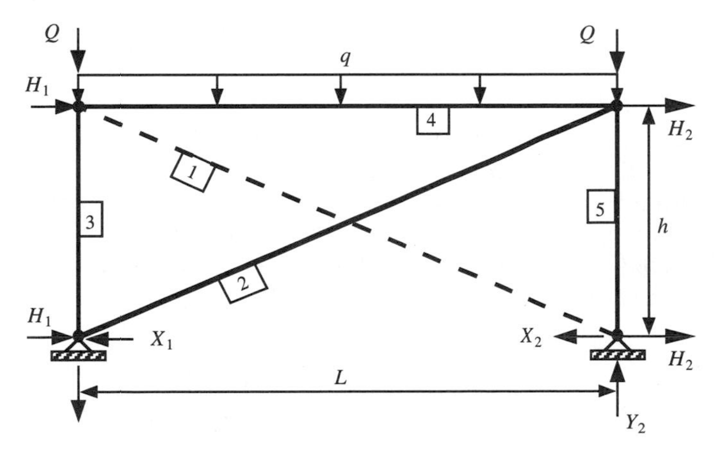

(b) Exterior walls attached to beams at hinge locations

Figure 6.2 Braced frame and $0.9D + 1.3W$ loading combination.

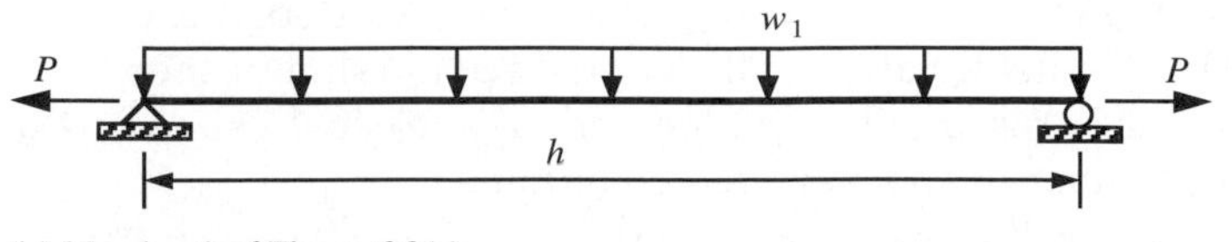

(a) Member 3 of Figure 6.2(a)

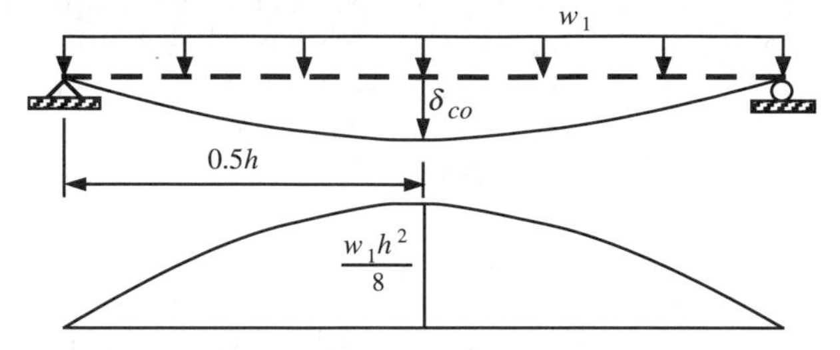

(b) Primary moment diagram is for $P = 0$

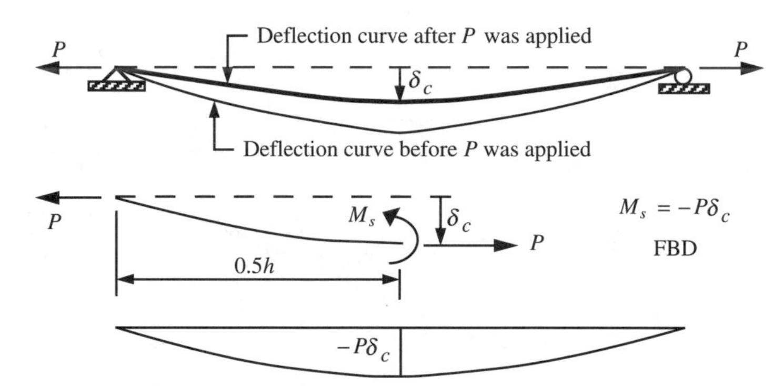

(c) Secondary moment diagram = *P(final deflected shape)*

(d) Combined moment diagram = (b) + (c); $M_{\max} = \frac{w_1 h^2}{8} - P\delta_c$

Figure 6.3 Secondary bending effects due to axial tension force.

effects of the axial tension force. Therefore, it is conservative to ignore the secondary effects on deflection and moment when the design requirements for strength and serviceability are checked for such a member.

Member 4 in Figure 6.2(b) is a beam-column. As shown in Figure 6.4, everywhere along the member the deflection and bending moment increase due to the effects of the axial compression force. For elastic behavior of a beam-column (from pp. 15 and 29 of [6]),

$$\delta_c \approx \frac{\delta_{co}}{1-\rho}$$

where

$$\rho = \frac{P}{P_e}$$

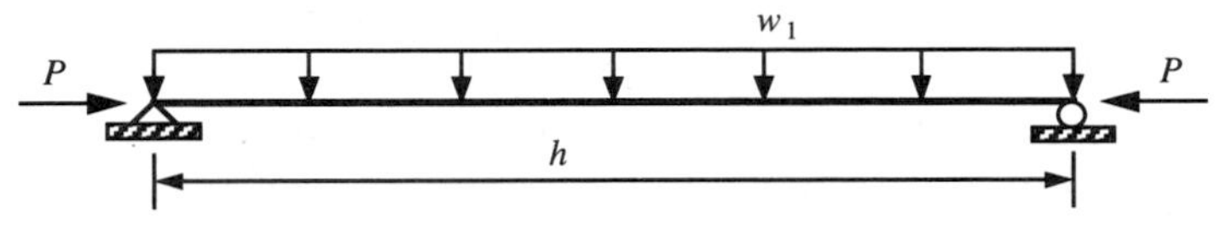

(a) Member 4 of Figure 6.2(b)

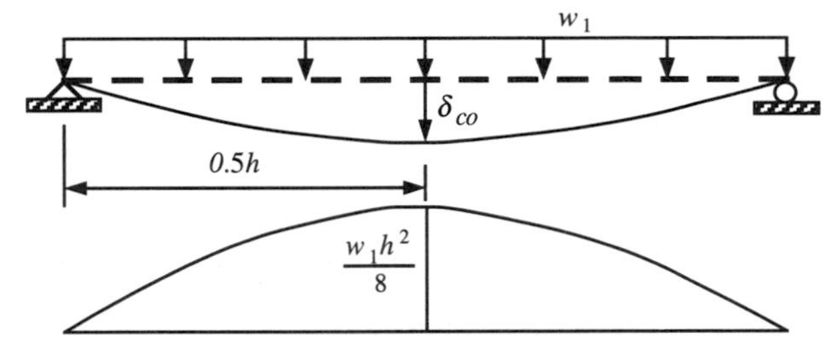

(b) Primary moment diagram is for $P = 0$

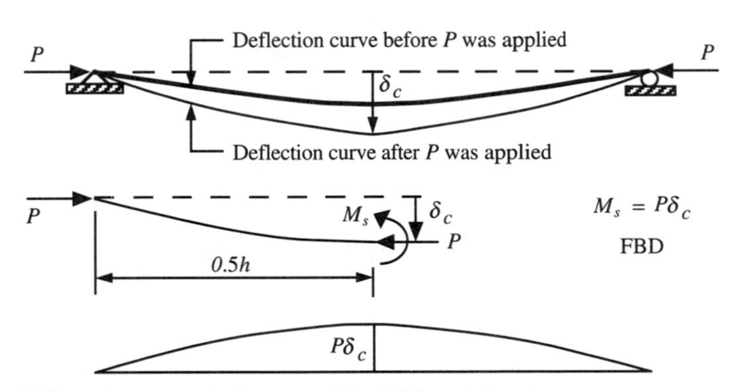

(c) Secondary moment diagram = *P(final deflected shape)*

(d) Combined moment diagram = (b) + (c); $M_{\max} = \frac{w_1 h^2}{8} + P\delta_c$

Figure 6.4 Secondary bending effects due to axial compression force.

$$P_e = \frac{\pi^2 EI}{L^2}$$

I and *L* are for the axis of bending

δ_{co} = maximum deflection when $P = 0$

δ_c = maximum deflection when $P > 0$

The maximum secondary bending moment [see Figure 6.4(c)] is

$$M_s = P\delta_c$$

and the maximum total bending moment [see Figure 6.4(d)] is

$$M_{\max} = \frac{qL^2}{8} + P\delta_c \approx \frac{qL^2}{8(1-\rho)}$$

The *member-secondary-bending* effects must be accounted for in checking the LRFD serviceability and strength design requirements for a beam-column. When the member-end moments are not zero, the member-secondary-bending effects are less pronounced. For other boundary conditions and/or other types of loads in Figure 6.4(a), the required bending strength can be obtained by using the information from LRFD Table C-C1.1 (p. 6-183) in LRFD Eq. (C1-2) (p. 6-41) and $M_{LT} = 0$ in LRFD Eq. (C1-1). That is, the required bending strength for a beam-column in a braced frame is

$$M_u = B_1 M_{NT}$$

where

$$M_{NT} = \text{required primary-bending strength}$$

B_1 = amplification factor that accounts for *member second-order* (Pδ) effects

$$B_1 = \text{larger of} \begin{cases} \dfrac{C_m}{(1 - P_u / P_e)} \\ 1.00 \end{cases}$$

C_m = factor that accounts for type of load and member-end moments

$$C_m = 1 + \psi\rho \text{ (from LRFD, p. 6-181)}$$

ψ is given in LRFD Table C-C1.1, p. 6-183.

$$\rho = \frac{P_u}{P_e}$$

$$P_u = \text{required column strength}$$

$$P_e = \frac{\pi^2 EI}{(KL)^2}$$

where I and KL are for the axis of bending. For a conservative approach, use $K = 1$. Otherwise, determine a value of $K < 1$ from either Eq. (4.26) (or from the sidesway inhibited nomograph on LRFD, p. 6-186) or LRFD Table C-C2.1 (p. 6-184). If the reader does not prefer to use the definition of $C_m = 1 + \psi\rho$ shown here for a beam-column subjected to a transverse loading, the C_m values given in LRFD C1(b) (p. 6-42) are applicable.

For a beam-column not subjected to a transverse loading, C_m is obtained from LRFD Eq. (C1-3), which is

$$C_m = 0.6 - 0.4(M_1 / M_2)$$

where M_1/M_2 is the ratio of the smaller and larger absolute-valued primary moments at the member's supports in the plane of bending. M_1/M_2 is positive when the member is bent in reverse curvature. M_1/M_2 is negative when the member is bent in single curvature.

The LRFD definitions applied to the member in Figure 6.4 are

$$M_u = B_1 M_{NT}$$

where

$$M_{nt} = \frac{w_1 h^2}{8}$$

$$C_m = 1$$

$$B_1 = \frac{1}{(1 - P_u / P_e)}$$

For the general case, when $M \neq 0$ on both member ends, the maximum moment occurs within the span of a beam-column when $[B_1 = C_m/(1 - P_u/P_e)] > 1$.

6.3 SYSTEM-SECOND-ORDER (PΔ) EFFECTS

Figure 6.5(a) shows an unbraced frame subjected to a factored load combination. The reaction directions assumed in Figure 6.5(a) are typical for the structure and the factored load combination shown. Note that all members are beam-columns. In the following discussion, as permitted by LRFD A5.1, p. 6-31, the required strength due to factored loads is obtained from elastic first-order analyses for which superposition is valid.

Figure 6.5(b) shows the primary moment diagram due only to the loads that cause *no* lateral *translation* of the frame joints (or no sidesway of the frame) to occur. The required primary bending strength for each member due to this portion of the factored load combination is denoted as M_{NT}. The *member-secondary-moment* (Pδ) associated with M_{NT} must be accounted for in checking the LRFD strength design requirements for a beam-column and the required bending strength is $M_u = B_1 M_{NT}$.

Figure 6.5(c) shows the primary moment diagram due to the loads that cause *lateral translation* of the frame joints to occur. The required primary bending strength for each member due to this portion of the factored load combination is denoted as M_{LT}.

Figure 6.5(d) shows the second-order effect due to sidesway of the entire structure. This *system-secondary-moment* (PΔ) must be accounted for in checking the LRFD strength design requirements for a beam-column.

Some of the commercially available structural analysis software packages can perform elastic P-DELTA analyses [9] that account for the system-secondary-moment (PΔ). In an elastic P-DELTA analysis, the solution for all the needed load combinations can be obtained in one computer run. For example, for the structure and loading shown in Figure 6.5, a *first-order analysis* (FOA) is performed. The system-secondary-moment in each member is obtained from the FOA results. For each column, PΔ is the product of the member-end axial force and the relative lateral movements of the member-end axial forces. As shown in Figure 6.6, equivalent member-end shears, PΔ/L, are computed and applied to create the PΔ effect in the next FOA. At each joint, the algebraic sum of the equivalent member-end shears gives an equivalent applied joint load that is added onto the original loads. Then, a FOA is performed for this modified loading. If any joint displacement changes by more than a specified amount (±2%, e.g.), new equivalent applied joint loads are computed and added onto the original loads. Again, a FOA is performed for the latest modified loading. The iterative process is terminated when no joint displacement changes by more than the specified amount in an iteration cycle. Unless an extra

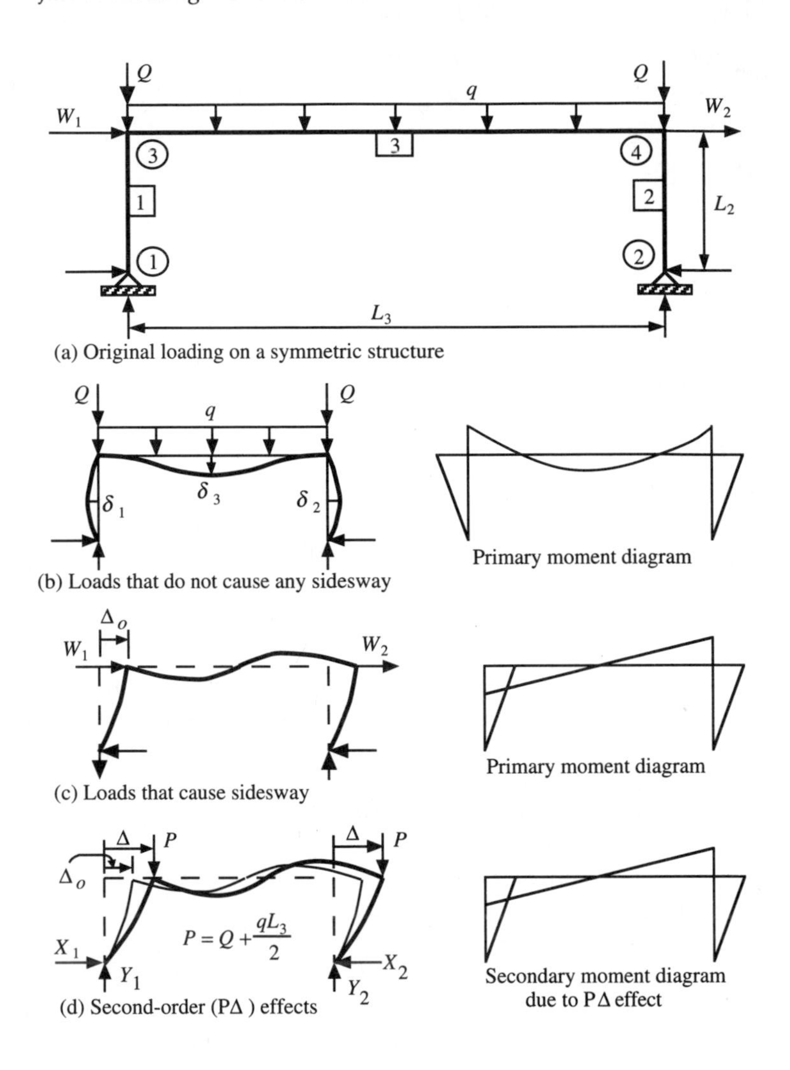

Figure 6.5 One-story unbraced frame.

joint is inserted at or close to the location of the maximum deflection in each member, the P-DELTA analyses do not account for the *member-secondary-moment* (Pδ).

Only a few of the available computer analysis software packages can perform a second-order analysis that can solve for the results due to only one load combination in each computer run. The secondary effects on joint displacements (rotations and translations), member-end forces, and member-end stiffness coefficients are a function of the axial force in each member of the structure.

Since not all available structural analysis software can perform P-DELTA analyses, LRFD C1 (p. 6-41) gives an approximate procedure that can be used in lieu of a second-order analysis. In the first edition, we dealt exclusively with this approximate procedure. However, in this edition, we will only use the approximate procedure to account for the *member-secondary-moment* (Pδ), and we will account for the *system-secondary-moment* (PΔ) by giving results from a P-DELTA analysis. Our

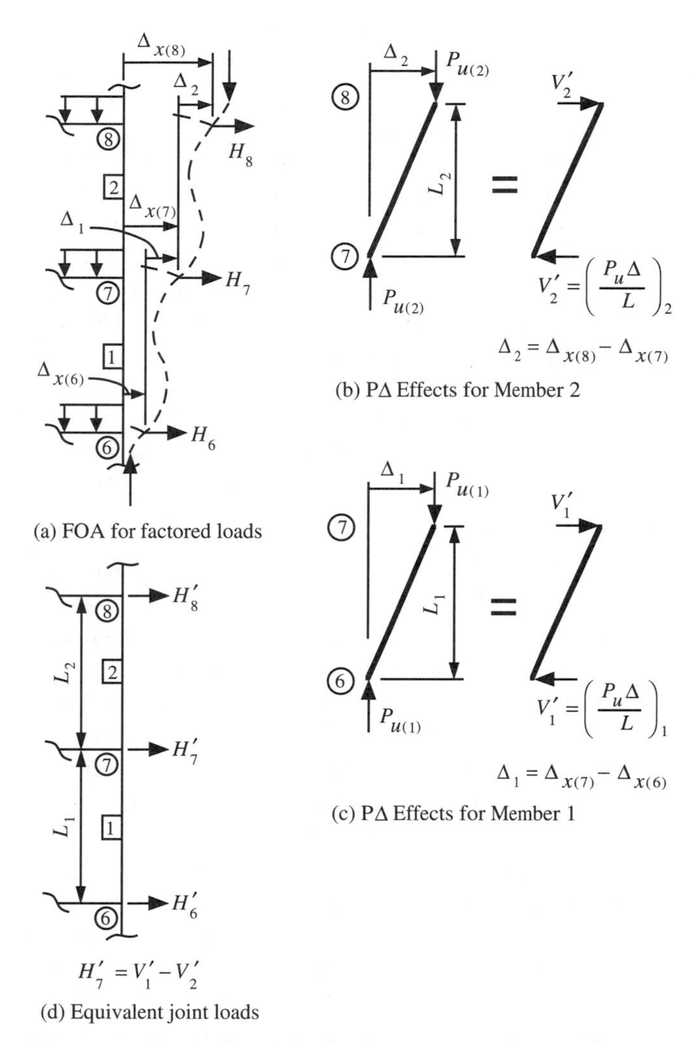

Figure 6.6 Equivalent joint loads to account for PΔ effects.

reason for doing this is that we assume that the reader has, or needs to obtain, a structural analysis package that performs a P-DELTA analysis of unbraced frames and accounts for the system-secondary-moments (PΔ).

6.4 ELASTIC FACTORED LOAD ANALYSES

Prior to performing the elastic factored load analyses of an unbraced frame, the structural designer performs approximate analyses for gravity loads only and for lateral loads (wind and/or earthquake) only. Approximate solutions for the required LRFD load combinations are obtained by superposing the results found in the approximate analyses. The approximate solutions are used for preliminary design purposes to obtain estimated member sizes to use in the first computer run of the displacement-method analyses. Then, drift and deflections are checked. If necessary, some member sizes are increased to improve serviceability. Also, for any member that does not satisfy the strength requirements given in LRFD Chapter H, a new member size is chosen and used in the next computer run. The described design process is continued, if necessary, until all serviceability and strength requirements are satisfied in the unbraced frame.

Some vendors of structural analysis software are including features that perform structural design tasks. These software packages also provide the capability of performing P-DELTA analyses (described in Section 6.2). The P-DELTA analyses account for the second-order (PΔ) effects required by LRFD C1 (p. 6-41). When such a package is being used, computing the LRFD interaction equations to check for each member in an unbraced frame for all required factored load combinations is a routine task. Each effective length factor is automatically computed, when needed, for each principal axis of each member. The LRFD column-strength and beam-strength definitions are automatically computed for each member using section properties from a built-in database of the hot-rolled steel sections.

Some observations:

1. LRFD C1 (p. 6-41) requires us to account for second-order effects.
2. Design features are being included in the structural software; we now have structural analysis and design software packages.
3. Consider the unbraced frame in Figure 6.7(a) subjected only to factored gravity direction loads on member 3. Since no sidesway occurs due to this loading, there are no *system-secondary-moments* (PΔ); that is, the results for this loading from a FOA and P-DELTA analysis are identical. Equivalent joint loads to account for the PΔ effects can be computed at each joint, but they have an appreciable effect only at joints that translate a significant amount in the direction of the equivalent joint loads.
4. Unless an extra joint is inserted at or close to the location of the maximum deflection in each member, a P-DELTA analysis does not account for the *member-secondary-moments* (Pδ). Figure 6.6 was prepared for columns, but a similar one can be prepared for each half of a subdivided beam. For that case, PΔ/(0.5L) is the equivalent end shear for each half of member 3 in Figure 6.7(a), and when an extra joint is inserted at midspan of member 3, the equivalent joint load is 2PΔ/(0.5L).
5. *Member-secondary-moments* (Pδ) are significant only in those beam-columns for which the maximum moment occurs within the span; that is, when $[B_1 = C_m/(1 - P_u/P_e)] > 1$.

Some recommendations:

1. After a structural design assumption is made, thereafter it must be used everywhere in the design process unless it is replaced by another assumption and the design process is restarted. For example, suppose that either $G = 10$ (for a hinged support) or $G = 1$ (for a fixed support) is used as recommended on LRFD p. 6-186 to obtain an effective length factor for a column. Then, a rotational spring of the appropriate stiffness must be included at that support in the P-DELTA analyses. Also, the connection between the column end and the support as well as the support itself must be designed for the reactions that account for the assumed value of G.

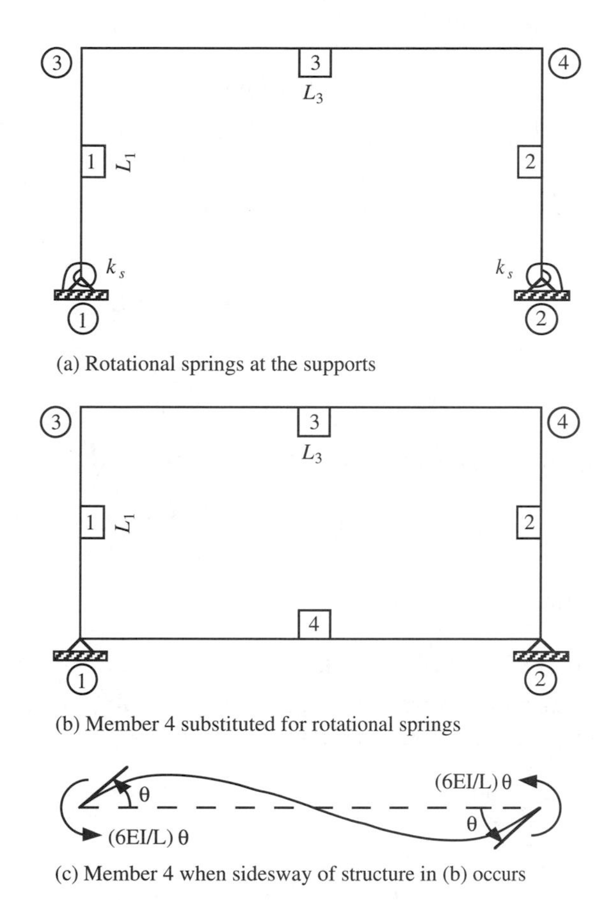

Figure 6.7 Rotational springs at the supports.

2. Perfect hinges at supports and completely fixed supports should not be acceptable boundary conditions in the P-DELTA analyses of an unbraced frame. Boundary condition effects on drift, deflections, and effective length are as important as accounting for the second-order (PΔ) effects. In some cases, the effects of boundary conditions are more important.

At the supports of a multistory building, $G = 2$ may be an appropriate choice to assume before the support conditions are completely known. Using a spring to represent $G = 2$ at a support attracts exactly half as much moment to a support as a completely fixed support when the loading consists only of gravity loads. We need to derive a rotational spring constant k_s to represent any desired value of G at a support. Due to the symmetry of Figure 6.7(a), Figure 6.7(b) is equivalent to Figure 6.7(a). When sidesway buckling of Figure 6.7(b) occurs, there is a point of inflection at midspan of member 4 as shown in Figure 6.7(c). Therefore,

$$k_s = \left(\frac{6EI}{L}\right)_4$$

$$\left(\frac{I}{L}\right)_4 = \frac{k_s}{6E}$$

$$G_1 = \frac{(I/L)_1}{(I/L)_4}$$

$$\left(\frac{I}{L}\right)_4 = \frac{(I/L)_1}{G_1} \equiv \frac{k_s}{6E}$$

$$k_s = \frac{1}{G_1}\left(\frac{6EI}{L}\right)_1$$

If we wish to account for inelastic column buckling, then

$$k_s = \frac{1}{G_1}\left(\frac{6E_t I}{L}\right)_1$$

and E_t is obtained from Eq. (4.23).

6.5 MEMBERS SUBJECT TO BENDING AND AXIAL TENSION

The strength design requirement for a W section subjected to an axial tension force and bending is given in LRFD H1.1:

1. When $\frac{P_u}{\phi P_n} \geq 0.2$,

$$\frac{P_u}{\phi P_n} + \frac{8}{9}\left(\frac{M_{ux}}{\phi M_{nx}} + \frac{M_{uy}}{\phi M_{ny}}\right) \leq 1.00$$

2. When $\frac{P_u}{\phi P_n} < 0.2$,

$$\frac{P_u}{2\phi P_n} + \left(\frac{M_{ux}}{\phi M_{nx}} + \frac{M_{uy}}{\phi M_{ny}}\right) \leq 1.00$$

where

P_u = required tensile strength
ϕP_n = design tensile strength (LRFD D1)
M_u = required bending strength
ϕM_n = design bending strength (LRFD F1)
x = subscript that denotes the strong bending axis
y = subscript that denotes the weak bending axis

As shown in Section 6.2, secondary bending effects can be ignored for a member subjected to an axial tension force and bending. Therefore, M_{ux}, M_{uy}, and P_u can be obtained directly from an elastic FOA.

Example 6.1

In Example 2.5, for A36 steel and a bearing-type bolted connection, a pair of L3 × 2 × 1/4 with long legs back to back was chosen as the trial section for the tension and bending members 34 and 43 in Figure 1.15. From Example 2.5, we also know:

$$P_u = 66.3 \text{ kips (tension)} \qquad \text{and} \qquad M_u = 0.18(12) = 2.16 \text{ in.-kips}$$

$$\phi P_n = 0.75F_u\,(A_n U) = 0.75(58)(1.9425)(0.9) = 76.0 \text{ kips}$$

Does the trial section satisfy the LRFD H1.1 strength design requirement?

Solution

$$L_b = (KL)_x = (KL)_y = L = 7.5 \text{ ft} = 90 \text{ in.}$$

1. For beam effects,

the double-angle section properties for $s = 3/8$ in. that will be needed are LRFD p. 1-98

$$S_x = 1.08 \text{ in.}^3 \qquad I_y = A(r_y)^2 = 2.38(0.891)^2 = 1.89 \text{ in.}^4$$

LRFD p. 1-160

$$J = 2(0.0270 \text{ in.}^4)$$

LRFD Table B5.1 (p. 6-38) does not give a local buckling criterion for flexural compression of a double-angle section. Take a conservative approach and use the local buckling criterion for axial compression, which is

$$\left(\frac{b}{t} = \frac{d-y}{t} = \frac{3-0.993}{0.25} = 8.03\right) \le \left(\frac{76}{\sqrt{36}} = 12.7\right)$$

Local buckling does not govern ϕM_{nx}. Therefore, LRFD F1.2c (p. 6-55) for double-angle beams with stems in compression is applicable for computing ϕM_{nx}:

$$B = -2.3\left(\frac{d}{L_b}\right)\sqrt{\frac{I_y}{J}} = -2.3\left(\frac{3}{90}\right)\sqrt{\frac{1.89}{2(0.0270)}} = -0.454$$

$$B + \sqrt{1+B^2} = -0.454 + \sqrt{1+(-0.454)^2} = 0.644$$

$$M_{cr} = \frac{\pi\sqrt{EI_y GJ}}{L_b}\left(B + \sqrt{1+B^2}\right)$$

$$M_{cr} = \frac{\pi\sqrt{29{,}000(1.89)(11{,}200)(0.054)}}{90}(0.644) = 129.4 \text{ in.-kips}$$

$$\phi M_{nx} = \text{smaller of} \begin{Bmatrix} 0.9 S_x F_y = 0.9(1.08)(36) = 35.0 \\ 0.9 M_{cr} = 0.9(129.4) = 116 \end{Bmatrix} = 35.0 \text{ in.-kips}$$

2. For axial tension effects,

$$[P_u/(\phi P_n) = 66.3/76.0 = 0.872] \geq 0.2$$

3. Check LRFD Eq. (H1-1a), p. 6-59:

$$\left[0.872 + \frac{8}{9}\left(\frac{2.16}{35.0}\right) = 0.927\right] \leq 1.00 \quad \text{(OK)}$$

Example 6.2

In Example 2.8, for A36 steel and welded connections, a WT7 × 15 was chosen as the trial section for the tension-and-bending members 5 to 14 in Figure 1.15. From Example 2.8, we also know:

$$P_u = 114.2 \text{ kips (tension)} \qquad \text{and} \qquad M_u = 1.62(12) = 19.4 \text{ in.-kips}$$

$$\phi P_n = 0.9FA_g = 143 \text{ kips}$$

Does the trial section satisfy the LRFD H1.1 strength design requirement?

Solution

$$L_b = (KL)_x = (KL)_y = L = 6 \text{ ft} = 72 \text{ in.}$$

1. For beam effects,

The section properties that will be needed are:

LRFD p. 1-78

$$d = 6.92 \text{ in.} \qquad S_x = 3.55 \text{ in.}^3 \qquad I_y = 9.79 \text{ in.}^4$$

LRFD p. 1-166

$$J = 0.190 \text{ in.}^4$$

LRFD Table B5.1 (p. 6-38) does not give a local buckling criterion for flexural compression of a WT stem. Take a conservative approach and use the local buckling criterion for axial compression, which is

$$\left(\frac{b}{t} = \frac{d-y}{t} = \frac{6.92 - 1.58}{0.270} = 19.8\right) \leq \left(\frac{127}{\sqrt{36}} = 21.2\right)$$

Local buckling does not govern ϕM_{nx}. Therefore, LRFD F1.2c (p. 6-55) for a WT with the stem in compression is applicable for computing ϕM_{nx}:

$$B = -2.3\left(\frac{d}{L_b}\right)\sqrt{\frac{I_y}{J}} = -2.3\left(\frac{6.92}{72}\right)\sqrt{\frac{9.79}{0.190}} = -1.59$$

$$B+\sqrt{1+B^2} = -1.59+\sqrt{1+(-1.59)^2} = 0.288$$

$$M_{cr} = \frac{\pi\sqrt{EI_y GJ}}{L_b}\left(B+\sqrt{1+B^2}\right)$$

$$M_{cr} = \frac{\pi\sqrt{29{,}000(9.79)(11{,}200)(0.190)}}{72}(0.288) = 808.9 \text{ in-kips}$$

$$\phi M_{nx} = \text{smaller of} \begin{Bmatrix} 0.9S_x F_y = 0.9(3.55)(36) = 115 \\ 0.9M_{cr} = 0.9(809) = 728 \end{Bmatrix} = 115 \text{ in-kips}$$

2. For axial tension effects,

$$[P_u/(\phi P_n) = 114.2/143 = 0.799] \geq 0.2$$

3. Check LRFD Eq. (H1-1a), p. 6-59:

$$\left[0.799+\frac{8}{9}\left(\frac{19.4}{115}\right) = 0.949\right] \leq 1.00 \quad \text{(OK)}$$

6.6 BEAM-COLUMNS

The strength design requirement for a W section subjected to bending and an axial compression force is given in LRFD H1.2:

1. When $\frac{P_u}{\phi P_n} \geq 0.2$,

$$\frac{P_u}{\phi P_n}+\frac{8}{9}\left(\frac{M_{ux}}{\phi M_{nx}}+\frac{M_{uy}}{\phi M_{ny}}\right) \leq 1.00$$

2. When $\frac{P_u}{\phi P_n} < 0.2$,

$$\frac{P_u}{2\phi P_n}+\left(\frac{M_{ux}}{\phi M_{nx}}+\frac{M_{uy}}{\phi M_{ny}}\right) \leq 1.00$$

where

P_u = required column strength

ϕP_n = column design strength (LRFD E2)

M_u = required bending strength (LRFD C1)

ϕM_n = design bending strength (LRFD F1)

x = subscript that denotes the strong bending axis

y = subscript that denotes the weak bending axis

M_{ux} and M_{uy} must be obtained from one of the following:

1. Elastic P-DELTA analysis
2. Elastic second-order analysis
3. Definition of M_u given in LRFD C1 for elastic FOA
4. Plastic second-order analysis

The design requirement for shear is given in LRFD F2 (p. 6-56) and, when P_u is a compression force in a column of a multistory building, in LRFD K1.7 (p. 6-95).

A beam-column is subjected to bending and an axial compression force. A brief review of the column design strength and the beam design strength for a W section is appropriate before the example problems are presented.

Consider a W section used as a beam-column. From LRFD Table B5.1 for column action only, local buckling does not govern ϕP_n when

$$\frac{0.5 b_f}{t_f} \leq \left(\lambda_r = \frac{95}{\sqrt{F_y}} \right) \quad \text{and} \quad \frac{h}{t_w} \leq \left(\lambda_r = \frac{253}{\sqrt{F_y}} \right)$$

and the design compressive strength is $\phi P_n = 0.85 A_g F_{cr}$ where F_{cr} is as defined in LRFD E2. When local buckling governs ϕP_n, then F_{cr} is defined in LRFD Appendix B.

KL, the effective length, for each principal axis of a column is needed to determine the compressive design strength. *KL* is the chord distance between two adjacent $M = 0$ points on the buckled shape. An $M = 0$ condition occurs at a real hinge and at a point of inflection on the buckled shape. Each principal axis has a slenderness ratio *KL/r*, where *r* is the radius of gyration for the principal axis associated with *KL*. When column flexural buckling occurs, the member bends about the principal axis having the larger slenderness ratio. If *KL* and *r* are different for each principal axis, *KL/r* must be computed for both principal axes to determine which axis has the larger *KL/r* value. For a built-up column, LRFD E4, E2, and Appendix E3 are applicable.

LRFD Table C-C2.1 (p. 6-183) gives *K* values for an isolated column having different boundary conditions. In an unbraced frame, lateral stability depends on the bending stiffness of the connected beams and columns. For design purposes, either use Eq. (4.27) or the *sidesway uninhibited* nomograph on LRFD p. 6-186 to determine *K* for a column in an unbraced frame. For a better estimate of *K*, the inelastic definition of *G* given in Eq.(4.29) can be used. In a braced frame, lateral stability is provided by diagonal bracing, shear walls, or equivalent means. For a column in a braced frame, *K*=1 can be conservatively chosen. Alternatively, Eq. (4.26) or the *sidesway inhibited* nomograph on LRFD p. 6-186 can be used to determine *K*.

Column design strength values are given in the LRFD column tables (pp. 3-16 to 3-116) for sections frequently used as a column.

For a W section used as a beam-column, the strong-axis design bending strength ϕM_{nx} is defined in LRFD F1 when all indicated restrictions obtained from LRFD Table B5.1 (pp. 6-38 to 6-39) are satisfied:

$$\frac{0.5 b_f}{t_f} \leq \left(\lambda_p = \frac{65}{\sqrt{F_y}} \right)$$

$$\frac{h}{t_w} \le \left[\lambda_p = \frac{640}{\sqrt{F_y}} \left(1 - \frac{2.75 P_u}{\phi_b P_y} \right) \right] \quad \text{when } \frac{P_u}{\phi_b P_y} \le 0.125$$

$$\frac{h}{t_w} \le \left\{ \left(\lambda_p = \frac{191}{\sqrt{F_y}} \left(2.33 - \frac{P_u}{\phi_b P_y} \right) \right) \ge \frac{253}{\sqrt{F_y}} \right\} \quad \text{when } \frac{P_u}{\phi_b P_y} > 0.125$$

where

$$\phi_b P_y = 0.9 A_g F_y$$

If any of the preceding restrictions is not satisfied, the design bending strength is defined in LRFD Appendix F (p. 6-111). Beam charts (ϕM_{nx} vs. L_b curves) for $C_b = 1$ are given on LRFD pp. 4-113 to 4-166 for W and M sections. For $C_b > 1$,

$$\phi M_{nx} = \text{smaller of } \begin{cases} C_b \overline{M}_1 \\ \phi M_{px} \end{cases}$$

$$\overline{M}_1 = (\phi M_{nx} \text{ for } C_b = 1)$$

The weak-axis design bending strength is $\phi M_{ny} = 0.9 Z_y F_y$ when

$$\frac{0.5 b_f}{t_f} \le \left(\lambda_p = \frac{65}{\sqrt{F_y}} \right)$$

Otherwise, flange local buckling governs ϕM_{ny} and either LRFD Eq. (A-F1-3) or (A-F1-4) is the applicable definition of M_{ny}.

6.7 BRACED FRAME EXAMPLES

An isolated beam-column is chosen for simplicity to illustrate the basic concepts involved in a member whose ends cannot translate perpendicularly to the member axis. However, these basic concepts are applicable to a floor member extracted from a multistory braced frame (see Figure 6.8) at the *i*th floor level, for example. Assume for discussion purposes that Figure 6.8 is only one bay of a multibay section on the interior of a building and that the lateral loads shown must be resisted only by the members shown. Suppose that a permanent partition wall is situated directly above each floor member. The wall only needs to extend barely above the ceiling level. An air gap exists between the bottom of the floor member and the top of the partition wall beneath the floor member. Each floor member is subjected to the uniformly distributed dead weight of the partition wall situated above the floor member. If the diagonal braces are designed to resist only tension, the axial compression force in the floor member at the *i*th floor level due to the factored wind loads shown is

$$P_{u(i)} = \sum_{j=i}^{n} W_j$$

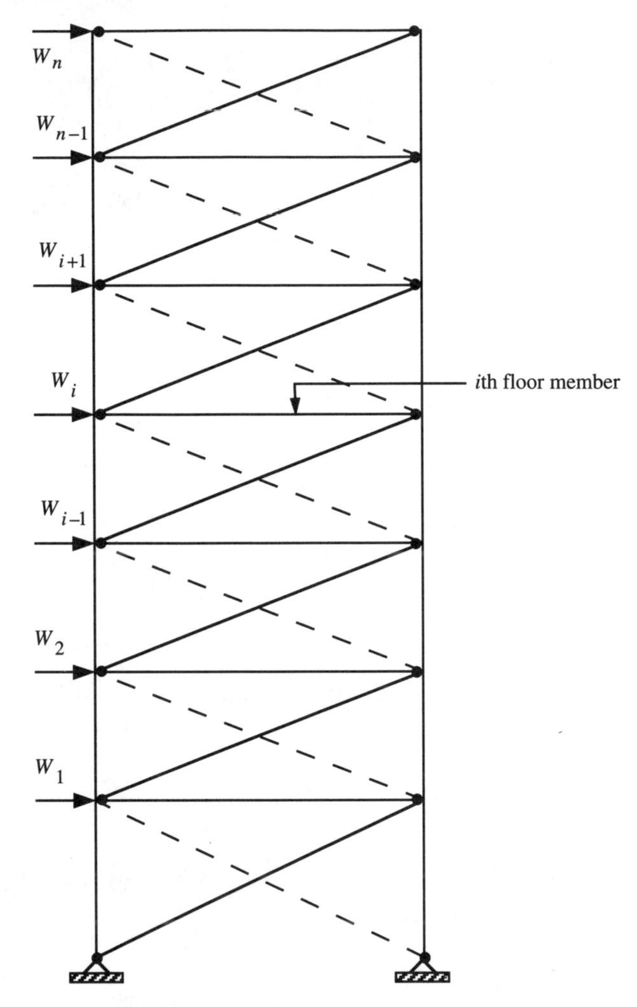

Figure 6.8 Multi-story braced frame.

In Part 4 of the LRFD Manual, any effects of local buckling were correctly accounted for in computing the beam design strength values given in the design aids. Therefore, we can use these values without having to check to see if local buckling governs.

In Part 3 of the LRFD Manual, any effects of flange local buckling were correctly accounted for in computing the column design strength values given in the design aids. Therefore, we can use these values without having to check to see if web buckling governs.

In Part 3 of the LRFD Manual, a footnote symbol is placed on each section for which the web must be checked when that section is used as a beam-column to see if web local buckling governs ϕM_{nx}.

Any section that is used as a beam-column and not listed in Part 3 of the LRFD Manual must be checked to see if web local buckling governs ϕM_{nx}. Example 6.7 has a member for which this web check is required.

Example 6.3

The W12 × 65 (F_y = 36 ksi) beam-column in Figure 6.9 is subjected to bending about the x-axis only due to the factored distributed load. Lateral braces are provided only at the member ends. Does the W12 x 65 satisfy the requirements of LRFD H1.2?

Solution

1. For beam effects,

the maximum primary moment is called M_{NT} since that is the name used for this parameter in LRFD Eq. (C1-1), which is applicable in item 2:

$$M_{NT} = q_u L^2/8 = 0.66(30)^2/8 = 74.3 \text{ ft-kips}$$

$C_b = 1.14$ (from the next-to-last case in LRFD Table 4-1, p. 4-9)

LRFD p. 4-19

$$\phi M_{px} = 261 \text{ ft-kips}$$

LRFD page 4-130

$$L_b = 30 \text{ ft} \qquad \overline{M}_1 = 212.5 \text{ ft-kips}$$

$$\phi M_{nx} = \text{smaller of} \begin{Bmatrix} C_b \overline{M}_1 & = 1.14(212.5) = 242 \\ \phi M_{px} & = 261 \end{Bmatrix} = 242 \text{ ft-kips}$$

2. For beam-column effects,

LRFD Eq. (C1-1) is applicable:

$$M_{ux} = B_1(M_{NT}) + B_2(0) = B_1(74.3 \text{ ft-kips})$$

$$C_{mx} = 1 \quad \text{(see LRFD, p. 6-183)}$$

$$P_{ex} = \frac{\pi^2 EI_x}{(KL)_x^2} = \frac{\pi^2 (29{,}000)(533)}{[30(12)]^2} = 1177 \text{ kips}$$

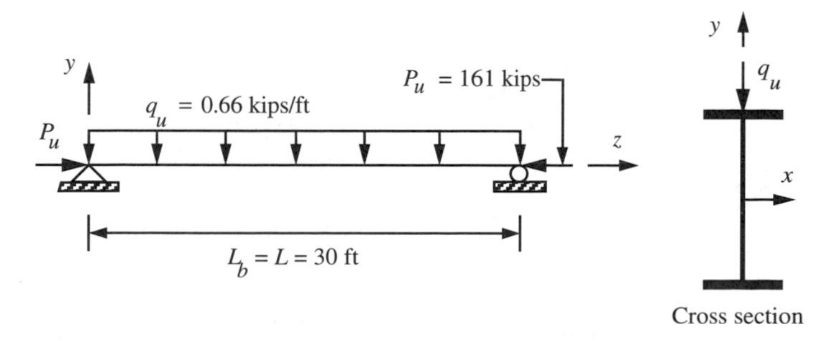

Figure 6.9 Example 6.3.

$$B_1 = \text{larger of} \begin{Bmatrix} \dfrac{C_m}{(1-P_u/P_e)} = \dfrac{1.0}{1-161/1177} = 1.1585 \\ 1.00 \end{Bmatrix} = 1.1585$$

$$M_{ux} = 1.1585(74.3) = 86.1 \text{ ft-kips}$$

3. For column effects:

LRFD p. 3-24 can be used to obtain ϕP_n:

$$\left[(KL)_y = 30 \text{ ft}\right] > \left[\frac{(KL)_x}{(r_x/r_y)} = \frac{30}{1.75} = 17.14 \text{ ft}\right]$$

$$\phi P_n = (\phi P_{ny} = 277 \text{ kips})$$

$$\left(\frac{P_u}{\phi P_n} = \frac{161}{277} = 0.581\right) \geq 0.2$$

4. Check LRFD Eq. (H1-1a), p. 6-59:

$$\left[0.581 + \frac{8}{9}\left(\frac{86.1}{242}\right) = 0.897\right] \leq 1.00 \qquad \text{(OK)}$$

Example 6.4

The W12 x 65 (F_y = 36 ksi) beam-column in Figure 6.10 is subjected only to x-axis bending due to the factored distributed load. Lateral braces are provided such that L_b = 15 ft; $(KL)_y$ = 15 ft. Does the W12 x 65 satisfy the requirements of LRFD H1.2?

Solution

1. For beam effects,

$$M_{NT} = 0.86(30)^2/8 = 96.8 \text{ ft-kips}$$

$$C_b = 1.30 \quad \text{(from the last case in LRFD Table 4-1, p. 4-9)}$$

LRFD p. 4-19

$$\phi M_{px} = 261 \text{ ft-kips}$$

LRFD p. 4-130

$$L_b = 15 \text{ ft} \qquad \overline{M}_1 = 254.5 \text{ ft-kips}$$

$$\phi M_{nx} = \text{smaller of} \begin{Bmatrix} C_b \overline{M}_1 = 1.30(254.5) = 331 \\ \phi M_{px} = 261 \end{Bmatrix} = 261 \text{ ft-kips}$$

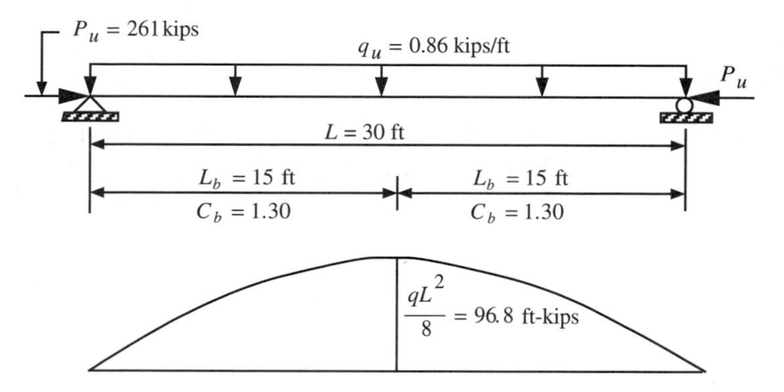

Figure 6.10 Example 6.4.

2. For beam-column effects

LRFD Eq. (C1-1) is applicable:

$$M_{ux} = B_1(M_{NT}) + B_2(0) = B_1(96.8 \text{ ft-kips})$$

$$C_{mx} = 1 \quad \text{(see LRFD, p.6-183)}$$

$$P_{ex} = \frac{\pi^2 EI_x}{(KL)_x^2} = \frac{\pi^2 (29{,}000)(533)}{[30(12)]^2} = 1177 \text{ kips}$$

$$B_1 = \text{larger of} \begin{Bmatrix} \dfrac{C_m}{(1 - P_u / P_e)} = \dfrac{1.0}{1 - 261/1177} = 1.285 \\ 1.00 \end{Bmatrix} = 1.285$$

$$M_{ux} = 1.285(96.8) = 124.4 \text{ ft-kips}$$

3. For column effects:

$$\left[\frac{(KL)_x}{(r_x / r_y)} = \frac{30}{1.75} = 17.14 \text{ ft} \right] > \left[(KL)_y = 15 \text{ ft} \right]$$

$$\phi P_n = (\phi P_{nx} = 458 \text{ kips})$$

$$\left(\frac{P_u}{\phi P_n} = \frac{261}{458} = 0.570 \right) \ge 0.2$$

4. Check LRFD Eq. (H1-1a), p. 6-59:

$$\left[0.571 + \frac{8}{9} \left(\frac{124}{261} \right) = 0.994 \right] \le 1.00 \qquad \text{(OK)}$$

Example 6.5

The beam-column in Figure 6.11 is subjected only to x-axis bending due to the factored concentrated load. Lateral braces for both flanges are provided only at the supports and midspan. Does a W12 × 72 (F_y = 50 ksi) satisfy LRFD H1.2?

Solution

1. For beam effects,
   LRFD p. 4-19

$$\phi M_{px} = 405 \text{ ft-kips}$$

The primary moment diagram from case 13 on LRFD p. 4-194 is shown in Figure 6.11.

For the first L_b region, M_{NT} = 150 ft-kips and

LRFD p. 4-196

$$L_b = 15 \text{ ft} \qquad \overline{M}_1 = 384 \text{ ft-kips}$$

$$C_b = 1.67 \quad \text{(from case 2 in LRFD Table 4-1, p. 4-9)}$$

$$\phi M_{nx} = \text{smaller of} \begin{Bmatrix} C_b \overline{M}_1 & = 1.67(384) = 642 \\ \phi M_{px} & = 405 \end{Bmatrix} = 405 \text{ ft - kips}$$

For the second L_b region, M_{NT} = 180 ft-kips and:

$$C_b = \frac{12.5\,M_{\max}}{2.5\,M_{\max} + 3\,M_A + 4\,M_B + 3\,M_C} = \frac{12.5(180)}{2.5(180) + 3(67.5) + 4(15) + 3(97.5)} = 2.24$$

$$\phi M_{nx} = \text{smaller of} \begin{Bmatrix} C_b \overline{M}_1 & = 2.24(384) = 860 \\ \phi M_{px} & = 405 \end{Bmatrix} = 405 \text{ ft - kips}$$

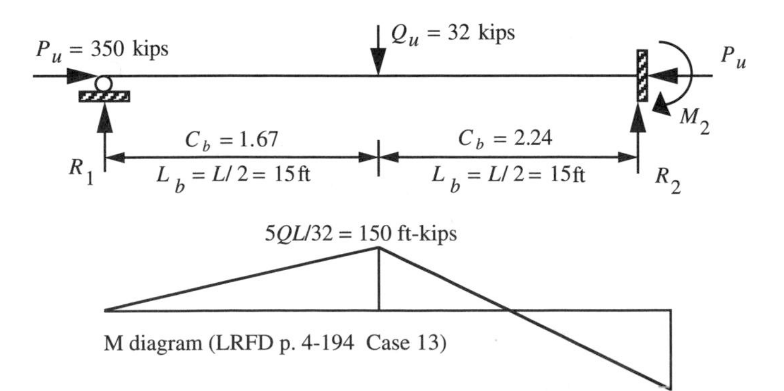

Figure 6.11 Example 6.5.

2. For beam-column effects,

$$P_{ex} = \frac{\pi^2 EI_x}{(KL)_x^2} = \frac{\pi^2 (29{,}000)(597)}{[24(12)]^2} = 2060 \text{ kips}$$

$$\rho = P_u/P_{ex} = 350/2060 = 0.170$$

$$C_{mx} = 1 - 0.3\rho \quad \text{(see LRFD p.6-183)}$$

$$C_{mx} = 1 - 0.3(0.170) = 0.949$$

$$B_1 = \text{larger of} \begin{Bmatrix} \dfrac{C_m}{(1 - P_u/P_e)} = \dfrac{0.949}{1 - 0.170} = 1.143 \\ 1.00 \end{Bmatrix} = 1.143$$

$$M_{ux} = 1.143(180) = 206$$

3. For column effects,

for y-axis buckling, it is conservative to use $(KL)_y = 15$ ft. For x-axis buckling, see case (e) on LRFD p. 6-184:

$$(KL)_x = 0.8(30) = 24 \text{ ft}$$

$$\left[(KL)_y = 15 \text{ ft}\right] > \left[\frac{(KL)_x}{(r_x/r_y)} = \frac{24}{1.75} = 13.7 \text{ ft}\right]$$

$$\phi P_n = (\phi P_{ny} = 694 \text{ kips})$$

$$\left[\frac{P_u}{\phi P_n} = \frac{350}{694} = 0.504\right] \geq 0.2$$

4. Check LRFD Eq. (H1-1a), p. 6-59, in all L_b regions.

Since P_u, ϕP_n, and ϕM_{px} are identical in both L_b regions, only the L_b region with the larger M_{ux} must be checked:

$$\left[0.504 + \frac{8}{9}\left(\frac{206}{405}\right) = 0.956\right] \leq 1.00 \qquad \text{(OK)}$$

6.8 UNBRACED FRAME EXAMPLES

The examples deal with the basic concepts involved in performing the LRFD interaction equation design check for a beam-column in an unbraced frame using the results from a P-DELTA structural analysis that accounts for the *system-secondary-moments* (PΔ), but does not account for the *member-secondary-moments* (Pδ). Therefore, when $[B_1 = C_m/(1 - P_u/P_e)] > 1$, the M_{ux} value obtained from the computer results will have to be increased to account for the *member-secondary-moments* (Pδ).

Example 6.6

For member 3 in Figure 1.15, assume that $L_b = (KL)_y = 21$ ft. See Appendix A for the P-DELTA solutions. Does a W12 × 40 (F_y = 36 ksi) satisfy LRFD H1.2?

Solution

1. For column effects,

   the lateral stability of the structure in Figure 1.15 is achieved by bending of members 1 to 4 and their interaction with the truss. To obtain an estimate of $(KL)_x$ for sidesway frame buckling, convert the truss to an equivalent beam (see Figure 6.12). From Appendix A,

   Members 5 to 14:

   $$A = 5.00 \text{ in.}^2 \qquad I = 20.9 \text{ in.}^4$$

   Members 15 to 24:

   $$A = 5.89 \text{ in.}^2 \qquad I = 23.3 \text{ in.}^4$$

   For the truss as an equivalent beam, try

   $$I = 0.85[44.2 + (5.00 + 5.89)(27)^2] = 6786 \text{ in.}^4$$

   The 0.85 factor was chosen based on experience. The assumed value of I = 6786 was verified by comparing the behavior of the assumed structure and the original structure subjected to the nominal wind loading in Appendix A:

   $$G_{24} = 1 \qquad G_{22} = \frac{310/21}{6786/60} = 0.1305$$

   $$K_x = \sqrt{\frac{1.6G_{24}G_{22} + 4(G_{24} + G_{22}) + 7.5}{G_{24} + G_{22} + 7.5}} = 1.19$$

   $$\left[(KL)_y = 21.0 \text{ ft}\right] > \left[\frac{(KL)_x}{(r_x/r_y)} = \frac{1.19(21)}{2.66} = 9.40 \text{ ft}\right]$$

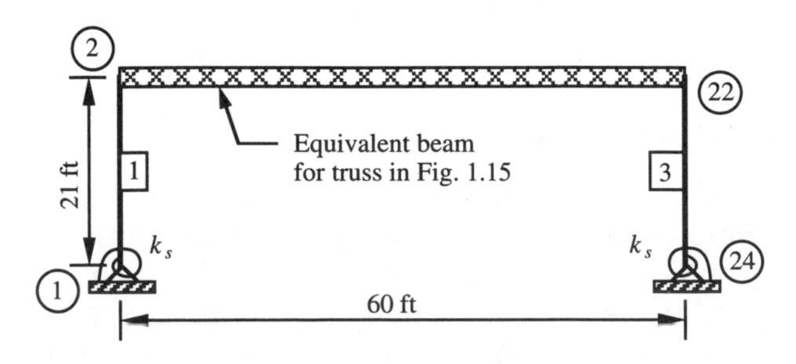

Figure 6.12 Example 6.6: Equivalent structure for Fig. 1.15.

$$\phi P_n = (\phi P_{ny} = 147.5 \text{ kips})$$

For member 3 in Figures 1.15 and 6.13, $L = 21$ ft. From loading 8 in Appendix A:

$$P_u = 64.6 \text{ kips}; \quad M_{22} = 44.1 \text{ ft-kips}; \quad M_{24} = 13.2 \text{ ft-kips}$$

2. For beam effects,

$$C_b = \frac{12.5(44.1)}{2.5(44.1) + 3(29.8) + 4(15.4) + 3(1.1)} = 2.08$$

$$\phi M_{nx} = \text{smaller of} \begin{Bmatrix} C_b \overline{M}_1 = 2.08(117.5) = 244 \\ \phi M_{px} = 155 \end{Bmatrix} = 155 \text{ ft-kips}$$

3. For beam-column effects,
Is LRFD Eq. (C1-1) applicable?

$$C_m = 0.6 - 0.4(M_1 / M_2) = 0.6 - 0.4(13.2/44.1) = 0.480$$

$$P_{ex} = \frac{\pi^2 EI_x}{(KL)_x^2} = \frac{\pi^2 (29{,}000)(310)}{[1.19(21)(12)]^2} = 987 \text{ kips}$$

$$B_1 = \text{larger of} \begin{Bmatrix} \dfrac{C_m}{(1 - P_u / P_e)} = \dfrac{0.480}{1 - 64.6/987} = 0.514 \\ 1.00 \end{Bmatrix} = 1.00$$

LRFD Eq. (C1-1) is not applicable; i.e., no *member-secondary-moment*(Pδ) needs to be computed, and $M_{ux} = 44.1$ ft-kips.

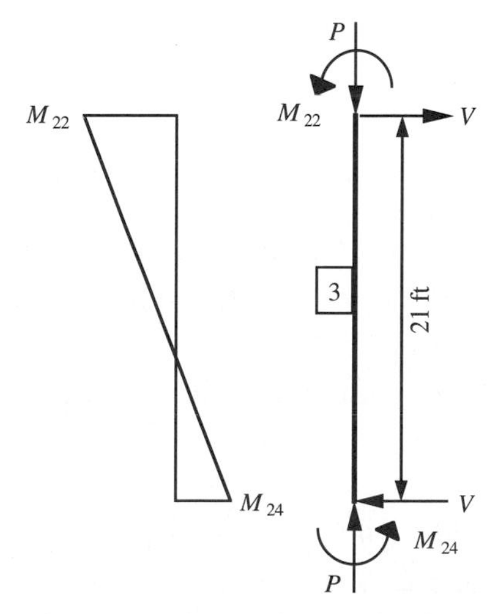

Figure 6.13 Example 6.6: Member 3 in Fig. 1.15.

$$\left[\frac{P_u}{\phi P_n} = \frac{64.6}{147.5} = 0.438\right] \geq 0.2$$

4. Check LRFD Eq. (H1-1a), p. 6-59 for loading 8:

$$\left[0.438 + \frac{8}{9}\left(\frac{44.1}{155}\right) = 0.690\right] \leq 1.00 \quad \text{(OK for loading 8)}$$

5. Check LRFD Eq. (H1-1a), p. 6-59 for loading 10
From Loading 10 in Appendix A,

$$P_u = 33.1 \text{ kips} \qquad M_{22} = 62.0 \text{ ft-kips} \qquad M_{24} = 44.1 \text{ ft-kips}$$

Obviously, $C_b >$ (C_b in loading 8); therefore,

$$\phi M_{nx} = (\phi M_{px} = 155 \text{ ft-kips})$$

From the loading 8 solution we know that no *member-secondary-moment* (Pδ) needs to be computed and $M_{ux} = 62.0$ ft-kips:

$$\left(\frac{P_u}{\phi P_n} = \frac{33.1}{147.5} = 0.224\right) \geq 0.2$$

$$\left[0.224 + \frac{8}{9}\left(\frac{62.0}{155}\right) = 0.580\right] \leq 1.00 \quad \text{(OK for loading 10)}$$

Only 69% of the W12 x 40 ($F_y = 36$ ksi) section's strength capacity is needed, and a lighter W section can be chosen for strength. However, when the serviceability requirements were checked at the end of Section 1.8, a W12 x 40 was found to be barely satisfactory for controlling drift. Therefore, a W12 x 40 is the lightest acceptable column section.

Example 6.7

For $F_y = 36$ ksi and the results given below from a P-DELTA analysis for Figure 6.14:

1. Check serviceability. Our limiting choices due to nominal loads are
Maximum drift = $L_1/350$ = (180 in.)/350 = 0.514 in.
Maximum live-load deflection = $L_3/360$ = (360 in.)/360 = 1.00 in.
2. Check LRFD strength requirements (H1.2, p. 6-60).

In Figure 6.14:

$k_s = 68.2$ ft-kips/degree (accounts for $G_1 = G_2 = 10$)

Members 1 and 2: W14 x 48 $\quad L = 15$ ft. $\quad L_b = (KL)_y = 7.5$ ft

Member 3: W21 x 44 $\quad L = 30$ ft. $\quad L_b = (KL)_y = 6$ ft

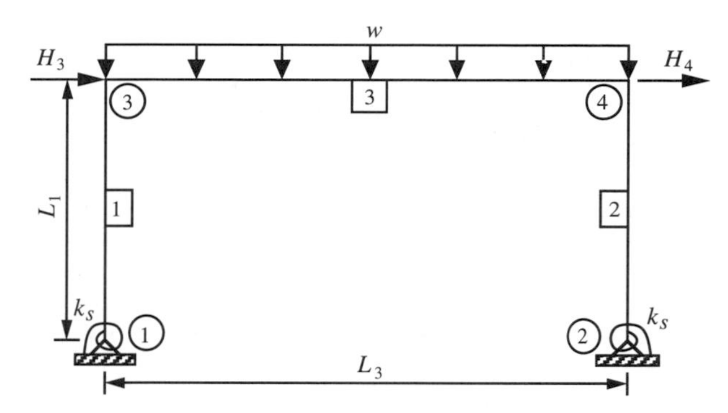

Figure 6.14 Example 6.7: Structure and loading.

Nominal loads are:

Dead: $w = 1.5$ kips/ft

Live: $w = 1$ kips/ft

Snow: $w = 0.6$ kips/ft

Wind: $w = -0.507$ kips/ft $\quad H_3 = 2.4$ kips $\quad H_4 = 1.5$ kips

P-DELTA analyses were performed for each of the required LRFD load combinations (p. 6-30) to check the LRFD strength requirements and for $0.9D + W$ to check drift. The results we need from the P-DELTA analyses are:

1. $\Delta_x = 0.300$ in. at joint 3 due to $0.9D + W$
2. $\Delta_y = -0.386$ in. at midspan of member 3 due to L
3. For member 2 (see Figure 6.15):
   (a) For $1.2D + 1.6L + 0.5S$,

$$Y_2 = 55.5 \text{ kips}; \; M_2 = 14.0 \text{ ft-kips}; \; M_4 = 176.7 \text{ ft-kips}$$

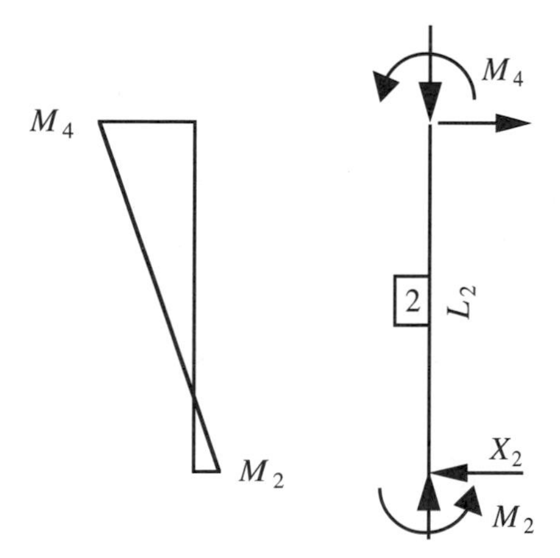

Figure 6.15 Example 6.7: Column

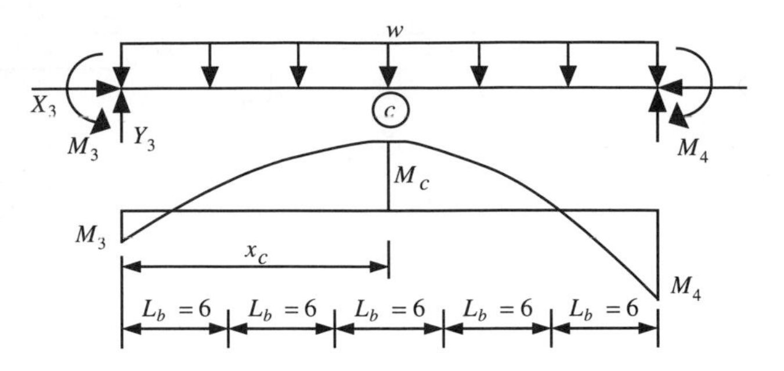

Figure 6.16 Example 6.7: Girder

(b) For $1.2D + 1.6S + 0.8W$

$$Y_2 = 36.5 \text{ kips} \qquad M_2 = 14.8 \text{ ft-kips} \qquad M_4 = 130.9 \text{ ft-kips}$$

4. For members 3 and 4 (see Figure 6.16) and $1.2D + 1.6L + 0.5S$,

$$X_3 = 12.71 \text{ kips} \qquad Y_3 = 55.5 \text{ kips}$$

$$M_3 = 176.7 \text{ ft-kips} \qquad M_4 = 176.7 \text{ ft-kips}$$

$$\text{At } x_c = Y_3 / w_3 = 15 \text{ ft}, M_c = 239.6 \text{ ft-kips}.$$

Solution

Check serviceability:

$$(\Delta_x = 0.300 \text{ in.}) \le (\text{maximum drift} = 0.514 \text{ in.}) \quad (\text{OK})$$

$$(\Delta_y = 0.386 \text{ in.}) \le (\text{maximum live-load deflection} = 1.00 \text{ in.}) \quad (\text{OK})$$

Check member 2 for strength (LRFD H1.2, p. 6-60). W14 × 48, $F_y = 36$ ksi. For $1.2D + 1.6L + 0.5S$ (governing load combination):

1. For column effects,

$$G_2 = 10 \qquad G_4 = (485/15)/(843/30) = 1.15$$

$$K_x = \sqrt{\frac{1.6G_2G_4 + 4(G_2 + G_4) + 7.5}{G_2 + G_4 + 7.5}} = 1.94$$

$$\left[\frac{(KL)_x}{(r_x/r_y)} = \frac{1.94(15)}{3.06} = 9.51 \text{ ft}\right] > \left[(KL)_y = 7.50 \text{ ft}\right]$$

$$\phi P_n = (\phi P_{nx} = 358 \text{ kips})$$

2. For beam effects,

LRFD p. 4-19

$$(L_b = 7.5 \text{ ft}) \le (L_p = 8.0 \text{ ft}), \qquad \phi M_{nx} = (\phi M_{px} = 212 \text{ ft-kips})$$

3. For beam-column effects,
Is LRFD Eq. (C1-1) applicable?

$$C_m = 0.6 - 0.4(M_1/M_2) = 0.6 - 0.4(14.0/176.7) = 0.568$$

$$P_{ex} = \frac{\pi^2 EI_x}{(KL)_x^2} = \frac{\pi^2(29{,}000)(485)}{[1.94(15)(12)]^2} = 1138 \text{ kips}$$

$$B_1 = \text{larger of} \begin{Bmatrix} \dfrac{C_m}{(1 - P_u/P_e)} = \dfrac{0.568}{1 - 55.5/1138} = 0.597 \\ 1.00 \end{Bmatrix} = 1.00$$

LRFD Eq. (C1-1) is not applicable; that is, no *member-secondary-moment* (Pδ) needs to be computed, and M_{ux} = 176.7 ft-kips:

$$\left(\frac{P_u}{\phi P_n} = \frac{55.5}{358} = 0.155\right) < 0.2$$

$$\left(\frac{0.155}{2} + \frac{176.7}{212} = 0.911\right) \le 1.00 \quad \text{(OK)}$$

Check member 3 for strength (LRFD H1.2, p. 6-60). W21 x 44 (F_y = 36 ksi):

1. For beam-column effects,

$$\left[\frac{P_u}{\phi P_y} = \frac{12.7}{0.9AF_y} = \frac{12.7}{0.9(13.0)(36)} = 0.0302\right] \le 0.125$$

$$\left(\frac{h}{t_w} = 53.6\right) \le \left\{\lambda_p = [1 - 2.75(0.0302)]\frac{640}{\sqrt{36}} = 97.8\right\}$$

WLB is not applicable.
Is LRFD Eq. (C1-1) applicable?

$$P_{ex} = \frac{\pi^2 EI_x}{(KL)_x^2} = \frac{\pi^2(29{,}000)(843)}{[(30)(12)]^2} = 1862 \text{ kips}$$

$$C_m = 1 - 0.4(P_u/P_{ex}) = 1 - 0.4(12.7/1862) = 0.997$$

$$B_1 = \text{larger of} \begin{Bmatrix} \dfrac{C_m}{(1 - P_u/P_e)} = \dfrac{0.997}{1 - 12.7/1862} = 1.004 \\ 1.00 \end{Bmatrix} = 1.004$$

$$M_{ux} = 1.004(239.6) = 240.6 \text{ ft-kips}$$

2. For beam effects,

$$(L_b = 6 \text{ ft}) > (L_p = 5.3 \text{ ft})$$

$$C_b = \frac{12.5(240.6)}{2.5(240.6) + 3(236.4) + 4(240.6) + 3(236.4)} = 1.008$$

$$\phi M_{nx} = \text{smaller of} \begin{cases} C_b \overline{M}_1 = 1.008(250) = 252 \\ \phi M_{px} = 258 \text{ ft - kips} \end{cases}$$

$$\phi M_{nx} = 252 \text{ ft-kips}$$

3. For column effects,

See Figure 6.14. Since the axial compression force in members 1 and 2 is not negligible, the braced frame formula or nomograph on LRFD p. 6-186 cannot be used to obtain K_x for the girder. Be conservative and use $K_x = 1$:

$$\left(\frac{KL}{r}\right)_x = \frac{360}{8.06} = 44.7$$

$$\left(\frac{KL}{r}\right)_y = \frac{72}{1.26} = 57.14$$

$$\left(\lambda_c = \frac{KL/r}{\pi}\sqrt{\frac{F_y}{E}} = \frac{57.14}{\pi}\sqrt{\frac{36}{29{,}000}} = 0.6409\right) < 1.5 \qquad \lambda_c^2 = 0.4107$$

$$\phi P_n = 0.85(13.0)(0.658^{0.4107})(36) = 335 \text{ kips}$$

$$\left[\frac{P_u}{(\phi P_n)} = \frac{12.7}{335} = 0.0379\right] < 0.2$$

4. Check LRFD Eq. (H1-1a), p. 6-59:

$$\left(\frac{0.0379}{2} + \frac{241}{250} = 0.983\right) \le 1.00 \qquad \text{(OK)}$$

All cited W sections are acceptable for strength and serviceability.

6.9 PRELIMINARY DESIGN

Consider the task of sizing the girders and columns in an unbraced multistory plane frame for an office building. See the required LRFD load combinations in LRFD A4.1. Assume that the building does not have to be designed to resist an earthquake loading.

When there are no more than about six stories above ground level in a typical unbraced multistory office building, one of the load combinations that is not a function of wind generally produces the axial force and bending moment values that cause the sum in the applicable interaction equation [LRFD Eq. (H1-1a) or (H1-1b)] to be a maximum for each member. We recommend that the ACI Code shear and moment coefficients [10] be used as the approximate method of analysis to obtain the girder-end shears and moments. However, to account for an estimate of the second-order effects, we recommend that the center-to-center span length be used instead of the clear span as stated in the ACI Code coefficients. At each joint in the plane frame that has two column ends, allocate half the unbalanced girder-end moment to each column end. The axial force in each column is obtained by summing the girder-end shears and beam-end shears at all floor levels for which the column is the vertical supporting member. Using this approach, we obtain estimated values of M_{ux} and P_u for each column. Then we use procedure 2 to select a trial column section to use in the first computer run of the P-DELTA analyses for all required LRFD load combinations.

Beam-Column Preliminary Design Procedure 1 (a beam section is desired)

This procedure is recommended when the member is predominantly a beam and the designer wants to use the beam charts to select a trial section.

Convert P_u to an *equivalent* $M_{ux} = P_u(d/2)$. Assume that the desired *span–depth* ratio is $L/d = 20$, which gives $d = L/20$ and equivalent $M_{ux} = P_uL/40$. Use the LRFD beam charts to select the lightest trial W section for which $\phi M_{nx} \geq (M_{ux} + P_uL/40)$.

Beam-Column Preliminary Design Procedure 2 (a column section is desired)

Use the LRFD column tables to select a trial section for which

$$\phi P_n \geq [P_u + m\,(M_{ux} + u\,M_{uy})]$$

where

1. m = the subsequent approximations value of m given on LRFD p. 3-12.
2. u = factor given in LRFD column tables. Let $u = 3$ to get started.

   Note: Each trial W section in Examples 6.6 to 6.7 was selected by using procedure 2.

Procedure 2 was obtained from LRFD Eq. (H1-1b):

$$\phi_c P_n \geq \left[P_u + \frac{8}{9}\left(\frac{\phi_c P_n}{\phi_b M_{nx}} M_{ux} + \frac{\phi_c P_n}{\phi_b M_{ny}} M_{uy} \right) \right]$$

$$\phi_c P_n \geq \left(P_u + m_x M_{ux} + m_y M_{uy} \right)$$

$$m_x = \frac{8}{9}\left(\frac{\phi_c P_n}{\phi_b M_{nx}} \right) \qquad m_y = \frac{8}{9}\left(\frac{\phi_c P_n}{\phi_b M_{ny}} \right)$$

Convert m_y to an equivalent m_x for later convenience purposes:

$$m_y = \frac{8}{9}\left(\frac{\phi_c P_n}{\phi_b M_{ny}} \right)\left(\frac{\phi_b M_{nx}}{\phi_b M_{nx}} \right) = \frac{8}{9}\left(\frac{\phi_c P_n}{\phi_b M_{nx}} \right)\left(\frac{\phi_b M_{nx}}{\phi_b M_{ny}} \right) = m_x \left(\frac{\phi_b M_{nx}}{\phi_b M_{ny}} \right)$$

To be conservative, assume that $\phi_b M_{nx} = \phi_b M_{px}$. Therefore,

$$u = \frac{\phi_b M_{nx}}{\phi_b M_{ny}} = \frac{Z_x}{Z_y}$$

where Z_x and Z_y are the plastic section moduli given in Part 1 of the LRFD Manual.
For the W sections listed in the column tables:

1. For W14, $2 \le u \le 4$. Only for the lightest three sections is $u > 3$.
2. For W12, $2.20 \le u \le 3.42$. Only for the lightest three sections is $u > 3$.
3. For W10, $2.12 \le u \le 2.77$.
4. For W8, $2.15 \le u \le 2.16$.

Therefore, we recommend $u = Z_x/Z_y = 3$ as the assumed value for the first trial section. If we replace m_x by m, we obtain

$$\phi_c P_n \ge \left[P_u + m\left(M_{ux} + uM_{uy} \right) \right]$$

where

$$m = \frac{8}{9}\left(\frac{\phi_c P_n}{\phi_b M_{nx}} \right)$$

which can be estimated as shown here.
For $C_b = 1$, $L_b = (KL)_y$, $[(KL)_x/(r_x/r_y)] \le (KL)_y$, $F_y = 36$ ksi, and $F_y = 50$ ksi:

1. $\phi_c P_n$ can be obtained from the column tables for a given $(KL)_y$.
2. $\phi_b M_{nx}$ can be obtained from the beam design charts for $L_b = (KL)_y$.

Example 6.8

Use Procedure 1 and select the lightest W section of A36 steel for

$$P_u = 12.8 \text{ kips} \qquad M_{ux} = 241 \text{ ft-kips}$$

$$L = 30 \text{ ft} \qquad C_b = 1 \qquad L_b = (KL)_y = 6 \text{ ft} \qquad (KL)_x = 30 \text{ ft}$$

Solution

$$\textit{Equivalent } M_{ux} = \left[M_{ux} + \frac{P_u L}{40} = 241 + 12.7\left(\frac{30}{40}\right) = 251 \text{ ft - kips} \right]$$

From LRFD p. 4-130,

$$C_b = 1 \qquad \text{plot point: } (L_b = 6,\ M_{ux} = 251)$$

$$\text{W21 x 44} \qquad (\phi M_{nx} = 250) \approx 251$$

This is how we chose the W21 x 44 for member 3 in Figure 6.14 for the loads stated in Example 6.7, where we checked a W21 x 44 for strength and found it was acceptable.

Example 6.9

Use Procedure 1 and select the lightest W section of A36 steel for

$$P_u = 12.8 \text{ kips} \qquad M_{ux} = 241 \text{ ft-kips}$$

$$L = 30 \text{ ft} \qquad C_b = 1.67 \qquad L_b = (KL)_y = 15 \text{ ft} \qquad (KL)_x = 30 \text{ ft}$$

Solution

$$\textit{Equivalent } M_{ux} = \left[M_{ux} + \frac{P_u L}{40} = 241 + 12.7\left(\frac{30}{40}\right) = 251 \text{ ft-kips} \right]$$

From LRFD p. 4-19,

$$\text{W21 x 44} \qquad (\phi M_{px} = 258) \geq 251$$

From LRFD p. 4-132,

$$C_b = 1.67 \qquad \text{plot point } (L_b, M_{ux}/C_b) = (15, 251/1.67 = 150)$$

W21 x 44 lies to right of and above this point. Choose W21 x 44 as the trial section.

Example 6.10

Use Procedure 2 and select the lightest W12 and lightest W14 of A36 steel for

$$P_u = 261 \text{ kips;} \quad M_{ux} = 96.8 \text{ ft-kips}$$

$$L = 30 \text{ ft;} \quad C_b = 1.67; \quad L_b = (KL)_y = 15 \text{ ft;} \quad (KL)_x = 30 \text{ ft}$$

Solution

1. For W12 and $F_y = 36$ ksi,
   LRFD p. 3-12 for $KL = 15$, $m = 1.55$

   $$\phi P_n \geq [P_u + mM_{ux} = 261 + (1.55)(96.8) = 411 \text{ kips}]$$

   LRFD p. 3-24

   $$(KL)_y = 15 \qquad \text{W12 x 65} \quad (\phi P_{ny} = 485) \geq 412$$

   $$[(KL)_x/(r_x/r_y) = 30/1.75 = 17.14 \text{ ft}] > [\,(KL)_y = 15]$$

   $$[\phi P_n = \phi P_{nx} = 460 - 0.14(14) = 458 \text{ kips}] \geq 412$$

   Try W12 x 65. (This is how we chose the W12 x 65 for Example 6.4.)

2. For W14 and $F_y = 36$ ksi
   LRFD p. 3-12 for $KL = 15$, $m = 1.4$

   $$\phi P_n \geq [P_u + mM_{ux} = 261 + (1.4)(96.8) = 397 \text{ kips}]$$

LRFD p. 3-21:

$$(KL)_y = 15 \qquad \text{W14 x 61}, \quad (\phi P_{ny} = 412) \geq 397$$

$$[(KL)_x/(r_x/r_y) = 30/2.44 = 12.3 \text{ ft}] < [(KL)_y = 15]$$

$$(\phi P_n = \phi P_{ny} = 412 \text{ kips}) \geq 397$$

Try W14 x 61.

Example 6.11

Use Procedure 2, LRFD column tables, $F_y = 36$ ksi, and select the lightest W14 and W12 for

$$P_u = 50.9 \text{ kips} \qquad M_{ux} = 186 \text{ ft-kips}$$

$$L = 15 \text{ ft} \qquad L_b = (KL)_y = 7.5 \text{ ft} \qquad (KL)_x/(r_x/r_y) = 9.51 \text{ ft}$$

Solution

1. For W14 and $F_y = 36$ ksi
LRFD p. 3-12 for $KL = 9.5$ ft, $m = 1.5$

$$\phi P_n \geq [P_u + mM_{ux} = 50.9 + (1.5)(186) = 330 \text{ kips}]$$

LRFD p. 3-21

$$\text{W14 x 48 } (\phi P_{nx} = 358) \geq 330$$

Try W14 × 48.

2. For W12 and $F_y = 36$ ksi
LRFD page 3-12: For $KL = 9.5$ ft., $m = 1.7$

$$\phi P_n \geq [P_u + mM_{ux} = 50.9 + (1.7)(186) = 367 \text{ kips}]$$

LRFD p. 3-25:

$$\text{W12 x 50} \qquad (\phi P_{nx} = 376) \geq 367$$

Try W12 x 50.

PROBLEMS

6.1 In Example 2.7, for A36 steel and a welded connection, a pair of L3 × 2 × 1/4 with long legs back to back was chosen as the trial section for members 34 and 43 in Figure 1.15. Also known from Example 2.7:

$$P_u = 66.3 \text{ kips (tension)} \qquad \text{and} \qquad M_u = 0.18(12) = 2.16 \text{ in-kips}$$

$$\phi P_n = 0.90 F_y A_g = 0.90(36)(2.38) = 77.1 \text{ kips}$$

Does the trial section satisfy the LRFD H1.1 strength design requirement?

6.2 See Figure P6.2. For $F_y = 36$ ksi, $q_u = 1.2$ kips/ft; $P_u = 170$ kips; $(KL)_y = L_b =$

$L = 10$ ft, assume that *yielding on* A_g governs ϕP_n and select the lightest acceptable W section that satisfies the LRFD H1.1 strength design requirement.

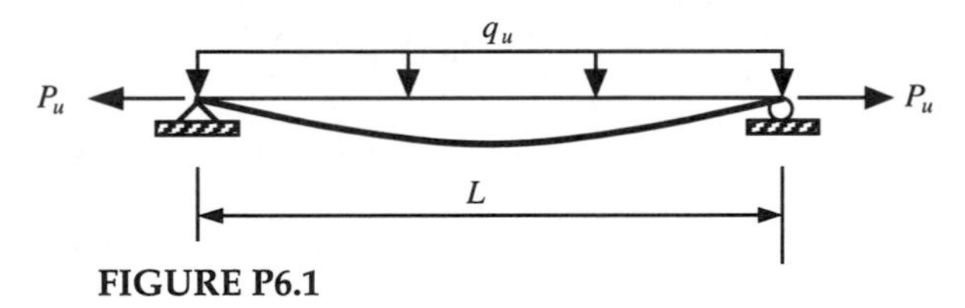

FIGURE P6.1

6.3 Solve Problem 6.2 for $F_y = 50$ ksi.

In Problems 6.4 to 6.10, does the W section indicated in each problem satisfy the LRFD C1 and H1.2 requirements for a beam-column?

6.4 See Figure P6.4. W12 x 45, $F_y = 50$ ksi.

$q_u = 1.0$ kips/ft $\quad P_u = 85$ kips $\quad (KL)_y = L_b = 15$ ft $\quad L = 30$ ft

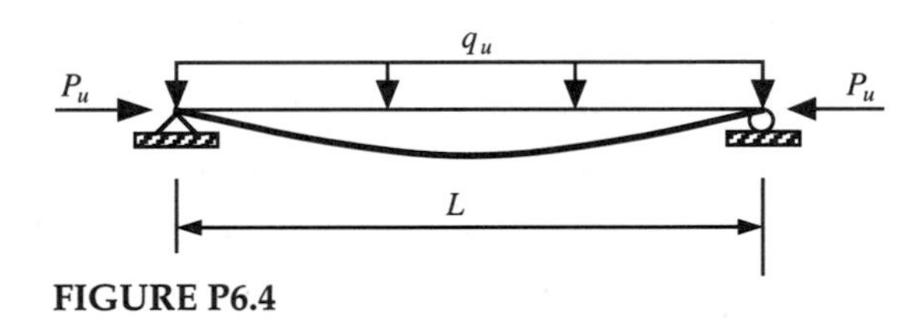

FIGURE P6.4

6.5 See Figure P6.5. W12 x 120, $F_y = 50$ ksi.

$Q_u = 60$ kips $\quad P_u = 360$ kips $\quad (KL)_y = L_b = 12$ ft $\quad L = 24$ ft

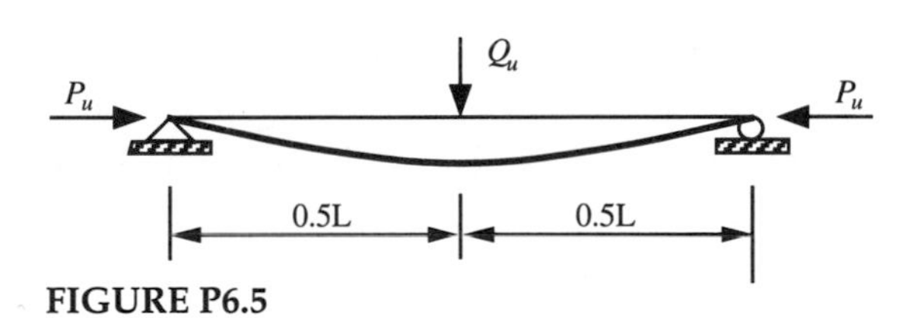

FIGURE P6.5

6.6 See Figure P6.6. W12 x 90 $F_y = 36$ ksi.

$M_1 = 130$ kips $\quad P_u = 600$ kips $\quad (KL)_y = L_b = L = 12$ ft

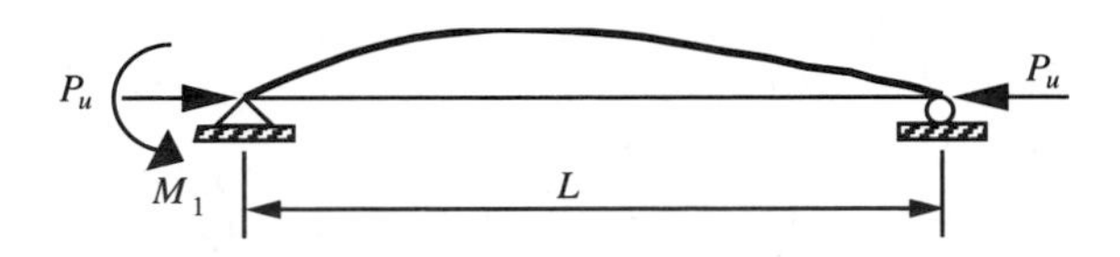

FIGURE P6.6

6.7 See Figure P6.7. W14 x 61 $F_y = 36$ ksi.

$M_1 = 170$ kips $M_2 = 60$ kips $P_u = 90$ kips $(KL)_y = L_b = 10$ ft $L = 20$ ft
$(KL)_x / (r_x / r_y) = 9.51$ ft

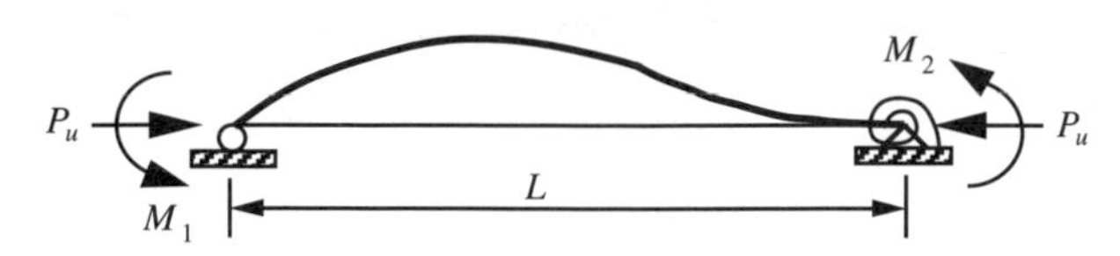

FIGURE P6.7

6.8 In Example 4.11 for A36 steel, a pair of L3 x 2 x 1/4 with long legs back to back and welded to a 3/80-in.-thick gusset plate was chosen as the trial section for members 5 to 14 in Figure 1.15 and in Appendix A. From Example 4.11, for these beam-columns we also know:

$$\phi P_n = 135 \text{ kips} \qquad P_u = 122.5 \text{ kips} \qquad M_u = 1.91 \text{ ft-kips}$$

Does the trial section satisfy the LRFD H1.2 strength design requirement?

6.9 Select the lightest acceptable WT section of A36 steel that satisfies LRFD H1.2 for members 5 to 14 in Figure 1.15 and in Appendix A.

6.10 Solve Problem 6.9 using $F_y = 50$ ksi steel.

In Problems 6.11 and 6.12, the member in each indicated figure is subjected to bending about the x-axis. Select the lightest acceptable W section from the indicated column section series that satisfies the requirements of LRFD C1 and H1.2 for the stipulated grade of steel.

6.11 For Figure P6.4, W12, $F_y = 50$ ksi, $q_u = 0.86$ kips/ft; $P_u = 261$ kips; $(KL)_y = L_b = 15$ ft; $L = 30$ ft.

6.12 For Figure P6.5, W12, $F_y = 50$ ksi, $Q_u = 32$ kips; $P_u = 350$ kips; $(KL)_y = L_b = 15$ ft; $L = 30$ ft.

6.13 See Figure 6.13. $F_y = 50$ ksi.

Members 1 and 2: W12 x 40 $L = 15$ ft; $L_b = (KL)_y = 15$ ft
Member 3: W18 x 35, $L = 30$ ft; $L_b = (KL)_y = 6$ ft
k_s accounts for $G = 10$ at joints 1 and 2.

Nominal loads are:

Dead: $w = 1.5$ kips/ft
Live: $w = 1$ kips/ft
Snow: $w = 0.6$ kips/ft
Wind: $w = -0.507$ kips/ft $H_3 = 2.4$ kips $H_4 = 1.5$ kips

P-DELTA analyses were performed for each of the required LRFD load combinations (p. 6-30) and the results at the ends of Member 3 are:

1. For $1.2D + 1.6L + 0.5S$,

$X_3 = 13.0$ kips $\quad Y_3 = 55.5$ kips; $M_3 = 180.5$ ft-kips; $M_4 = -180.5$ ft-kips

2. For $1.2D + 1.6S + 0.5L$

$X_3 = 11.45$ kips $\quad Y_3 = 48.9$ kips $\quad M_3 = 159$ ft-kips $\quad M_4 = -159$ ft-kips

Does the design satisfy LRFD H1.2 (p. 6-60)?

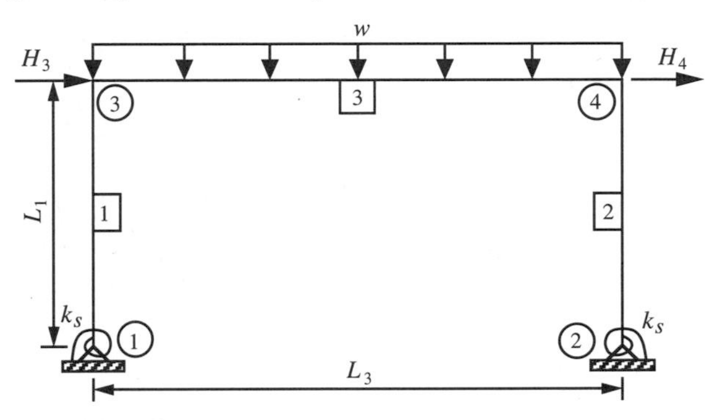

FIGURE P6.13

6.14 See Figure 6.14. $F_y = 50$ ksi.

Members 1 and 2: W14 x 48, $L = 15$ ft; $\quad L_b = (KL)_y = 15$ ft
Member 3: W21 x 44, $L = 36$ ft; $\quad L_b = (KL)_y = 6$ ft
k_s accounts for $G = 10$ at joints 1 and 2.

Nominal loads are :

Dead: $w = 1.5$ kips/ft
Live: $w = 1$ kips/ft
Snow: $w = 0.6$ kips/ft
Wind: $w = -0.507$ kips/ft $\quad H_3 = 2.4$ kips $\quad H_4 = 1.5$ kips

P-DELTA analyses were performed for each of the required LRFD load combinations (p. 6-30) and the results at the ends of member 3 are:

1. For $1.2D + 1.6L + 0.5S$

$X_3 = 19.5$ kips $\quad Y_3 = 66.6$ kips $\quad M_3 = 270.8$ ft-kips $\quad M_4 = -270.8$ ft-kips

2. For $1.2D + 1.6S + 0.5L$

$X_3 = 17.2$ kips $\quad Y_3 = 58.7$ kips $\quad M_3 = 239$ ft-kips $\quad M_4 = -239$ ft-kips

Does the design satisfy LRFD H1.2 (p. 6-60)?

6.15 See Figure 6.15. $F_y = 50$ ksi.

Members 1 and 2: W10 x 33, $L = 15$ ft; $\quad L_b = (KL)_y = 15$ ft
Member 3: W10 x 39, $L = 15$ ft; $\quad L_b = (KL)_y = 15$ ft
Members 4 and 5: W24 x 62, $L = 36$ ft; $\quad L_b = (KL)_y = 6$ ft.

k_s accounts for $G = 10$ at joints 1 to 3.

Nominal loads are:

Dead: $w = 1.5$ kips/ft

Live: $w = 1$ kips/ft

Snow: $w = 0.9$ kips/ft

Wind: $w = -0.507$ kips/ft $\quad H_2 = 2.4$ kips $\quad H_6 = 1.5$ kips

P-DELTA analyses were performed for each of the required LRFD load combinations (p. 6-30) and the results at the ends of members 4 and 5 are:

1. For $1.2D + 1.6L + 0.5S$,

$X_2 = 5.22$ kips $\quad Y_2 = 55.2$ kips $\quad M_2 = 72.4$ ft-kips $\quad M_4 = -580.7$ ft-kips

$X_4 = 5.22$ kips $\quad Y_6 = 55.2$ kips $\quad M_4 = 580.7$ ft-kips $\quad M_6 = -72.4$ ft-kips

2. For $1.2D + 1.6S + 0.5L$,

$X_2 = 5.07$ kips $\quad Y_2 = 53.6$ kips $\quad M_2 = 70.3$ ft-kips $\quad M_4 = -564.1$ ft-kips

$X_4 = 5.07$ kips $\quad Y_6 = 81.0$ kips $\quad M_4 = 564.1$ ft-kips $\quad M_6 = -70.3$ ft-kips

Does the design satisfy LRFD H1.2 (p. 6-60)?

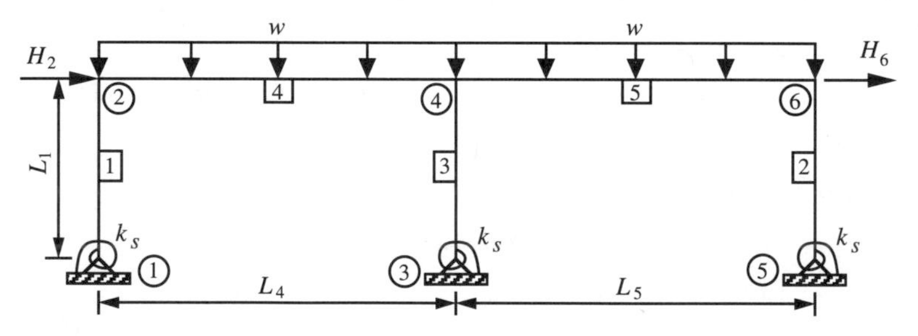

FIGURE P6.15

6.16 See Figure 6.16. $F_y = 50$ ksi.

Members 1 to 4: W14 x 43, $L = 12.5$ ft; $\quad L_b = (KL)_y = 12.5$ ft.

Member 5: W21 x 44, $L = 30$ ft; $\quad L_b = (KL)_y = 6$ ft.

Member 6: W18 x 35, $L = 30$ ft; $\quad L_b = (KL)_y = 6$ ft.

k_s accounts for $G = 2$ at joints 1 and 2.

Nominal loads are:

Dead: $w_5 = 2.61$ kips/ft $\quad w_6 = 2.1$ kips/ft

Live (reduced): $w_5 = 1.54$ kips/ft

Snow: $w_6 = 0.9$ kips/ft

Wind: $w_6 = -0.507$ kips/ft

$H_3 = 3.25$ kips $\quad H_4 = 2.03$ kips

$H_5 = 1.62$ kips $\quad H_6 = 1.02$ kips

P-DELTA analyses were performed for each of the required LRFD load combinations (p. 6-30) and the results at the ends of members 5 and 6 are:

1. For $1.2D + 1.6L + 0.5S$,

$X_3 = -18.2$ kips $Y_3 = 83.9$ kips $M_3 = 349.3$ ft-kips $M_4 = -349.3$ ft-kips

$X_5 = 31.9$ kips $Y_5 = 44.55$ kips $M_5 = 190$ ft-kips $M_6 = -190$ ft-kips

2. For $1.2D + 1.6S + 0.5L$,

$X_3 = -25.7$ kips $Y_3 = 58.5$ kips $M_3 = 254$ ft-kips $M_4 = -254$ ft-kips

$X_5 = 33.3$ kips $Y_5 = 59.4$ kips $M_5 = 241$ ft-kips $M_6 = -241$ ft-kips

Does the design satisfy LRFD H1.2 (p. 6-60)?

6.17 See Figure 6.17. $F_y = 50$ ksi.

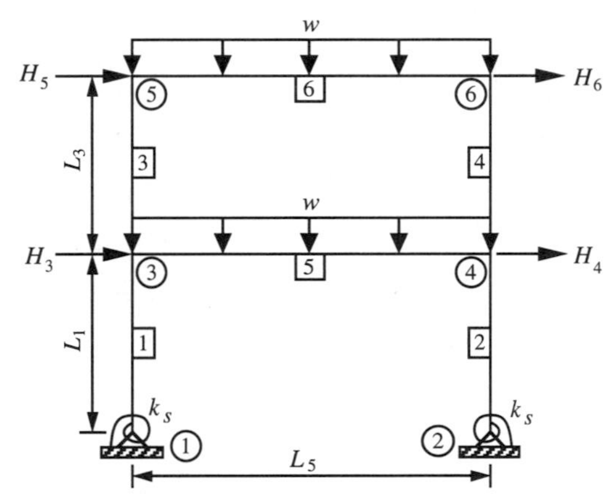

P6.16

Members 1 to 4: W10 x 33, $L = 12.5$ ft $L_b = (KL)_y = 12.5$ ft
Members 5 to 6: W10 x 39, $L = 12.5$ ft $L_b = (KL)_y = 12.5$ ft
Members 7 to 8: W24 x 62, $L = 30$ ft $L_b = (KL)_y = 6$ ft
Members 9 to 10: W21 x 50, $L = 30$ ft $L_b = (KL)_y = 6$ ft
k_s accounts for $G = 2$ at joints 1 to 3.

Nominal loads are:

Dead: $w_7 = w_8 = 2.61$ kips/ft $w_9 = w_{10} = 2.1$ kips/ft
Live (reduced): $w_7 = w_8 = 1.54$ kips/ft
Snow: $w_9 = w_{10} = 0.9$ kips/ft
Wind: $w_9 = w_{10} = -0.507$ kips/ft

$H_4 = 3.25$ kips $H_6 = 2.03$ kips

$H_7 = 1.62$ kips $H_9 = 1.02$ kips

P-DELTA analyses were performed for each of the required LRFD load combinations (p. 6-30) and the results at the ends of members 7 to 10 are:

1. For $1.2D + 1.6L + 0.5S$

$X_4 = -8.88$ kips $\quad Y_4 = 71.1$ kips $\quad M_4 = 157.5$ ft-kips $\quad M_5 = -541.6$ ft-kips

$X_5 = -8.88$ kips $\quad Y_5 = 96.7$ kips $\quad M_5 = 541.6$ ft-kips $\quad M_6 = -157.5$ ft-kips

$X_7 = 14.87$ kips $\quad Y_7 = 38.1$ kips $\quad M_7 = 89.6$ ft-kips $\quad M_8 = -282.7$ ft-kips

$X_8 = 14.87$ kips $\quad Y_8 = 51.0$ kips $\quad M_8 = 282.7$ ft-kips $\quad M_9 = -89.6$ ft-kips

2. For $1.2D + 1.6S + 0.5L$

$X_4 = -11.0$ kips $\quad Y_4 = 50.4$ kips $\quad M_4 = 124.0$ ft-kips $\quad M_5 = -369.3$ ft-kips

$X_5 = -11.0$ kips $\quad Y_5 = 66.7$ kips $\quad M_5 = 369.3$ ft-kips $\quad M_6 = -124.0$ ft-kips

$X_7 = 14.9$ kips $\quad Y_7 = 49.9$ kips $\quad M_7 = 102.7$ ft-kips $\quad M_8 = -386.8$ ft-kips

$X_8 = 14.9$ kips $\quad Y_8 = 68.9$ kips $\quad M_8 = 386.8$ ft-kips $\quad M_9 = -102.7$ ft-kips

3. For $1.2D + 1.6S + 0.8W$

$X_4 = -6.95$ kips $\quad Y_4 = 39.5$ kips $\quad M_4 = 83.5$ ft-kips $\quad M_5 = -309.1$ ft-kips

$X_5 = -8.34$ kips $\quad Y_5 = 52.3$ kips $\quad M_5 = 280.5$ ft-kips $\quad M_6 = -121.2$ ft-kips

$X_7 = 14.5$ kips $\quad Y_7 = 44.5$ kips $\quad M_7 = 86.3$ ft-kips $\quad M_8 = -351.7$ ft-kips

$X_8 = 13.4$ kips $\quad Y_8 = 61.7$ kips $\quad M_8 = 344.6$ ft-kips $\quad M_9 = -94.7$ ft-kips

Does the design satisfy LRFD H1.2 (p. 6-60)?

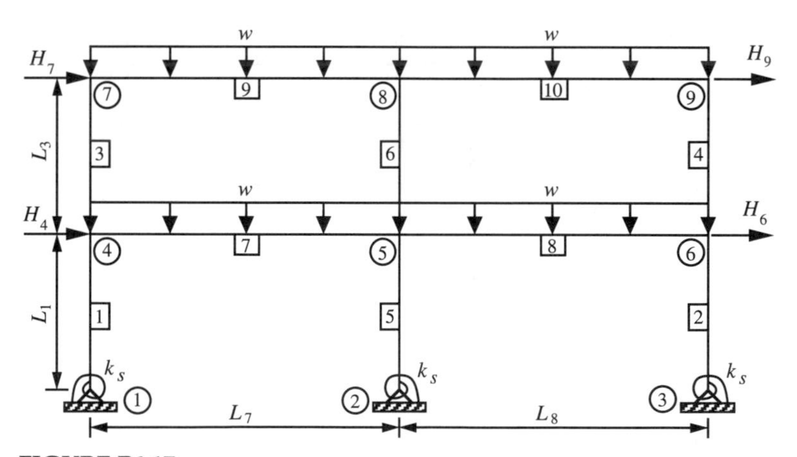

FIGURE P6.17

CHAPTER 7

Bracing Requirements

7.1 INTRODUCTION

This chapter discusses bracing requirements and provides conservative guidelines adapted from Winter [18], McGuire [19], and Galambos [20] for the preliminary design of the bracing. The following types of bracing are considered:

1. Diagonal bracing in a braced frame (LRFD C1 and C2.1)
   (a) Cross braces (each truss diagonal is designed as a tension member)
   (b) K braces (one K-truss diagonal is in compression; other diagonal is in tension)
2. Weak-axis column braces (LRFD B4)
3. Compression flange braces of a beam (to prevent lateral-torsional buckling)

7.2 STABILITY OF A BRACED FRAME

In the following discussion, we assume that lateral stability for the braced frame direction of a structure is to be provided by diagonal bracing in vertical, cantilever, plane trusses (see Figures 7.1 to 7.5).

The structure in Figure 7.1 is composed of unbraced frames in the *Y*-direction and braced frames in the *X*-direction. Only a two-bay by three-bay structure was chosen to simplify the graphical presentation. The flat roof slab for this structure is assumed to be rigid in-plane and adequately fastened to the roof framing members such that the joints at the top end of all columns translate the same amount when the structure is subjected to wind independently in each of the *X*-and *Y*-directions. Therefore, each of the three diagonal braces provides one-third of the total bracing required for the sum of the weak-axis column buckling loads and the total factored wind load in the *X*- direction. If a structure has more bays in the *X*-direction than shown in Figure 7.1, the structural designer probably will choose to locate some of the required diagonal bracing in two or more bays (see Figure 7.2).

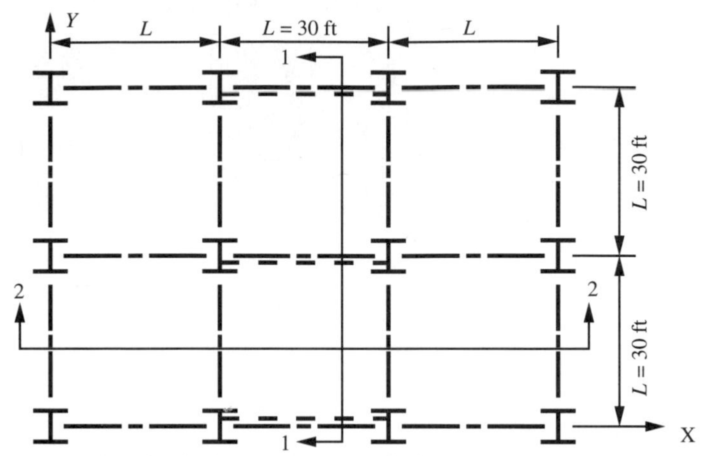

A36 steel; All columns: W14 x 48; All girders (Y direction): W24 x 55

(a) Roof framing plan

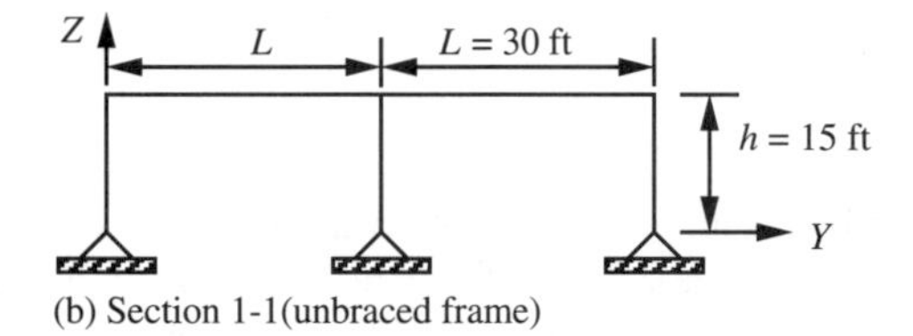

(b) Section 1-1(unbraced frame)

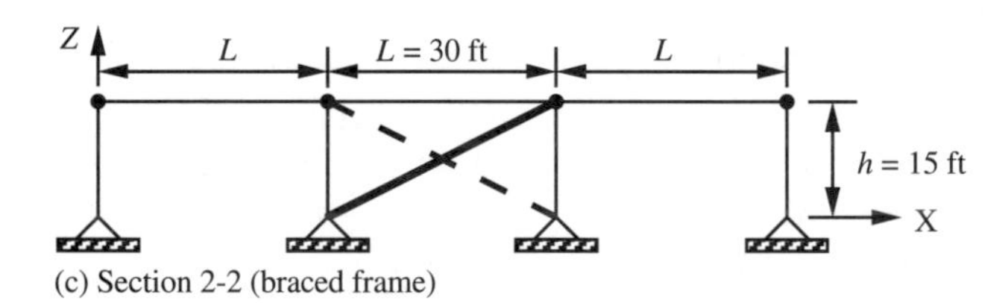

(c) Section 2-2 (braced frame)

Figure 7.1 One-story building.

The lateral bracing in the Y-direction of Figure 7.1 must provide adequate stiffness to control drift and adequate strength to prevent overturning for each of the required loading combinations [LRFD Eqs. (A4-1) to (A4-6)]. For simplicity in the discussion, lateral bracing is discussed only for each of the loading combinations defined by LRFD Eqs. (A4-3) and (A4-4). The size of the diagonal braces can be chosen using the preliminary design guidelines given here. Then, accounting for second-order effects (LRFD C1 and C2.1), a structural analysis of the vertical,

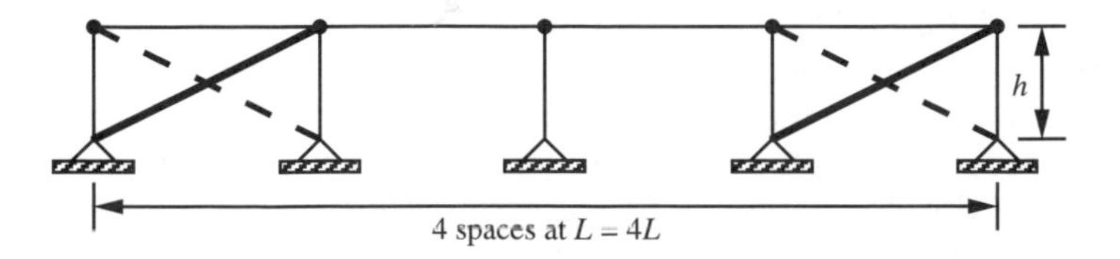

Figure 7.2 Bracing in more than one bay.

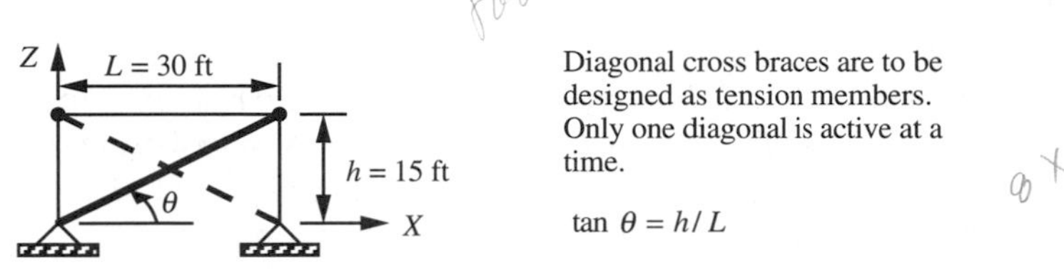

(a) Details of the ideal braced bay in Figure 7.1

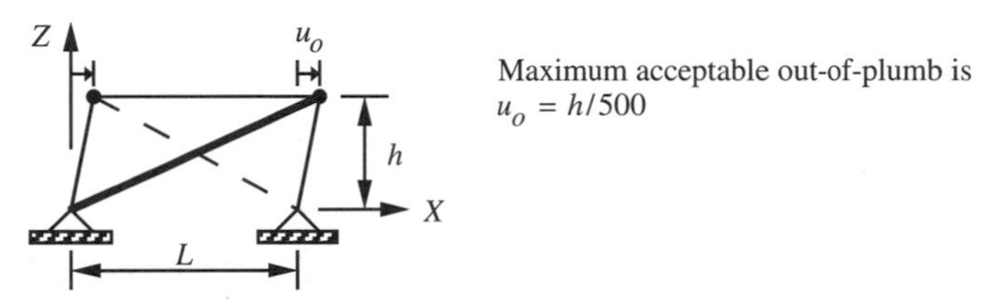

(b) Actual erection position

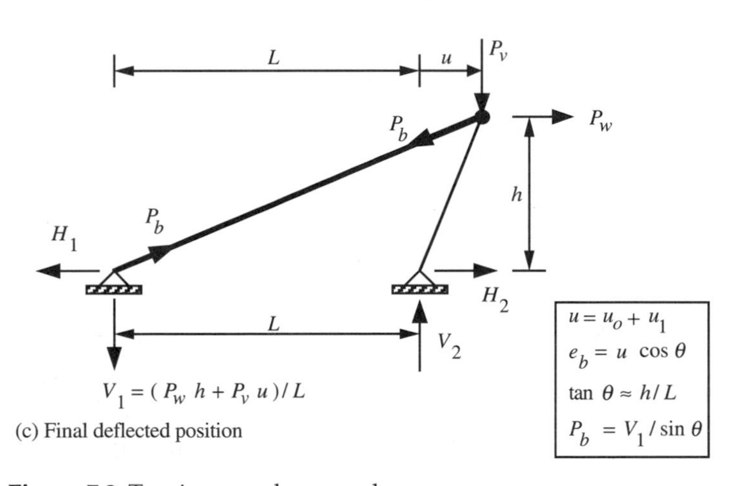

(c) Final deflected position

Figure 7.3 Tension-member cross braces.

cantilever, and plane trusses containing the diagonal braces must be performed to ensure lateral stability of the structure in the Y-direction. In this lateral stability analysis, axial deformation of all members in the vertical bracing system shall be included (LRFD C2.1). In the preliminary design guidelines given for sizing the diagonal braces, axial deformation of only the diagonal braces is included.

As shown in Figure 7.3, during construction the columns cannot be perfectly plumbed and are erected with an acceptable out-of-plumbness $u_o \leq h/500$ (see LRFD, p. 6-254). The diagonal braces are installed in this out-of-plumb position and subsequently subjected to nominal loads that increase as construction progresses to completion.

7.2.1 Required Stiffness and Strength of Cross Braces

The derivations for the required stiffness and strength of a typical diagonal brace in Figure 7.3(c) are made for a loading combination that contains a term for wind. The solutions for a loading combination that does not contain a wind term are obtained by deleting the term involving the factored wind force P_w from the solutions that contain a term for wind. P_v is the sum of the column buckling loads to be resisted by one diagonal brace. P_w is the total factored load to be resisted by one diagonal brace due to wind on the end of the building and is applied in the X- direction. P_v and P_w are applied to the joint at the top end of the diagonal brace.

At reaction 2 in Figure 7.2(c), $\Sigma M_Y = 0$ gives

$$V_1 = \frac{P_w h + P_v u}{L}$$

where

$$u_o = \text{original out-of-plumbness}$$

$$u_1 = \text{drift that occurs due to } P_w \text{ and } P_v$$

$$u = (u_o + u_1) = \text{the final deflected position of the column tops}$$

The force in the brace is

$$P_b = \frac{V_1}{\sin\theta} = \frac{P_w h + P_v u}{L\sin\theta}$$

and the elongation of the brace is

$$(e_b = u\cos\theta) = [(PL/EA)_b = P_b/S_b]$$

where $S_b = (EA/L)_b$ is the actual stiffness of the brace. Thus, for this loading combination, the required stiffness is

$$S_b = \frac{P_b}{e_b} = \frac{V_1/\sin\theta}{u\cos\theta}$$

$$S_b = \frac{P_w h + P_v u}{uL\sin\theta\,\cos\theta} = \frac{P_v + P_w h/u}{L\sin\theta\,\cos\theta}$$

To be conservative, assume that the maximum out-of-plumbness $u_o = h/500$ exists and the maximum acceptable value of additional drift u_1 occurs. Note that u_1 is due to factored loads and must be chosen by the structural designer since the LRFD Specification does not give any guidelines on the maximum acceptable drift. For serviceability of a steel framework structure, the drift index is customarily chosen to be in the range of $h/667$ to $h/200$. For $1.2D + 1.6L$ and $L/D = 3$, this corresponds to $1.5(D + L)$. To be conservative, we recommend that $u = 1.5(h/200) = 3h/400 = h/133$ be used in the preliminary design of the brace. Thus, the required stiffness is

$$\left[S_b = \left(\frac{EA}{L}\right)_b\right] \geq \frac{(P_v + 133\,P_v)}{L\sin\theta\,\cos\theta}$$

For a tension member, the LRFD strength requirement is

$$\phi P_n \geq \left[P_b = \frac{(P_w h + P_v u)}{(L\sin\theta)}\right]$$

Thus, the strength requirement is

$$\phi P_n \geq \frac{\left[h\left(P_w + P_v / 133\right)\right]}{\left(L \sin\theta\right)}$$

Summary

The design requirements for each diagonal brace are

1. For stiffness,

$$\left[S_b = \left(\frac{EA}{L}\right)_b\right] \geq \frac{\left(P_v + 133\,P_w\right)}{\left(L\sin\theta\ \cos\theta\right)}$$

2. For strength

$$\phi P_n \geq \frac{\left[h\left(P_w + P_v / 133\right)\right]}{\left(L \sin\theta\right)}$$

The preceding preliminary design guidelines for a one-story building can be applied in each story of a multistory building (see Figure 7.4). Then, in the story for which the diagonal brace is being designed, the definitions of P_v and P_w are

$$P_v = \sum_{j=1}^{nc} P_j$$

where

$$nc = \text{total number of columns in the story}$$

$$P_j = \phi P_{ny} \text{ of the } j\text{th column in the story}$$

$$P_w = \sum_{j=i+1}^{ns} W_j$$

where

$$ns = \text{total number of stories}$$

$$i = \text{number of the floor level in the story}$$

$$W_j = \text{total factored wind load applied at the } j\text{th floor level}$$

Consider a brace that is perpendicular to the members being braced. As in the preceding discussion, let P_v denote the compressive force being braced. Temporarily, assume that the brace does not have to resist lateral loads due to wind or earthquakes. The long-standing conservative rule of thumb is that the required strength of this brace is $0.02P_v$. If the brace is inclined to the members being braced as in Figure 7.1(c), $0.02P_v$ is the horizontal component of the required strength of the brace. However, the brace must have *adequate stiffness* and *strength* to qualify as lateral bracing.

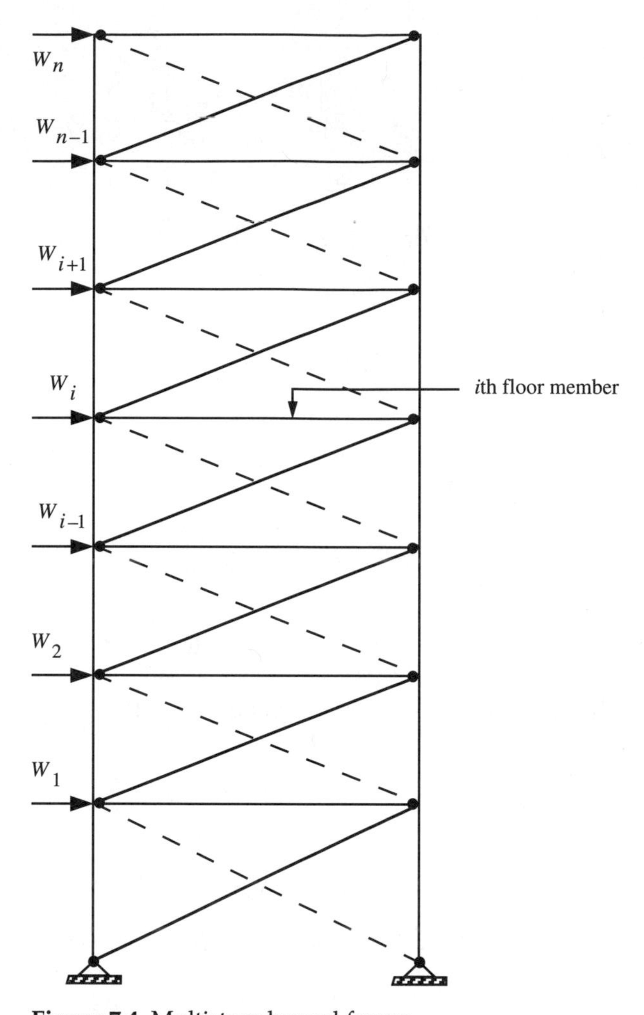

Figure 7.4 Multistory braced frame.

Example 7.1

The diagonal braces in Figure 7.1(a) are to be designed for the bracing configuration in Figure 7.1(c) and the following conditions:

$$F_y = 36 \text{ ksi} \qquad F_u = 58 \text{ ksi} \qquad h = 15 \text{ ft}; \; L = 30 \text{ ft};$$

$$\tan\theta = h/L = 15/30 = 0.5 \qquad \theta = 26.57°; \qquad \sin\theta = 0.447$$

Select the minimum acceptable diameter A36 threaded rod to serve as the diagonal brace in Figure 7.1(b) for the following LRFD load combinations:

Loading 1

$$1.2D + 1.6S + 0.5L \quad \text{which is LRFD (A4-3)}$$

Loading 2

$$1.2D + 1.3W + 0.5L + 0.5S \quad \text{which is LRFD (A4-4)}$$

For all loadings, use $P_v = \Sigma P_n$, where P_n = nominal strength of each column. In loading 2, assume that $P_w = 3.30$ kips.

Solution

Compute $P_v = \Sigma P_n$ for the W14 × 48 columns of A36 steel. For each exterior column,

$$\left[(KL)_y = 15 \text{ ft}\right] > \left[\frac{(KL)_x}{(r_x/r_y)} = \frac{1.85(15)}{3.06} = 9.07 \text{ ft}\right]$$

For each interior column,

$$\left[(KL)_y = 15 \text{ ft}\right] > \left[\frac{(KL)_x}{(r_x/r_y)} = \frac{1.77(15)}{3.06} = 8.68 \text{ ft}\right]$$

Therefore, weak-axis column strength governs. From the LRFD column tables, we find $\phi P_n = \phi P_{ny} = 270$ kips for each column. Each diagonal brace must provide the weak-axis bracing for four columns. Therefore, each diagonal brace must be satisfactorily designed for:

Loading 1

$$P_v = \Sigma P_n = 4(270/0.85) = 1271 \text{ kips}$$

Loading 2

$$P_v = 1271 \text{ kips} \qquad \text{and} \qquad P_w = 3.30 \text{ kips}$$

The strength design requirement is

$$\phi P_n \geq \left[P_b = \frac{h(P_w + P_v/133)}{(L\sin\theta)}\right]$$

From LRFD Table J3.2 (p. 6-81), the design tensile strength for an A36 threaded rod is

$$\phi P_n = 0.75\,(0.75F_uA_b)$$
$$= 0.75\,(0.75)(58\text{ ksi})A_b = 32.625A_b$$

where A_b = gross area of the threaded rod.

Loading 2 governs P_b:

$$P_b = \frac{h(P_w + P_v/133)}{L\sin\theta}$$

$$= \frac{(15)(3.30 + 1271/133)}{(30)(0.447)} = 14.4 \text{ kips}$$

To satisfy the strength requirement, we need

$$(\phi P_n = 32.625A_b) \geq (P_b = 14.4 \text{ kips})$$
$$A_b \geq (14.4/32.625 = 0.441 \text{ in.}^2)$$
$$(\pi d^2/4) \geq 0.441 \text{ in.}^2$$
$$d \geq 0.7496 \text{ in.}$$

Try $d = 3/4$ in. threaded rod: $(A_b = 0.442 \text{ in.}^2) \geq 0.441 \text{ in.}^2$

Note: By the 2% rule-of-thumb method,

$$P_b = 0.02(1271 + 3.30)/\cos\theta = 32.1 \text{ kips}$$

which would require 32.1/14.4 = 2.22 times more strength than we actually need.

Check the stiffness design requirement, which is

$$\left[S_b = \left(\frac{EA}{L}\right)_b\right] \geq \left[\frac{(P_v + 133P_w)}{(L\sin\theta\ \cos\theta)}\right]$$

$$L_b = \frac{h}{\sin\theta} = \frac{15}{0.447} = 33.56 \text{ ft}$$

$$S_b = \frac{(29{,}000)(0.442)}{33.56} = 382 \text{ k/ft}$$

For loading 1,

$$\frac{P_v}{L\sin\theta\ \cos\theta} = \frac{1271}{(30)(0.447)(0.894)} = 106 \text{ k/ft}$$

For loading 2,

$$\frac{P_v + 133P_w}{L\sin\theta\ \cos\theta} = \frac{1271 + 133(3.30)}{(30)(0.447)(0.894)} = 143 \text{ ft}$$

Since 382 > 143, we would use a 3/4-in.-diameter threaded rod as the diagonal brace. For this choice of threaded rod, note that $L/d = 33.56(12)/0.75 = 537$, and for threaded rods we prefer to use $d \geq 5/8$ in. and $L/d \leq 500$, which corresponds to $L/r \leq 2000$.

Instead of the threaded rod chosen in Example 7.1, suppose that either a single-angle or double-angle tension member with field-welded end connections is to be selected for the diagonal brace. If the welds are adequately designed such that the tension member governs the design strength, ϕP_n is the least of $0.90F_yA_b$, $0.75F_uA_bU$, and ϕR_n for block shear rupture. A_b of the diagonal brace must be chosen such that

$$\phi P_n \geq \left[P_b = \frac{h(P_w + P_v/133)}{(L\sin\theta)}\right]$$

$$\left[S_b = \left(\frac{EA}{L}\right)_b\right] \geq \left[\frac{(P_v + 133P_w)}{(L\sin\theta\ \cos\theta)}\right]$$

are satisfied for strength and stiffness.

Instead of the threaded rod chosen in Example 7.1, suppose that either a single-angle or double-angle tension member with bolted end connections is to be selected for the diagonal brace. If the bolts are adequately designed such that the tension member governs the design strength, ϕP_n is the least of $0.90F_yA_b$, $0.75F_uU(A_b - A_{\text{holes}})$, and ϕR_n for block shear rupture. A_b of the diagonal brace must be chosen such that

$$\phi P_n \geq \left[P_b = \frac{h(P_w + P_v/133)}{(L \sin\theta)} \right]$$

$$\left[S_b = \left(\frac{EA}{L}\right)_b \right] \geq \left[\frac{(P_v + 133P_w)}{(L \sin\theta \cos\theta)} \right]$$

are satisfied for strength and stiffness.

7.2.2 Required Stiffness and Strength of K Braces

Figure 7.5 shows how the diagonals of a K truss participate to provide lateral bracing for a braced plane frame. The axial force in each diagonal is P_b. One diagonal is in tension and the other diagonal is in compression. Thus, the strength requirement of $\phi P_n \geq P_b$ must be satisfied for both diagonals and ϕP_n of the compression diagonal usually is smaller than ϕP_n of the tension diagonal. However, when the wind direction is reversed, the original tension diagonal becomes a compression diagonal. Therefore, both diagonals must be identical.

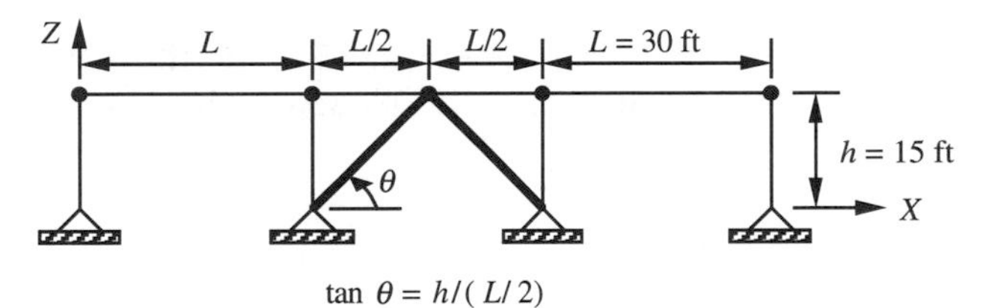

(a) Section 2-2 (braced frame) of Figure 7.1(a)

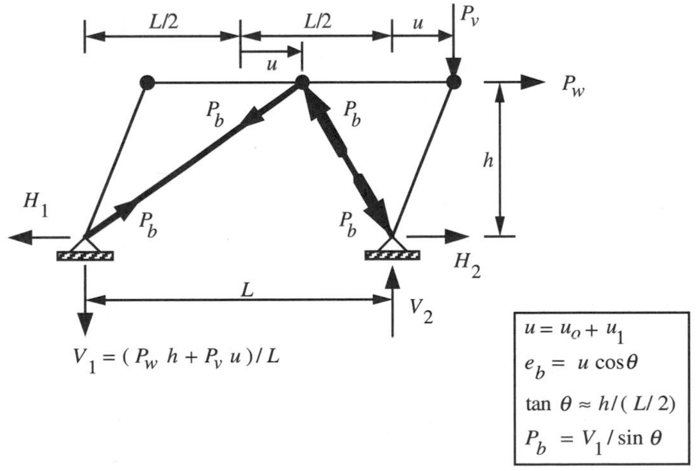

(b) Final deflected position

Figure 7.5 K-truss braces.

The derivations for the required stiffness and strength of each diagonal brace in Figure 7.5(b) are similar to those for the cross braces. The design requirements are:

1. For stiffness,

$$\left[S_b = \left(\frac{EA}{L} \right)_b \geq \frac{(P_v + 133 P_w)}{(L \sin \theta \cos \theta)} \right]$$

2. For strength,

$$\phi P_n \geq \left[\frac{h(P_w + P_v / 133)}{(L \sin \theta)} \right]$$

Example 7.2

The diagonal braces in Figure 7.1(a) are to be designed for the bracing configuration in Figure 7.5 and the following conditions:

$$F_y = 36 \text{ ksi} \qquad F_u = 58 \text{ ksi} \qquad h = 15 \text{ ft} \qquad L = 30 \text{ ft}$$

$$\tan \theta = h/(L/2) = 15/15 = 1 \qquad \theta = 45^\circ \qquad \sin \theta = 0.707$$

Select the minimum acceptable pair of A36 steel angles that can serve as the diagonal brace in Figure 7.5 for the following LRFD load combinations:

Loading 1

$$1.2D + 1.6S + 0.5L \quad \text{which is LRFD (A4-3)}$$

Loading 2

$$1.2D + 1.3W + 0.5L + 0.5S \quad \text{which is LRFD (A4-4)}$$

For all loadings, use $P_v = \Sigma P_n$, where P_n = nominal strength of each column.
In loading 2, assume that $P_w = 3.30$ kips.
Select the pair of angles for:

1. Welded-end connections
2. Bolted-end connections

Solution

Note that Example 7.2 is Example 7.1 with the following changes:

1. K braces (Figure 7.5) are to be designed instead of X braces [Figure 7.1(c)]
2. A pair of angles is to be selected instead of a tie rod.

Therefore, from Example 7.1, each diagonal brace must be satisfactorily designed for

Loading 1

$$P_v = \Sigma P_n = 4(270/0.85) = 1271 \text{ kips}$$

Loading 2

$$P_v = 1271 \text{ kips} \qquad \text{and} \qquad P_w = 3.30 \text{ kips}$$

The strength design requirement is:

$$\phi P_n \geq \left[P_b = \frac{h(P_w + P_v/133)}{(L\sin\theta)} \right].$$

For loading 1, $P_w = 0$, and

$$P_b = \frac{h(P_v h/133)}{(L\sin\theta)} = \frac{(1271)(15)/133}{(30)(0.707)} = 6.76 \text{ kips}$$

For loading 2, $P_w = 3.30$, and

$$P_b = \frac{(15)(3.30 + 1271/133)}{(30)(0.707)} = 9.09 \text{ kips}$$

Assume that $\phi P_n \geq 9.09$ kips for the diagonal as a compression member will govern the size. $(KL)_x = (KL)_y = 15/0.707 = 21.2$ ft. From LRFD p. 3-70, a pair of L4 x 3 x 0.25 with two intermediate connectors and $A = 3.38$ in.^{2} is chosen. ($\phi P_n = 17$ kips) ≥ 9.09 kips.

Check the stiffness design requirement, which is

$$\left[S_b = \left(\frac{EA}{L}\right)_b \right] \geq \left[\frac{(P_v + 133 P_w)}{(L\sin\theta \; \cos\theta)} \right]$$

$$S_b = \frac{(29{,}000)(3.38)}{21.2} = 4624 \text{ kips/ft}$$

For loading 1,

$$\frac{P_v}{L\sin\theta \; \cos\theta} = \frac{1271}{(30)(0.707)(0.707)} = 84.8 \text{ kips/ft}$$

For loading 2,

$$\frac{P_v + 133 P_w}{(L\sin\theta \; \cos\theta)} = \frac{1271 + 133(3.30)}{(30)(0.707)(0.707)} = 114 \text{ kips/ft}$$

Since 4624 > 114, more than adequate stiffness is provided.

Now, we must check the diagonal brace as a tension member:

1. Check the strength of the tension diagonal with welded-end connections:

$$\phi F_y A_g = 0.9(36)(3.38) = 110 \text{ kips}$$

$$\phi F_u A_e = 0.75(58)(0.85)(3.38) = 125 \text{ kips}$$

$$(\phi P_n = 110 \text{ kips}) \geq 17 \text{ kips}$$

Therefore, the compression design strength governed as assumed.

The welds will be designed for $\phi P_n = 17$ kips instead of $P_b = 9.09$ kips. That is, the design compressive strength of the member will be developed in the design of the welds. For E70 electrodes and 3/16-in. fillet welds, the strength of the transverse welds on the ends of the pair of angles is

$$[2(0.75)(0.6)(70)(0.707)(3/16)(4)(1.5) = 50.1 \text{ kips}] \geq 17 \text{ kips}$$

Thus, no longitudinal welds are needed for strength on a tension member and there is no block shear rupture condition to check. However, ductility will be improved by using longitudinal welds with end returns instead of the transverse welds.

2. Check the strength of the tension diagonal with bolted-end connections.

The bolts will be designed for $\phi P_n = 17$ kips instead of $P_b = 9.09$ kips. For one 5/8-in.-diameter A325N bolt in double shear, $(\phi R_n = 22.1 \text{ kips/bolt}) \geq 17$ kips is adequate at each end of the member.

Check bearing:

2 (0.25) > 0.375 in.; the 3/8-in.-thick separator plate governs.

$$\phi R_n = 0.75(2.4dtF_u) = 0.75(2.4)(0.625)(0.375)(58)$$

$$[\phi R_n = 24.5 \text{ kips/(bolt location)}] \geq 17 \text{ kips} \quad \text{(OK)}$$

Check design tension strength:

When there is only one bolt, we use the recommendation on LRFD page 6-172 which is

$$A_e = \text{net area of the connecting element in each angle}$$

$$[\phi F_u A_e = 0.75(58)(2)(0.25)(4 - 0.75) = 70.7 \text{ kips}] \geq 17 \text{ kips} \quad \text{(OK)}$$

Check block shear rupture:

The minimum end distance is the larger of 1.125 in. and 1.5(0.625) = 0.9375 in

$$A_{\text{hole}} = 0.5[2(0.25)(0.75)] = 0.1875 \text{ in.}^2$$

$$A_{gv} = 2\,(1.125)(0.25) = 0.5625 \text{ in.}^2$$

$$A_{nv} = A_{gv} - A_{\text{hole}} = 0.375 \text{ in.}^2$$

$$A_{gt} = 2\,(0.25)(1.5) = 0.75 \text{ in.}^2$$

$$A_{nt} = A_{gt} - A_{\text{hole}} = 0.5625 \text{ in.}^2$$

$$\phi R_n = 0.75(F_u A_{nt} + 0.6F_y A_{gv}) = 0.75[58 + 0.6(36)](0.5625)$$

$$(\phi R_n = 33.6 \text{ kips}) \geq 17 \text{ kips} \quad \text{(OK)}$$

7.3 WEAK AXIS STABILITY OF A COLUMN

For a pinned-ended, W section column (see Figure 7.6), ϕP_{ny} governs the column design strength. A perfectly straight and perfectly plumb column does not exist. The LRFD definition for ϕP_n contains the assumption that the initial crookedness is a half sine wave with an amplitude of $u_o = L/1500$ [see Figure 7.6(c)]. The maximum acceptable out-of-plumbness is $U_o = L/500$ [see Figure 7.6(d)]. Figure 7.6(e) shows the buckled configuration of an imperfect, pinned-ended, W section column.

Let n denote an integer ($n \geq 2$), and $h = L/n$ denote the distance along the column length between laterally braced points. Then $(KL)_y = (h = L/n)$. Figures 7.7(c–e) are for the case $n = 2$. For a pinned-ended column, $(KL)_x = L$. For some W sections used as columns, $r_x/r_y > 3$. The larger of $(KL)_x/(r_x/r_y)$ and $(KL)_y$ governs the column design strength ϕP_n. In Figure 7.7(c), when

$$\left[(KL)_y = h = \frac{L}{2}\right] > \frac{(KL)_x}{(r_x/r_y)}$$

and when elastic buckling occurs, ϕP_n for this column is four times ϕP_n of the pinned-ended column in Figure 7.7(a). By bracing the weak axis of a column [see Figures 7.7(c) to 7.9] at a sufficient number of points such that

$$\left[(KL)_y = h = \frac{L}{n}\right] < \frac{(KL)_x}{(r_x/r_y)}$$

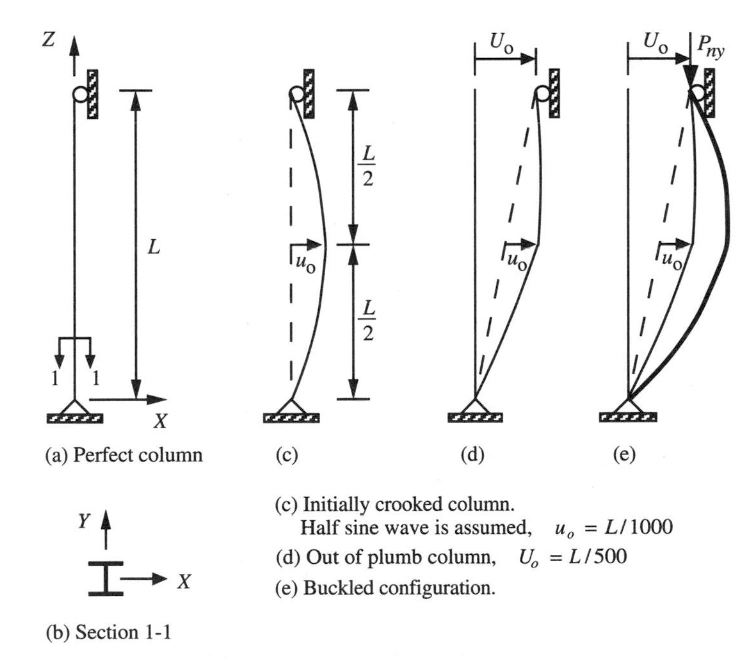

Figure 7.6 Pinned-ended column.

then $\phi P_n = \phi P_{nx}$. In Figure 7.7(c), the roller support at midheight of the column is shown to indicate that a weak-axis column brace exists at this location. Using a roller support to indicate column-braced points at intermediate locations along the length is customary in textbooks until more specific details are required. In the following discussion, an elastic spring [see Figure 7.7(c)] is used to indicate a weak-axis column brace. The elastic spring may be thought of as an axially loaded member whose axial stiffness is

$$S_b = (PL/EA)_b$$

where subscript b denotes *brace*, and

$$P_b = \text{axial force in the brace}$$
$$L_b = \text{length of the brace}$$
$$A_b = \text{gross area of the brace}$$
$$E_b = \text{modulus of elasticity of the brace}$$

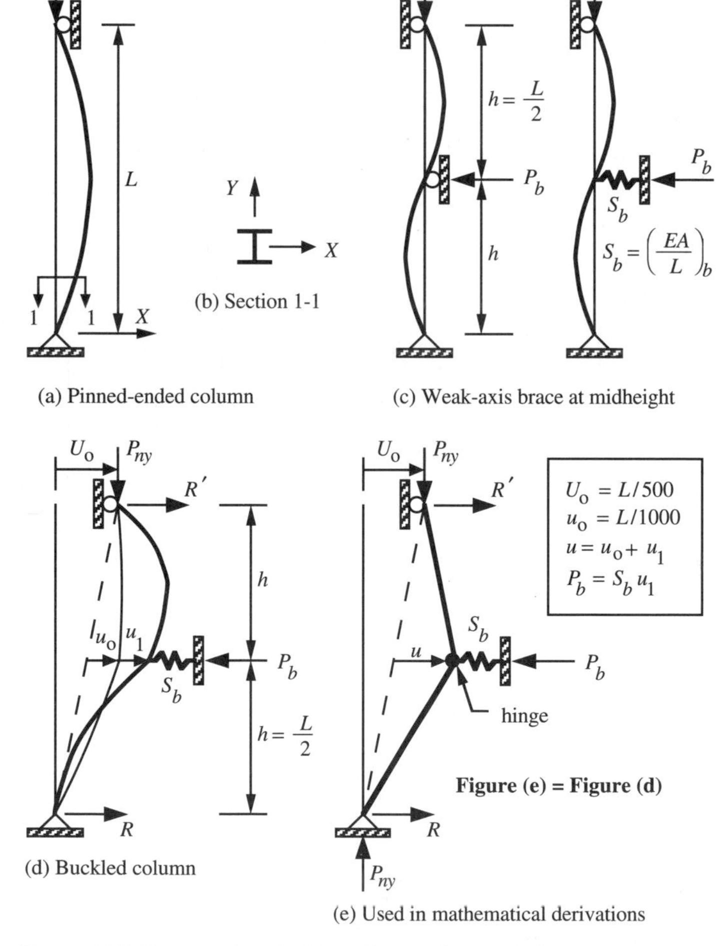

Figure 7.7 One weak-axis brace for a column.

The axial strength of the brace is $(\phi P_n)_b$ and $(\phi P_n)_b \geq P_b$ is required, where P_b is the force in the brace when weak-axis column buckling occurs. If a brace with adequate stiffness and strength is located at the L/n points along the height of a pinned-ended column, then $(KL)_y = (h = L/n)$. If $h < (KL)_x/(r_x/r_y)$, then we have strengthened the column such that $\phi P_n = \phi P_{nx}$. The purpose of the following discussion is to provide guidelines on what is *adequate stiffness* and *strength* for an intermediate column brace.

The braces must be installed such that they prevent lateral translation and twisting. If twisting is not prevented at the braces, the governing column design strength will be due to torsional buckling for $(KL)_z = L$.

7.3.1 Bracing Stiffness and Strength Requirements when $h = L/2$

In Figure 7.7(c), the buckled shape is two half sine waves. There is a point of inflection on the sine curve at the midheight support. Therefore, the internal bending moment $M = 0$ at this location. This is important to remember since the following discussion is based on $M = 0$ at the laterally braced points. Since $M = 0$ at each braced point, for mathematical and graphical convenience we use pinned-ended, straight-column segments between the braced points as shown in Figures 7.7(e) to 7.9. In Section 7.3.5, we will show how to cope with the situation when the point of inflection on the buckled shape does not occur at a braced point.

If less stiffness is provided than that shown in the following derivation, the weaker symmetrical mode shape [Figure 7.7(a) with a dimple at the brace location] will develop. Thus, it is essential that the brace have at least the required stiffness shown here.

In Figure 7.7(e), the maximum out-of-plumbness $U_o = L/500$ and the maximum out-of-straightness $u_o = L/1000$ have been assumed for the weak-axis buckled shape with one intermediate lateral brace P_b. Note that P_{ny} (LRFD nominal column strength for the y-axis) is the buckling load. Also, note that we have chosen to be more conservative by using $u_o = L/1000$ than the $u_o = L/1500$ assumed in the LRFD column strength definition.

For the FBD shown in Figure 7.7(e), from $\Sigma M_y = 0$ at the bottom support,

$$LR' + U_o P_{ny} = hP_b$$

$$R' = \frac{P_b}{2} - \frac{P_{ny}}{500}$$

From $\Sigma F_x = 0$,

$$R = P_b - R' = \frac{P_b}{2} + \frac{P_{ny}}{500}$$

From $\Sigma M_y = 0$ at the top end of the bottom column segment as a FBD,

$$hR = \left(\frac{U_o}{2} + u\right) P_{ny}$$

As recommended by Winter [18], assume that $u_1 = u_o$ in $u = u_o + u_1$; then,

$$P_b = 0.008 P_{ny}$$

P_b is the required bracing strength to produce P_{ny} for the buckling shape of two half sine waves, each of length $h = L/2$.

Since $U_o/2$ and u_o existed when the brace was installed, then

$$P_b = S_b u_1$$

$$S_b = \frac{8P_{ny}}{L}$$

S_b is the required bracing strength to produce P_{ny} for the buckling shape of two half sine waves, each of length $h = L/2$.

The design requirements of an axially loaded member serving as the brace are

$$(\phi P_n)_b \geq \left(P_b = 0.008 P_{ny}\right)$$

$$\left[S_b = \left(\frac{EA}{L}\right)_b\right] \geq \frac{8P_{ny}}{L}$$

7.3.2 Bracing Stiffness and Strength Requirements when $h = L/3$

The buckled shape for Figure 7.8(a) is three half sine waves, and this antisymmetric mode shape is the one desired. If less stiffness is provided than that shown in the following derivation, the weaker symmetrical mode shape (only one half sine wave with a dimple between the braces) shown in Figure 7.8(b) will develop. Thus, it is essential that the brace have at least the required stiffness shown here.

From $\Sigma M_y = 0$ at the bottom support,

$$LR = U_o P_{ny} + hP_b = 2hP_b$$

$$R' = \frac{P_b}{3} - \frac{P_{ny}}{500}$$

From $\Sigma F_x = 0$,

$$R = \frac{P_b}{3} - \frac{P_{ny}}{500}$$

From $\Sigma M_y = 0$ at the top end of the bottom column segment,

$$hR = \left(u - \frac{U_o}{3}\right) P_{ny}$$

Assume that $u_1 = u_o$ in $u = u_o + u_1$; then

$$P_b = 0.018 P_{ny}$$

Since $U_o/3$ and u_o existed when the brace was installed, then

$$P_b = S_b u_1$$

$$S_b = \frac{18 P_{ny}}{L}$$

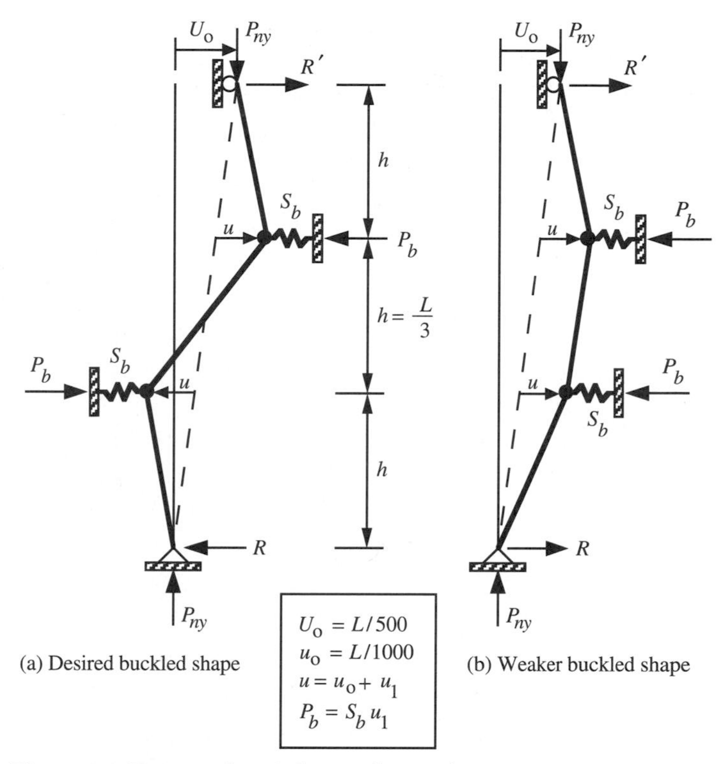

Figure 7.8 Two weak-axis braces for a column.

7.3.3 Bracing Stiffness and Strength Requirements when $h = L/4$

The buckled shape for Figure 7.9 is four half sine waves, and this antisymmetric mode shape is the one desired. If less stiffness is provided than that shown in the following derivation, the weaker symmetrical mode shape (only one half sine wave with dimples at the brace locations) will develop. Thus, it is essential that the brace have at least the required stiffness shown here.

Even though all braces are assumed to have the same stiffness in Figure 7.9, the theoretical mode shape relation is that the displacement at the center brace is maximum and the displacement at the other two braces is 0.707 times the maximum displacement. These relations are shown in Figure 7.9.

From $\Sigma M_y = 0$ at the bottom support,

$$LR' + U_o P_{ny} + 2hP_b = 0.707(h + 3h)P_b$$

$$R' = \frac{P_b}{4.83} - \frac{P_{ny}}{500}$$

From $\Sigma F_x = 0$,

$$R = \frac{P_b}{4.83} + \frac{P_{ny}}{500}$$

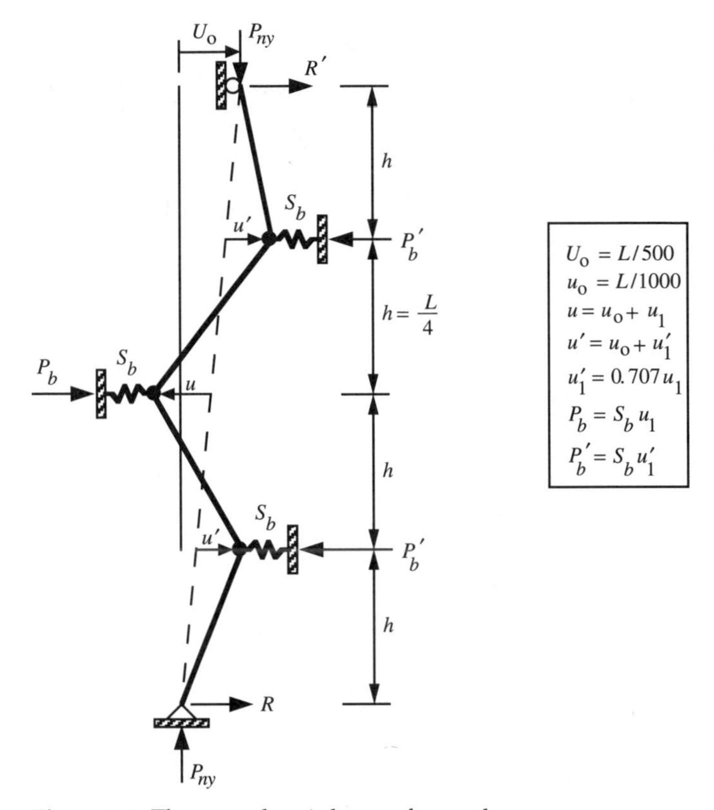

Figure 7.9 Three weak-axis braces for a column.

From $\Sigma M_y = 0$ at the top end of the bottom column segment,

$$hR = \left(u + \frac{U_o}{4}\right)P_{ny}$$

Assume that $u_1 = u_o$ in $u = u_o + u_1$; then,

$$P_b = 0.0273\,P_{ny}$$

Since $U_o/4$ and u_o existed when the brace was installed, then,

$$P_b = S_b u_1$$

$$S_b = \frac{27.3\,P_{ny}}{L}$$

7.3.4 Bracing Stiffness and Strength Requirements When $h = L/n$ for Large n

The buckled shape is $n + 1$ half sine waves and is antisymmetric. If less stiffness is provided than that shown below, the weaker symmetrical mode shape (only one half sine wave with dimples at the brace locations) will develop. Thus, it is essential that the brace have at least the following required stiffness and strength:

$$P_b = 0.008\, nP_{ny}$$

$$S_b = \frac{8\, nP_{ny}}{L}$$

7.3.5 When Point of Inflection Does Not Occur at a Braced Point

For the structure in Figure 7.10, we showed [1, pp. 651–653] that the $M = 0$ point is at $0.5179L$ from the bottom support, whereas the brace is at $0.6L$ from the bottom support.

From $\Sigma\, M_y = 0$ at the bottom support,

$$LR' + U_o P_{ny} = 0.6\, P_b L$$

$$R' = 0.6\, P_b - \frac{P_{ny}}{500}$$

From $\Sigma\, F_x = 0$,

$$R = 0.4\, P_b + \frac{P_{ny}}{500}$$

The ordinate of the half sine wave at the braced point is $0.951u_o$, where u_o is the ordinate at midheight of the column.

For mathematical convenience, assume that $u_1 = 0.951u_o$; thus, $u = 0.951\,(u_o + u_1)$.

From $\Sigma\, M_y = 0$ at the top end of the bottom column segment,

$$0.5179\, LR = \left(0.998\, u / 0.951 + 0.5179\, U_o\right) P_{ny}$$

$$P_b = 0.0101\, P_{ny}$$

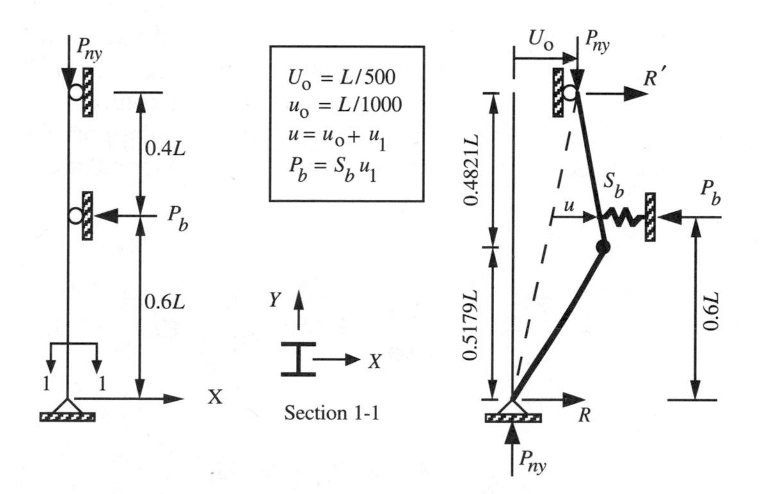

Figure 7.10 Weak-axis column brace does not occur at M = 0 point.

Since U_o and u_o existed when the brace was installed, then

$$P_b = S_b u_1$$

$$S_b = \frac{10.7 P_{ny}}{L}$$

If elastic buckling occurs, according to the LRFD column strength definition,

$$P_{ny} = 0.877\left[\frac{\pi^2 EI}{(0.5179L)^2}\right] = 0.877\left(\frac{3.73\pi^2 EI}{L^2}\right)$$

Since $0.5179L$ is very close to $0.5L$, we can use the solutions for P_b and S_b obtained in Section 7.3.1 for $h = L/2$ and compare the requirements. Again, if elastic buckling occurs when the brace is placed at midheight of the column,

$$P_{ny} = 0.877\left[\frac{\pi^2 EI}{(L/2)^2}\right] = 0.877\left(\frac{4\pi^2 EI}{L^2}\right)$$

The strength required for a brace at $0.6L$ from the bottom end of the column as shown in Figure 7.10 is 0.010(3.73)/ [0.008(4)] = 1.17 times the strength required for a brace at midheight of the column. The stiffness required for a brace at $0.6L$ from the bottom end of the column as shown in Figure 7.10 is 10.7(3.73)/ [8(4)] = 1.25 times the strength required for a brace at midheight of the column. Thus, the reader should not blindly use the stiffness and strength guidelines given in Sections 7.3.1 to 7.3.4 if the braces are not placed at $h = L/n$ points along the column length.

7.3.6 Example Problem

Example 7.3

See Figure 7.11 for the structure in which weak-axis column bracing is to be provided. Member 1 must be designed as a tension member to brace the top end of the four columns. See Section 7.2.1 for the design requirements of this member. Member 2 must be designed as a tension member to brace the weak axis of the four columns at midheight and to transfer the tension force in member 1 to the foundation support. Select the minimum acceptable diameter A36 threaded rod to serve as the diagonal braces.

Solution

$$\left[\frac{(KL)_x}{(r_x/r_y)} = \frac{1.85(15)}{3.06} = 9.07\text{ ft}\right] > \left[(KL)_y = 7.5\text{ ft}\right]$$

For each column,

$$\phi P_n = \phi P_{nx} = 364\text{ kips}$$

$$\phi P_{ny} = 384\text{ kips}$$

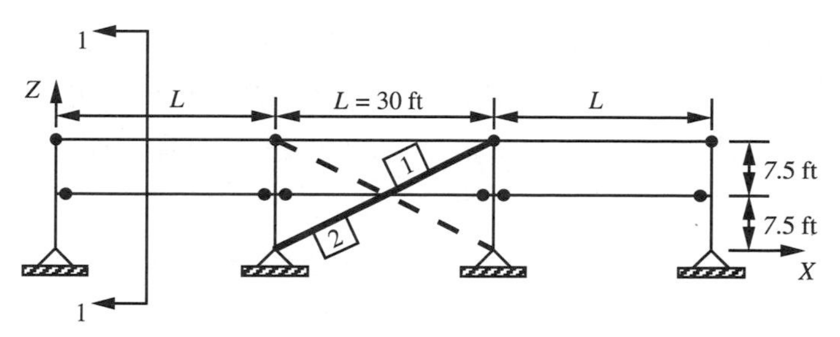

(a) Side elevation view (braced frame)

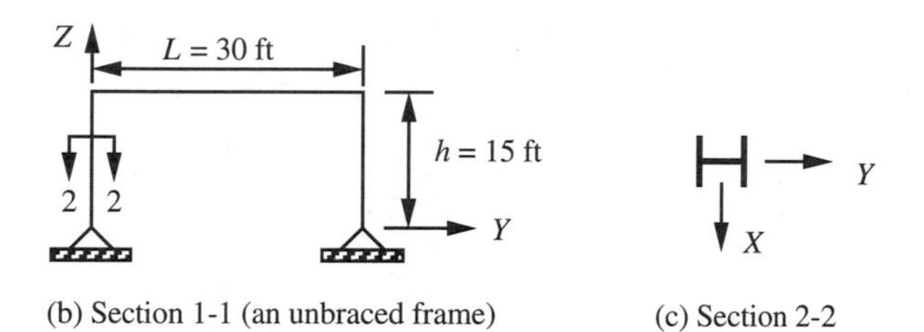

(b) Section 1-1 (an unbraced frame)

(c) Section 2-2

A36 steel; All columns: W14 x 48; All girders (Y direction): W24 x 55

Figure 7.11 Example 7.3: one-story building.

For Member 1,

$$P_v = 4\ (364/0.85) = 1713 \text{ kips}$$

$$P_w = 2.48 \text{ kips (an assumed value for illustration purposes)}$$

The strength design requirement is:

$$\phi P_n \geq \left[P_b = \frac{h(P_w + P_v / 133)}{(L \sin \theta)} \right]$$

From LRFD Table J3.2 (p. 6-81), the design tensile strength for an A36 threaded rod is

$$\phi P_n = 0.75\ (0.75 F_u A_b)$$

$$\phi P_n = 0.75\ (0.75)(58 \text{ ksi})\ A_b = 32.625 A_b$$

where A_b = gross area of the threaded rod.

For Member 1,

Loading 2 governs P_b:

$$P_b = \frac{h(P_w + P_v / 133)}{L \sin \theta}$$

$$= \frac{15(2.48 + 1713/133)}{30(0.707)} = 10.9 \text{ kips}$$

To satisfy the strength requirement, we need

$$(\phi P_n = 32.625A_b) \geq (P_b = 10.9 \text{ kips})$$

$$A_b \geq (10.9/32.625 = 0.334 \text{ in.}^2)$$

$$(\pi d^2/4) \geq 0.334 \text{ in.}^2$$

$$d \geq 0.652 \text{ in.}$$

Try $d = 3/4$ in. threaded rod: $(A_b = 0.442 \text{ in.}^2) \geq 0.334 \text{ in.}^2$

Check the stiffness design requirement, which is

$$\left[S_b = \left(\frac{EA}{L}\right)_b\right] \geq \left[\frac{(P_v + 133P_w)}{(L\sin\theta \ \cos\theta)}\right]$$

$$L_b = \frac{h}{\sin\theta} = \frac{15}{0.707} = 21.2 \text{ ft}$$

$$S_b = \frac{(29{,}000)(0.442)}{21.2} = 605 \text{ kips/ft}$$

$$\frac{P_v + 133P_w}{L\sin\theta \ \cos\theta} = \frac{1713 + 133(2)}{30(0.707)(0.707)} = 136 \text{ kips/ft}$$

Since 605 > 136 and $[L/d = 21.2(12)/0.75 = 339] \leq 500$, a 3/4-in.-diameter A36 steel threaded rod is acceptable for member 1.

Member 2

Since $h = L/2 = 7.5$ ft, Section 7.3.1 gives the design requirements for a weak-axis column brace at midheight:

$$(\phi P_n)_b \geq (P_b = 0.008P_{ny})$$

$$\left[S_b = \left(\frac{EA}{L}\right)_b\right] \geq \frac{8P_{ny}}{L}$$

We must add the axial force from member 1 to the strength requirement since member 2 transfers the force in member 1 to the foundation support:

$$P_b = 10.9 + (0.008)(1713)/0.707 = 30.3 \text{ kips}$$

To satisfy the strength requirement, we need

$$(\phi P_n = 32.625A_b) \geq (P_b = 30.3 \text{ kips})$$

$$A_b \geq (30.3/32.625 = 0.929 \text{ in.}^2)$$

$$(\pi d^2/4) \geq 0.929 \text{ in.}^2$$

$$d \geq 1.09 \text{ in.}$$

Try $d = 1.125$-in. threaded rod: $(A_b = 0.994 \text{ in.}^2) \geq 0.929 \text{ in.}^2$

Check the stiffness design requirement:

$$S_b = 29{,}000(0.994)/21.2 = 1360 \text{ kips/ft}$$

We must add the vertical component of the axial force from member 1 to the stiffness requirement since member 2 transfers the force in member 1 to the foundation support:

$$8\,[1713 + 0.707(10.9)]/15 = 918 \text{ kips/ft}$$

Since $1360 > 918$ and $[L/d = 21.2(12)/1.125 = 226] \leq 500$, a 1.125-in.-diameter threaded rod is acceptable as the diagonal brace.

7.4 LATERAL STABILITY OF A BEAM COMPRESSION FLANGE

For lateral bracing purposes, the compression flange and the top half of the web of a W section beam can conservatively be treated as a column. Thus, the cross-sectional area of the equivalent column is $A_c = A/2$, where A is the gross area of the W section used as a beam. Then, the design requirements given in Section 7.3 can be used to design the lateral braces located at intervals of $h = L/n$ along the length of a beam to prevent lateral-torsional buckling.

Example 7.4

Each flume girder spanning 30 ft in Figure 7.12 is a W24 × 55, $F_y = 36$ ksi that was designed assuming lateral braces were to be provided at intervals of $h = L/n = 30/4 = 7.5$ ft. The lateral braces are to be provided by a pair of angles as shown in Figures 7.12(b) and (c) in conjunction with the cross braces shown in Figure 7.12(b). Design the bracing members.

Solution

We have a simply supported W24 × 55, $F_y = 36$ ksi that spans 30 ft with lateral bracing provided at intervals of 7.5 ft. From the LRFD beam charts, $\phi M_{nx} = 337$ ft-kips for $C_b = 1$ and $L_b = 7.5$ ft:

$$C_b = \frac{12.5\,M_{\max}}{2.5\,M_{\max} + 3\,M_A + 4\,M_B + 3\,M_C}$$

$$C_b = \frac{12.5}{2.5 + 3(55/64) + 4(15/16) + 3(63/64)} = 1.06$$

$$\phi M_{nx} = \text{smaller of} \begin{cases} C_b \overline{M}_1 = 1.06(337) = 357 \\ \phi M_{px} = 362 \text{ ft-kips} \end{cases}$$

$$\phi M_{nx} = 357 \text{ ft-kips}$$

Treat $\phi M_{nx} = 357$ ft-kips as a couple with a lever arm of

$$a = 0.95d = 0.95(23.57) = 22.4 \text{ in.} = 1.87 \text{ ft}$$

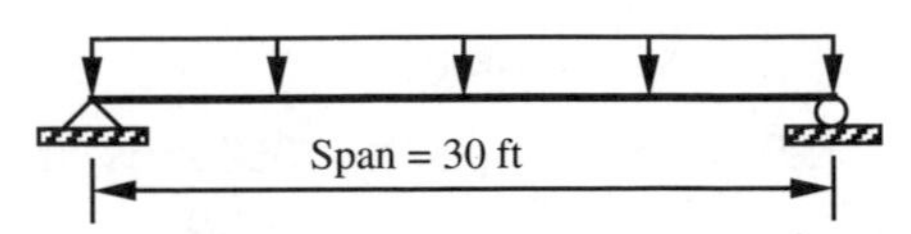

(a) Each flume girder: W24 x 55

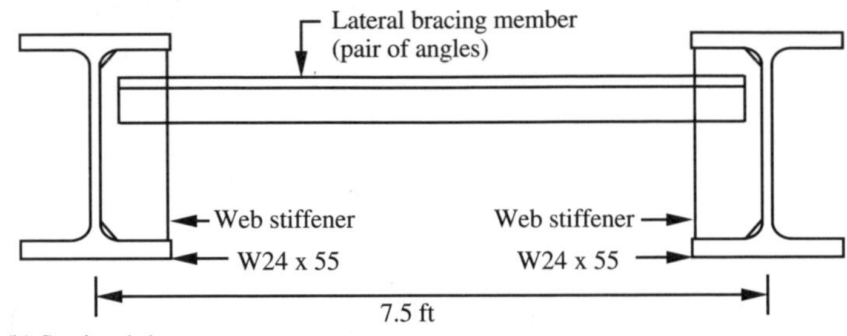

(b) Section 1-1

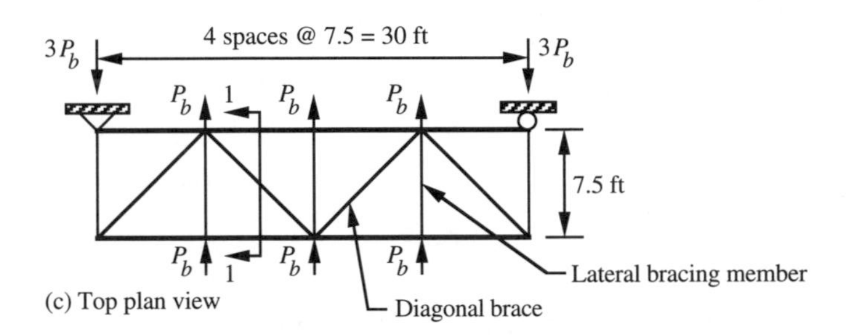

(c) Top plan view

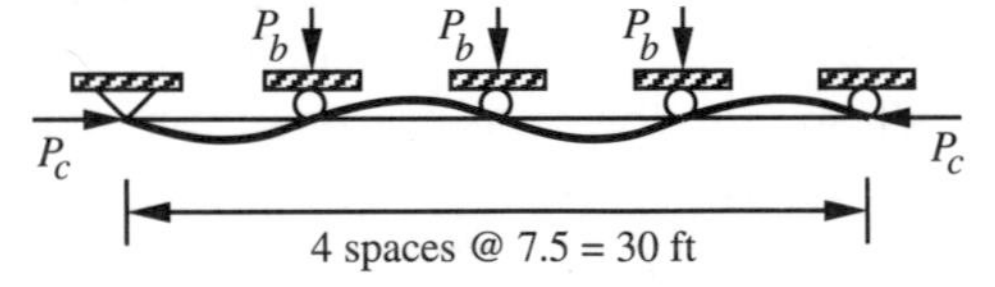

(d) Plan view of buckled W24 x 55 flume girder compression flange

Figure 7.12 Example 7.4: lateral braces for two girders.

Then, the axial compressive force in the equivalent column [see Figure 7.12(d)] is

$$P_c = 357/1.87 = 191 \text{ kips}$$

The force for which lateral bracing is to be provided is

$$P_{ny} = P_c/\phi = 191/0.85 = 225 \text{ kips}$$

From Section 7.3.3 for $h = L/4$, the design requirements for the brace are

$$P_b = 0.0273\,P_{ny}$$

$$S_b = \frac{27.3\,P_{ny}}{L}$$

$$P_b = 0.0273(225) = 6.14 \text{ kips}$$

From LRFD p. 3-64, for $(KL)_x = (KL)_y = 7.5$ ft and $P_u = 6.14$ kips, try a pair of L2 × 2 × 1/4 with two intermediate spacers:

$$(\phi P_n = 10.0 \text{ kips}) \geq (P_b = 6.14 \text{ kips})$$

Check the stiffness requirement:

$$[S_b = 29{,}000(0.960)/7.5 = 3712 \text{ kips/ft}] \geq [27.3(228)/30 = 207] \quad \text{(OK)}$$

Use a pair of L2 x 2 x 1/4 with two intermediate spacers as the lateral brace between the top flanges of the W24 x 55 flume girders. In Figure 7.12(b), each web stiffener serves as the end-connector plate for the pair of L2 x 2 x 1/4 and prevents twisting of the W24 x 55 at each brace location.

Choose a pair of angles welded at their ends to the underside of the top flange of each W24 x 55 to serve as the bracing diagonals in Figure 7.12(c). For the end diagonals that must resist an axial compression force:

$$P_d = (3P_b)/0.707 = 3(6.22)/0.707 = 40.0 \text{ kips}$$

$$L = 7.5/0.707 = 10.6 \text{ ft}$$

From LRFD p. 3-64, try a pair of L3.5 x 3 x 1/4:

$$(\phi P_n = 46 \text{ kips}) \geq (P_b = 40.0 \text{ kips})$$

Check the stiffness requirement:

$$[S_b = 29{,}000(3.13)/7.5/0.707 = 17{,}120 \text{ kips/ft}] \geq [27.3(3)(228)/30 = 622] \quad \text{(OK)}$$

Use a pair of L3.5 x 3 x 1/4.

CHAPTER 8

Connections

8.1 INTRODUCTION

The theoretical analysis techniques for a bolted and a welded connection of the same type usually contain some assumed behavioral features that are very similar. Consequently, we believe that the most appropriate presentation is to discuss bolting and welding for a particular type of connection and then to move on to the discussion of another type of connection. Connectors subjected to concentric shear were discussed in Chapter 2.

8.2 CONNECTORS SUBJECTED TO ECCENTRIC SHEAR

In this type of connection (see Figure 8.1), the resultant force acting on the connectors does not coincide with the center of gravity of the connectors. The alternate terminology is that the resultant force acting on the connectors is eccentric with respect to the center of gravity of the connectors. Figure 8.1 shows the following cases of the connectors being subjected to eccentric shear:

1. A bracket plate is fastened to the flange of a column to support a load [see Figures 8.1(a) and (b)]. LRFD pp. 12-5 to 12-10 contain some tabular information that is useful in the design of a bracket plate.
2. A plate is shop-welded either to the flange or to the web of a column and either bolted or field-welded to a beam web at the ends of a simply supported beam [see Figure 8.1(c)]. The connectors on the beam end are subjected to eccentric shear and a negligible moment. However, the groove weld connecting the plate to the column flange is subjected to shear and eccentric tension (see Section 8.5).
3. A shear splice of a beam [see Figure 8.1(d)] is connected only to the ends of each beam web. Therefore, each spliced member end has a shear and a negligible moment that must be transferred through the splice to the adjacent spliced member end.

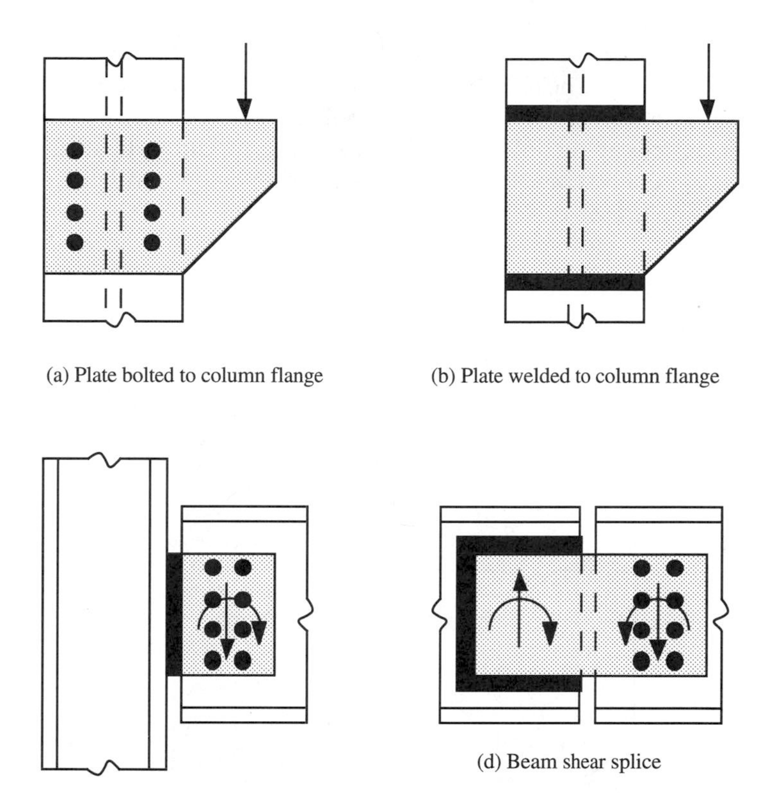

(a) Plate bolted to column flange

(b) Plate welded to column flange

(c) Plate welded to column flange and bolted to beam web

(d) Beam shear splice

FIGURE 8.1 Connectors subjected to eccentric shear.

Two methods of analysis for connectors subjected to eccentric shear will be discussed: the *ultimate strength method* and the *elastic method*. The ultimate strength method allows more economical fastener groups to be used, provides a more consistent factor of safety, and was used in the preparation of the tabular information in LRFD Tables 8-18 (p. 8-40) through 8-25 (p. 8-87) for bolt groups and in LRFD Tables 8-38 (p. 8-163) through 8-45 (p. 8-210) for weld groups. However, the ultimate strength method is an iterative procedure. If a structural designer has a connection configuration that does not conform to one of those shown in the LRFD tabular information, the elastic method is simpler and more conservative, but in some cases it is excessively conservative. We choose to describe the methods of analysis for connector groups subjected to an eccentrically applied failure load that is inclined to the vertical axis of the connector groups.

8.2.1 A Bolt Group Subjected to Eccentric Shear

Ultimate Strength Method

The following discussion is applicable for a bearing-type connection and is based on the description given by Crawford and Kulak [21].

Figure 8.2(a) shows the free-body diagram of a bracket plate subjected to an inclined, factored load that is applied at a point whose eccentric location with respect to the center of gravity of the bolt group is denoted by e_x and e_y. At a point called the instantaneous center of rotation (ICR), only a pure rotation β of the bracket plate is assumed to occur. The ICR point is located at an initially unknown distance of e_{ox} with respect to the center of gravity of the bolt group.

The forces at the holes in the bracket plate are due to bearing of the bolts on the bracket plate. Each bearing force is assumed to be perpendicular to its radius from ICR. For the ith force,

$$F_i = \phi r_n \left(1 - e^{10\, r_i \beta}\right)^{0.55}$$

where

$$\phi r_n = \text{least of} \begin{cases} \text{DSSB} \\ \text{DBSBH} \\ \text{DSCSB} \end{cases}$$

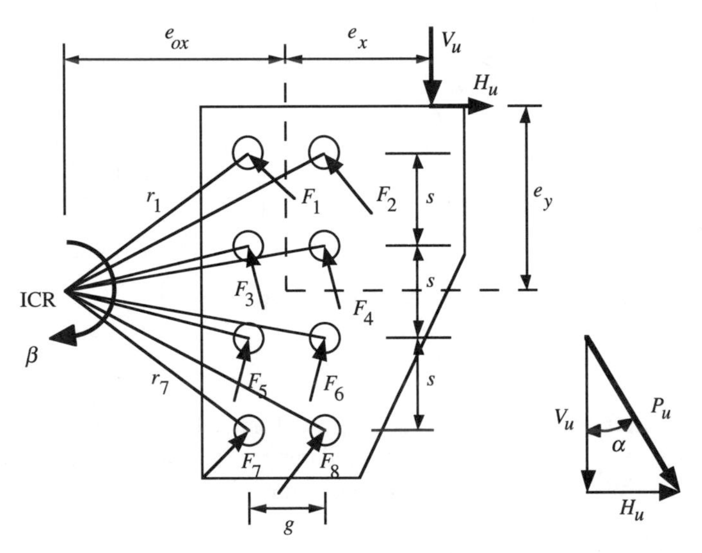

(a) FBD of bracket plate (b) Components of eccentric force

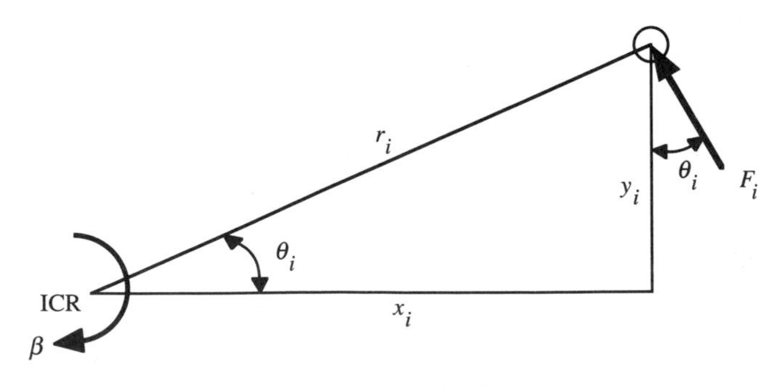

(c) Details for ith bearing force

FIGURE 8.2 Ultimate strength method for eccentric shear of a bolt group

DSSB = design shear strength of a single bolt

DBSBH = design bearing strength at a single bolt hole

DSCSB = design slip-critical strength of a single bolt (if applicable)

$e = 2.718$ (base of natural logarithm)

r_i = radius from ICR to the center of the ith bolt hole

$$\beta = \Delta_{max}/r_{max}$$

$\Delta_{max} = 0.34$ in. (from LRFD p. 8-30)

r_{max} = radius from ICR to the center of the most remote bolt hole

The three equations of statics are

$$\sum_{i=1}^{n} F_i \sin\theta_i - P_u \sin\alpha = 0$$

$$\sum_{i=1}^{n} F_i \cos\theta_i - P_u \cos\alpha = 0$$

$$\left[(e_{ox} + e_x)\cos\alpha + e_y \sin\alpha\right]P_u - \sum_{i=1}^{n} r_i F_i = 0$$

and the parameters are defined in Figure 8.2.

In these equations, the known parameters are the bolt spacings (g and s), e_x, e_y, a, and ϕr_n. We want to find the value of P_u from the last equilibrium equation that also satisfies the first two equilibrium equations. An iterative procedure must be used to find this value of P_u. For an assumed value of e_{ox}, the ICR point is located, r_{max} is determined, and β, r_i, F_i, and θ_i are computed. P_u is then determined from the last equilibrium equation. If this value of P_u satisfies the first two equilibrium equations, the assumed value of e_{ox} was correct. Otherwise, we must assume another value of e_{ox} and repeat the process. Obviously, a computer program needs to be written to conduct the iterative search for the correct value of e_{ox}. Let F_x and F_y, respectively, be the absolute value obtained on the left-hand side of the first and second equilibrium equations. Then, we can start by assuming $e_{ox} = 0.10$ in. and increase e_{ox} by an increment of 0.02 in. until $F_x/(P_u \sin\alpha) \leq 0.01$ and $F_y/(P_u \cos\alpha) \leq 0.01$, which means we ensure that no more than a 1% discrepancy in equilibrium can occur. This type of procedure was used to obtain the C values given in LRFD Tables 8-18 (p. 8-40) through 8-25 (p. 8-87), where $C = P_u/(\phi r_n)$. If we use the LRFD tabular values, the LRFD strength-design requirement for the bolt group is

$$\left[\phi R_n = C(\phi r_n)\right] \geq P_u$$

Note: LRFD p. 8-32 tells us for these tables that:

1. Linear interpolation on any given page is acceptable.
2. Linear interpolation between pages is not acceptable.

Elastic Method

The eccentric force components in Figure 8.3(a) are transferred to the center of gravity of the bolt group that necessitates, as shown in Figure 8.3(c), that we also must have $M = e_x V_u + e_y H_u$ at the center of gravity of the bolt group. Then, the following assumptions are made:

1. At each bolt hole location due to V_u only, there is an upward force of V_u/n, where n = total number of bolts.
2. At each bolt hole location due to H_u only, in the negative x-direction there is a force of H_u/n, where n = total number of bolts.
3. Due only to $M = e_x V_u + e_y H_u$, the plate is assumed to rotate about CG (the center of gravity of the bolt group) and in the direction of β. A radius is connected from CG to the center of each bolt hole. As shown in Figure 8.3(d), F_i due only

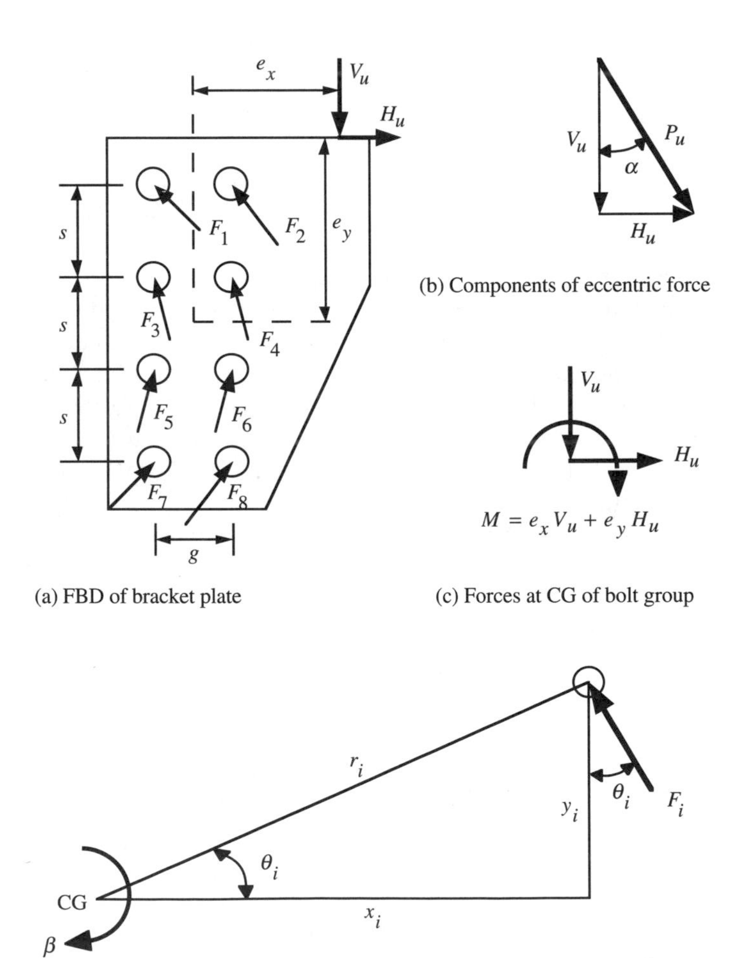

FIGURE 8.3 Elastic method for eccentric shear of a bolt group

to M is assumed to be perpendicular to r_i. The moment of F_i about CG is r_iF_i, and this moment is opposite to the direction of M. Also, we assume that F_i is proportional to r_i, which can be mathematically written as $F_i = r_i\beta$.

Since β is unknown, we can arbitrarily choose $\beta = F_1/r_1$. Then

$$F_i = r_i\left(\frac{F_1}{r_1}\right)$$

and the equilibrium condition that must be satisfied is

$$\left(M = e_x V_u + e_y H_u\right) \equiv \left[\sum_{i=1}^{n} r_i F_i = \left(\frac{F_1}{r_1}\right)\sum_{i=1}^{n} r_i^2\right]$$

which enables us to obtain

$$\frac{F_1}{r_1} = \frac{\left(e_x \cos\alpha + e_y \sin\alpha\right)P_u}{\sum_{i=1}^{n} r_i^2}$$

and

$$F_i = r_i\left(\frac{F_1}{r_1}\right)$$

Let

$$F_{xi} = F_i \sin\theta_i$$
$$F_{yi} = F_i \cos\theta_i$$

Then, the shear in the most heavily loaded bolt is

$$F_j = \sqrt{\left(\frac{H_u}{n} + F_{xj}\right)^2 + \left(\frac{V_u}{n} + F_{yj}\right)^2}$$

The LRFD strength-design requirement for the most heavily loaded bolt is

$$\phi R_n \geq F_j$$

where

$$\phi R_n = \text{least of} \begin{cases} \text{DSSB} \\ \text{DBSBH} \\ \text{DSCSB} \end{cases}$$

DSSB = design shear strength of a single bolt

DBSBH = design bearing strength at a single bolt hole

DSCSB = design slip-critical strength of a single bolt (if applicable)

The elastic method is not iterative, but it gives an excessively conservative value of F_j in some cases.

Example 8.1

For the bearing-type connection shown in Figure 8.4, we are given the following numerical information: A36 steel bracket plate, $g = 3$ in., $s = 6$ in., $e_x = 5$ in., minimum bearing thickness = 0.5 in., and four 7/8-in.-diameter A325X bolts in single shear.

The objectives of this example are to obtain the maximum acceptable value of P_u for this bolt group by using:

1. LRFD p. 8-46
2. The ultimate strength method equations
3. The elastic method equations

Solution 1

From LRFD p. 8-24,

$$\phi r_n = 28.1 \text{ kips/bolt} = (\text{single shear of one bolt})$$

From LRFD p. 8-26,

$$\phi r_n = (91.4 \text{ kips/in.})(0.5 \text{ in.}) = 45.7 \text{ kips/(bolt hole)}$$

From LRFD p. 8-46, for $g = 3$ in., $s = 6$ in., $e_x = 5$ in., and $n = 2$ bolts in each vertical row of the bolt group, we find $C = 2.10$, which enables us to compute the LRFD strength-design requirement for the bolt group:

$$\left[\phi R_n = C(\phi r_n) = 2.10(28.1) = 59.0 \text{ kips}\right] \geq P_u$$

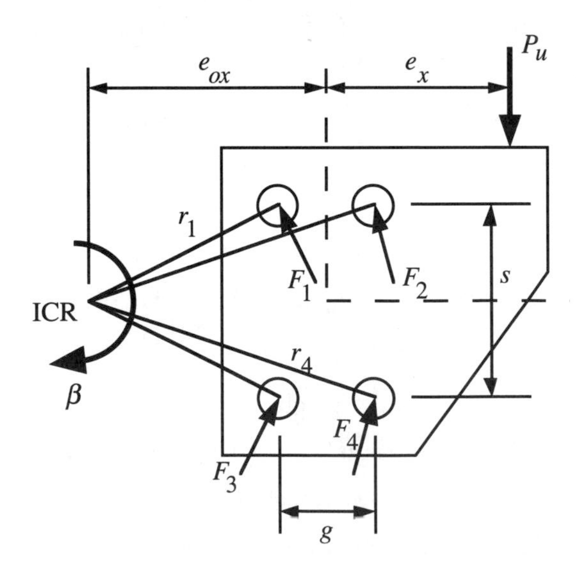

FIGURE 8.4 Example 8.1

Solution 2

Try $e_{ox} = 2.40$ in.

$$x_1 = 2.40 - 1.5 = 0.9 \qquad y_1 = 3 \qquad r_3 = r_1 = 3.1321$$

$$x_2 = 2.40 + 1.5 = 3.9 \qquad y_2 = 3 \qquad r_4 = r_2 = 4.9204$$

$$r_{\max} = 4.9204$$

$$\beta = \frac{\Delta_{\max}}{r_{\max}} = \frac{0.34}{4.9204} = 0.06910$$

$$F_i = \phi r_n \left(1 - e^{10\, r_i \beta}\right)^{0.55}$$

$$F_3 = F_1 = 26.28$$

$$F_4 = F_2 = 27.58$$

$$\sum_{i=1}^{4} r_i F_i = 2[3.1231(2628) + 4.9204(27.58)] = 463.03$$

$$P_u = \frac{\sum_{i=1}^{4} r_i F_i}{e_{ox} + e_x} = \frac{436.03}{2.40 + 5.00} = 58.92 \text{ kips}$$

$$F_Y = \sum_{i=1}^{4} F_i \cos\theta_i = 2\left[\frac{26.28(0.9)}{3.1321} + \frac{27.58(3.9)}{4.9204}\right] = 58.82$$

$$\left(\frac{P_u - F_Y}{P_u} = \frac{58.92 - 58.82}{58.92} = 0.0017\right) \le 0.01$$

Vertical equilibrium is only violated by 0.17% and another iteration cycle is not needed. $P_u = 58.9$ kips is very nearly the same as $P_u = 59.0$ kips found in Solution 1.

Solution 3

Try $P_u = 59.0$ kips:

$$M = e_x P_u = 5(59) = 295 \text{ in.-kips}$$

At each bolt hole due only to $P_u = 59.0$ kips, there is an upward force of 59.0/4 = 14.75 kips.

At each bolt hole due only to $M = 295$ in.-kips, all radii are equal and

$$r_1^2 = x_1^2 + y_1^2 = (1.5)^2 + (3)^2 = 11.25$$

$$\sum_{i=1}^{4} r_i^2 = 4(11.25) = 45.0$$

$$\frac{F_1}{r_1} = \frac{e_x P_u}{\sum_{i=1}^{n} r_i^2} = \frac{295}{45.0} = 6.56$$

For $i = 1$ to 4,

$$F_i = 3.354(6.56) = 22.0 \text{ kips}$$

$$F_{xi} = F_i \sin\theta_i = \frac{22.0(3)}{3.3541} = 19.7$$

$$F_{yi} = F_i \cos\theta_i = \frac{22.0(1.5)}{3.3541} = 9.84$$

The resultant force on the most heavily loaded bolt is

$$F_j = \sqrt{\left(F_{xj}\right)^2 + \left(\frac{P_u}{n} + F_{yj}\right)^2} = \sqrt{(19.7)^2 + \left(\frac{59.0}{4} + 9.84\right)^2} = 31.51$$

which is 31.51/27.58 = 1.14 times the force in the most heavily loaded bolt by the ultimate strength method.

Since the LRFD strength requirement is that $\phi R_n \geq F_j$, by the elastic method the maximum acceptable value of $P_u = (28.1/31.51)(59.0) = 52.6$ kips.

By the ultimate strength method, the maximum acceptable value of $P_u = 59.0$ kips. Therefore, by using the ultimate strength method in this example, we can have 59.0/52.6 = 1.12 times more applied load than we can have according to the elastic method.

Example 8.2

For the bearing-type connection shown in Figure 8.5, we are given the following numerical information: A36 steel bracket plate; $g = 5.5$ in.; $s = 3$ in.; $e_x = 15$ in.; minimum bearing thickness = 0.5 in.; 7/8-in.-diameter A325X bolts in single shear; $P_u = 100$ kips; and $\alpha = 45°$. Using LRFD p. 8-55, find the number of bolts required.

Solution

From LRFD p. 8-24,

$$\phi r_n = 28.1 \text{ kips/bolt} = (\text{single shear of one bolt})$$

From LRFD p. 8-26,

$$\phi r_n = (91.4 \text{ kips/in.})(0.5 \text{ in.}) = 45.7 \text{ kips/(bolt hole)}$$

On LRFD p. 8-55, for $g = 5.5$ in., $s = 3$ in., $e_x = 15$ in., and $n = 5$ bolts in each vertical row of the bolt group, we need

$$\left[\phi R_n = C\left(\phi r_n\right) = C(28.1)\right] \geq \left(P_u = 100 \text{ kips}\right)$$

$$C \geq 3.56$$

For $n = 5$ bolts in each vertical row of the bolt group, we find $C = 3.64$ and $[\phi R_n = 3.64(28.1) = 102 \text{ kips}] \geq (P_u = 100 \text{ kips})$.

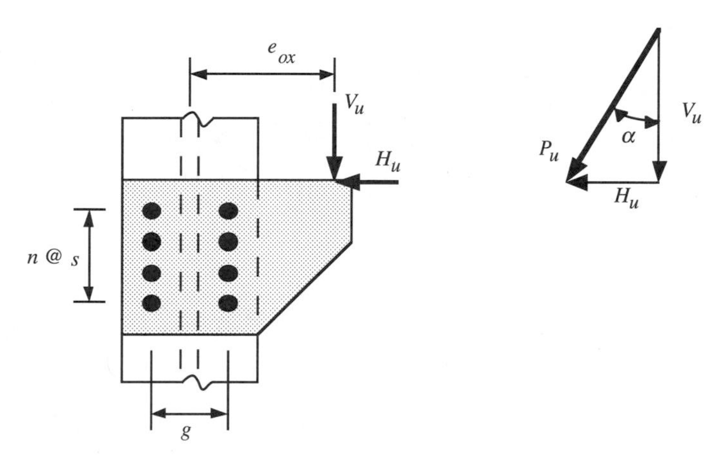

FIGURE 8.5 Example 8.2

8.2.2 A Weld Group Subjected to Eccentric Shear

Ultimate Strength Method

The following discussion is based on the descriptions given by Butler et al. [22] and Lesik and Kennedy [33].

Figure 8.6(a) shows the free-body diagram of a fillet weld group fastened to a bracket plate subjected to an inclined, factored load that is applied at a point whose eccentric location with respect to the center of gravity of the weld group is denoted by e_x and e_y. At the ICR point, only a pure rotation b of the bracket plate is assumed to occur. The ICR point is located at an initially unknown distance of e_{ox} with respect to the center of gravity of the bolt group.

The shear force on each of two typical differential elements in the weld group is shown in Figure 8.6(a). Each shear force is assumed to be perpendicular to its radius from ICR. For the ith force,

$$F_i = \left\{ \phi r_n = 0.75(0.6 F_{Exx} A_w)\,(1.0 + 0.50 \sin^{1.5} \theta)[\rho(1.9 - 0.9\rho)]^{0.3} \right\}_i$$

where

$$F_{Exx} = \text{weld electrode strength}$$

$$(A_w)_i = (0.707 S_w L_w)_i$$

$$(S_w)_i = \text{weld size}$$

$$(L_w)_i = \text{weld length chosen for the } i\text{th element}$$

$$\theta_i = \text{angle measured in degrees as shown in Figure 8.6(c)}$$

$$\rho_i = \left[\frac{r_i \Delta_u / r_{cr}}{3.34(\theta + 2)^{-0.32} S_w} \right]_i$$

$$r_i = \text{radius from ICR to center of } i\text{-th element}$$

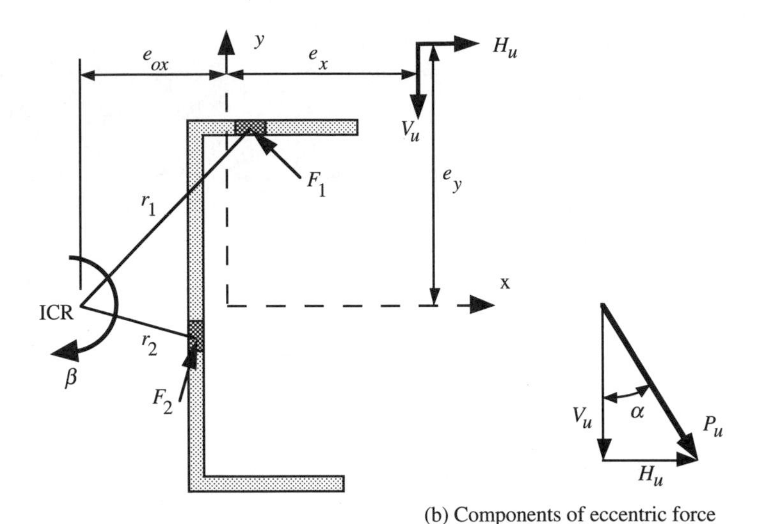

(b) Components of eccentric force

(a) FBD of weld group

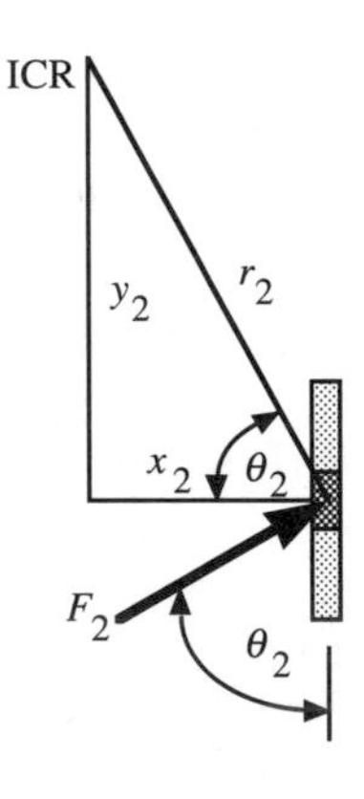

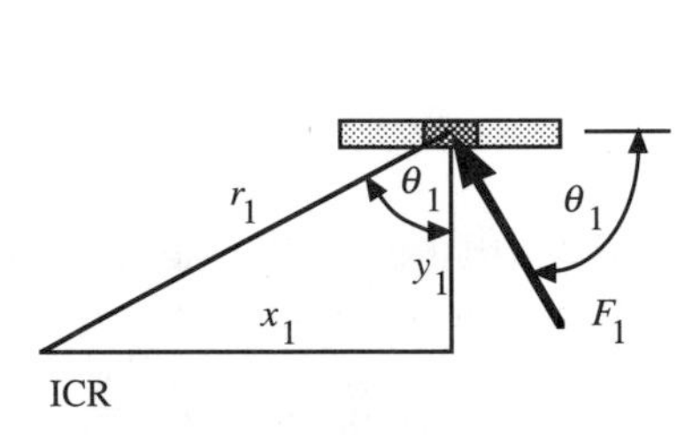

Each theta angle is measured in degrees from the length-direction axis of a fillet weld to the force on the differential weld element.

(c) Details for the force on a weld element

FIGURE 8.6 Ultimate strength method for eccentric shear of a weld group

r_{cr} = radius from ICR to center of element with minimum Δ_u/r_i

$$\left[\Delta_u = 17.39(\theta+6)^{-0.65} S_w\right]_i \leq 2.72(S_w)_i$$

The three equations of statics are

$$\sum_{i=1}^{n} F_{xi} - P_u \sin\alpha = 0$$

$$\sum_{i=1}^{n} F_{yi} - P_u \cos\alpha = 0$$

$$\left[(e_{ox} + e_x)\cos\alpha + e_y \sin\alpha\right]P_u - \sum_{i=1}^{n} r_i F_i = 0$$

where F_{xi} and F_{yi}, respectively, denote the x and y components of F_i. In these equations, the known parameters are the dimensions and properties of each fillet weld in the weld group, e_x, e_y, and α. We want to find the value of P_u from the last equilibrium equation that also satisfies the first two equilibrium equations. An iterative procedure must be used to find this value of P_u. For an assumed value of e_{ox}, the ICR point is located, F_i is computed for an appropriately chosen $(L_w)_i$, and P_u is then determined from the last equilibrium equation. If this value of P_u satisfies the first two equilibrium equations, the assumed value of e_{ox} was correct. Otherwise, we must assume another value of e_{ox} and repeat the process. Obviously, a computer program needs to be written to conduct the iterative search for the correct value of e_{ox}. Let F_X and F_Y, respectively, be the absolute value obtained on the left-hand side of the first and second equilibrium equations. Then, we can start by assuming e_{ox} = 0.10 in. and increase e_{ox} by an increment of 0.02 in. until $F_X/(P_u \sin\alpha) \le 0.005$ and $F_Y/(P_u \cos\alpha) \le 0.005$. This type of procedure was used to obtain the C values given in LRFD Tables 8-38 (p. 8-163) through 8-45 (p. 8-210). If we use the LRFD tabular values, the LRFD strength-design requirement for the weld group is

$$(\phi R_n = 16CC_1S_wL_c) \ge P_u$$

where

$$C = \text{tabular value} \quad (\text{which includes } \phi = 0.75)$$

$$C_1 = \text{electrode coefficient from LRFD Table 8-37}$$

$$S_w = \text{weld size (in.)}$$

$$L_c = \text{length of connection (in.)}$$

$$P_u = \text{required strength of weld group (kips)}$$

Note: LRFD p. 8-157 tells us for these tables that:

1. Linear interpolation on any given page is acceptable.
2. Linear interpolation between pages is not acceptable.
3. C_1 accounts for
   (a) The C values being done for E70 electrodes.
   (b) Additional strength reduction factors of 0.9 for E70 and E80 electrodes, and 0.85 for E100 and E110 electrodes; these strength reductions account for the uncertainty of the extrapolation from E70 test results to higher strength electrodes.

Elastic Method

Figure 8.7(a) shows a weld group subjected to eccentric shear. For mathematical convenience, each weld is considered to be a line coincident with the edge of material to be welded. Thus, each weld is treated as having an effective throat thickness of t_e = unity. For the weld group in Figure 8.7(a):

1. The area is $A = 2b + d$.
2. The center of gravity of the weld group is located at

$$\bar{x} = \frac{b^2}{A}$$

3. The polar moment of inertia is

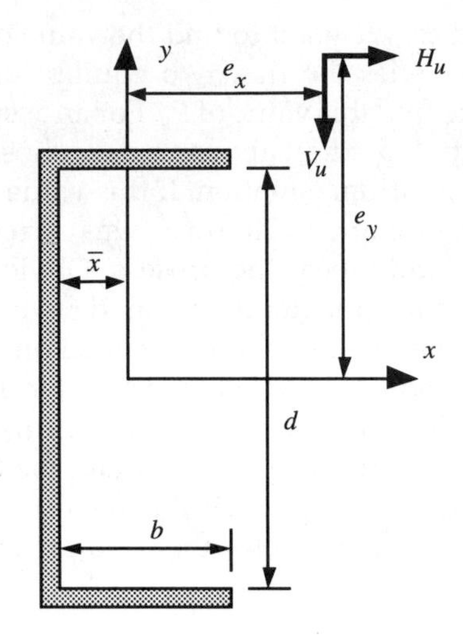

(a) FBD of weld group

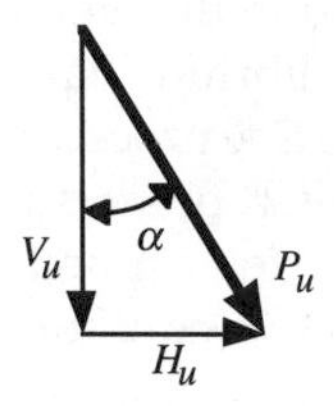

(b) Components of eccentric force

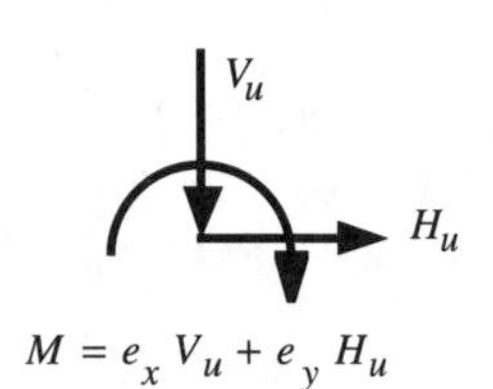

(c) Forces at CG of weld group

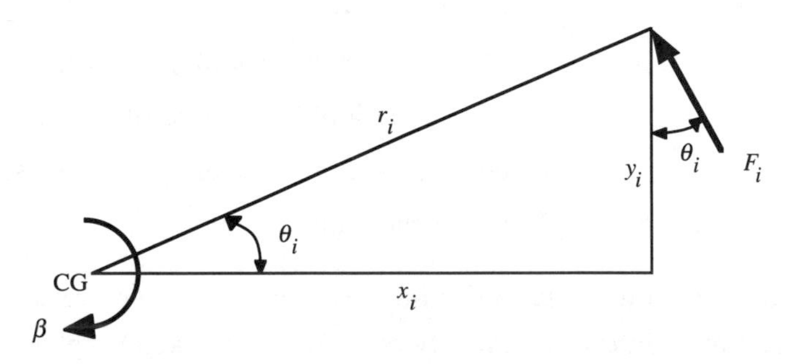

(d) Details for ith force due to M shown in (c)

FIGURE 8.7 Elastic method for eccentric shear of a weld group

$$I_p = I_x + I_y = \frac{2b^3 + 6bd^2 + d^3 + 6\bar{x}bd}{12}$$

The eccentric force components in Figure 8.7a are transferred to the center of gravity of the weld group that necessitates, as shown in Figure 8.7(c), that we also must have $M = e_x V_u + e_y H_u$ at the center of gravity of the weld group. Then, the assumptions made are:

1. Due to V_u only, there is an upward force per inch of weld length of $q_y = V_u/A$.
2. Due to H_u only, in the negative x-direction there is a force per inch of weld length of $q_x = H_u/A$.
3. Due only to $M = e_x V_u + e_y H_u$, the plate is assumed to rotate about the center of gravity of the weld group and in the direction of β.

A radius is connected from CG to the center of each differential weld element. As shown in Figure 8.7(d), F_i due only to M is assumed to be perpendicular to r_i. The moment of F_i about CG is $r_i F_i$, and this moment is opposite to the direction of M. Also, we assume that F_i is proportional to r_i, which can be mathematically written as $F_i = r_i \beta$. Then, at a point whose coordinates are (x, y), the force per inch of weld length in the x- and y- directions, respectively, is

$$q'_x = \frac{My}{I_p}$$

$$q'_y = \frac{Mx}{I_p}$$

Let q_j denote the maximum value due to the sum of the force components found in items 1 to 3. Then, the shear per inch of weld length in the most heavily loaded differential weld element is

$$q_j = \sqrt{(q_x + q'_x)^2 + (q_y + q'_y)^2}$$

Note that for the jth differential weld element, q_x and q'_x are in the same direction; also, q_y and q'_y are in the same direction. The LRFD strength design requirement is

$$\phi F_w t_e \geq q_j$$

where

$$\phi F_w = 0.75\ (0.60 F_{Exx})$$

$$F_{Exx} = \text{weld electrode strength}$$

$$t_e = 0.707 S_w$$

The elastic method is not iterative , but it gives an excessively conservative value of q_j in some cases.

Example 8.3

For the fillet-welded connection shown in Figure 8.8, we are given the following numerical information: A36 steel bracket plate; b = 4 in.; d = 8 in.; e_x = 12 in.; E70 electrodes ($C_1 = 1$); and 5/16-in. fillet welds. Obtain the maximum acceptable value of P_u for this weld group by using:

1. LRFD p. 8-187
2. The elastic method equations.

Solution 1

From LRFD p. 8-187, for $kL = (b = 4$ in.$)$, $L = (d = 8$ in.$)$, and $aL = (e_x = 12$ in.$)$, which means that $k = 0.5$ and $a = 1.5$, we find C = 0.706 and

$$P_u = 16CC_1S_wL = 16(0.706)(1)(5/16)(8) = 28.2 \text{ kips}$$

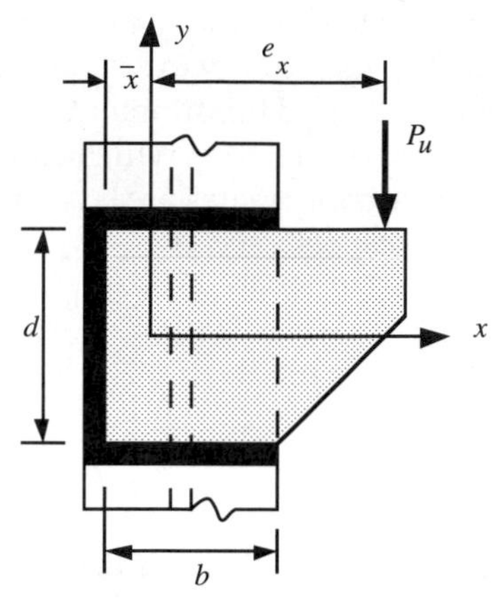

FIGURE 8.8 Example 8.3

Solution 2

$$A = 2b + d = 2(4) + 8 = 16 \text{ in.}^2$$

$$\bar{x} = \frac{b^2}{A} = \frac{(4)^2}{16} = 1.00$$

$$I_p = \frac{2b^3 + 6bd^2 + d^3 + 6\bar{x}bd}{12}$$

$$I_p = \frac{2(4)^3 + 6(4)(8)^2 + (8)^3 + 6(1)(4)(8)}{12} = 197.33 \text{ in.}^4$$

$$\phi F_w t_e = 0.75(0.60)(70)(0.707)(5/16) = 6.96 \text{ kips/in.}$$

Try $P_u = 28.2$ kips:

$$M = e_x P_u = 12(28.2) = 338 \text{ in.-kips}$$

Due to P_u only, there is an upward force per inch of weld length of

$$q_y = P_u/A = 28.2/16 = 1.76 \text{ kips/in.}$$

At $(x_j, y_j) = (3, 4)$ due only to $M = 338$ in.-kips,

$$q'_x = \frac{My}{I_p} = \frac{338(4)}{197} = 6.86 \text{ kips/in.}$$

$$q'_y = \frac{Mx}{I_p} = \frac{338(3)}{197} = 5.15 \text{ kips/in.}$$

The resultant force on the most heavily loaded point in the weld group is

$$q_j = \sqrt{(q'_x)^2 + (q_y + q'_y)^2} = \sqrt{(6.86)^2 + (1.76 + 5.156)^2} = 9.74 \text{ kips/in.}$$

By the elastic method, the shear/in. at the most heavily loaded point is $q_j = 9.10$ kips/in. Since the LRFD strength requirement is that $\phi F_w t_e \geq q_j$, by the elastic method

$$P_u = (6.96/9.74)(28.2) = 20.2 \text{ kips}$$

Recall that $P_u = 28.2$ kips for the ultimate strength method for which, in this example, we can have 28.2/20.2 = 1.40 times more applied load than we can have according to the elastic method.

Example 8.4

For the fillet-welded connection shown in Figure 8.8, we are given the following numerical information: A36 steel bracket plate; $b = 4$ in.; $d = 8$ in.; $e_x = 12$ in.; E70 electrodes ($C_1 = 1$); and $P_u = 50$ kips. Using LRFD p. 8-187, find the required weld size.

Solution

From LRFD p. 8-46, for $kL = (b = 4$ in.), $L = (d = 8$ in.), $\alpha L = (e_x = 12$ in.), which means that $k = 0.5$ and $a = 1.5$, we find $C = 0.706$ and

$$S_w \geq \{P_u/(16CC_1L) = 50/[16(0.706)(1)(8)] = 0.553 \text{ in.}\}$$

Use $S_w = 9/16$ in.

8.3 BOLTS SUBJECTED TO TENSION AND PRYING ACTION

Figure 8.9 shows a hanger connection in which the bolts are subjected to tension. Also, the top bolt in the angle section of Figure 1.12(c) is subjected to tension. The bolts in the web angle of Figure 1.12(c) are subjected to shear and tension due to a small bending moment (see Section 8.7).

As shown in Figures 8.9(c) and (d), each bolt is subjected to a direct tension force $T + Q$, where $T = P_u/4$ and Q is a prying force. For this type of connection, according to LRFD J3.3:

1. A high-strength bolt must be used.
2. The required bolt strength is $T + Q$.

Fully tensioned bolts should be used in bolt groups subjected to prying action in order to reduce the deformations of the connection parts and the prying forces. The empirical analysis and design procedures given on LRFD pp. 11-5 to 11-11 may be used for hanger connections. The minimum feasibly possible value of b in Figure 8.9 should be chosen to reduce the prying force. See LRFD p. 11-11 for an example problem.

As shown in Figure 8.10, preferably the hanger should be stiffened to eliminate prying action. If more than one bolt on each side of the hanger web is used, hanger stiffeners should be used.

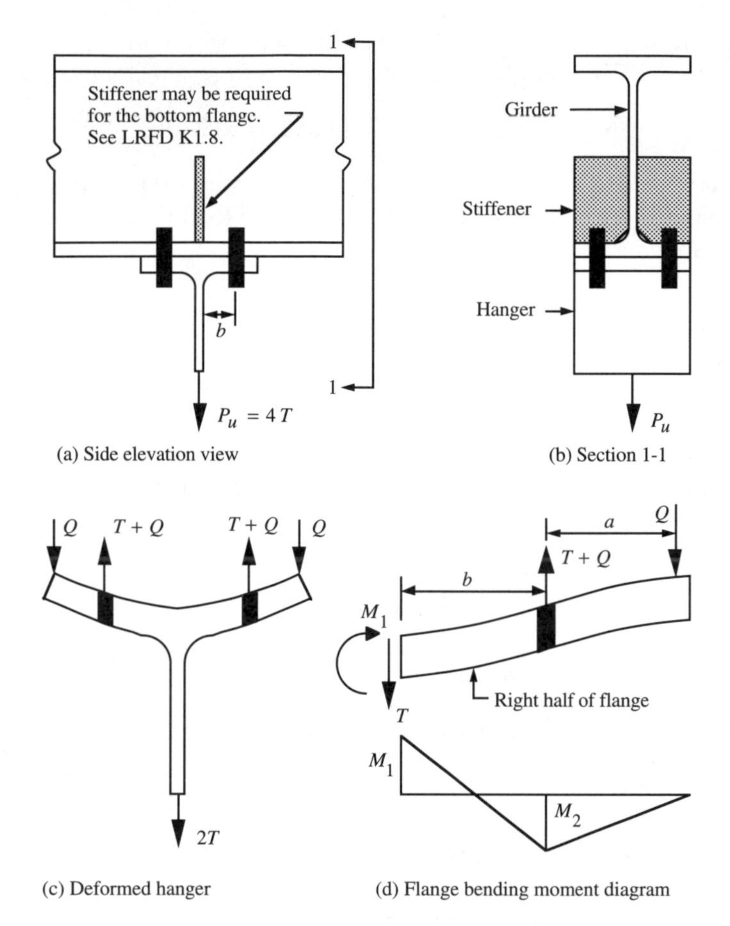

FIGURE 8.9 T section hanger connection

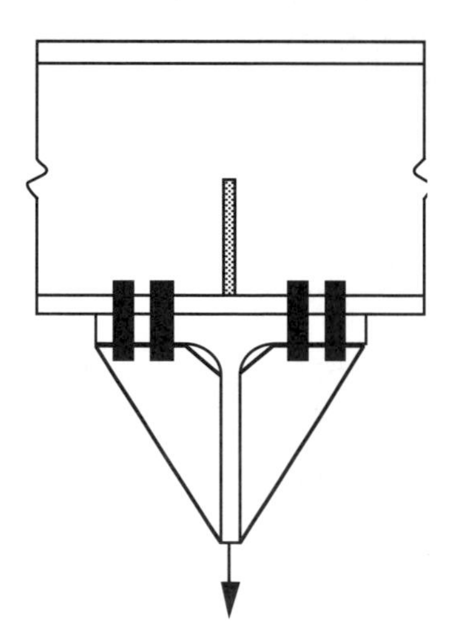

FIGURE 8.10 Stiffened T section hanger connection

8.4 BOLTS SUBJECTED TO TENSION AND SHEAR

Figure 8.11 shows a truss diagonal member used to provide tension bracing for a braced frame. The vertical component of P_u subjects the bolt group at the end of the tension member to uniform shear. The horizontal component of P_u subjects the bolt group at the end of the tension member to uniform tension. Each bolt in this bolt group is assumed to resist an equal share of each component of P_u. LRFD J3.7 and LRFD Table J3.5 (p. 6-84) give the needed design information for these bolts in a bearing-type connection. LRFD Figure C-J3.1 (p. 6-226) shows the interaction curve as three straight lines approximating an ellipse. This figure is the basis for the limiting tensile stress equations given in LRFD Table J3.5. If a slip-critical connection is desired, LRFD J3.8 and LRFD Tables J3.1 and J3.6 are applicable.

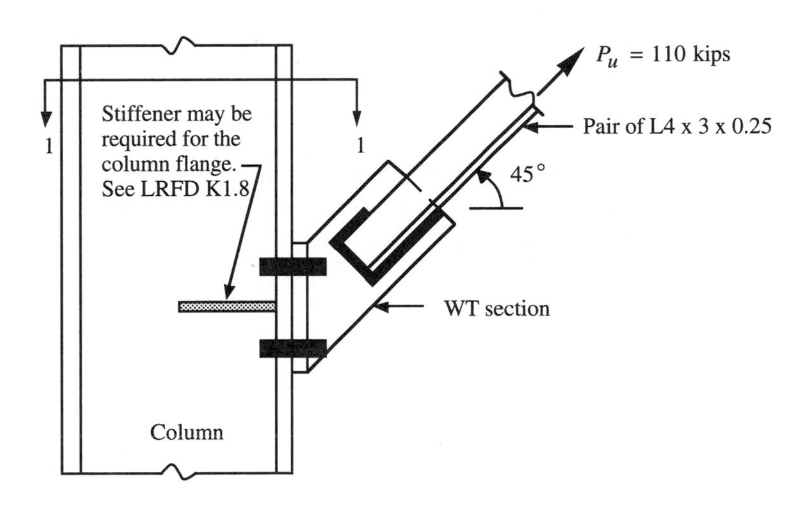

(a) Side elevation view

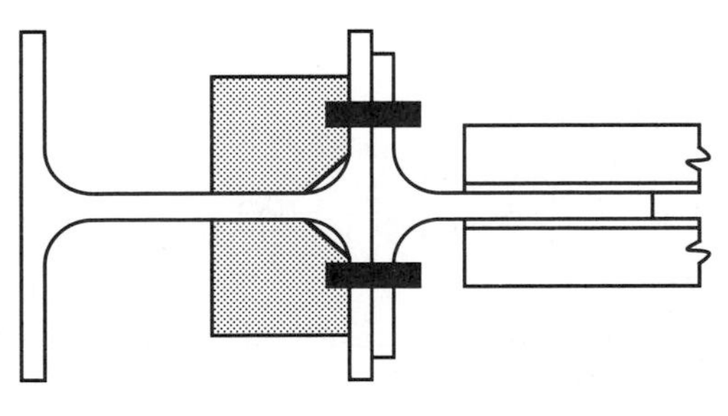

(b) Section 1-1

FIGURE 8.11 Bolts subjected to shear and tension

Example 8.5

For the bearing-type connection shown in Figure 8.11, determine the number of 7/8-in.-diameter A325X bolts required.

Solution

Try four bolts:

$$f_t = \left[f_v = \frac{0.707\,P_u}{nA_b} = \frac{0.707(110)}{4(0.6013)} = 32.3 \text{ ksi} \right]$$

$$\left[F_t = 117 - 1.5 f_v = 117 - 1.5(32.3) = 68.55 \text{ ksi} \right] \leq 90 \text{ ksi}$$

Since (F_t = 68.6) ≥ (f_t = 32.3), four bolts are acceptable if the bearing strength is adequate.

Example 8.6

If a slip-critical connection is desired in Example 8.5, are four 1-in.-diameter A325X bolts adequate? Assume standard bolt holes, D = 30.6 kips, L = 45.8 kips, and the governing factored loading combination is 1.2D + 1.6L. Therefore, P_u = 110 kips (the same as in Example 8.5) and $D + L$ = 76.4 kips.

Solution

The service tension force in each of the four bolts is

$$T = \frac{0.707(D+L)}{4} = \frac{0.707(76.4)}{4} = 13.5 \text{ kips}$$

From LRFD Table J3.1, the minimum pretension force is T_b = 51 kips:

$$f_t = \left[f_v = \frac{0.707\,T}{A_b} = \frac{0.707(13.5)}{0.7854} = 12.15 \text{ ksi} \right]$$

The limiting shear strength is

$$F_v = \left(1 - \frac{T}{T_b}\right)(17 \text{ ksi}) = \left(1 - \frac{13.5}{51}\right)(17) = 12.5 \text{ ksi}$$

Since (F_v = 12.5) ≥ (f_v = 12.15), four bolts are adequate for a slip-critical connection if the bearing strength is adequate.

8.5 CONNECTORS SUBJECTED TO ECCENTRIC TENSION AND SHEAR

Figures 8.12 and 8.13 show examples of beam-to-column connections in which the connectors are subjected to eccentric tension (due to a bending moment) and shear. Since we preferred to show the eccentric force (reversed beam reaction) in Figures

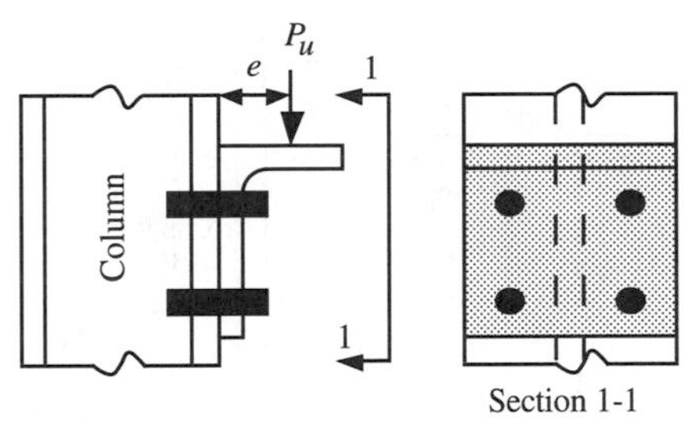

(a) Beam seat angle bolted to column flange

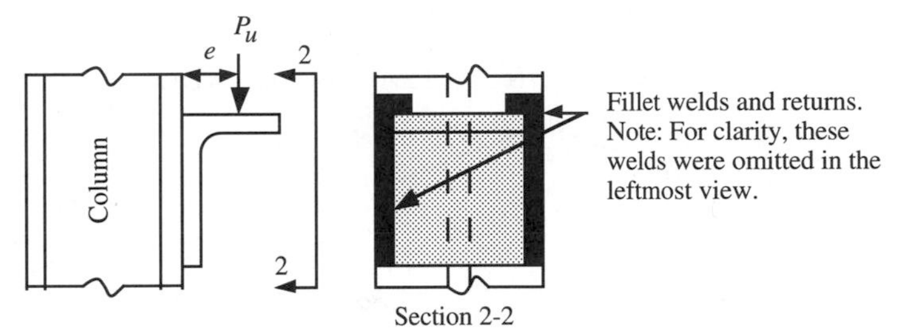

(b) Beam seat angle welded to column flange

FIGURE 8.12 Connectors for beam-seat angles

8.12(a) and (b), the beam sitting on the beam seat and the required top angle to provide lateral bracing of the top beam flange were not shown. These omitted details are shown in the figures on LRFD pp. 9-128 and 9-138.

8.5.1 Weld Groups

The design strength of the welds on a beam seat [see Figure 8.12(b) and the figure on LRFD p. 9-128] is given in LRFD Table 9-7 (p. 9-137) for some seat angle sizes. Entries in LRFD Table 9-7 (p. 9-137) were computed by the elastic method for this weld group. In the elastic method for this weld group treated as line elements, the welds are subjected to a shear/in. of

$$q_v = \frac{P_u}{\Sigma L_w}$$

where ΣL_w = sum of the line element lengths (including the returns) in the weld group. Also, the weld returns are subjected to a tension/in. of

$$q_t = \frac{(P_u e)\bar{y}}{I}$$

where

$\bar{y}$ = distance from top of weld group to neutral axis

I = moment of inertia of weld group

e = distance from weld group to beam reaction P_u

Note: Examples 8.10 and 8.11 illustrate how to determine the value of e used in computing any entry in LRFD Table 9-7 (p. 9-137).

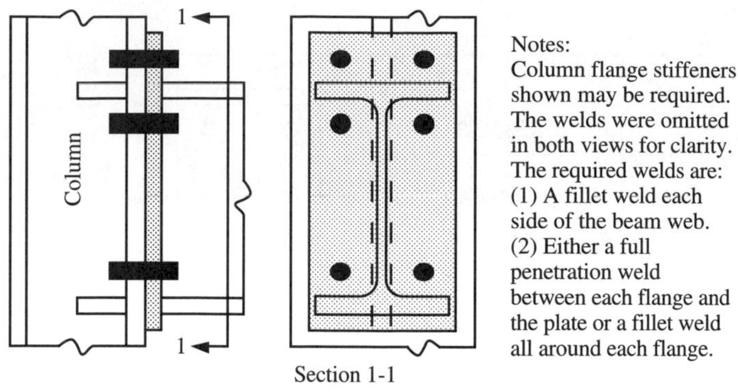

(a) Plate welded to beam end and bolted to column flange

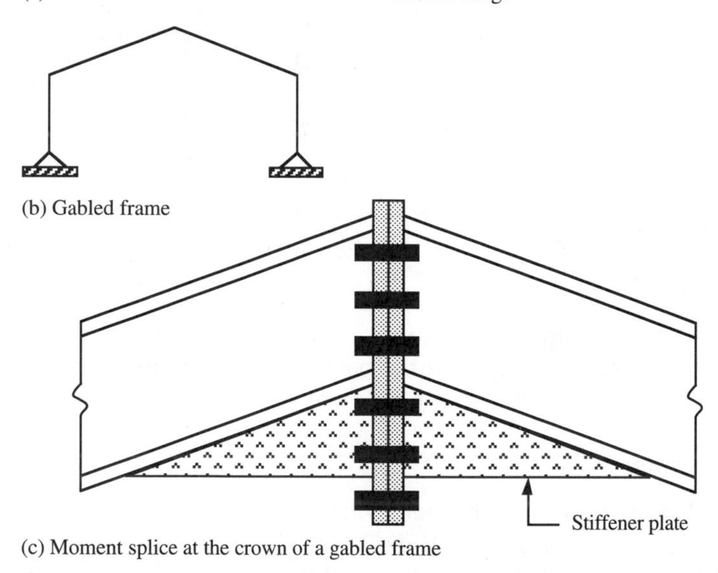

(b) Gabled frame

(c) Moment splice at the crown of a gabled frame

FIGURE 8.13 Connectors for beam-end plates

The design requirement is

$$\left[\phi F_w t_e = 0.75(0.6 F_{Exx})(0.707 S_w)\right] \geq \left(q = \sqrt{q_v^2 + q_t^2}\right)$$

Alternatively, if we conservatively ignore the weld returns, the special case on LRFD p. 8-163 for the ultimate strength method of a weld group can be used to determine the design strength of the welds on a beam seat.

For the weld group described in Figure 8.13(a), the procedure recommended on LRFD p. 10-24 can be used. In this procedure, if only fillet welds are used, the eccentric tension is assumed to be resisted by the fillet weld group surrounding the beam tension flange and the shear is assumed to be resisted by the fillet weld group on the beam web. Therefore, in this procedure the eccentric tension and shear are uncoupled; that is, they are treated independently.

8.5.2 Bolt Groups

For the elastic method and the ultimate strength method, respectively, Figures 8.14(a) and 8.14(b) schematically show the bolt tension force distributions that without much thought appear reasonable to assume in a bearing-type bolt group subjected to a bending moment and a shear. If either of these bolt tension force distributions is assumed, Figure 8.14 is applicable for Figures 8.12 and 8.13, and LRFD J3.7 is applicable for the bolts since they are subjected to tension and shear. We will show how involved the elastic method and the ultimate strength method are for the bolt tension force distributions assumed in Figure 8.14. Then, other bolt tension force distributions that may be assumed in certain cases to simplify the mathematics will be noted.

Consider a bolt group that has m vertical lines of bolts and n horizontal rows of bolts. The total number of bolts in the bolt group is $n_b = mn$. For example, in Figure 8.14, $m = 2$, $n = 5$, and $n_b = 10$. Other parameters common to both methods are

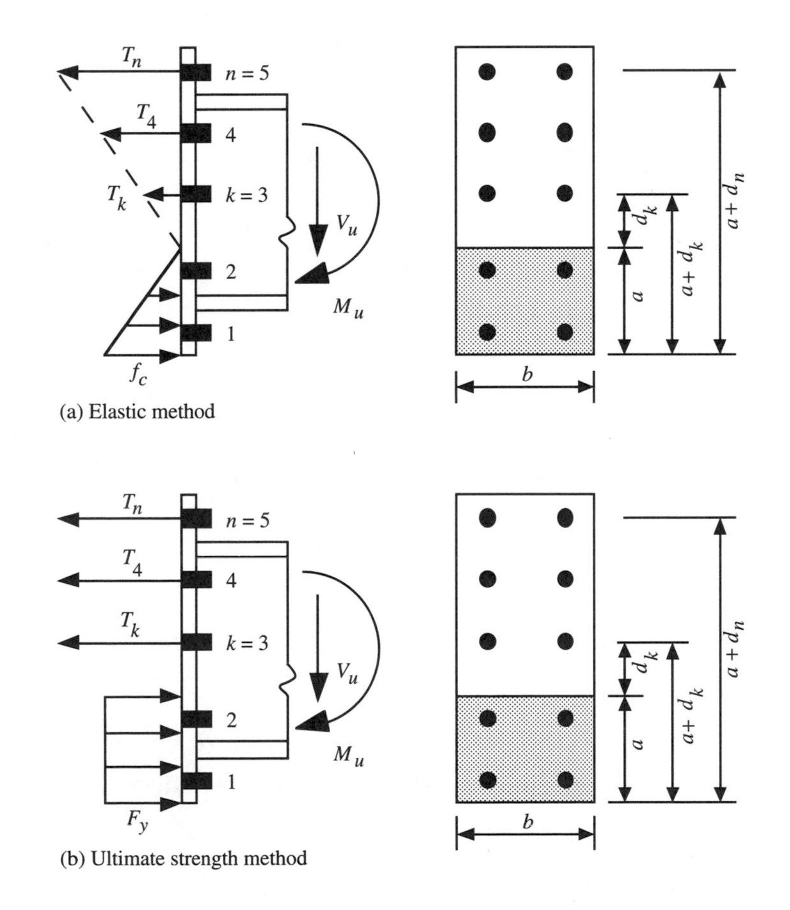

FIGURE 8.14 Bolts subjected to shear and eccentric tension

a = depth of compression stress block

b = width of compression stress block

d_i = distance from neutral axis to ith row of bolts in tension

$A_i = mA_b$ = total area of bolts in ith row of bolts in tension

A_b = area of one bolt

k = number of first row of bolts above neutral axis (in Figure 8.14, $k = 3$)

Elastic Method

The neutral axis is found by taking moments about the bottom edge of the compression zone:

$$\frac{a}{2}(ab) = \sum_{i=k}^{n}(a + d_i)A_i$$

An iterative procedure must be used to determine a, and then the moment of inertia is

$$I = \frac{a^3 b}{3} + \sum_{i=k}^{n} A_i d_i^2$$

and the tension force in the ith row of bolts is

$$T_i = \frac{M_u A_i d_i}{I}$$

which is valid for $i = k$ to n. The tension stress in each bolt in the ith row is

$$(f_t)_i = \frac{T_i}{mA_b}$$

and must not exceed the tension stress limit F_t given by the formulas in LRFD Table J3.5. If we assume that each bolt in the bolt group resists an equal amount of shear, then the bolt shear stress f_v is

$$f_v = \frac{V_u}{n_b A_b}$$

where

V_u = factored shear force acting on the bolt group

n_b = total number of bolts in the bolt group

A_b = area of one bolt

Ultimate Strength Method

Equilibrium of horizontal forces requires that

$$F_y ab = \sum_{i=k}^{n} T_i$$

where

$$F_y = \text{yield stress of the plate}$$

$$T_i = mA_bF_t$$

$$A_b = \text{area of one bolt}$$

F_t = tension stress limit specified in LRFD Table J3.5 is a function of f_v.

The usual assumption for the bolt shear stress f_v is that

$$f_v = \frac{V_u}{n_b A_b}$$

where V_u = the factored shear force acting on the bolt group.

Since k is unknown, an iterative procedure must be used to determine

$$a = \frac{\sum_{i=k}^{n} T_i}{F_y b}$$

and then the design bending strength of the connection is

$$\phi M_n = \frac{0.9F_y a^2 b}{2} + \sum_{i=k}^{n} d_i T_i$$

The design requirement is that $\phi M_n \geq M_u$.

As we noted prior to the presentation of the elastic method and the ultimate strength method, there are other bolt tension force distributions that may be assumed in certain cases to simplify the mathematics. Bolt force distributions that are assumed in the LRFD Manual procedures for beam seats and end plates will be noted.

The various bolt group types customarily used for beam seats are shown on LRFD p. 9-128. The shear design strength of the bolts for each bolt group type is given in LRFD Table 9-6 (p. 9-136). In preparing this table, the tension in the bolts due to $M_u = P_u e$ was ignored. For example, from LRFD p. 8-24 for a 0.75-in.-diameter A325N bolt in single shear, we find that f_v = 36.0 ksi and ϕR_n = 15.9 kips/bolt. For six bolts, ϕR_n = 6(15.5) = 93.0 kips, which agrees with the shear design strength of 93.0 kips given on LRFD p. 9-136 for connection type C (six bolts). The LRFD Manual does not give a justification for ignoring $M_u = P_u e$ in a bolt group on a beam seat and accounting for $M_u = P_u e$ in a weld group on a beam seat.

For an end plate welded to an entire section of a beam end and bolted to a column flange, the bolt force distributions assumed in the LRFD Manual design procedure are stated in the first paragraph on LRFD p. 10-24. The bolts located an equal distance above and below the beam tension flange are assumed to resist tension only. The tension force in each of these bolts is assumed to be equal. ϕF_t from LRFD Table J3.2 (p. 6-81) or, alternatively, LRFD 8-15 (p. 8-27) is used in choosing a sufficient number of these bolts to develop the beam flange force.

$$F_f = \frac{M_u}{d - t_f}$$

where

$$M_u = \text{required factored beam end moment}$$
$$d = \text{depth of the chosen beam section}$$
$$t_f = \text{flange thickness of the chosen beam section}$$

The other bolts in the bolt group are assumed to resist only shear. The shear force in each of these bolts is assumed to be equal. LRFD Table I-D (p. 8-24) is used in choosing a sufficient number of these bolts to develop the required factored beam end shear V_u.

8.6 TRUSS MEMBER CONNECTIONS AND SPLICES

An adequate number of bearing-type connection and splice examples for tension members was given in Chapter 2. For compression members in trusses, the member-end and splice end forces must be developed by the connectors. Therefore, the discussion in Chapter 2 for connectors in tension member ends is also applicable for the connectors in truss compression member ends. If a truss compression member needs to be field spliced, the moment of inertia for each principal axis of the splice connection parts should not be less than those of the compression member being spliced. Also, the buckling strength of the splice connection parts must not be less than the column design strength.

Since Chapter 2 contains only bearing-type connections, a discussion of slip-critical connections is given. When high-strength bolts are fully tensioned, the parts being connected are clamped together by the tension force in the bolts. A friction force equal to the clamping force times the coefficient of friction on the clamped surfaces is developed. If the maximum axial force in the member due to service loads does not exceed the available friction force, the connection is classified as being slip-resistant. According to LRFD Commentary J3.8 (p. 6-226), slippage occurs at about 1.4 to 1.5 times the maximum service load. LRFD J3.8 (p. 6-83) and LRFD Tables J3.5 and J3.6 give the information needed for slip-critical connections, which may be designed either at service loads or at factored loads. In LRFD Table J3.6, the nominal slip-critical shear strength of a high-strength bolt is really the nominal friction force per bolt that can be counted on to occur in a slip-resistant connection. Slip-critical connections must also satisfy the design requirements as a bearing-type connection.

Example 8.7

See Figure 2.6 and Example 2.6 where, due to factored loads, the governing design strength of the connection was found to be 50.6 kips. If we want this connection to be satisfactory as a slip-critical connection, what is the maximum acceptable force due to service loads? Assume that $L/D = 1.5$, where L is live load and D is dead load, and assume the loading combination that governs is $1.2D + 1.6L$.

Solution

$$1.2D + 1.6(1.5D) = 50.6 \text{ kips}$$
$$D = 50.6/3.6 = 14.1 \text{ kips}$$
$$L = 1.5(14.1) = 21.2 \text{ kips}$$

Required maximum service load = $D + L = 35.3$ kips.

See LRFD p. 8-29. Due to service loads for one 0.75-in.-diameter A325 bolt in a standard-sized hole and subjected to double shear, the maximum acceptable slip-critical shear = 15.0 kips/bolt. In Figure 2.6(a), there are three bolts in the connection. Therefore, the maximum acceptable service load = 3(15) = 45.0 kips. Since $45.0 \geq 35.3$, the connection satisfies all LRFD design requirements when the three bolts are fully tensioned.

If the required maximum service load = $D + L = 52.1$ kips, for example, then the number of bolts needed for the connection to qualify as being slip-resistant would be 52.1/15 = 3.47 bolts and four bolts would have to be used.

8.7 COLUMN BASE PLATES

Suggested column base plate details are given on LRFD p. 11-55. The suggested design procedure on LRFD p. 11-57 to 11-60 for a column base plate is applicable when the base plate is required to transfer only an axial compression force from a column to a reinforced concrete footing or to a reinforced concrete pedestal (pier). Two example problems are given on LRFD p. 11-60 and 11-64. Suggested anchor bolt details are shown on LRFD pp. 8-88 to 8-91.

The LRFD Manual does not suggest a design procedure for a column base plate when the column is subjected to an axial compression and bending. For the two eccentrically loaded column cases shown in Figure 8.15, we recommend the following design procedure based on the ultimate strength method.

In Figure 8.15(a), the base plate dimensions are denoted as

$$B = \text{width}$$
$$H = d + 2h'$$
$$t = \text{thickness}$$

where

$$d = \text{depth of the column section}$$
$$h' \geq (w_e + C_1)$$
$$w_e = \text{minimum edge distance} \quad \text{(see LRFD pp. 6-82, 11-57)}$$
$$C_1 = \text{minimum clearance for socket wrench head} \quad \text{(see LRFD p. 8-13)}$$

H is estimated prior to entering the base plate design procedure. Consequently, H is known in the base plate design procedure. For welding purposes, we need $B \geq b_f + 2(S_w + 1/16 \text{ in.})$, where b_f = flange width of the column section. B_m = minimum B required for bearing strength is computed as shown for each case. The required values in inches for B and H are rounded up to either the next integer or the next even integer depending on the availability of the desired plate size. See LRFD p. 1-133 for the preferred increment of thickness.

Our discussion assumes that the design bearing strength of the concrete is

$$\phi_c P_p = 0.6(1.7 f_c' A_1) = 1.02 f_c' A_1$$

which was obtained from LRFD Eq. (J9-2) for $A_2 \geq 4A_1$ where

$$A_1 = \text{area of base plate resting on the concrete support}$$
$$A_2 = \text{area of concrete support on which the base plate sits}$$

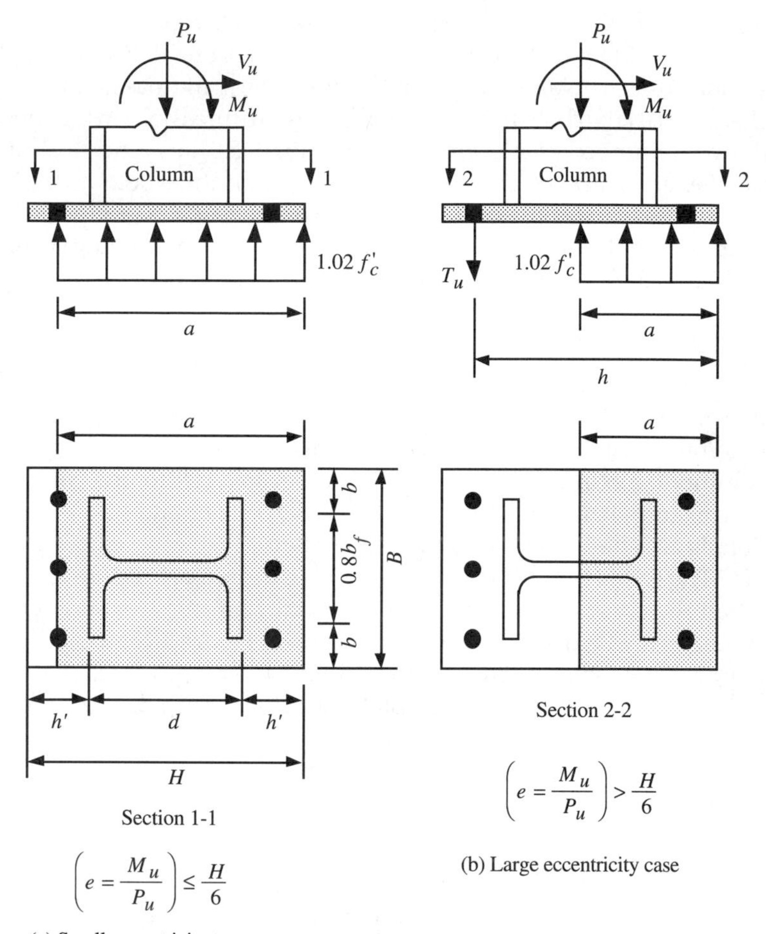

FIGURE 8.15 Eccentrically-loaded column base plates

Case 1: $(e = \mathbf{M_u}/\mathbf{P_u}) \leq H/6$

For the assumption that a plane section remains plane when subjected to an axial compression plus bending, when $e = H/6$ the neutral axis is located at the left end of the base plate. Therefore, no anchor bolt is required to resist any tension due to M_u when $e = H/6$.

The depth of the rectangular compression stress block is conservatively assumed to be $a = H - 2e$ [see Figure 8.15(a)]. Equilibrium of the vertical forces requires that

$$\left[\phi_c P_p = (1.02\, f_c' B_m)(H - 2e)\right] = P_u$$

where B_m = minimum B required for bearing strength

$$B_m = \frac{P_u}{1.02 f_c' (H - 2e)}$$

At the face of the most heavily loaded column flange, the design bending strength requirement is

$$\left[\phi M_n = 0.9F_y B\left(\frac{t}{2}\right)^2\right] \geq \left[1.02 f_c' B_m h'\left(\frac{h'}{2}\right)\right]$$

where B = actual value chosen for base-plate width

$$t \geq h'\sqrt{\frac{2.27 f_c' B_m}{F_y B}}$$

If B is large enough, the critical sections are in the vicinity of the column flange tips [see Figure 8.15(a)]. Note that the location of these sections is identical to the assumption on LRFD p. 2-101, but we chose to use b instead of n to denote the distance from these sections to the edge of the plate. To be conservative, the plate design bending strength is computed only for the length a:

$$\left[\phi M_n = 0.9F_y a\left(\frac{t}{2}\right)^2\right] \geq \left[1.02 f_c' ab\left(\frac{b}{2}\right)\left(\frac{B_m}{B}\right)\right]$$

which requires that

$$t \geq b\sqrt{\frac{2.27 f_c' B_m}{F_y B}}$$

Case 2: (e = $\mathbf{M_u}/\mathbf{P_u}$) > H/6

As shown in Figure 8.15(b), the anchor bolts on the left end of the base plate are required to resist tension. This case is considerably more complex than case 1. However, the base plate thickness formulas derived in case 1 are applicable for case 2 and are not repeated here. An additional base plate thickness requirement for plastic bending of the plate due to T_u [see Figure 8.15(b)] is

$$t \geq 2.108\sqrt{\frac{T_u(h' - w_e)}{F_y B}}$$

where T_u = required tension strength of anchor bolt group.

Equilibrium of the vertical forces requires

$$T_u = C_u - P_u$$

where

$$C_u = 1.02 f_c' Ba$$

in which B and a are unknown.

Moment equilibrium requires

$$T_u\left(h - \frac{a}{2}\right) + P_u\left(\frac{H-a}{2}\right) = M_u$$

Substitution of $T_u = C_u - P_u$ and then substituting for C_u we find that

$$a = h - \sqrt{h^2 - \frac{P_u(2h - H) + 2M_u}{1.02 f_c' B}}$$

in which B is unknown and h is approximately known. For example, initially we can assume that $h = H$ - (minimum w_e). Alternatively, we can assume that $h = H - 0.5h'$.

For a trial value of $B \geq [b_f + 2\,(S_w + 1/16 \text{ in.})]$, we can compute a and T_u. Then, we can use the following procedure adapted from Shipp and Haninger [23] to determine the size and the number of anchor bolts required to produce T_u. An identical anchor bolt group will be used on the compression end of the base plate. If the governing loading combination includes wind, an identical anchor bolt group on each end of the base plate will be necessary since the direction of M_u reverses when wind reverses.

Let n denote the total number of anchor bolts; thus, there are $n/2$ anchor bolts on each end of the base plate. Each anchor bolt is assumed to resist an equal amount of shear. The design requirement for the anchor bolts that produce T_u is

$$\frac{n}{2}\left(\phi R_n - C_v \frac{V_u}{n}\right) \geq T_u$$

where

ϕR_n = design tension strength of one anchor bolt (p. 8-27)

C_v = shear coefficient (accounts for the effects of various shear failure surfaces)

C_v = 1.10 when the base plate is embedded in the concrete support and the top surface of the base plate is flush with the support surface

C_v = 1.25 when the base plate is recessed in grout

C_v = 1.85 when the base plate is supported on, but not recessed in, grout.

If a structural designer wants to try a particular anchor bolt group having $n/2$ bolts of a certain size on each end of the base plate, the required value of B can be computed by using the following formulas:

$$T_u = \frac{n}{2}\left(\phi R_n - C_v \frac{V_u}{n}\right)$$

$$a = 2\left[h - \frac{M_u + P_u\left(h - \frac{H}{2}\right)}{P_u + T_u}\right]$$

$$B \geq \frac{P_u + T_u}{1.02 f'_c a}$$

Example 8.8

Select an A36 steel base plate for the following reaction requirements of a W14 × 99 column: P_u = 524 kips, M_u = 120 ft-kips = 1440 in.-kips, and V_u = 24 kips due to 1.2D + 1.6L + 0.5(L_r or S or R). For the support, assume that the concrete grade is 3 ksi.

Solution

For a W14 x 99, b_f = 14.57 in. and d = 14.16 in.

For an anchor bolt diameter d_b, the minimum value of $h' \approx 4.25d_b$. We need an estimated anchor bolt diameter to begin our design. When case 1 is applicable, the anchor bolt diameter required for strength is zero since $T_u = 0$. However, for erection purposes we would not use less than a 0.75-in.-diameter for the column size used in this example. Larger anchor bolt diameters may be required when case 2 is applicable. From LRFD p. 8-13 for a 0.75-in.-diameter bolt, the minimum socket wrench head clearance is C_1 = 1.25 in. From LRFD p. 6-82 for a 0.75-in.-diameter bolt, (minimum w_e) = 1.25 in. However, LRFD Table 11-3 (p. 11-57) indicates the hole diameter in a base plate for a 0.75-in.-diameter bolt should be 1.3125 in., which is larger than what was assumed on LRFD p. 6-82, and we find a revised estimate of (minimum w_e) = 1.25 in. + (1.3125 - 1.25) = 1.3125 in. Assume that $H \geq$ [14.16 + 2 (1.25 + 1.3125) = 19.3 in.]. Try H = 20 in. and h' = (20 - 14.16)/2 = 2.92 in.

Since ($e = M_u/P_u$ = 1440/524 = 2.75 in.) $\leq$ ($H/6$ = 20/6 = 3.37 in.), Case 1 is applicable and $a = H - 2e$ = 20 – 2(2.75) = 14.5 in. For strength, we need

$$B_m = \frac{P_u}{1.02 f_c' (H - 2e)} = \frac{524}{1.02(3)(14.5)} = 11.81 \text{ in}$$

Since (b_f = 14.57 in.) > 11.8 in., choose B = 16 in., which gives

$$b = [16 - 0.8(14.57)]/2 = 2.17 \text{ in.}$$

The required base plate thickness is

$$t \geq \left(h' \sqrt{\frac{2.27 f_c' B_m}{F_y B}} = 2.92 \sqrt{\frac{2.27(3)(11.81)}{36(16)}} = 1.09 \text{ in.} \right)$$

For the investigated loading combination, a 16 × 20 × 1.25 baseplate is needed.
If we only want to use an anchor bolt group with n = 4 bolts and a base plate embedded in grout, the required tension design strength of one bolt is

$$\phi R_n \geq \left[\frac{2T_u + C_v V_u}{n} = \frac{2(0) + 1.25(24)}{4} = 7.50 \text{ kips} \right]$$

From LRFD p. 8-27 for a 0.75-in.-diameter A325 bolt, we find that (ϕR_n = 29.8 kips/bolt) $\geq$ 7.5 kips/bolt.

Example 8.9

Select an A36 steel base plate for the following reaction requirements of a W14 × 99 column: P_u = 280 kips, M_u = 240 ft-kips = 2880 in.-kips, and V_u = 48 kips due to 1.2D + 1.3W+ 0.5L + 0.5(L_r or S or R). For the support, assume the concrete grade is 3 ksi.

Solution

For a W14 × 99, b_f = 14.57 in. and d = 14.16 in. Try d_b = 0.875 in.; $h' \approx 4.25d_b$ = 3.72 in.; $H \geq [d + 2h'$ = 14.16 +2(3.72) = 21.6 in.]. In Example 8.8, for a different loading combination, we needed B = 16 in. Try a base plate size of B = 16 in. and H = 22 in., which gives h' = (22 -14.16)/2 = 3.92 in. and b = 2.17 in.

Since ($e = M_u/P_u = 2880/280 = 10.29$ in.) > ($H/6 = 22/6 = 3.67$ in.), Case 2 is applicable. Try $h = H - 2.25$ in. = 23 - 2.25 = 20.75 in:

$$a = h - \sqrt{h^2 - \frac{P_u(2h - H) + 2M_u}{1.02 f_c' B}}$$

$$a = 20.75 - \sqrt{(20.75)^2 - \frac{280(41.5 - 23) + 2(2880)}{1.02(3)(16)}} = 6.36 \text{ in.}$$

$$T_u = 1.02 f_c' Ba - P_u = 1.02(3)(16)(6.36) - 280 = 31.4 \text{ kips}$$

If we only want to use an anchor bolt group with n = 4 bolts and a base plate embedded in grout, then the required tension design strength of one bolt is

$$\phi R_n \left[\frac{2T_u + C_v V_u}{n} = \frac{2(31.4) + 1.25(48)}{4} = 30.7 \text{ kips}\right]$$

From LRFD p. 8-27 for a 7/8-in.-diameter A325 bolt, we find that

$$(\phi R_n = 40.6 \text{ kips/bolt}) \geq 430.7 \text{ kips/bolt}$$

$$t \geq \left[2.108\sqrt{\frac{T_u(h' - w_e)}{F_y B}} = 2.108\sqrt{\frac{31.4(3.92 - 2.25)}{36(16)}} = 0.64 \text{ in.}\right]$$

$$t \geq \left[h'\sqrt{\frac{2.27 f_c'}{F_y}} = 3.92\sqrt{\frac{2.27(3)}{36}} = 1.70 \text{ in.}\right]$$

For the investigated loading combination, a 16 × 23 × 1.75 base plate embedded in grout with four headed 7/8-in.-diameter A325 anchor bolts is adequate.

8.8 COLUMN SPLICES

Suggested column splice details for columns in a multistory building are given on LRFD pp. 11-64 to 11-91. Column splices are usually occur at 4 ft above the finish floor level, which means that the same W section is used in two stories.

An axial compressive force usually can be totally transferred by bearing of the upper column on the lower column at the splice location. This requires that the column ends at the splice locations be milled to provide a smooth bearing surface. The section depth of the upper column is usually less than the section depth of the lower column. This requires that filler plates or shims be inserted between the splice plates and the upper column flanges. In some cases, as shown in case III on LRFD p. 11-75 and in case IX on LRFD p. 11-89, a butt plate must be welded to the bottom end of the upper column end at a column splice location to provide an adequate bearing seat on the top end of the lower column.

The column splice plates must be designed to resist the required factored moment and shear in the column at a splice location. When all the required factored axial compression force in a column cannot be resisted by bearing, the remainder of this force must be resisted by the splice plates and their connectors. All the required factored axial tension force in a column must be resisted by the splice plates and their connectors.

8.9 SIMPLE SHEAR CONNECTIONS FOR BEAMS

The connections for this category are designed to transfer only the required factored shear at the end of the beam from the beam end to the column.

Definitions of fully restrained (FR) and partially restrained (PR) connection types are given on LRFD p. 6-25. In the structural analysis of the structure shown in Figure 1.10, joints 2, 5, 7, and 8 are rigid (nondeformable). FR connections must be designed at these joints. An FR connection must have adequate strength and stiffness to transfer the required shear and bending strengths at the beam ends to the column without any appreciable change in the angle between each beam and column. At joint 10 in Figure 1.10, the beam end is hinged to the column. A PR connection is designed at joint 10 to transfer only the required beam end shear to the column. At joint 4 in Figure 1.10, there is a rotational spring between the end of member 1 and the column. A PR connection is designed to transfer the required beam end shear and the required rotational spring moment from member 1 to the column. When loads are applied to the structure, there is a significant change in the angle between the left end of member 1 and the columns. The moment at the end of member 1 is on the order of 20 to 60% of the moment that would exist if the left end of member 1 were rigidly attached (fully restrained) at joint 4.

When the moment capacity of the connected parts is negligible in PR construction, the terminology "simple framing" is often used to describe the type of construction. When no moment is to be transferred by a connection, structural designers refer to the connection as a simple connection, a shear connection, or a no-moment connection. Simple connections are designed to be flexible enough to allow the beam end essentially to rotate freely as assumed at a hinge in structural analysis.

Simple shear connections commonly used for beams are *web framing angles, beams seats, shear end-plate connections,* and *single plates*. Information for each of these connections is given in LRFD Part 9. The simple connection choice is often a matter of personal preference, but there are situations when the choice is dictated by framing conditions, fabrication costs, and convenience in field erection.

Suggested details for beam-to-beam connections are given on LRFD pp. 9-179 and 10-66. The top flanges of all beams in a floor system must be at the same elevation. For two mutually perpendicular sets of beams in a floor system, the beam ends in one set must be coped to eliminate the otherwise interfering flanges (see LRFD pp. 8-226, 10-66). Tabular information that is useful in the required checking of block shear rupture at coped beam ends is given on LRFD pp. 8-214 to 8-224.

8.9.1 Beam Web Connections

Recommended design procedure information for all-bolted double-angle connections and example problems are given on LRFD pp. 9-11 to 9-14. Tables of these connections are given on LRFD pp. 9-22 to 9-87. Although a gage of 3 in. is shown in these tables, other gages may be used, provided that the design bearing strength is reduced for a gage in the range of $2.67d$ to $3d$, where d = bolt diameter. As permitted by footnote c in LRFD Table J3.4 (p. 6-82), the minimum end distance of 1.25 in. is used. These connections are designed on the assumption that the effects due to eccentric shear are negligible. Therefore, each bolt in these connections is assumed to resist only an equal share of the required beam end shear (see LRFD Table 9-2, p. 9-22). The design strength of the connection angles is assumed to be governed by shear through the net section. In a bearing-type connection, the design strength of the connection may be governed by bearing on the beam web or on the connection angles. The length of the connection angles should be at least $T/2$, where T is the flat height of the beam web. If the connection angle length is less than T, the angles should be attached on the web as close as possible to the compression flange of the beam in order to provide lateral stability for that flange.

Recommended design procedure information for one-sided all-bolted double-angle connections and example problems are given on LRFD pp. 9-161 to 163. A bolt group in each leg of the single angle is subjected to an eccentric shear that must not be ignored. The tabular information can be used to design a bolt group subjected to an eccentric shear. In welded web connections, either a WT section or a flat plate may be less expensive than a web framing angle and serves the same structural purpose. However, as shown in the figure on LRFD p. 9-128, we would install either a top angle or a flat plate to provide lateral support for the compression beam flange.

Recommended design procedure information for bolted/welded and all-welded double-angle connections and example problems are given on LRFD pp. 9-15 to 9-20. Tables of these connections are given on LRFD pp. 9-88 to 9-90. The weld group on each side of the beam web is subjected to an eccentric shear that is not ignored in the design of these connections.

8.9.2 Unstiffened Beam Seats

Recommended design procedure information for all-bolted unstiffened beam seats, example problems, and tables for these connections are given on LRFD pp. 9-128 to 9-136.

Recommended design procedure information for all-welded unstiffened beam seats, example problems, and tables for these connections are given on LRFD pp. 9-132, 9-135, and 9-137.

A single angle may serve satisfactorily as an unstiffened beam seat when the required factored beam end shear is small. For a beam supported by a seat angle, as shown in the figure on LRFD p. 9-128, either a light angle or plate must be attached either to the top flange of the beam or to the upper portion of the beam web to provide lateral support for the compression beam flange. Usually, an $L4 \times 4 \times 0.25$ is chosen for this top angle.

Since the entries in Table 9-7 for a 4-in. outstanding leg of an angle for a welded beam seat are identical to the corresponding entries in Table 9-6 for a bolted beam seat, the critical section for bending of the angle leg was assumed to be the same whether a bolted beam seat or a welded beam seat is used. This is consistent with the assumption that the bolt group is subjected only to shear. In Figure 8.16(b), we show the assumed deformation of the seat angle. Only the outstanding leg of the angle is subjected to bending. The inclined bearing force between the beam and the beam seat has a horizontal component that pins the heel of the seat angle against the column flange. Therefore, the following design bending requirement of the outstanding angle leg must be satisfied:

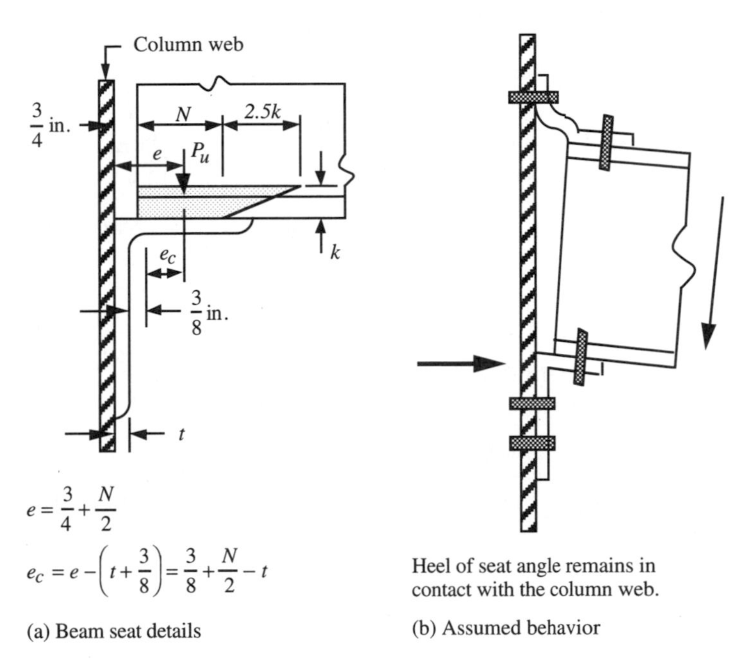

(a) Beam seat details

(b) Assumed behavior

FIGURE 8.16 Bolted beam-seat connection

$$\left[\phi M_n = 0.9\frac{t}{2}\left(F_y B\frac{t}{2}\right) = 0.225 F_y B t^2\right] \geq (M_u = P_u e_c)$$

where

B = length of beam seat angle section

t = thickness of beam seat angle section

e_c = eccentricity of the beam reaction P_u with respect to the critical section

e = eccentricity of the beam reaction P_u with respect to the heel of the angle

$$e = 0.75 + N/2$$

N = bearing length defined in the next paragraph

$$e_c = e - (t + 0.375) = 0.375 + N/2 - t$$

Note: $t + 0.375$ is assumed to be representative of the value of k for an angle section.

In preparing LRFD Tables 9-6 and 9-7, a beam setback of 0.75 in. was assumed and is made in the following discussion. The critical section in bending is at the toe of the fillet, which is assumed to be located at $t + 0.375$ in. from the heel of the angle.

If beam web yielding [LRFD Eq. (K1-3)] governs the minimum bearing length N_{wy} required on the angle leg,

$$N_{wy} = \frac{P_u}{F_{yw} t_w} - 2.5k$$

If beam web crippling [LRFD Eq. (K1-5)] governs the minimum bearing length N_{wc} required on the angle leg,

$$N_{wc} = \left(\frac{P_u}{51 t_w^2 \sqrt{F_{yw} t_f / t_w}} - 1 \right) \frac{d}{3} \left(\frac{t_f}{t_w} \right)^{1.5}$$

The actual bearing length used in preparing the LRFD Manual tables was

$$N_m = \frac{P_u}{2 F_{yw} t_w}$$

The bearing length N is the larger of N_{wy}, N_{wc}, and N_m. Assume that the bearing force is uniformly distributed on the bearing length N:

$$e_c = 0.75 + N/2 - (t + 0.375) = 0.375 + N/2 - t$$

The value of t that satisfies the design bending strength requirement for the angle leg is

$$t \geq \frac{-P_u + \sqrt{\left[P_u + 0.9 F_y B(0.375 + N/2)\right] P_u}}{0.45 F_y B}$$

Compute t for the desired value of B; the usual values of B are 6 and 8 in.

Let n be the required number of bolts in a beam seat. The design requirement for a bearing-type bolt group in a beam seat is that $n \geq P_u/R_b$, where R_b is the smaller of the single shear design strength of one bolt and the bearing design strength at one bolt location.

Example 8.10

See figures on LRFD p. 9-128. Verify entries LRFD Table 9-6 for the following: The beam seat is L4 x 4 x 0.75 x 8 $F_y = 36$ ksi., and a type D bolted connectionis used with three A325N bolts; $d = 7/8$ in. The beam is a W16 x 50:

$t_w = 0.380$ in. $t_f = 0.630$ in. $d = 16.26$ in. $k = 21/16$ in. $F_y = 36$ ksi.

Solution

Bolt shear strength

The bolts are assumed to be subjected only to shear. For single shear,

$$\phi R_n = 3(21.6) = 64.8 \text{ kips}$$

which agrees with the tabulated value on LRFD p. 9-136.

Bending strength of the outstanding leg

As noted, N_m was used for the bearing length N in preparing the LRFD tables. Also, $t_w = 3/8$ in. was used in preparing the LRFD tables:

$$N_m = \frac{P_u}{2F_{yw}t_w} = \frac{55}{2(36)(0.375)} = 2.04 \text{ in.}$$

For illustration purposes, we will also compute N_{wy} and N_{wc}.

For the prevention of web yielding,

$$N_{wy} = \frac{P_u}{F_{yw}t_w} - 2.5k = \frac{55}{36(0.380)} - 2\left(\frac{21}{16}\right) = 0.739 \text{ in.}$$

For the prevention of web crippling:

$$N_{wc} = \frac{d}{3}\left(\frac{P_u}{51t_w^2\sqrt{F_{yw}t_f/t_w}} - 1\right)\left(\frac{t_f}{t_w}\right)^{1.5}$$

$$N_{wc} = \left(\frac{16.26}{3}\right)\left[\frac{55}{51(0.380)^2\sqrt{36(0.630)/0.380}} - 1\right]\left(\frac{0.630}{0.380}\right)^{1.5} = -0.385 \text{ in.}$$

The bearing length N is the larger of N_{wy}, N_{wc}, and N_m:

$$N = N_m = 2.04 \text{ in.}$$

For the bending strength requirement of the seat angle, we need:

$$t \geq \frac{-P_u + \sqrt{\left[P_u + 0.9F_yB(0.375 + N/2)\right]P_u}}{0.45F_yB}$$

$$t \geq \frac{-55 + \sqrt{[55 + 0.9(36)(8)(0.375 + 2.04/2)](55)}}{0.45(36)(8)} = 0.744 \text{ in.}$$

Use $t = 0.75$ in. If we wish to verify the design bending strength tabulated in the LRFD Manual, we proceed as follows:

$$\left[\phi M_n = 0.9\left(\frac{F_yBt^2}{4}\right) = 0.225F_yBt^2\right] \geq (M_u = P_ue_c)$$

$$e_c = e - (t + 0.375) = 0.375 + \frac{N}{2} - t$$

$$N = N_m = \frac{\phi R_n}{2F_{yw}t_w} = \frac{55.6}{2(36)(0.375)} = 2.06 \text{ in.}$$

$$e_c = 0.375 + \frac{2.06}{2} - 0.75 = 0.655 \text{ in.}$$

$$\phi R_n = P_u = \frac{0.225(36)(8)(0.75)^2}{0.655} = 55.6 \text{ kips}$$

and the tabulated bending strength on LRFD p. 9-136 is $\phi R_n = 55.6$ kips.

Alternatively, we could have treated the design bending strength as unknown, which requires the following approach for the determination of the design bending strength:

$$\left[\phi M_n = 0.9\left(\frac{F_y Bt^2}{4}\right) = 0.225 F_y Bt^2\right] \geq (M_u = P_u e_c)$$

$$e_c = e - (t + 0.375) = 0.375 + \frac{N}{2} - t$$

$$N = N_m = \frac{P_u}{2F_{yw} t_w}$$

Substitute N into e_c and e_c into M_u to find

$$\phi R_n = P_u = b + \sqrt{b^2 + c}$$

where

$$b = 2F_{yw} t_w (t - 0.375) = 2(36)(0.375)(0.75 - 0.375) = 10.125$$

$$c = 0.9 F_{yw} t_w F_y Bt^2 = 0.9(36)(0.375)(36)(0.75)^2 = 1968.3$$

Thus, we obtain the design bending strength:

$$\phi R_n = P_u = 10.125 + \sqrt{(10.125)^2 + 1968.3} = 55.6 \text{ kips}$$

Example 8.11

See the figures on LRFD p. 9-128. Verify the entries in LRFD Table 9-7 for the following.
The beam seat is L7 x 4 x 0.75 x 8, F_y = 36 ksi. with 5/16 in. E70xx fillet welds. The beam is a W21 x 62:

t_w = 0.400 in. t_f = 0.615 in. d = 20.99 in. k = 11/8 in. F_y = 36 ksi

Solution

Weld design strength

The welds are assumed to be subjected to shear and bending. According to LRFD p. 9-132, this design strength was computed using the elastic method. Include the length of the weld returns in q_v due to shear. Ignore the length of the weld returns in q_M due to bending moment (see Figure 8.17):

$$S_w = 5/16 \text{ in.}$$

$$B = 2S_w = \text{weld return length} = 0.625 \text{ in.}$$

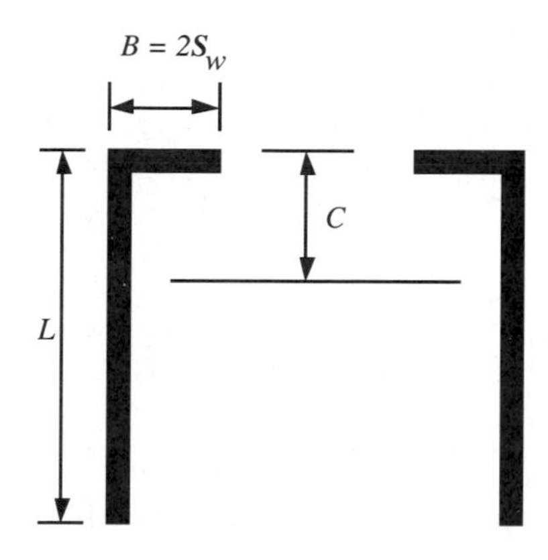

FIGURE 8.17 Example 8.11

$$\phi R_n = P_u = 10.125 + \sqrt{(10.125)^2 + 1968.3} = 55.6 \text{ kips}$$

$$q_M = \frac{Mc}{I} = \frac{(P_u e)\frac{L}{2}}{2(L^3/12)} = \frac{3(P_u e)}{L^2}$$

$$e = 0.75 + N/2$$

$$N = N_m = \frac{P_u}{2F_{yw}t_w} = \frac{P_u}{2(36)(0.375)} = 0.0370\,P_u$$

$$q = \sqrt{q_v^2 + q_M^2}$$

$[\phi\,(0.6F_{Exx}t_e) = 0.75(0.6)(70)(0.707S_w)] \geq q$ is required.

Solve for P_u, which is renamed as the design strength ϕR_n:

$$\phi R_n = 55.0 \text{ kips}$$

and the tabulated value on LRFD p. 9-137 is $\phi R_n = 53.4$ kips.

Bending strength of the outstanding leg

As noted, N_m was used for the bearing length N in preparing the LRFD tables. Also, $t_w = 0.375$ in. was used in preparing the LRFD tables:

$$N_m = \frac{P_u}{2F_{yw}t_w} = \frac{50}{2(36)(0.375)} = 1.85 \text{ in.}$$

For illustration purposes, we will also compute N_{wy} and N_{wc}. For the prevention of web yielding,

$$N_{wy} = \frac{P_u}{F_{yw}t_w} - 2.5k = \frac{50}{36(0.400)} - 2\left(\frac{11}{8}\right) = 0.722 \text{ in.}$$

For the prevention of web crippling,

$$N_{wc} = \frac{d}{3}\left(\frac{P_u}{51t_w^2\sqrt{F_{yw}t_f/t_w}} - 1\right)\left(\frac{t_f}{t_w}\right)^{1.5}$$

$$N_{wc} = \left(\frac{20.99}{3}\right)\left[\frac{50}{51(0.400)^2\sqrt{36*0.615/0.400}} - 1\right]\left(\frac{0.615}{0.400}\right)^{1.5} = -2.35 \text{ in.}$$

The bearing length N is the larger of N_{wy}, N_{wc}, and N_m:

$$N = (N_m = 1.85 \text{ in.})$$

For the bending strength requirement of the seat angle, we need

$$t \geq \frac{-P_u + \sqrt{[P_u + 0.9F_yB(0.375 + N/2)]P_u}}{0.45F_yB}$$

$$t \geq \frac{-50 + \sqrt{[50 + 0.9(36)(8)(0.375 + 1.85/2)]50}}{0.45(36)(8)} = 0.687 \text{ in.}$$

Use $t = 0.75$ in. The computations for the verification and determination of the design bending strength are identically the same as in Example 8.10.

8.9.3 Stiffened Beam Seats

When the required factored beam end shear exceeds the design strength values listed in the unstiffened beam seat table, the beam seat angle is stiffened by either vertical angles or plates to eliminate bending in the outstanding leg of the seat angle.

Recommended design procedure information for all-bolted stiffened beam seats, example problems, and tables for these connections are given on LRFD pp. 9-138 to 9-144.

Recommended design procedure information for bolted/welded stiffened beam seats, example problems, and tables for these connections are given on LRFD p. 9-140, 9-143, and 9-146.

The paired stiffener angles shown in the figure on LRFD p. 9-138 can be separated to accommodate column gages. A filler or spacer plate should be inserted in the separation gap and stitch-bolted as one does in a double-angle column when the angles are not in contact.

Example 8.12

See the figures on LRFD p. 9-138. Verify the entries in LRFD Table 9-8 for the following. Use A36 steel.

The beam seat is a 6 × 10 × 0.375 plate stiffened by a pair of L5 x 5 x 5/16 x 8.625, and attached by a type A bolted connection (six A325N bolts; $d = 7/8$ in.) to a W30×99 beam.

Solution

Bolt shear strength

The bolts are assumed to be subjected only to shear. For single shear,

$$\phi R_n = 6(21.6) = 129.6 \text{ kips}$$

which agrees with the tabulated value on LRFD p. 9-144.

Pair of angle stiffeners

For the bearing strength, LRFD J8.1 (p. 6-89):

$$\phi R_n = 0.75\,(1.8 F_y A_{pb})$$

The effective bearing length of the stiffener is the outstanding leg length minus 0.75 in:

$$A_{pb} = 2(5 - 0.75)(5/16) = 2.66 \text{ in.}^2$$

$$\phi R_n = 0.75(1.8)(36)(2.66) = 129 \text{ kips}$$

which agrees with the tabulated value on LRFD p. 9-144.

Check width–thickness ratio

When the angle legs are in continuous contact,

$$\left(\frac{b}{t} = \frac{5}{5/16} = 16\right) \cong \left(\lambda_r = \frac{95}{\sqrt{36}} = 15.8\right)$$

If the angles are separated, a thicker angle is required to satisfy the limiting width–thickness ratio.

Example 8.13

See the figures on LRFD p. 9-138. Verify the entries in LRFD Table 9-9 for the following. Use A36 steel.
The beam seat is a $6 \times 12 \times 0.375$ plate stiffened by a $6 \times 15 \times 0.625$ plate and attached by 5/16 in. fillet welds (E70XX) to a W30 × 116 ($t_w = 0.565$ in.)

Solution

For the stiffener plate,

$$(t = 0.625 \text{ in.}) > (t_w = 0.565 \text{ in.}) \quad \text{as required}$$

$$(t = 0.625 \text{ in.}) \geq [2(5/16) = 0.625 \text{ in.}] \quad \text{as required}$$

Check the width–thickness ratio:

$$\left(\frac{b}{t} = \frac{6}{5/8} = 9.6\right) \leq \left(\lambda_r = \frac{95}{\sqrt{36}} = 15.8\right) \qquad \text{as required}$$

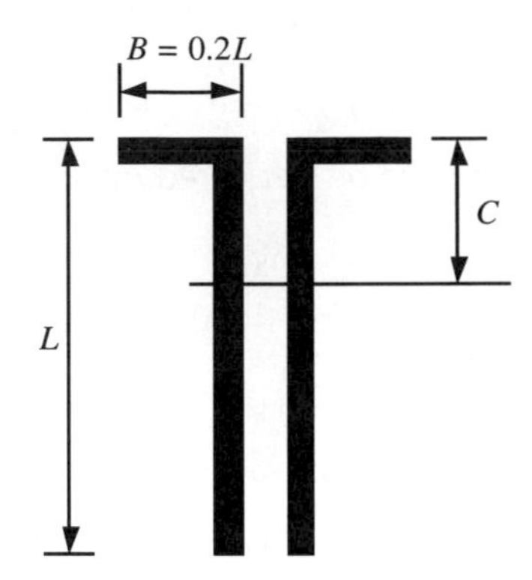

FIGURE 8.18 Example 8.13

Design strength of the $S_w = 5/16$ in. welds:
The welds are assumed to be subjected to shear and bending. This design strength was computed using the elastic method (see Figure 8.18):

$$q_v = \frac{P_u}{A} = \frac{P_u}{2(L+B)} = \frac{P_u}{2(15+3)} = 0.0278\,P_u$$

From the top of the weld group to the centroid of the weld group is

$$c = \frac{2(15)(7.5)}{2(15+3)} = 6.25 \text{ in.}$$

$$I = 2\left[\frac{(15)^3}{12} + 15(7.5-6.25)^2 + 3(6.25)^2\right] = 844$$

$$W = 6 \text{ in.}$$

Since the only variables on LRFD p. 9-145 are W, L, and the weld size, assume that

$$N = W - 0.75 = 6 - 0.75 = 5.25 \text{ in.}$$

$$e = W - \frac{N}{2} = 6 - \frac{5.25}{2} = 3.375 \text{ in.}$$

$$q_M = \frac{Mc}{I} = \frac{P_u(3.375)(6.25)}{844} = 0.0250\,P_u$$

$$q = \sqrt{q_v^2 + q_M^2}$$

$$[\phi(0.6F_{Exx}t_e) = 0.75(0.6)(70)(0.707S_w)] \geq q \quad \text{is required}$$

Solve for P_u, which is renamed as the design strength ϕR_n:

$$\phi R_n = 186 \text{ kips}$$

and the tabulated value on LRFD p. 9-145 is $\phi R_n = 154$ kips.

8.9.4 Shear End-Plate Connections

At the end of a beam, as shown in the figure on LRFD p. 9-91, a plate can be shop-welded to the beam web and field-bolted to a column flange to serve as the means to transfer the required factored beam end shear to the column. This type of connection is classified as a *shear end-plate connection*. Recommended design procedure information for these connections is given on LRFD pp. 9-91 to 9-127. The end plate should be attached on the beam web as close as possible to the compression flange of the beam in order to provide lateral stability for that flange.

8.9.5 Bracket Plates

On LRFD p. 12-5, there are two figures of bracket plates. Design information for these bracket plates is given on LRFD pp. 12-5 to 12-10. LRFD Table 12-1 gives the critical net section indicated in the figure for that table. For the bracket plate, the design bending requirement for the extreme fiber in tension is

$$\left(\phi M_n = 0.9 F_y S_{net}\right) \geq P_u e$$

where

S_{net} = net section modulus of the critical section

e = eccentricity of the beam reaction P_u with respect to the critical section

Examples 8.1 and 8.2 discussed the design strength of a bolt group for a bracket plate. For design purposes (to select the required number of bolts), the tables on LRFD pp. 8-40 to 8-87 can be used.

8.10 MOMENT CONNECTIONS FOR BEAMS

The connections for this category are designed to transfer the required factored shear and moment at the beam end to the supporting member.

8.10.1 Beam-to-Beam Connections and Splices

Suggested design details for moment connections and splices of a beam are shown on LRFD pp. 10-57 and 10-66. In a floor system, one may desire that the lighter loaded set of intersecting beams be continuous at their supports (the heavier loaded set of intersecting beams, which are usually called girders). As shown on LRFD pp. 10-57 and 10-66, the beam end shears at the girder supports are resisted by either web connections or by beam seats. The continuous beam end moments are transferred either through top and bottom splice plates or through top splice plates and the beam seats.

The moment to be transferred can be treated as a couple whose lever arm is $d + t_p$, where d is the beam depth being connected and t_p is the thickness of the connection plate. The top splice plate is subjected to a tension force of $T_u = M_u/(d + t_p)$ where M_u is the required moment at the continuous beam support. If a tension splice plate is bolted to the beam flange being connected, the chosen splice plate must satisfy LRFD J5.2.

For a splice plate that transfers a compression force of $C_u = M_u/(d + t_p)$, the limiting width–thickness ratios on LRFD p. 6-39 for *flange cover plates* are applicable.

If the continuous beam is designed by assuming that

$$\frac{0.5b_f}{t_f} \leq \left(\lambda_p = \frac{65}{\sqrt{F_y}} \right)$$

then, for the compression splice plate we must require that

$$\frac{b}{t} \leq \left(\lambda_p = \frac{190}{\sqrt{F_y}} \right)$$

where b is the plate width between the lines of bolts in the splice plate and t is the thickness of the splice plate. At the splice gap location, the design compressive strength requirement of a compression splice plate is that

$$\left(\phi_c P_n = 0.85 A_g F_y \right) \geq C_u$$

where A_g is the gross area of the splice plate.

8.10.2 Beam-to-Column Connections

Suggested design details for this type of connection are shown on LRFD pp. 10-10, 10-22, and 10-23. Only two of the suggested possibilities (see Figures 8.19 and 8.20) are mentioned in this text.

For the extended end-plate connection shown in Figures 8.13 and 8.19, the suggested design procedure information is given on LRFD pp. 10-21 to 10-35. There are comprehensive example problems in the LRFD Manual, and we see no need to provide any additional examples.

The criteria given on LRFD pp. 10-35 to 10-42 must be used to determine when column web stiffeners aligned with the beam flange ends are required. A paragraph on LRFD p. 10-25 encourages the avoidance of such stiffeners. According to research by Curtis and Murray [24], excessive column flange bending at the beam tension flange does not occur when the column flange thickness satisfies the following criterion:

$$t_{fc} > \sqrt{\frac{4M_{eu}}{0.9F_{yc}b_s}}$$

where

b_s = effective column flange length region subjected to bending (in.)

$b_s = 2.5s_4$ for four-bolt arrangement (in.) [see Figure 8.19(a)]

$b_s = 3.5p_b + s_4$ for eight-bolt arrangement (in.) [see Figure 8.19(b)]

$$s_4 = 2p_f + t_{fb}$$

The other parameters are defined on LRFD p. 10-24.

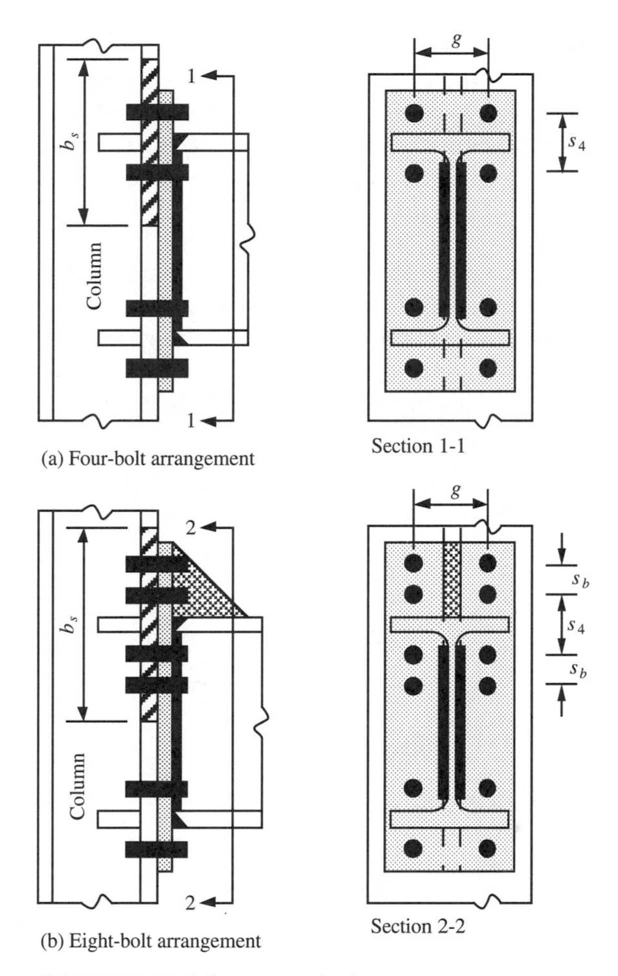

FIGURE 8.19 Moment end-plate connection

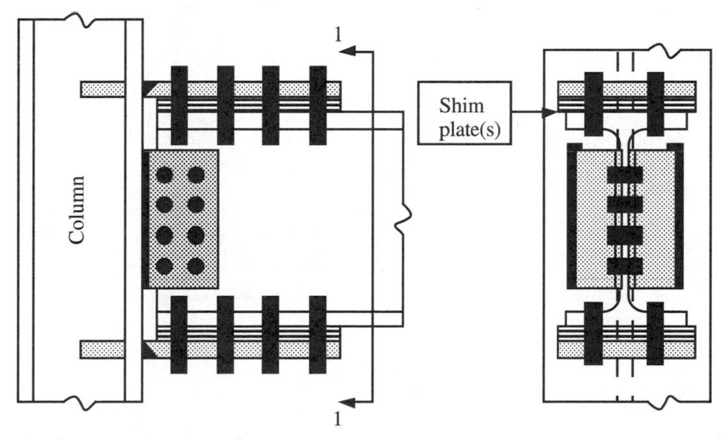

FIGURE 8.20 Top and bottom plate connection

8.11 KNEE OR CORNER CONNECTIONS

LRFD Figure 10-29 (p. 10-67) shows examples of knee or corner connections frequently used in one-story frames designed using FR connections.

Figure 8.21(a) shows a square-knee connection. The beam extends to the edge of the exterior column flange and sits on a base plate (cross-hatched element) welded to the top end of the column. A plate is welded to the end of the beam and to the exterior column flange.

As shown in Figure 8.21(b), the tension flange forces must be transferred by shear into the beam web. Usually, a pair of beam web stiffeners is required from the points *C* to *D* to transfer some of the column flange force C_v via fillet welds subjected to shear into the web of the girder. If either the design shear strength or the design buckling strength of the beam web is not adequate, a pair of diagonal web stiffeners must be provided from points *A* to *C*.

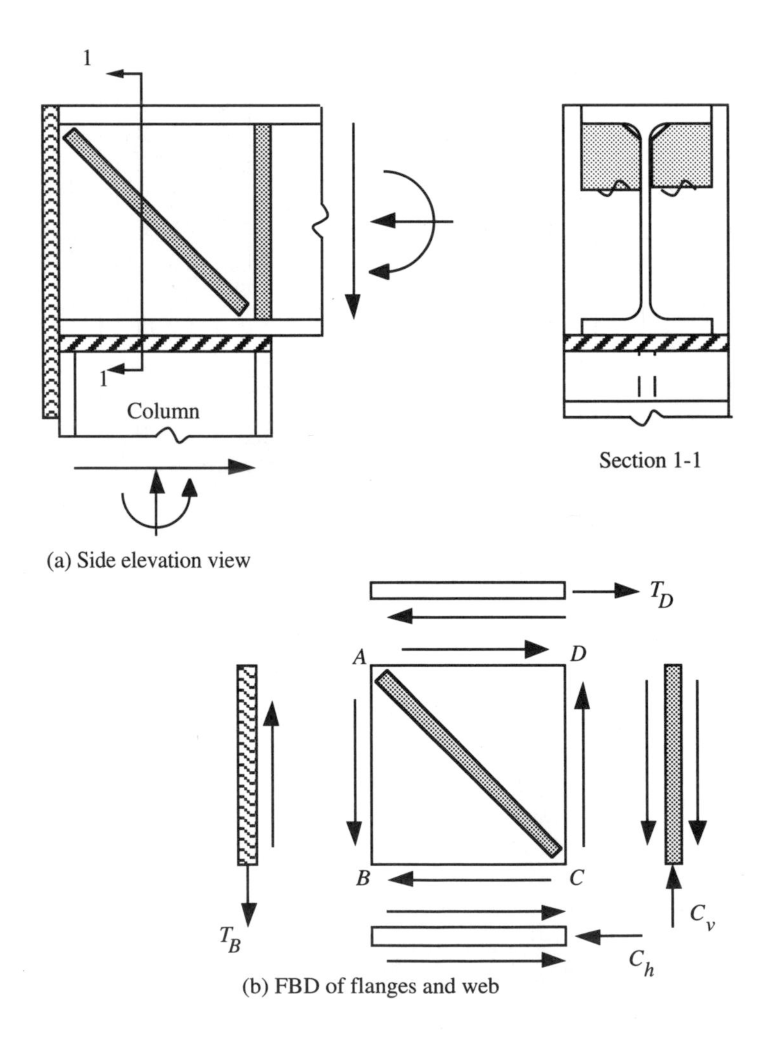

FIGURE 8.21 Corner connection

Example 8.14

Design the square-knee connection in Figure 8.21 for a W24 × 55 beam and a W14 x 82 column. For the W24 x 55 section, the required strengths are $V_u = 60$ kips, $P_u = 24$ kips, and $M_u = 320$ ft-kips. For the W14 × 82 section, the axial compressive design strength is $\phi P_n = 558$ kips and the required strengths are $V_u = 24$ kips, $P_u = 60$ kips, and $M_u = 320$ ft-kips. Use A36 steel and E70 electrodes.

Solution

In Figure 8.21(b), assume:

1. The beam flange forces are

$$T_D = \frac{M_u}{0.95d} - \frac{P_u}{2} = \frac{320(12)}{0.95(23.57)} - \frac{24}{2} = 160 \text{ kips}$$

$$C_h = \frac{M_u}{0.95d} + \frac{P_u}{2} = \frac{320(12)}{0.95(23.57)} + \frac{24}{2} = 184 \text{ kips}$$

2. The column flange forces are

$$T_B = \frac{M_u}{0.95d} - \frac{P_u}{2} = \frac{320(12)}{0.95(14.31)} - \frac{60}{2} = 253 \text{ kips}$$

$$C_v = \frac{M_u}{0.95d} + \frac{P_u}{2} = \frac{320(12)}{0.95(14.31)} + \frac{60}{2} = 313 \text{ kips}$$

Design the beam end plate:

The beam flange width is 7.01 in. and the column flange width is 10.13 in. Choose the end plate width as 8 in. to accommodate fillet welds on the column flange. Since the end plate is in tension, the design strength requirement for yielding on A_g is

$$[0.9(36)(8t)] \geq (T_B = 253)$$

$$t \geq 0.976 \text{ in. is required. Use } t = 1 \text{ in.}$$

The column flange thickness is 0.855 in. and the end-plate thickness is 1 in. The thicker part joined is 1 in., for which the minimum acceptable $S_w = 5/16$ in. and the maximum acceptable $S_w = 15/16$ in. Try $S_w = 5/8$ in. The design requirement for the weld group on the column flange is

$$[0.75(0.6)(70)(0.707)(0.625)(8 + 2L_w)] \geq (T_B = 253)$$

$L_w \geq 10.2$ in. is required for $S_w = 0.625$ in. to develop the column flange force $T_B = 253$ kips. On the end of the beam web, the design requirement for the weld group is

$$[0.75(0.6)(70)(0.707S_w)(2)(21)] \geq (T_B = 253)$$

For strength, $S_w \geq 0.277$ in. is required. Use $S_w = 5/16$ in., which is the minimum acceptable weld size.

Check the beam web:

$$\left(h/t_w = 54.6\right) \leq \left(187\sqrt{5/36} = 69.7\right)$$

From B to A in Figure 8.21, the design shear strength of the web is

$$\phi V_n = 0.9(0.6)(36)(21)(0.395) = 161 \text{ kips}$$

which is less than the required strength $T_B = 253$ kips. Therefore, on the beam end, there is an excess force of 253 - 161 = 92 kips. If the web buckling strength from LRFD K1.6 in the A to C direction is inadequate, a pair of diagonal web stiffeners must be designed.

From D to A, the design shear strength of the web is

$$\phi V_n = 0.9(0.6)(36)(11)(0.395) = 84 \text{ kips}$$

which is less than the required strength $T_D = 160$ kips. Therefore, parallel to the top beam flange, there is an excess force of 160 - 84 = 76 kips. If the web buckling strength from LRFD K1.6 in the A to C direction is inadequate, a pair of diagonal web stiffeners must be designed. The diagonal length of the web in compression is

$$d_c = \sqrt{(21)^2 + (11)^2} = 23.7 \text{ in.}$$

From LRFD K1.6 (p. 6-94), the web buckling design strength is

$$\phi R_n = 0.9\left[\frac{4100(0.395)^3\sqrt{36}}{23.7}\right] = 57.6 \text{ kips}$$

and the required strength is

$$P_u = \sqrt{(92)^2 + (76)^2} = 119 \text{ kips}$$

Since $\phi R_n < P_u$, a pair of diagonal web stiffeners must be designed for the region from A to C for $P_u = 119$ kips.

In Figure 8.20, the ends of the diagonal stiffener need to be mitered to fit the corners in bearing since the diagonal compression force exists at points A and C. The length of the diagonal stiffener is

$$L_s = \sqrt{(23.57 - 20.505)^2 + [14.32 - 2(0.855)]^2} = 25.85 \text{ in.}$$

From LRFD K1.9 (p. 6-96), $KL = 0.75\,(25.85) = 19.4$ in. is the effective length of the pair of diagonal web stiffeners to be installed between points A and C.

Try a pair of $3.25 \times 23.75 \times t$ plates:

$$\left(\frac{b}{t} = \frac{3.25}{0.375} = 8.67\right) \leq \left(\frac{95}{\sqrt{36}} = 15.8\right) \quad \text{as required.}$$

Since the ends of the diagonal stiffener are mitered to fit the corners in bearing, LRFD

J8.1 (p. 6-89) requires that

$$[\phi R_n = 0.75(1.8)(36A_b)] \geq (119 - 57.6 = 61.4 \text{ kips})$$

$$\left[A_b = 2\left(3.25 + \frac{0.395}{2} - \frac{15}{16}\right)t\right] \geq 1.26 \text{ in.}^2$$

$t \geq 0.251$ in. is required. Use $t = 0.25$ in.

From LRFD K1.9 (p. 6-96), the effective area of the diagonal web stiffener is

$$A_s = 2(3.25)(0.25) + (25)(0.395)(0.395) = 5.53 \text{ in.}^2$$

The moment of inertia for column buckling perpendicular to the web plane is

$$I_{st} = \frac{0.25[2(3.25) + 0.395]^3}{12} = 6.83 \text{ in.}^4$$

$$r_{st} = \sqrt{\frac{I_{st}}{A_{st}}} = \sqrt{\frac{6.83}{5.53}} = 1.11 \text{ in.}$$

$$\left(\frac{KL}{r}\right)_{st} = \frac{0.75(25.85)}{1.11} = 17.5$$

From LRFD E2 (p. 6-39):

$$\lambda_c = 17.5\sqrt{36/29000} = 0.617 \qquad \lambda_c^2 = 0.380$$

$$\left[\phi P_n = 0.85(5.53)(0.658^{0.380})(36) = 144 \text{ kips}\right] \geq \left(P_u = 119 \text{ kips}\right)$$

Therefore, a pair of 3.25 x 0.25 x 23.75 plates is adequate.

Minimum fillet welds along the stiffener must be provided to prevent buckling of each individual web stiffener plate as a column in the direction of the plane of the beam web. The thicker part joined is 0.395 in. and the minimum fillet weld size is 3/16 in.

Check the beam web at point *C* in Figure 8.21(b). Since the diagonal stiffener is required, we would automatically provide a pair of web stiffeners from *C* to *D*. From the FBD of this stiffener in Figure 8.20(b), the required compressive strength of the stiffener from *C* to *D* is

$$P_u = 313 - (92 + 60) = 161 \text{ kips}$$

LRFD J8.1 (p. 6-89) requires

$$[\phi R_n = 0.75(1.8)(36A_b] \geq 161 \text{ kips}$$

$$\left[A_b = 2\left(3.25 + \frac{0.395}{2} - \frac{15}{16}\right)t\right] \geq 3.31 \text{ in.}^2$$

$$t \geq 0.660 \text{ in.}$$

Use $t = 0.75$ in.

From C to D, the design shear strength of the beam web is

$$\phi V_n = 0.6(36)(23.57)(0.395) = 181 \text{ kips}$$

$$(\phi V_n = 181 \text{ kips}) \geq (P_u = 161 \text{ kips})$$

Since this design strength is so close to the required strength, we would extend the stiffener to the toe of the top web fillet. Therefore, the strength requirement of the weld group (4 welds) is

$$4\,[0.75(0.6)(70)(0.707S_w)(21)] \geq (P_u = 161)$$

The weld leg size required for strength is $S_w \geq 0.086$ in., and the minimum acceptable $S_w = 0.25$ in. for the 0.75-in. thicker part joined. Use $S_w = 0.25$ in.

PROBLEMS

8.1 Select an A36 steel base plate for the following reaction requirements of a W14 × 82 column: $P_u = 400$ kips, $M_u = 120$ ft-kips, and $V_u = 30$ kips. For the support, assume that the concrete grade is 3 ksi.

8.2 Select an A36 steel base plate for the following reaction requirements of a W14 x 82 column: $P_u = 200$ kips, $M_u = 300$ ft-kips, and $V_u = 60$ kips. For the support, assume that the concrete grade is 3 ksi.

8.3 A pair of L4 x 3.5 x 0.5 is fillet welded to a column web ($t_w = 0.510$ in.) and bolted to a beam end ($t_w = 0.510$ in.). Five 0.75-in.-diameter A325N bolts in a bearing-type connection with standard size bolt holes are in the 3.5-in. angle legs and the beam web. The pair of angles, the beam, and the column are A36 steel.

Use LRFD Table 9-3 (p. 9-88) and determine the required weld size for the weld group. Verify the design strength tabulated for weld group B. This design strength was computed using the elastic method. Include the length of the weld returns in q_v due to shear. Ignore the length of the weld returns in q_M due to bending moment (see Figure P8.3):

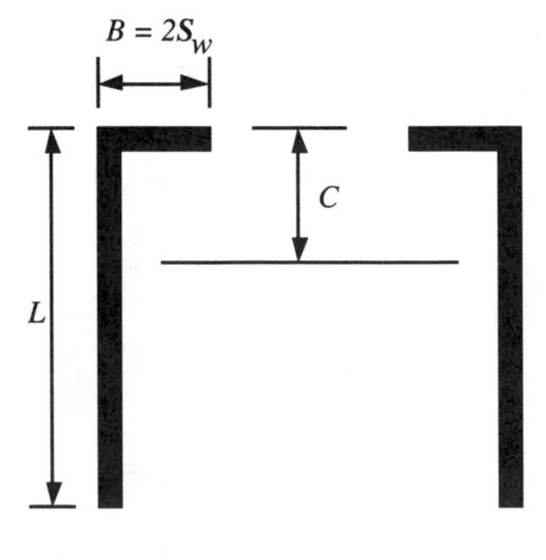

FIGURE P8.3

$$q_v = \frac{P_u}{A} = \frac{P_u}{2(L+2S_w)}$$

$$S_w = \text{weld leg size}$$

$$2S_w = \text{weld return length}$$

$$q_M = \frac{Mc}{I} = \frac{(P_u e)(L/2)}{2(L^3/12)} = \frac{3(P_u e)}{L^2}$$

$e = 2.25$ in. (from LRFD, p. 8-271)

$$q = \sqrt{q_v^2 + q_M^2}$$

$$[\phi(0.6F_{Exx}t_e) = 0.75(0.6)(70)(0.707S_w)] \geq q \quad \text{is required}$$

Solve for P_u, which is renamed as the design strength ϕR_n.

8.4 Solve Problem 8.3 for 7/8-in.-diameter A325N bolts.

8.5 Solve Problem 8.3 for 1-in.-diameter A325N bolts.

8.6 Solve Problem 8.3 for A307 bolts.

8.7 Using LRFD Tables 9-2 and 9-3, select a pair of angles bolted to beam web and welded to column flange. The beam is a W24 × 55 and the column is a W14 x 82. The weld group on the pair of angles is also on the column web. Use A36 steel for the angles, the beam, and the column. Use 0.75-in.-diameter A325N bolts in a bearing-type connection. The required shear strength of the beam end is 100 kips.

8.8 Solve Problem 8.7 for A307 bolts.

8.9 As shown on LRFD p. 9-89, a pair of L3 x 3 x 5/16 is fillet-welded to a column web and to a beam end. The pair of angles, the beam, and the column are A36 steel. Verify the entries given in LRFD Table 9-4, using:

(a) LRFD Table 8-42 for weld group A.
(b) The elastic method for weld group B. Include the length of the weld returns in q_v due to shear. Ignore the length of the weld returns in q_M due to bending moment (see Figure P8.9):

$$q_v = \frac{P_u}{A} = \frac{P_u}{2(L+2S_w)}$$

$$S_w = \text{weld leg size}$$

$$2S_w = \text{weld return length}$$

$$q_M = \frac{Mc}{I} = \frac{(P_u e)(L/2)}{2(L^3/12)} = \frac{3(P_u e)}{L^2}$$

$e = 3$ in.

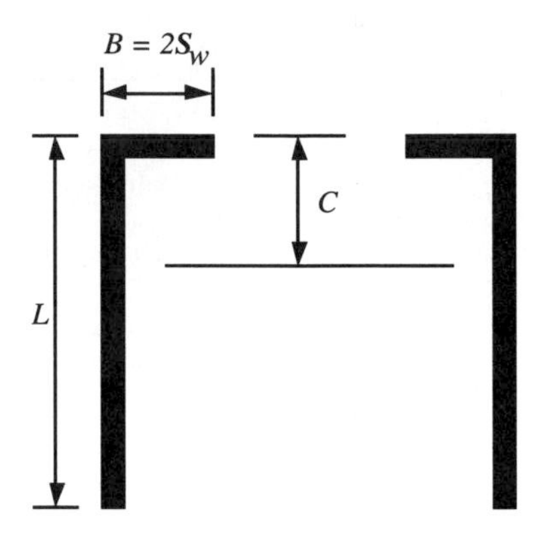

FIGURE P8.9

$$q = \sqrt{q_v^2 + q_M^2}$$

$[\phi(0.6F_{Exx}t_e) = 0.75(0.6)(70)(0.707S_w)] \geq q$ is required

Solve for P_u, which is renamed as the design strength ϕR_n.

8.10 If the web thickness of the beam and the column in Problem 8.9 is $t_w = 0.510$ in., do the weld sizes for Problem 8.9 satisfy LRFD J2.2b (p. 6-75)?

8.11 Use LRFD Table 9-4 to select a pair of angles and the weld sizes to be used for the following conditions. The beam is a W24 × 55 and the column is a W14 × 82. Weld group B is on the column web. Use A36 steel for the angles, the beam, and the column. The connection must develop the design shear strength of the beam end.

8.12 Determine the block shear rupture strength of the connection in Example 9-1 (p. 9-16) by using:

(a) LRFD equations for BSR
(b) LRFD Tables 8-47 and 8-48 (pp. 8-215 to 8-224)

8.13 Determine the block shear rupture strength of the connection in Example 9-2 (p. 9-18) by using:

(a) LRFD equations for BSR
(b) LRFD Tables 8-47 and 8-48 (pp. 8-215 to 8-224).

8.14 See the figure on LRFD p. 9-91. Verify the entries in Table 9-5 for the end-plate connection designed in Example 9-6 (p. 9-92).

8.15 Repeat Example 9-6 and Problem 8.14 for $F_y = 36$ ksi.

8.16 See LRFD pp. 9-161 and 9-162. Verify the entries in Table 9-11 for the connection designed in Example 9-13 (p. 9-163).

8.17 See LRFD p. 9-163. Verify the entries in Table 9-12 for the connection designed in Example 9-14 (p. 9-166).

8.18 For the bearing-type connection shown in Figure P8.18, we are given the following numerical information: A36 steel bracket plate and column, $g = 5.5$ in., $s =$ 3 in., $e_x = 6$ in., column flange thickness = 0.855 in., 0.75-in.-diameter A325X bolts in single shear, and $P_u = 50$ kips. Use LRFD p. 8-52 to determine the number of bolts required. Use LRFD pp. 12-5 to 12-10 to determine the required bracket plate thickness for $M = P_u e$.

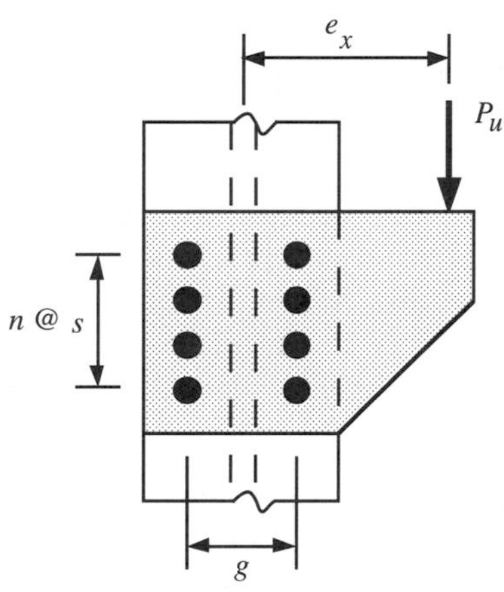

FIGURE P8.18

8.19 For the bearing-type connection shown in Figure P8.18, we are given the following numerical information: A36 steel bracket plate and column, $g = 8$ in., $s =$ 3 in., $e_x = 14$ in., column flange thickness = 0.710 in., 7/8-in.-diameter A325X bolts in single shear, and $P_u = 75$ kips. Use LRFD p. 8-52 to determine the number of bolts required. Use LRFD pp. 12-5 to 12-10 to determine the required bracket plate thickness for $M = P_u e$.

8.20 For the welded connection shown in Figure P8.20, we are given the following numerical information: A36 steel bracket plate and column, $b = 4$ in., $d =$ 10 in., $e_x = 12$ in., column flange thickness = 0.855 in., E70 electrodes for the fillet

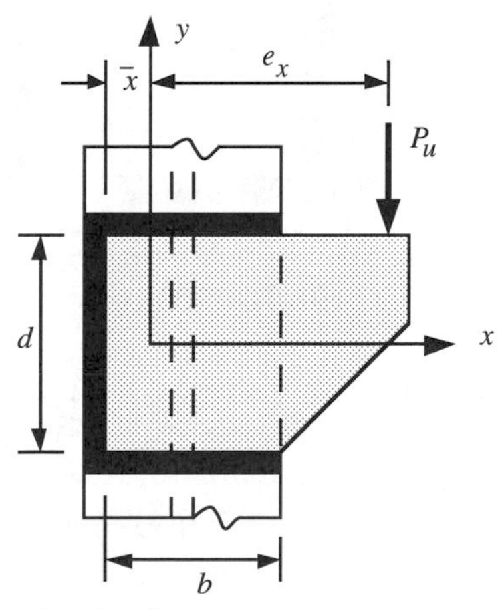

FIGURE P8.20

welds, and P_u = 50 kips. Use LRFD p. 8-187 to determine the required weld size. Determine the required bracket plate thickness such that $M/S_x \geq 0.9F_y$, where

$$M = P_u e$$

$$S_x = \frac{I_x}{d/2} = \text{elastic section modulus of the bracket plate}$$

Also, the cantilevered plate length divided by the thickness must not exceed $\lambda_r = 95/\sqrt{F_y}$.

8.21 For the welded connection shown in Figure P8.20, we are given the following numerical information: A36 steel bracket plate and column, b = 12 in., d = 14 in., e_x = 14 in., column flange thickness = 0.710 in., E70 electrodes for the fillet welds, and P_u = 100 kips. Use LRFD p. 8-187 to determine the required weld size. Determine the required bracket plate thickness such that $M/S_x \geq 0.9F_y$, where

$$M = P_u e$$

$$S_x = \frac{I_x}{d/2} = \text{elastic section modulus of the bracket plate}$$

Also, the cantilevered plate length divided by the thickness must not exceed $\lambda_r = 95/\sqrt{F_y}$.

8.22 Design the beam splice shown in Figure P8.22 Use A36 steel for the pair of splice plates and the W24 × 55 beam. The required shear strength of the beam ends is V_u = 121 kips. Use 0.75-in.-diameter A325X bolts in a bearing-type connection and E70 electrodes. Determine the required number of bolts, the weld size, and the size of each splice plate.

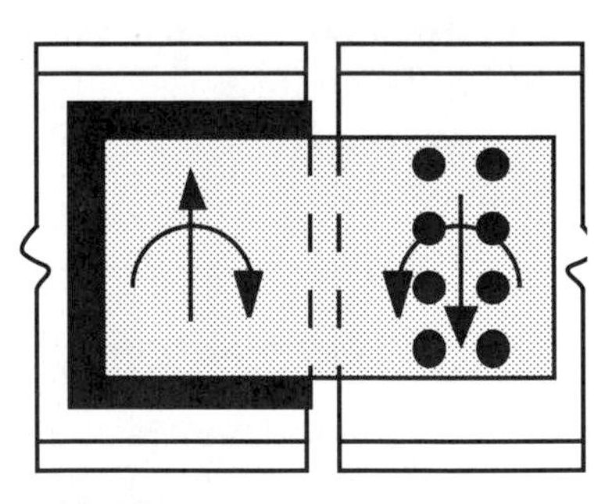

FIGURE P8.22

8.23 Design the beam splice shown in Figure P8.22. Use F_y = 50 ksi steel for the pair of splice plates and the W24 × 55 beam. The required shear strength of the beam ends is V_u = 167 kips. Use 1-in.-diameter A325X bolts in a bearing-type connection and E80 electrodes. Determine the required number of bolts, the weld size, and the size of each splice plate.

8.24 Solve Example 8.5 for the tension member inclined at tan α = 0.5.

8.25 Solve Example 8.6 for the tension member inclined at tan $\alpha = 0.5$.

8.26 Design a beam splice as shown in LRFD Figure 10-20 (p. 10-57). Use A36 steel for the splice plates and the W24 × 55 beam. The connection must develop ϕM_{px} = 330 ft-kips and ϕV_n = 121 kips. Use 7/8-in.-diameter A325X bolts in a bearing-type connection and E70 electrodes. Determine the required number of bolts, the weld size, and the size of each splice plate. The girder is a W30 x 90.

8.27 Design an all-welded beam splice similar to the welded half of LRFD Figure 10-20 (p. 10-57). Use A36 steel for the splice plates and the W24 × 55 beam. The connection must develop ϕM_{px} = 362 ft-kips and ϕV_n = 121 kips. Use E70 electrodes. Determine the required weld sizes, the beam seat size, and the size of the splice plate. The girder is a W36 x 135.

8.28 Solve LRFD Example 10-2 (p. 10-16) for M_u = 362 ft-kips, a factored beam end shear of 66 kips, and a factored column web shear of 36 kips. Also, LRFD Example 10-6 Solution A is applicable.

8.29 Solve Problem 8.28 for a W24 x 55 beam framed to a W14 x 82 column. M_u = 503 ft-kips. The factored beam shear is 92 kips. The factored column web shear is 50 kips. Use F_y = 50 ksi steel for the connection parts. Use 1-in.-diameter A325X bolts and E80 electrodes.

CHAPTER 9

Plate Girders

9.1 INTRODUCTION

In LRFD Chapter G (p. 6-58), we find that flexural members are classified as either beams or plate girders. Some of the characteristics of beams and plate girders need to be illustrated. Therefore, we choose to state the characteristics of beams and plate girders in separate figures.

The characteristics of a *beam* are given in Figure 9.1(d), where F_{yf} is the yield strength of the flanges. *Intermediate web stiffeners*, as shown in Figure 9.2(a) for a plate girder, are not allowed in the definition of a beam. However, *bearing web stiffeners* as shown in Figure 9.1(a) may be required by LRFD K1.9 (p. 6-96). A beam may be a built-up hybrid section (F_y of the web and flanges may be different). LRFD Appendix G (p. 6-203) states that inelastic web buckling in a hybrid flexural member is dependent on the flange strain.

The characteristics of a *plate girder* are given in Figure 9.2(c), where F_{yf} is the yield strength of the flanges. Bearing web stiffeners as shown in Figure 9.2(a) may be required by LRFD K1.9 (p. 6-96). Intermediate web stiffeners may be required by LRFD F2.3 (p. 6-113) or, when *tension field action* is utilized (see Figure 9.4), by LRFD G4 (p. 6-125). Intermediate web stiffeners [see Figure 9.2(a)] are not required when either of the following conditions is satisfied:

1. $h / t_w \leq 418 / \sqrt{F_{yw}}$
2. $\phi V_n \geq V_u$

where

$$F_{yw} = \text{yield strength of the web}$$

$$V_u = \text{required shear strength}$$

$$\phi V_n = \text{design shear strength given in LRFD F2.2 (p. 6-56)}$$

h and t_w are defined in Figure 9.1(b).

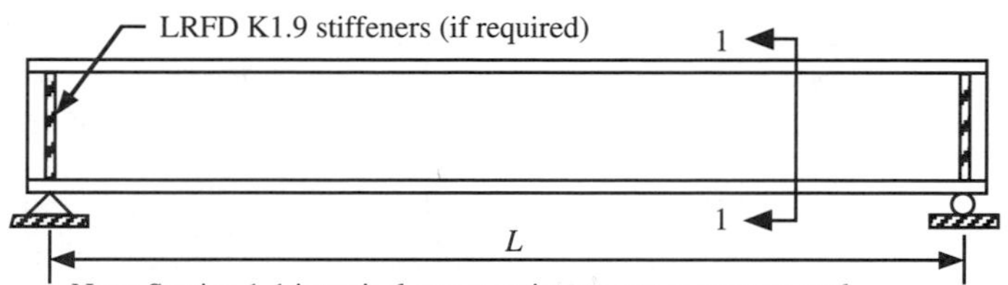

Note: Section 1-1 is typical cross section except at supports and concentrated load points.

(a) Side elevation view

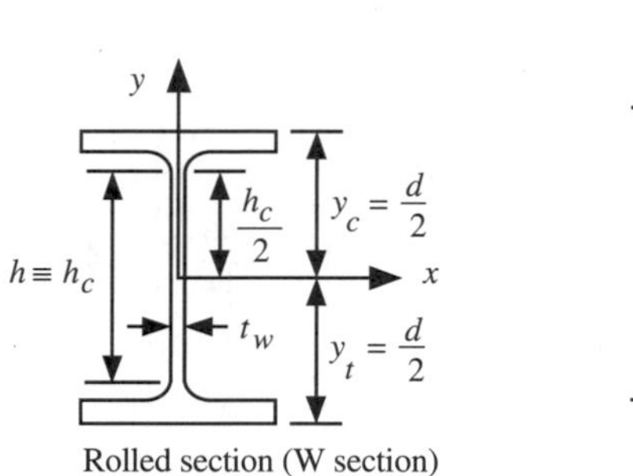

Rolled section (W section)

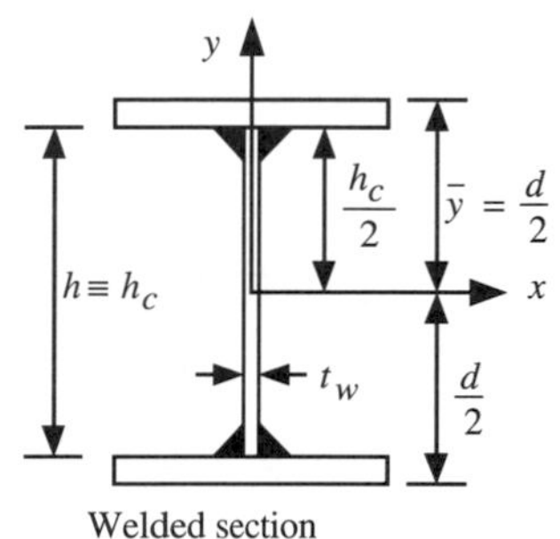

Welded section

(b) Section 1-1 when cross section is doubly symmetric

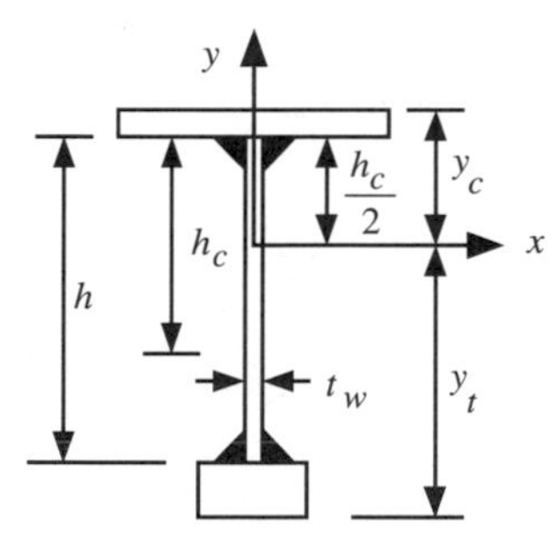

(d) Beam characteristics are:
Rolled or welded section.

$$\frac{h}{t_w} \leq \frac{970}{\sqrt{F_{yf}}}$$

Intermediate web stiffeners [see LRFD F2.3, (p. 6-56)] are an **excluded** characteristic.

(c) Section 1-1 when cross section is singly symmetric

FIGURE 9.1 Beam.

If intermediate web stiffeners are required, the member is a plate girder and the maximum permissible value of h/t_w is given on LRFD p. 6-122:

1. For $\frac{a}{h} \leq 1.5$: $\quad \text{Maximum}\,\frac{h}{t_w} = \frac{2000}{\sqrt{F_{yf}}}$

2. For $\frac{a}{h} > 1.5$: $\quad \text{Maximum}\,\frac{h}{l_w} = \frac{14{,}000}{\sqrt{F_{yf}\left(F_{yf} + 16.5\right)}}$

where

F_{yf} = yield strength of the flanges

a = clear distance between intermediate web stiffeners

h and t_w are defined in Figure 9.2(b).

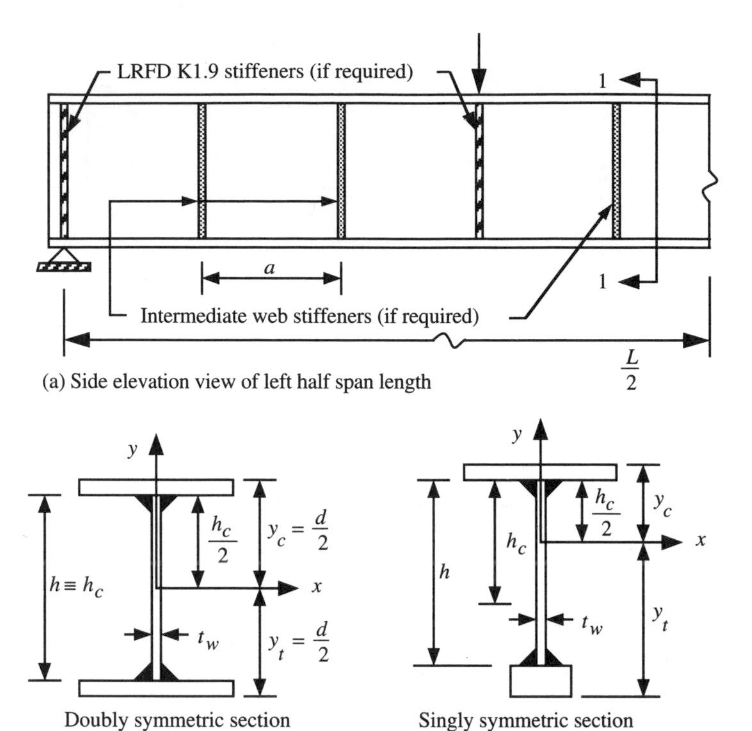

(b) Section 1-1 when LRFD G4 stiffeners are not used

(c) Plate Girder characteristics are:
Rolled or welded section. Either or both of the following:

$$\frac{h}{t_w} > \frac{970}{\sqrt{F_{yf}}}$$

Intermediate web stiffeners are required by LRFD F2.3 or G4.

FIGURE 9.2 Plate girder.

As shown in Figure 9.2(b), the most commonly used built-up section for a plate girder is composed of two flange plates fillet-welded to a web plate. Some other cross sections that have been used for a plate girder are shown in Figure 9.3. Any of the sections in Figures 9.2 and 9.3 may be a hybrid section; that is, for example, the web(s) may be $F_y = 36$ ksi steel and the flanges may be $F_y = 50$ ksi steel.

In a multistory office or hotel building, large column-free areas (ballroom, dining room, lobby) may be needed or desirable. Either a built-up plate girder or a truss may be the economical choice to span these column-free rooms and to support the columns in the stories above these rooms. Students want design rules without any exceptions to be given for them to use in deciding the most economical type of flexural member for a specific design situation. Such rules are impossible for us to give, but the following guidelines may be helpful:

1. For rolled beams, an economical depth is $L/20$ to $L/50$, where L is the span length. See LRFD Table 4-2 (p. 4-30).

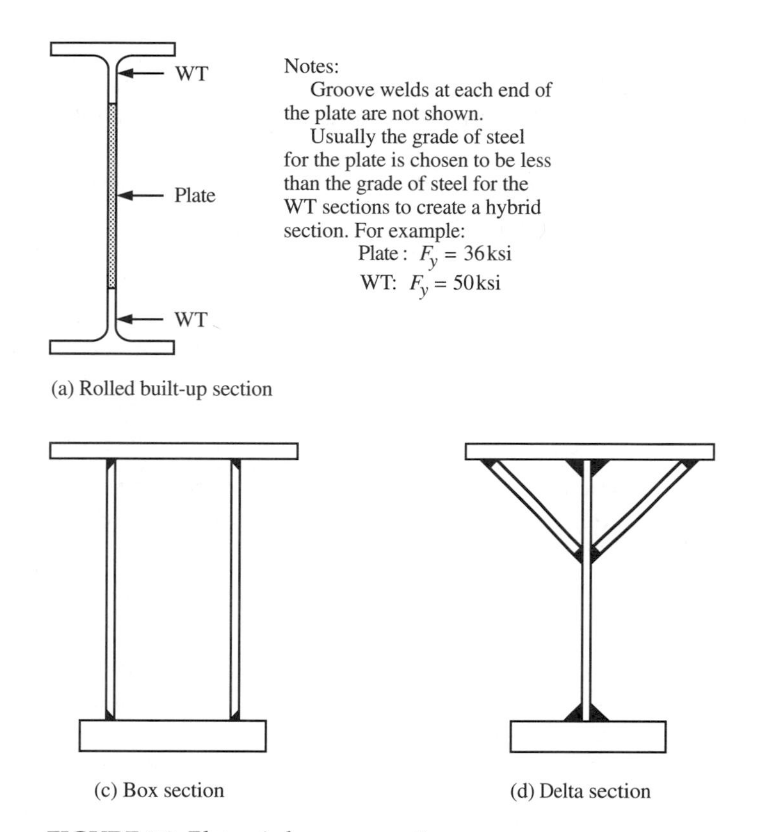

FIGURE 9.3 Plate girder cross sections.

2. For built-up beams and plate girders, an economical depth is $L/15$ to $L/8$. Simple spans of 70 to 150 feet are typical for plate girders in buildings and highway bridges. Shorter simple spans are economical in unusual cases and railway bridges. Longer continuous spans of 90 to 400 feet are feasible, but for the longer spans the section depth varies from a maximum at the supports to a minimum at midspan.
3. For parallel chord trusses, an economical depth is $L/12$ to $L/8$.

A built-up plate girder usually has the following advantages when compared to a truss:

1. Lower fabrication cost.
2. Fewer field erection problems, but a larger crane capacity may be required.
3. Fewer critical points in the member at which design requirements may govern. In a truss, if an overload occurs in any member or connection, the result may be a disaster. However, if an overload occurs only at one point in a plate girder, some of the neighboring material may help prevent a disaster.

We restrict the discussion to the case where the required strengths are determined from an elastic factored analysis. In the LRFD Specification, there are two design procedures for plate girders. We choose to restrict the following discussion to the case for which the cross sections shown in Figures 9.1(b) to (c) and 9.2(b) are applicable. The two design procedures for this case are referred to as:

1. *Conventional Design Method* A brief description of the plate girder behavior for this design method is given in Section 9.2. See Section 5.10 and Example 5.8 for the design of bearing web stiffeners.
2. *Tension Field Design Method* A brief description of the plate girder behavior for this design method is given in Section 9.3. See Section 5.10 and Example 5.8 for the design of bearing web stiffeners. The Tension Field Method is not applicable for
   (a) Web-tapered girders
   (b) Hybrid girders
   (c) The member-end panels in a nonhybrid girder
   (d) Any panels of a nonhybrid girder for which

$$\frac{a}{h} > 3 \qquad \text{or} \qquad \frac{a}{h} > \left(\frac{260}{h/t_w} \right)^2$$

According to a statement on LRFD p. 6-249, plate girders usually are more economical when they are designed by the Conventional Method. The reason cited for this is that if a thicker web is used, fewer intermediate web stiffeners are required (possibly none are required) by the Conventional Method than for the Tension Field Method. Consequently, considerably less fabrication time is required when the number of intermediate web stiffeners can be held to a minimum or to none. By making the web thicker (adds more steel cost), which requires fewer intermediate web stiffeners, skilled labor costs can be significantly reduced, and the overall fabrication cost may be less than for a minimum girder weight design by the Tension Field Method.

9.2 CONVENTIONAL DESIGN METHOD

LRFD F2 (p. 6-53) is applicable for shear. LRFD F2.3 is applicable for the intermediate (transverse) stiffeners. For bending, when LRFD F1.1 to F1.2 are not applicable, LRFD G (p. 6-58) invokes:

1. LRFD Appendix F1 (p. 6-111) when $h/t_w \leq 970/\sqrt{F_{yf}}$
2. LRFD G2 (p. 6-122) when $h/t_w > 970/\sqrt{F_{yf}}$

Except for having to design intermediate web stiffeners (LRFD F2.3), this method involves the familiar design requirements for a beam when $h/t_w \leq 970/\sqrt{F_{yf}}$. However, the determination of the design shear strength (LRFD F2) is more complex than for a beam, and the design bending strength may have to be determined from LRFD Appendix F1 for which there are no available design aids similar to the $C_b = 1$ beam charts. When $h/t_w > 970/\sqrt{F_{yf}}$, previously undiscussed definitions must be used to determine the design bending strength (LRFD, Appendix G2). Since there are no available design aids similar to the $C_b = 1$ beam charts for

LRFD Appendices F1 and G2, the design process is iterative (a trial section is checked and, if necessary, revised, checked, and so on until the design requirements are satisfied). However, on LRFD pp. 4-183 to 4-185, some tabular information is provided as a guide for selecting a trial section whose depth is in the range of 45 to 92 in. Also, the tables on LRFD pp. 6-155 to 6-158 may be useful in choosing the intermediate stiffener spacings such that the design requirement for shear is satisfied.

9.2.1 Design Strength Definitions

The definitions that follow are not valid for a web-tapered girder for which the reader should see LRFD Appendix F3 (p. 6-118).

For a plate girder subjected to bending about the strong axis only, the design requirements are

1. $\phi_b M_{nx} \geq M_{ux}$
2. $\phi_v V_n \geq V_u$

where $\phi_b M_{nx}$ and $\phi_v V_n$ are the design bending and shear strengths, M_{ux} and V_u the required bending and shear strengths

For the Conventional Design Method, the design strength definitions are:

1. Design bending strength $\phi_b M_{nx}$

$$\text{When} \quad \frac{0.5 b_f}{t_f} \leq \frac{65}{\sqrt{F_y}} \quad \text{and} \quad \frac{h}{t_w} \leq \frac{640}{\sqrt{F_{yf}}}$$

for nonhybrid girders, $\phi_b M_{nx}$ is defined in LRFD F1. (p. 6-52 to 6-56).

$$\text{When} \quad \frac{0.5 b_f}{t_f} > \frac{65}{\sqrt{F_y}} \quad \text{and} \quad \frac{970}{\sqrt{F_{yf}}} \geq \frac{h}{t_w} > \frac{640}{\sqrt{F_{yf}}}$$

$\phi_b M_{nx}$ is defined in LRFD Appendix F1 (pp. 6-111).

$$\text{When} \quad \frac{h}{t_w} > \frac{970}{\sqrt{F_{yf}}}$$

$\phi_b M_{nx}$ is defined in LRFD Appendix G2 (p. 6-122), which states that the governing definition is the smaller value obtained from:

(a) Tension-flange yield:

$$\phi M_{nx} = 0.9\left(S_{xt} R_e F_{yt}\right)$$

(b) Compression-flange buckling:

$$\phi M_{nx} = 0.9\left(S_{xc} R_{PG} R_e F_{cr}\right)$$

where

$$\left[R_{PG} = 1 - \frac{a_r}{1200 + 300 a_r}\left(\frac{h_c}{t_w} - \frac{970}{\sqrt{F_{cr}}}\right)\right] \leq 1.00$$

$$R_e = \frac{12 + a_r(3m - m^3)}{12 + 2a_r} \quad \text{for hybrid girders}$$

$$R_e = 1 \quad \text{for nonhybrid girders}$$

a_r = ratio of web area to compression flange area

h_c and t_w are defined in Figure 9.2(b)

$$S_{xc} = I_x/y_c \quad \text{and} \quad S_{xt} = I_x/y_t$$

I_x = moment of inertia for strong-axis bending (in.^{4})

y_t = distance from x-axis to the extreme fiber in tension (in.)

F_{yt} = yield stress of tension flange (ksi)

y_c = distance from x-axis to the extreme fiber in compression (in.)

F_{cr} = critical compression flange stress (ksi) (see definitions in next paragraph)

$$m = F_{yw}/F_{cr}$$

In the following equations and conditional relations, the slenderness parameters λ, λ_p, λ_r, and C_{PG} are as defined subsequently for *lateral-torsional buckling* (LTB) and *flange local buckling* (FLB). F_{cr} must be computed for LTB and FLB; the smaller F_{cr} governs:

(a) When $\lambda \le \lambda_p$,

$$F_{cr} = F_{yf}$$

(b) When $\lambda_p < \lambda \le \lambda_r$,

$$\left\{F_{cr} = C_b F_{yf}\left[1 - \frac{1}{2}\left(\frac{\lambda - \lambda_p}{\lambda_r - \lambda_p}\right)\right]\right\} \le F_{yf}$$

(c) When $\lambda > \lambda_r$,

$$F_{cr} = \frac{C_{PG}}{\lambda^2}$$

For LTB,

$$\lambda = \frac{L_b}{r_t}$$

L_b = laterally unbraced length of the compression flange (in.)

r_T = radius of gyration of a T section comprised of the compression flange and one-third of the compression web area

$$\lambda_p = \frac{300}{\sqrt{F_{yf}}}$$

$$\lambda_r = \frac{756}{\sqrt{F_{yf}}}$$

$$C_{PG} = 286{,}000 C_b$$

C_b = a bending parameter defined in LRFD Eq. (F1-3), p. 6-53

For FLB:

$$\lambda = \frac{0.5 b_f}{t_f}$$

$$\lambda_p = \frac{65}{\sqrt{F_{yf}}}$$

$$\lambda_r = \frac{230}{\sqrt{F_{yf} / k_c}}$$

$$C_{PG} = 26{,}200 k_c$$

$$k_c = \frac{4}{\sqrt{h / t_w}}$$

$$0.35 \le k_c \le 0.763$$

2. Design shear strength $\phi_v V_n$

From LRFD F2 (p. 6-56), we find:

(a) If intermediate stiffeners are not used, the design requirements are

$$\frac{h}{t_w} \le 260 \quad \text{and either} \quad \frac{h}{t_w} \le \frac{418}{\sqrt{F_{yw}}} \quad \text{or} \quad \phi V_n \ge V_u$$

where ϕV_n is defined in LRFD F2 (pp. 6-56 and 6-113).

When

$$\frac{h}{t_w} \le \frac{418}{\sqrt{F_{yw}}}$$

web shear yielding governs the design shear strength:

$$\phi V_n = 0.90\left(0.6 F_{yw} A_w\right)$$

When

$$\frac{418}{\sqrt{F_{yw}}} < \frac{h}{t_w} \le \frac{523}{\sqrt{F_{yw}}}$$

inelastic web buckling governs the design shear strength:

$$\phi V_n = 0.90\left(0.6 F_{yw} A_w \frac{418 / \sqrt{F_{yw}}}{h / t_w}\right)$$

When

$$\frac{523}{\sqrt{F_{yw}}} < \frac{h}{t_w} \le 260$$

elastic web buckling governs the design shear strength:

$$\phi V_n = 0.90\left(\frac{132{,}000 A_w}{(h/t_w)^2}\right)$$

(b) If intermediate stiffeners are used, the design requirements are as follows. When $a/h \le 1.5$,

$$k_v = 5 + \frac{5}{(a/h)^2} \quad \text{and} \quad \text{Maximum}\frac{h}{t_w} = \frac{2000}{\sqrt{F_{yf}}}$$

When $3.0 \ge a/h > 1.5$,

$$k_v = 5 + \frac{5}{(a/h)^2} \quad \text{and} \quad \text{Maximum}\frac{h}{t_w} = \frac{14{,}000}{\sqrt{F_{yf}\left(F_{yf} + 16.5\right)}}$$

When $a/h > 3$ or $a/h > (260/h/t_w)^2$,

$$k_v = 5 \quad \text{and} \quad \text{Maximum}\frac{h}{t_w} = \frac{14{,}000}{\sqrt{F_{yf}\left(F_{yf} + 16.5\right)}}$$

When

$$\frac{h}{t_w} \le 187\sqrt{\frac{k_v}{F_{yw}}}$$

web shear yielding governs the design shear strength:

$$\phi V_n = 0.90\left(0.6 F_{yw} A_w\right)$$

When

$$187\sqrt{\frac{k}{F_{yw}}} < \frac{h}{t_w} \le 234\sqrt{\frac{k}{F_{yw}}}$$

inelastic web buckling governs the design shear strength:

$$\phi V_n = 0.90\left(0.6 F_{yw} A_w \frac{418/\sqrt{F_{yw}}}{h/t_w}\right)$$

When

$$\frac{h}{t_w} > 234\sqrt{\frac{k}{F_{yw}}}$$

elastic web buckling governs the design shear strength:

$$\phi V_n = 0.90\left(\frac{26{,}400 A_w k_v}{(h/t_w)^2}\right)$$

9.2.2 Intermediate Stiffener Requirements

Intermediate stiffeners are not required when $h/t_w \le 418/\sqrt{F_{yw}}$ or when $\phi_v V_n \ge V_u$, where $\phi_v V_n$ is determined from the formulas given in item 2a of Section 9.2.1.

The design requirement for intermediate stiffeners is

$$I_{st} \ge a t_w^3 j$$

where

$$\left[j = \frac{2.5}{(a/h)^2} - 2 \right] \ge 0.5$$

a = clear distance between intermediate web stiffeners

I_{st} = moment of inertia about an axis in the web center for a pair of stiffeners or about the face in contact with the web plate for a single stiffener

h and t_w are defined in Figure 9.2(b).

See LRFD F2.3 (p. 6-118) for some clearly stated, restrictive conditions for intermediate stiffeners and their fasteners.

9.2.3 Design Examples

LRFD Example 4-10 (p. 4-168) is an example of the Tension Field Design Method. LRFD Example 4-11 (p. 4-176) is an example of the Conventional Design Method and illustrates the following items:

1. Design of a hybrid girder for $F_{yw} = 36$ ksi and $F_{yf} = 50$ ksi
2. $\left[\frac{h}{t_w} = \frac{54}{(9/16)} = 96 \right] \le \left(\frac{970}{\sqrt{F_{yf}}} = \frac{970}{\sqrt{50}} = 137 \right)$
3. Determination of the design bending strength for $L_b = 0$
4. Design of the intermediate stiffeners

LRFD Example 4-12 (p. 4-180) is for a nonhybrid member that does not require any intermediate stiffeners, and since

$$\left(\frac{h}{t_w} = \frac{50}{0.5} = 100 \right) \le \left(\frac{970}{\sqrt{F_y}} = \frac{970}{\sqrt{36}} = 162 \right)$$

this member is a built-up beam and not a plate girder.

The general design procedure for the Conventional Design Method when $L_b = 0$ is illustrated in LRFD Example 2 (p. 4-176). Therefore, in the following examples, we choose to illustrate some strength calculations for a trial section.

Example 9.1

The built-up W57 × 18 × 160 section on LRFD p. 4-184 is chosen as a trial section and needs to be investigated for the following conditions:

$$L_b = 20 \text{ ft} \qquad C_b = 1.15 \qquad F_{yw} = F_{yf} = 36 \text{ ksi}$$

1. Find ϕV_n when intermediate web stiffeners are not used.
2. Find ϕV_n in a panel for which $a/h = 60/56 = 1.07$ when intermediate web stiffeners are used.
3. Find ϕM_{nx}.

Solution 1

$$A_w = (56)(7/16) = 24.5 \text{ in.}^2$$

For an unstiffened web in a plate girder, from LRFD F2.2 (p. 6-113) we find that

$$\left[\frac{h}{t_w} = \frac{56}{(7/16)} = 128\right] \le 260 \text{ as required}$$

$$k_v = 5$$

$$\left(\frac{h}{t_w} = 128\right) > \left(234\sqrt{\frac{k_v}{F_{yw}}} = 234\sqrt{\frac{5}{36}} = 87.2\right)$$

$$\phi V_n = 0.9\left[\frac{26,400 A_w k_v}{(h/t_w)^2}\right] = 0.9\left[\frac{26400(24.5)(5)}{(128)^2}\right] = 178 \text{ kips}$$

These computations illustrate how a ϕV_n value listed in the table on LRFD p. 4-184 was computed.

If intermediate web stiffeners are not used, from LRFD F2.3 (p. 6-113) we find for our example that

$$\left(\frac{h}{t_w} = 128\right) > \left(\frac{418}{\sqrt{F_{yw}}} = \frac{418}{\sqrt{36}} = 69.7\right)$$

$$(\phi V_n = 178 \text{ kips}) \ge V_u \quad \text{is required.}$$

Solution 2

From LRFD F2.2 (p. 6-113), we find that

$$\left(\frac{a}{h} = \frac{60}{56} = 1.07\right) \le 3.0 \le \left[\left(\frac{260}{h/t_w}\right)^2 = \left(\frac{260}{128}\right)^2 = 4.13\right]$$

$$k_v = 5 + \frac{5}{(a/h)^2} = 5 + \frac{5}{(60/56)^2} = 9.356$$

$$\left(\frac{h}{t_w} = 128\right) > \left(234\sqrt{\frac{k_v}{F_{yw}}} = 234\sqrt{\frac{9.356}{36}} = 119\right)$$

$$\phi V_n = 0.9\left[\frac{26{,}400 A_w k_v}{(h/t_w)^2}\right] = 0.9\left[\frac{26400(24.5)(9.356)}{(128)^2}\right] = 332\,\text{kips}$$

Solution 3

LRFD Appendix F (p. 6-111) is applicable. The three limit states that may need to be investigated to determine the governing ϕM_{nx} are:

- **a.** Flange local buckling (FLB)
- **b.** Web local buckling (WLB)
- **c.** Lateral-torsional buckling (LTB)

ϕM_{nx} must be computed for each applicable limit state, and the least of these computed values is the governing value of ϕM_{nx}.

From LRFD p. 4-184,

$$b_f = 18\text{ in.} \qquad t_f = 5/8\text{ in.} \qquad h_w = 56\text{ in.}$$

$$t_w = 7/16\text{ in.} \qquad Z_x = 980\text{ in.}^3 \qquad S_x = 854\text{ in.}^3$$

For a nonhybrid member, since $(Z_x/S_x = 980/854 = 1.15) < 1.5$,

$$\phi M_{px} = 0.9F_y Z_x = 0.9\,(36)(980) = 31{,}752\text{ in.-kips} = 2646\text{ ft-kips}$$

which is applicable for all three limit states.

a. For FLB (see LRFD pp. 6-111 and 6-114):

$$\lambda = \frac{0.5b_f}{t_f} = \frac{0.5(18)}{(5/8)} = 14.4$$

$$\lambda_p = \frac{65}{\sqrt{F_{yf}}} = \frac{65}{\sqrt{36}} = 10.8$$

$$F_L = \text{smaller of}\begin{cases}(F_{yf} - F_r) = (36 - 16.5) = 19.5\ \text{ksi}\\ F_{yw} = 36\ ksi\end{cases}$$

$$0.35 \le \left(k_c = \frac{4}{\sqrt{h/t_w}} = \frac{4}{\sqrt{128}} = 0.3536\right) \le 0.763$$

$$\lambda_r = \frac{162}{\sqrt{F_L / k_c}} = \frac{162}{\sqrt{19.5/0.3536}} = 21.8$$

$$(\lambda_p = 10.8) < (\lambda = 14.4) \le (\lambda_r = 21.8)$$

$$\phi M_{rx} = 0.9\, F_L\, S_x = 0.9\,(19.5)(854) = 14{,}988 \text{ in.-kips} = 1249 \text{ ft-kips}$$

$$\phi M_{nx} = \phi M_{px} - (\phi M_{px} - \phi M_{rx})\left(\frac{\lambda - \lambda_p}{\lambda_r - \lambda_p}\right)$$

$$\phi M_{nx} = 2646 - (2646 - 1249)\left(\frac{14.4 - 10.8}{21.8 - 10.8}\right) = 2189 \text{ ft-kips}$$

b. For WLB (See LRFD pp. 6-111 and 6-114):

$$\left(\lambda_p = \frac{640}{\sqrt{F_{yf}}} = \frac{640}{\sqrt{36}} = 107\right) < \left[\lambda = \frac{h}{t_w} = \frac{56}{(7/16)} = 128\right] \le \left(\lambda_r = \frac{970}{\sqrt{F_{yf}}} = \frac{970}{\sqrt{36}} = 162\right)$$

$$R_e = 1.0 \quad \text{(nonhybrid member)}$$

$$\phi M_{rx} = 0.9 R_e F_{yf} S_x = 0.9(1.0)(36)(854) = 27{,}670 \text{ in.-kips} = 2306 \text{ ft-kips}$$

$$\phi M_{nx} = \phi M_{px} - (\phi M_{px} - \phi M_{rx})\left(\frac{\lambda - \lambda_p}{\lambda_r - \lambda_p}\right)$$

$$\phi M_{nx} = 2646 - (2646 - 2306)\left(\frac{128 - 107}{162 - 107}\right) = 2516 \text{ ft - kips}$$

c. For LTB (See LRFD pp. 6-111 and 6-114): $C_b = 1.15$ and $L_b = 20$ ft, Also, we will need the following section properties:

$$I_y = 2\left[\frac{(5/8)(18)^3}{12}\right] + \frac{56(7/16)^3}{12} = 607.9 \text{ in.}^4$$

$$r_y = \sqrt{\frac{I_y}{A}} = \sqrt{\frac{607.9}{47.0}} = 3.60 \text{ in.}$$

$$J = \sum\left(\frac{bt^3}{3}\right) = 2\left[\frac{18(5/8)^3}{3}\right] + \frac{56(7/16)^3}{3} = 4.49 \text{ in.}^4$$

$$C_w = \frac{t_f b_f^3 (h + t_f)^2}{24} = \frac{(5/8)(18)^3 [56 + (7/16)]^2}{24} = 486{,}971 \text{ in.}^6$$

$$\lambda = \frac{L_b}{r_y} = \frac{20(12)}{3.60} = 66.7$$

$$\lambda_p = \frac{300}{\sqrt{F_{yf}}} = \frac{300}{\sqrt{36}} = 50.0$$

$$X_1 = \frac{\pi}{S_x}\sqrt{\frac{EAGJ}{2}} = \frac{\pi}{854}\sqrt{\frac{29,000(47.0)(11,200)(4.49)}{2}} = 681$$

$$X_2 = \frac{4C_w}{I_y}\left(\frac{S_x}{GJ}\right)^2 = \frac{4(486,971)}{607.9}\left[\frac{854}{11,200(4.49)}\right]^2 = 0.924$$

$$\lambda_r = \frac{X_1}{F_L}\sqrt{1+\sqrt{1+X_2F_L^2}} = \frac{681}{19.5}\sqrt{1+\sqrt{1+0.924(19.5)^2}} = 155$$

$$(\lambda_p = 50.0) < (\lambda = 66.7) \le (\lambda_r = 155)$$

$$\phi M_{rx} = 0.9\, F_L\, S_x = 0.9\,(19.5)(854) = 14,988 \text{ in.-kips} = 1249 \text{ ft-kips}$$

$$\phi M_{nx} = C_b\left[\phi M_{px} - (\phi M_{px} - \phi M_{rx})\left(\frac{\lambda - \lambda_p}{\lambda_r - \lambda_p}\right)\right]$$

$$0.1 < (I_{yc}/I_y = 0.5) < 0.9$$

$$C_b = 1.15 \quad \text{is applicable}$$

$$\phi M_{nx} = 1.15\left[2646 - (2646 - 1249)\left(\frac{66.7 - 50.0}{155 - 50.0}\right)\right] = 2787 \text{ ft-kips}$$

Summary for the determination of ϕM_{nx}:

a. For FLB:

$$\phi M_{nx} = 2189 \text{ ft-kips}$$

b. For WLB:

$$\phi M_{nx} = 2516 \text{ ft-kips}$$

c. For LTB:

$$\phi M_{nx} = 2787 \text{ ft-kips}$$

The governing value is $\phi M_{nx} = 2189$ ft-kips.

Example 9.2

The built-up W57 × 18 × 160 section on LRFD p. 4-184 is chosen as a trial section and needs to be investigated for the following conditions:

$$L_b = 20 \text{ ft} \qquad C_b = 1.15 \qquad F_{yw} = 36 \text{ ksi} \qquad F_{yf} = 50 \text{ ksi}$$

1. Find ϕV_n when intermediate web stiffeners are not used.
2. Find ϕV_n in a panel for which a/h = 60/56 = 1.07 when intermediate web stiffeners are used.
3. Find ϕM_{nx}.

Solution 1 and 2

These solutions for ϕV_n are the same as in Example 9.1.

Solution 3

LRFD Appendix F (p. 6-111) is applicable. The three limit states that may need to be investigated to determine the governing ϕM_{nx} are:

a. FLB
b. WLB
c. LTB

ϕM_{nx} must be computed for each applicable limit state, and the least of these computed values is the governing value of ϕM_{nx}.
From LRFD p. 4-184:

$$b_f = 18 \text{ in.} \qquad t_f = 5/8 \text{ in.} \qquad h_w = 56 \text{ in.}$$
$$t_w = 7/16 \text{ in.} \qquad Z_x = 980 \text{ in.}^3 \qquad S_x = 854 \text{ in.}^3$$

From LRFD pp. 6-111 and 6-114, we find that for a hybrid member, M_{px} is to be computed from the fully plastic stress distribution and is applicable for all three limit states. For a doubly symmetric hybrid section,

$$M_{px} = \text{smaller of} \begin{cases} Z_{xf}F_{yf} + Z_{xw}F_{yw} \\ 1.5S_xF_{yf} \end{cases}$$

where

$$Z_{xf} = 2A_f\left(\frac{d-t_f}{2}\right) = 2(18)(0.625)\left(\frac{57.25-0.625}{2}\right) = 637 \text{ in.}^3$$

$$Z_{xw} = 2\left[\frac{A_w}{2}\left(\frac{h}{4}\right)\right] = \frac{A_wh}{4} = \frac{(56)(7/17)(56)}{4} = 343 \text{ in.}^3$$

$$M_{px} = \text{smaller of} \begin{cases} (637)(50)+(343)(36) = 44,198 \text{ in.-kips} \\ 1.5(854)(50) = 64,050 \text{ in.-kips} \end{cases}$$

$$\phi M_{px} = 0.9\ (44,198) = 39,778 \text{ in.-kips} = 3315 \text{ ft-kips}$$

a. For FLB:

$$\lambda = \frac{0.5b_f}{t_f} = \frac{0.5(18)}{(5/8)} = 14.4$$

$$\lambda_p = \frac{65}{\sqrt{F_{yf}}} = \frac{65}{\sqrt{50}} = 9.19$$

$$F_L = \text{smaller of} \begin{cases} (F_{yf} - F_r) = (50-16.5) = 33.5 \text{ ksi} \\ F_{yw} = 36 \text{ ksi} \end{cases}$$

$$0.35 \le \left(k_c = \frac{4}{\sqrt{h/t_w}} = \frac{4}{\sqrt{128}} = 0.3536 \right) \le 0.763$$

$$\lambda_r = \frac{162}{\sqrt{F_L / k_c}} = \frac{162}{\sqrt{33.5/0.3536}} = 16.6$$

$$(\lambda_p = 9.19) < (\lambda = 14.4) \le (\lambda_r = 16.6)$$

$$\phi M_{rx} = 0.9\, F_L\, S_x = 0.9\,(33.5)(854) = 25{,}748 \text{ in.-kips} = 2146 \text{ ft-kips}$$

$$\phi M_{nx} = \phi M_{px} - (\phi M_{px} - \phi M_{rx})\left(\frac{\lambda - \lambda_p}{\lambda_r - \lambda_p}\right)$$

$$\phi M_{nx} = 3315 - (3315 - 2146)\left(\frac{14.4 - 9.19}{16.6 - 9.19}\right) = 2493 \text{ ft - kips}$$

b. For WLB,

$$\left(\lambda_p = \frac{640}{\sqrt{F_{yf}}} = \frac{640}{\sqrt{50}} = 90.5 \right) < \left[\lambda = \frac{h}{t_w} = \frac{56}{(7/16)} = 128 \right] \le \left(\lambda_r = \frac{970}{\sqrt{F_{yf}}} = \frac{970}{\sqrt{50}} = 137 \right)$$

From Appendix G2 (p. 6-123) we find for a hybrid member that

$$a_r = \frac{A_w}{A_{cf}} = \frac{56(7/16)}{18(0.625)} = 2.18$$

$$m = \frac{F_{yw}}{F_{yf}} = \frac{36}{50} = 0.72$$

$$\left\{ R_e = \frac{12 + a_r(3m - m^3)}{12 + 2a_r} = \frac{12 + 2.18[3(0.72) - (0.72)^3]}{12 + 2(2.18)} = 0.9716 \right\} \le 1.0$$

$$\phi M_{rx} = 0.9 R_e F_{yf} S_x = 0.9\,(0.9716)(50)(854) = 37{,}339 \text{ in.-kips} = 3112 \text{ ft-kips}$$

$$\phi M_{nx} = \phi M_{px} - (\phi M_{px} - \phi M_{rx})\left(\frac{\lambda - \lambda_p}{\lambda_r - \lambda_p}\right)$$

$$\phi M_{nx} = 3315 - (3315 - 3112)\left(\frac{128 - 90.5}{137 - 90.5}\right) = 3151 \text{ ft - kips}$$

c. For LTB, $C_b = 1.15$ and $L_b = 20$ ft.

From Example 9.1:

$I_y = 607.9$ in.^{4} $\quad r_y = 3.60$ in. $\quad J = 4.49$ in.^{4} $\quad C_w = 486{,}971$ in.^{6}

$X_1 = 681 \quad X_2 = 0.924 \quad \lambda_r = 155$

$$\lambda = \frac{L_b}{r_y} = \frac{20(12)}{3.60} = 66.7$$

$$\lambda_p = \frac{300}{\sqrt{F_{yf}}} = \frac{300}{\sqrt{50}} = 42.4$$

$$X_1 = \frac{\pi}{S_x}\sqrt{\frac{EAGJ}{2}} = \frac{\pi}{854}\sqrt{\frac{29,000(47.0)(11,200)(4.49)}{2}} = 681$$

$$X_2 = \frac{4C_w}{I_y}\left(\frac{S_x}{GJ}\right)^2 = \frac{4(486,971)}{607.9}\left[\frac{854}{11,200(4.49)}\right]^2 = 0.924$$

$$\lambda_r = \frac{X_1}{F_L}\sqrt{1+\sqrt{1+X_2F_L^2}} = \frac{681}{33.5}\sqrt{1+\sqrt{1+0.924(33.5)^2}} = 117$$

$$(\lambda_p = 42.4) < (\lambda = 66.7) \leq (\lambda_r = 117)$$

$\phi M_{rx} = 0.9\, F_L\, S_x = 0.9\,(33.5)(854) = 25{,}748$ in.-kips = 2146 ft-kips

$$\phi M_{nx} = C_b\left[\phi M_{px} - \left(\phi M_{px} - \phi M_{rx}\right)\left(\frac{\lambda - \lambda_p}{\lambda_r - \lambda_p}\right)\right]$$

$$0.1 < \left(I_{yc}/I_y = 0.5\right) < 0.9$$

$C_b = 1.15$ is applicable

$$\phi M_{nx} = 1.15\left[3315 - (3315 - 2146)\left(\frac{66.7 - 42.4}{117 - 42.4}\right)\right] = 3374 \text{ ft - kips}$$

Summary for the determination of ϕM_{nx}:

a. For FLB:

$$\phi M_{nx} = 2493 \text{ ft-kips}$$

b. For WLB:

$$\phi M_{nx} = 3151 \text{ ft-kips}$$

c. For LTB:

$$\phi M_{nx} = 3374 \text{ ft-kips}$$

The governing value is $\phi M_{nx} = 2493$ ft-kips

Example 9.3

The built-up W57 × 18 × 160 section on LRFD p. 4-184 is chosen as a trial section and needs to be investigated for the following conditions: $L_b = 20$ ft, $C_b = 1.15$, $F_{yw} = F_{yf} =$

36 ksi, and $\phi V_n \geq (V_u = 360$ kips) in the end panel. Note that except for the $\phi V_n \geq (V_u = 360$ kips) requirement in the end panel, the other conditions are identical to those in Example 9.1. Find the intermediate web stiffener spacing required to satisfy $\phi V_n \geq (V_u = 360$ kips).

Solution

From Example 9.1 or from LRFD page 4-184, $\phi V_n = 178$ kips when intermediate web stiffeners are not provided. Since $\phi V_n \geq (V_u = 360$ kips) is required, we must provide intermediate web stiffeners.

Enter LRFD Table 9-36 (p. 6-155) with

$$\frac{h}{t_w} = \frac{56}{(7/16)} = 128$$

$$\frac{\phi V_n}{A_w} = \frac{V_u}{A_w} = \frac{360}{56(7/16)} = 14.7 \text{ ksi}$$

and find a trial value of $a/h = 0.98$ for which $a = 0.98(56) = 54.9$. Try $a = 55$ in. and $a/h = 55/56 = 0.982$ in LRFD F2.2 (p. 6-113):

$$k_v = 5 + \frac{5}{(a/h)^2} = 5 + \frac{5}{(0.982)^2} = 10.18$$

$$\left(\frac{h}{t_w} = 128\right) > \left(234\sqrt{\frac{k_v}{F_{yw}}} = 234\sqrt{\frac{10.18}{36}} = 124.5\right)$$

$$A_w = (56)(7/16) = 24.5 \text{ in.}^2$$

$$\phi V_n = 0.9\left[\frac{26{,}400 A_w k_v}{(h/t_w)^2}\right] = 0.9\left[\frac{26{,}400(24.5)(10.18)}{(128)^2}\right] = 362 \text{ kips}$$

($\phi V_n = 362$ kips) $\geq$ ($V_u = 360$ kips) as required for $a/h = 0.982$ and $a = 55$ in.

Use $a = 55$ in.

Example 9.4

The built-up W57 × 18 × 206 section on LRFD page 4-184 is chosen as a trial section and needs to be investigated for the following conditions:

$$L_b = 20 \text{ ft} \qquad C_b = 1.15 \qquad F_{yw} = F_{yf} = 36 \text{ ksi}$$

1. Find ϕV_n when intermediate web stiffeners are not used.
2. Find ϕM_{nx}.

Solution 1

For an unstiffened web in a plate girder, from LRFD F2.2 (p. 6-113) we find that

$$\left[\frac{h}{t_w} = \frac{56}{(7/16)} = 128\right] \le 260 \quad \text{as required}$$

From LRFD p. 4-184 and footnote c on page 4-185, $\phi V_n = 177.6$ kips when intermediate web stiffeners are not used.

Solution 2

From LRFD p. 4-184, for a W57 × 18 × 206:

$$b_f = 18 \text{ in.} \qquad t_f = 1 \text{ in.} \qquad h_w = 56 \text{ in.}$$

$$t_w = 7/16 \text{ in.} \qquad Z_x = 1370 \text{ in.}^3 \qquad S_x = 1230 \text{ in.}^3$$

For a nonhybrid member, since $(Z_x / S_x = 1370/1230 = 1.11) < 1.5$,

$$\phi M_{px} = 0.9F_yZ_x = 0.9\,(36)(1370) = 44{,}388 \text{ in.-kips} = 3699 \text{ ft-kips}$$

which is applicable for all three limit states.

a. For FLB,

$$\left[\lambda = \frac{0.5b_f}{t_f} = \frac{0.5(18)}{1} = 9.00\right] \le \left(\lambda_p = \frac{65}{\sqrt{F_{yf}}} = \frac{65}{\sqrt{36}} = 10.8\right)$$

$$\phi M_{nx} = (\phi M_{px} = 3699 \text{ ft-kips})$$

b. For WLB:

$$\left(\lambda_p = \frac{640}{\sqrt{F_{yf}}} = \frac{640}{\sqrt{36}} = 107\right) < \left[\lambda = \frac{h}{t_w} = \frac{56}{(7/16)} = 128\right] \le \left(\lambda_r = \frac{970}{\sqrt{F_{yf}}} = \frac{970}{\sqrt{36}} = 162\right)$$

From LRFD p. 6-123, for a nonhybrid member we find that $R_e = 1.0$:

$$\phi M_{rx} = 0.9R_eF_{yf}S_x = 0.9\,(1.0)(36)(1230) = 39{,}852 \text{ in.-kips} = 3321 \text{ ft-kips}$$

$$\phi M_{nx} = \phi M_{px} - \left(\phi M_{px} - \phi M_{rx}\right)\left(\frac{\lambda - \lambda_p}{\lambda_r - \lambda_p}\right)$$

$$\phi M_{nx} = 3699 - (3699 - 3321)\left(\frac{128 - 107}{162 - 107}\right) = 3555 \text{ ft} - \text{kips}$$

c. For LTB, $C_b = 1.15$ and $L_b = 20$ ft. Also, we will need the following section properties:

$$I_y = 2\left[\frac{1(18)^3}{12}\right] + \frac{56(7/16)^3}{12} = 972.4 \text{ in.}^4$$

$$r_y = \sqrt{\frac{I_y}{A}} = \sqrt{\frac{972.4}{60.5}} = 4.01 \text{ in.}$$

$$J = \sum\left(\frac{bt^3}{3}\right) = 2\left[\frac{18(1)^3}{3}\right] + \frac{56(7/16)^3}{3} = 13.56 \text{ in.}^4$$

$$C_w = \frac{t_f b_f^3 (h+t_f)^2}{24} = \frac{1(18)^3(56+1)^2}{24} = 789{,}507 \text{ in.}^6$$

$$\lambda = \frac{L_b}{r_y} = \frac{20(12)}{4.01} = 59.85$$

$$\lambda_p = \frac{300}{\sqrt{F_{yf}}} = \frac{300}{\sqrt{36}} = 50.0$$

$$X_1 = \frac{\pi}{S_x}\sqrt{\frac{EAGJ}{2}} = \frac{\pi}{1230}\sqrt{\frac{29{,}000(60.5)(11200)(13.56)}{2}} = 932.3$$

$$X_2 = \frac{4C_w}{I_y}\left(\frac{S_x}{GJ}\right)^2 = \frac{4(789{,}507)}{972.4}\left(\frac{1230}{11{,}200(13.56)}\right)^2 = 0.213$$

$$F_L = \text{smaller of } \begin{cases} (F_{yf} - F_r) = (36 - 16.5) = 19.5 \text{ ksi} \\ F_{yw} = 36 \text{ ksi} \end{cases}$$

$$\lambda_r = \frac{X_1}{F_L}\sqrt{1+\sqrt{1+X_2F_L^2}} = \frac{932.3}{19.5}\sqrt{1+\sqrt{1+0.213(19.5)^2}} = 151.6$$

$$(\lambda_p = 50.0) < (\lambda = 59.85) \le (\lambda_r = 151.6)$$

$$\phi M_{rx} = 0.9\, F_L\, S_x = 0.9\,(19.5)(1230) = 21{,}587 \text{ in-kips} = 1799 \text{ ft-kips}$$

$$\left\{\phi M_{nx} = C_b\left[\phi M_{px} - (\phi M_{px} - \phi M_{rx})\left(\frac{\lambda - \lambda_p}{\lambda_r - \lambda_p}\right)\right]\right\} \le \phi M_{px}$$

$$1.15\left[3699 - (3699 - 1799)\left(\frac{59.85 - 50.0}{151.6 - 50.0}\right)\right] = 4042 \text{ ft-kips} > \phi M_{px}$$

$$\phi M_{nx} = (\phi M_{px} = 3699 \text{ ft-kips})$$

Summary for the determination of ϕM_{nx}:

a. For FLB:

$$\phi M_{nx} = 3699 \text{ ft-kips}$$

b. For WLB:

$$\phi M_{nx} = 3555 \text{ ft-kips}$$

c. For LTB:

$$\phi M_{nx} = 3699 \text{ ft-kips}$$

The governing value is $\phi M_{nx} = 3555$ ft-kips.

Example 9.5

The built-up section used in Example 9.4 is chosen as a trial section and is investigated for the following conditions:

$$L_b = 60 \text{ ft} \qquad C_b = 1 \qquad F_{yw} = F_{yf} = 36 \text{ ksi}$$

Find ϕM_{nx}.

Solution

See Example 9.4 for the FLB and WLB solutions.
From Example 9.4,

$$r_y = 4.01 \text{ in.,} \qquad \lambda_p = 50.0, \qquad X_1 = 932.3, \qquad X_2 = 0.213,$$

$$\lambda_r = 151.6, \qquad S_x = 1230 \text{ in.}^3 \qquad \phi M_{rx} = 1799 \text{ ft-kips} \qquad \phi M_{px} = 3699 \text{ ft-kips}$$

c. For LTB,

$$\left[\lambda = \frac{L_b}{r_y} = \frac{60(12)}{4.01} = 179.6\right] > (\lambda_r = 151.6)$$

From LRFD pp. 6-111 and 6-114, we find that

$$\phi M_{nx} = (0.9 M_{cr} = 0.9 S_x F_{cr})$$

$$F_{cr} = \frac{C_b X_1 \sqrt{2}}{\lambda}\sqrt{1+\frac{X_1^2 X_2}{2\lambda^2}} = \frac{1.0(932.3)\sqrt{2}}{179.6}\sqrt{1+\frac{(932.3)^2(0.213)}{2(179.6)^2}} = 14.4 \text{ ksi}$$

$$\phi M_{nx} = [\phi M_{cr} = 0.9(1230)(14.4) = 15{,}941 \text{ in.-kips} = 1328 \text{ ft-kips}]$$

Summary for the determination of ϕM_{nx}:

a. For FLB:

$$\phi M_{nx} = 3699 \text{ ft-kips}$$

b. For WLB:

$$\phi M_{nx} = 3555 \text{ ft-kips}$$

c. For LTB:

$$\phi M_{nx} = 1328 \text{ ft-kips}$$

The governing value is $\phi M_{nx} = 1328$ ft-kips

9.3 TENSION FIELD DESIGN METHOD

LRFD Appendix G (p. 6-122) must be used for shear and the intermediate stiffeners. None of the design strength definitions in LRFD Appendix G have been previously discussed in this text, and they are more complex than those in LRFD Appendix F1. The design procedure for this method is also iterative as described in Section 9.2 for the Conventional Design Method.

For the Tension Field Method, a structurally efficient web chosen to barely satisfy the design requirement for shear is so slender that web buckling occurs in each panel between the intermediate web stiffeners. However, the post buckling strength of the web can be utilized if the intermediate web stiffeners are properly designed to keep the flanges in their original location as described in the next paragraph.

Consider Figure 9.4, which shows a simply supported plate girder subjected to a factored loading that is symmetric with respect to midspan. The maximum shear occurs at the supports and the maximum moment occurs at midspan. In the vicinity of the maximum shear, *shear web buckling* occurs in the diagonal compression direction in each panel between the intermediate web stiffeners. The length direction of the ripples due to shear web buckling is parallel to the long direction of the shaded areas in Figure 9.4. After shear web buckling occurs, the following analogy is valid when the intermediate web stiffeners are properly designed. The intermediate web stiffeners are analogous to the vertical compression members in a truss. The diagonal tension region (shaded in Figure 9.4) of each panel between the intermediate web stiffeners is analogous to a diagonal tension member in a truss. In the vicinity of maximum moment, *flexural web buckling* occurs in the top region of each panel between the intermediate web stiffeners. The design bending strength of the plate girder cross section is affected by flexural web buckling. As shown in Figure 9.4, the compression flange line is curved due to the factored loading. Curvature of the compression flange causes compression in the web, and the compression is perpendicular to the flange. Therefore, in the region of maximum moment, the post buckling strength of the web must be sufficient when assisted by the intermediate web stiffeners to prevent local buckling of the compression flange in the plane of the web direction. In the region of maximum moment, the intermediate web stiffeners are analogous to the vertical compression members in a truss.

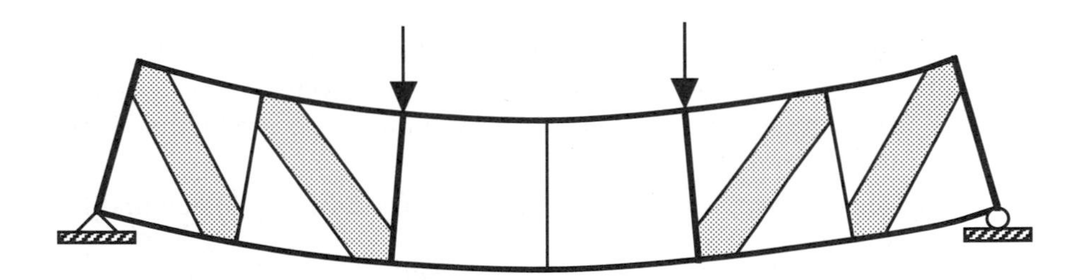

Notes:
Bearing stiffeners are shown at the reactions and the concentrated loads.
Intermediate stiffeners are shown between the bearing stiffeners.
The shaded areas of the web are "diagonal tension members."

FIGURE 9.4 Tension-field action in a plate girder.

9.3.1 Design Strength Definitions

According to LRFD Appendix G3 (p. 6-103), tension field action is not permitted for:

1. Hybrid and web-tapered girders.
2. The end panels in a nonhybrid girder.
3. The panels of a non-hybrid girder for which $a/h > 3$ or $a/h > (260 / h/t_w)^2$. This restriction is to prevent a girder from being too flexible perpendicular to the web for handling purposes during fabrication, shipping, and field erection.

For these cases, the Conventional Design Method must be used.

For a plate girder subjected to bending about the strong axis only, the design requirements are:

1. $\phi_b M_{nx} \geq M_{ux}$
2. $\phi_v V_n \geq V_u$
3. When intermediate stiffeners are required and when

$$0.60 \leq \frac{0.9 V_n}{V_u} \leq 1.00 \text{ and } 0.75 \leq \frac{0.9 M_{nx}}{M_u} \leq 1.00$$

the following flexure-shear interaction equation must be satisfied at each stiffener location:

$$\frac{M_{ux}}{\phi M_{nx}} + 0.625 \frac{V_u}{\phi V_n} \leq 1.375$$

where $\phi_b M_{nx}$ and $\phi_v V_n$, respectively, are the design bending and shear strengths, and

$$\phi = 0.9$$

$$M_{ux} = \text{required bending strength}$$

$$V_u = \text{required shear strength}$$

For the Tension Field Design Method, the design strength definitions are:

1. Design bending strength $\phi_b M_{nx}$ The design bending strength definitions are the same as those given for the Conventional Design Method in Section 9.2.1, and they are not repeated here.
2. Design shear strength $\phi_v V_n$ From LRFD G3 (p. 6-103), we find that:
   (a) When

$$\frac{h}{t_w} \leq 187 \sqrt{\frac{k_v}{F_{yw}}}$$

web shear yielding governs the design shear strength:

$$\phi V_n = 0.90\left(0.6 F_{yw} A_w\right)$$

(b) When

$$\frac{h}{t_w} > 187\sqrt{\frac{k_v}{F_{yw}}}$$

we have

$$\phi V_n = 0.90\left[0.6F_{yw}A_w\left(C_v + \frac{1-C_v}{1.15\sqrt{1+(a/h)^2}}\right)\right]$$

$$k_v = 5 + \frac{5}{(a/h)^2}$$

C_v = ratio of "critical" web stress, according to linear buckling theory, to the shear yield stress of the web. C_v is determined as follows. When

$$187\sqrt{\frac{k_v}{F_{yw}}} < \frac{h}{t_w} \le 234\sqrt{\frac{k_v}{F_{yw}}}$$

inelastic web buckling governs the design shear strength for which:

$$C_v = \frac{187\sqrt{k_v/F_{yw}}}{h/t_w}$$

When

$$\frac{h}{t_w} > 234\sqrt{\frac{k_v}{F_{yw}}}$$

elastic web buckling governs the design shear strength for which:

$$C_v = \frac{44{,}000\,k_v}{(h/t_w)^2 F_{yw}}$$

Note: In the member-end panels, the Conventional Design Method must be used to determine ϕV_n.

9.3.2 Intermediate Stiffener Requirements

Intermediate stiffeners are not required when

$$\frac{h}{t_w} \le \frac{418}{\sqrt{F_{yw}}} \quad \text{nor when} \quad \left[\phi V_n = 0.90\left(0.6F_{yw}A_w\right)C_v\right] \ge V_u$$

where

$$C_v = \frac{187\sqrt{5/F_{yw}}}{h/t_w} \quad \text{when } 187\sqrt{\frac{5}{F_{yw}}} < \frac{h}{t_w} \le 234\sqrt{\frac{5}{F_{yw}}}$$

or

$$C_v = \frac{220,000}{(h/t_w)^2 F_{yw}} \quad \text{when} \quad \frac{h}{t_w} > 234\sqrt{\frac{5}{F_{yw}}}$$

In order to satisfy LRFD G1 (p. 6-122), stiffeners may be required regardless of any other LRFD Specification. For such cases, Appendix F2.3 is applicable for the design of the stiffeners.

When tension field action is utilized, the design requirements for intermediate stiffeners are

$$I_{st} \geq a t_w^3 j$$

$$\left\{ A_{st} \geq \frac{F_{yw}}{F_{yst}} \left[0.15 D h t_w (1 - C_v) \frac{V_u}{\phi V_n} - 18 t_w^2 \right] \right\} \geq 0$$

$$j = 0.5 \quad \text{when } a/h \geq 1$$

$$\text{otherwise } j = \frac{2.5}{(a/h)^2} - 2$$

where

a = clear distance between intermediate web stiffeners (in.)

h and t_w are defined in Figure 9.2(b)

I_{st} = moment of inertia about an axis in the web center for a pair of stiffeners or about the face in contact with the web plate for a single stiffener (in.^{4})

A_{st} = stiffener area (in.^{2})

F_{yst} = yield stress of the stiffeners (ksi)

D = 1 when a pair of stiffeners is used

D = 1.8 when a single angle stiffener is used

D = 2.4 when a single plate stiffener is used

C_v and ϕV_n are as defined in item 2b of Section 9.3.1

V_u = required shear at the stiffener location

The intermediate stiffeners are load-bearing and in compression when tension field action is utilized. These stiffeners must be designed such that the following requirements from LRFD Table B5.1 (p. 6-38) are satisfied:

$$\frac{b}{t} \leq \left(\lambda_r = \frac{95}{\sqrt{F_{yst}}} \right) \quad \text{for plate web stiffeners}$$

$$\frac{b}{t} \leq \left(\lambda_r = \frac{76}{\sqrt{F_{yst}}} \right) \quad \text{for angle web stiffeners}$$

Note: When the Conventional Design Method must be used to determine fV_n, LRFD F2.3 (p. 6-113) is applicable for the design of the intermediate stiffeners.

9.3.3 Design Examples

LRFD Example 4-10 (p. 4-168) is an example of the Tension Field Design Method and illustrates:

1. Design of a nonhybrid girder: $F_y = 50$ ksi.
2. $\left(\dfrac{h}{t_w} = \dfrac{70}{(5/16)} = 224\right) > \left(\dfrac{970}{\sqrt{F_{yf}}} = \dfrac{970}{\sqrt{50}} = 137\right)$
3. Determination of the design bending strength when lateral braces for the compression flange are provided only at the supports and at the two concentrated load points:

   (a) For the two end regions: $L_b = 17$ ft and $C_b = 1.67$.
   (b) For the center region: $L_b = 14$ ft and $C_b = 1$.
4. Design of the intermediate stiffeners.
5. In the end panels, the Conventional Design Method is required and is used, but for the other panels the Tension Field Design Method is applicable and is used.

The general design procedure for the Tension Field Method when $L_b \leq L_p$ is illustrated in LRFD Example 4-10 (p. 4-168). Therefore, in the following examples we choose to illustrate some strength calculations for a trial section.

Example 9.6

Change the web of the section investigated in Example 9.4 to a 56 x 1/4 plate. The new section properties are: $A_w = 56(1/4) = 14.0$ in.^{2}, $S_x = 1135$ in.^{3}, and $Z_x = 1222$ in.^{3} Using the Tension Field Design Method wherever applicable, perform the following design tasks:

1. In the end panels (at the supports of a simply supported plate girder), the Tension Field Design Method is not permitted. Using the Conventional Design Method, find the required *a/h* such that $\phi V_n = 177.6$ kips, which is the same design shear strength in Example 9.4 when intermediate web stiffeners were not used.
2. In all other panels the Tension Field Design Method is permitted since

$$\left[\frac{h}{t_w} = \frac{56}{(1/4)} = 224\right] \geq \left(\frac{970}{\sqrt{F_{yf}}} = \frac{970}{\sqrt{36}} = 161.7\right)$$

Find the required *a/h* for an interior panel such that $\phi V_n \geq (V_u = 160$ kips) and design a single-plate intermediate stiffener.

3. Find ϕM_{nx}.

Solution 1

To find *a/h* of the end panels, enter LRFD page 6-155 with:

$$\frac{h}{t_w} = 224$$

$$\frac{\phi V_n}{A_w} = \left(\frac{V_u}{A_w} = \frac{177.6}{14.0} = 12.7 \text{ ksi} \right)$$

which indicates that we should try $a/h \approx 0.48$; $a \approx 0.48(56) = 26.88$ in. Try $a = 27$ in. and $a/h = 27/56 = 0.482$ in. for the determination of ϕV_n in LRFD F2.3 (p. 6-113).

$$k_v = 5 + \frac{5}{(a/h)^2} = 5 + \frac{5}{(0.482)^2} = 26.52$$

$$\left(\frac{h}{t_w} = 224 \right) > \left(234\sqrt{\frac{k_v}{F_{yw}}} = 234\sqrt{\frac{26.52}{36}} = 201 \right)$$

$$\phi V_n = 0.9\left[\frac{26400\, A_w k_v}{(h/t_w)^2} \right] = 0.9\left[\frac{26,400(14.0)(26.52)}{(224)^2} \right] = 175.8 \text{ kips}$$

$V_u = 177.6$ kips exceeds $\phi V_n = 175.8$ kips by 1% for $a = 27$ in. and $a/h = 27/56 = 0.482$ in. Use $a = 26$ in. and $a/h = 26/56 = 0.464$ in. for which ($\phi V_n = 187$ kips) $\geq$ ($V_u = 177.6$ kips) as required.

Solution 2

To find *a/h* for an interior panel in which $\phi V_n \geq (V_u = 160$ kips) is required for *tension field action*, enter LRFD p. 6-157 with

$$\frac{h}{t_w} = 224$$

$$\frac{\phi V_n}{A_w} = \left(\frac{V_u}{A_w} = \frac{160}{14.0} = 11.4 \text{ ksi} \right)$$

which indicates that we should try $a/h \approx 1.3$; $a \approx 1.3(56) = 72.8$ in. Try $a = 73$ in. for the determination of ϕV_n in LRFD Appendix G3 (p. 6-124), which is applicable since

$$\left(\frac{a}{h} = \frac{73}{56} = 1.3036 \right) \leq \left[\left(\frac{260}{h/t_w} \right)^2 = \left(\frac{260}{224} \right)^2 = 1.347 \right] \leq 3.0$$

$$k_v = 5 + \frac{5}{(a/h)^2} = 5 + \frac{5}{(1.3036)^2} = 7.94$$

$$\left(\frac{h}{t_w} = 224\right) > \left(234\sqrt{\frac{k_v}{F_{yw}}} = 234\sqrt{\frac{7.94}{36}} = 110\right)$$

$$C_v = \frac{44{,}000k_v}{F_{yw}(h/t_w)^2} = \frac{44{,}000(7.94)}{36(224)^2} = 0.1934$$

$$\phi V_n = 0.9\left(0.6A_w F_{yw}\right)\left[C_v + \frac{1-C_v}{1.15\sqrt{1+(a/h)^2}}\right]$$

$$\phi V_n = 0.9(0.6)(14.0)(36)\left(0.1934 + \frac{1-0.1934}{1.15\sqrt{1+(1.3036)^2}}\right) = 170 \text{ kips}$$

Use $a = 73$ in. and $a/h = 73/56 = 1.3036$ in. for which $(\phi V_n = 170 \text{ kips}) \geq (V_u = 160 \text{ kips})$ as required.

Design a single-plate intermediate stiffener. When tension field action is utilized, the design requirements for intermediate stiffeners are

$$\left\{A_{st} \geq \frac{F_{yw}}{F_{yst}}\left[0.15Dht_w(1-C_v)\frac{V_u}{\phi V_n} - 18t_w^2\right]\right\} \geq 0$$

$$\left\{A_{st} \geq \frac{36}{36}\left[0.15(2.4)(56)(0.25)(1-0.1934)\frac{160}{170} - 18(0.25)^2\right] = 2.70 \ in.^2\right\} \geq 0$$

$$\frac{b}{t} \leq \left(\lambda_r = \frac{95}{\sqrt{F_{yst}}} = \frac{95}{\sqrt{36}} = 15.8\right)$$

Try $6 \times 1/2$ plate. $(A_{st} = 3.00) \geq 2.70$ in.^{2} and $(b/t = 12.0) \leq 15.8$ as required. Check required I_{st}. Since $(a/h = 1.3036) \geq 1$, $j = 0.5$.

$$I_{st} \geq \left[at_w^3 j = 73(0.25)^3(0.5) = 0.570 \text{ in.}^4\right] \text{ is required}$$

$$\left[I_{st} = \frac{0.5(6)^3}{3} = 36.0\right] \geq 0.570 \text{ in.}^4 \quad \text{as required}$$

Use a single 6 x 1/2 plate as the intermediate web stiffener.

Solution 3

a. For tension flange yield,

$$\phi M_{nx} = 0.9\left(S_{xt}R_e F_{yt}\right) = 0.9(1135)(1.0)(36) = 36{,}774 \text{ in - kips} = 3064 \text{ ft - kips}$$

b. F_{cr} for FLB,

$$\left(\lambda = \frac{0.5b_f}{t_f} = \frac{0.5(18)}{1} = 9.00\right) \leq \left(\lambda_p = \frac{65}{\sqrt{F_{yf}}} = \frac{65}{\sqrt{36}} = 10.8\right)$$

$$F_{cr} = F_{yf}$$

c. F_{cr} for LTB,

$$C_b = 1.15 \qquad \text{and} \qquad L_b = 20 \text{ ft}$$

Also, we will need the following section properties of a T section composed of the compression flange and $A_{wc}/3$, where $A_{wc} = t_w(h_c/2) = 0.25(56/2) = 7.00 \text{ in.}^2$:

$$A_T = A_{cf} + \frac{A_{wc}}{3} = 18.0 + \frac{7.00}{3} = 20.3$$

$$I_T = I_y = \frac{1(18)^3}{12} + \frac{(56.0/6)(0.25)^3}{12} = 486 \text{ in.}^4$$

$$r_T = r_y = \sqrt{\frac{I_y}{A_T}} = \sqrt{\frac{486}{20.3}} = 4.89 \text{ in.}$$

$$\left(\lambda = \frac{L_b}{r_T} = \frac{20(12)}{4.89} = 49.08\right) \le \left(\lambda_p = \frac{300}{\sqrt{F_{yf}}} = \frac{300}{\sqrt{36}} = 50.0\right)$$

$$F_{cr} = F_{yf}$$

d. For compression flange buckling,

$$a_r = \frac{A_w}{A_{cf}} = \frac{14.0}{18.0} = 0.778$$

$$F_{cr} = F_{yf}$$

$$\left[R_{PG} = 1 - \frac{a_r}{1200 + 300a_r}\left(\frac{h_c}{t_w} - \frac{970}{\sqrt{F_{cr}}}\right)\right] \le 1.0$$

$$\left[R_{PG} = 1 - \frac{0.778}{1200 + 300(0.778)}\left(224 - \frac{970}{\sqrt{36}}\right) = 0.966\right] \le 1.0$$

$$\phi M_{nx} = 0.9(S_{xc} R_{PG} R_e F_{cr}) = 0.9(1135)(0.966)(1.0)(36) = 35{,}518 \text{ in-kips} = 2960 \text{ ft-kips}$$

Summary for the determination of ϕM_{nx} by the Tension Field Design Method:

a. For tension-flange yield:

$$\phi M_{nx} = 3064 \text{ ft-kips}$$

d. For compression-flange buckling:

$$\phi M_{nx} = 2960 \text{ ft-kips}$$

The governing value is $\phi M_{nx} = 2960$ ft-kips.
Since M_{ux} was not given in the problem statement, we are unable to perform the *flexure-shear interaction* check required by LRFD G5 (p. 6-125).

The solution for this example was obtained for the purpose of comparison to the solution in Example 9.4 by the Conventional Design Method for the built-up W57 ×

18 × 206 section on LRFD p. 4-184 chosen as a trial section and investigated for the following conditions: L_b = 20 ft, C_b = 1.15, and $F_{yw} = F_{yf}$ = 36 ksi.

The solutions obtained in Example 9.4 were:

1. ϕV_n = 177.6 kips when intermediate web stiffeners were not used.

2. Summary for the determination of ϕM_{nx}:

a. For FLB: ϕM_{nx} = 3699 ft-kips

b. For WLB: ϕM_{nx} = 3555 ft-kips

c. For LTB: ϕM_{nx} = 3699 ft-kips

The governing value is ϕM_{nx} = 3555 ft-kips.

A comparison of the results for Examples 9.4 and 9.6 is as follows:

1. In Example 9.4, t_w = 7/16 in. and the Conventional Design Method was applicable. In Example 9.6, t_w = 1/4 in. and the Tension Field Design Method was applicable except for the web in the end panels.
2. In Example 9.4, the web plate weight = (7/16)(56)(490/144) = 83.4 lb/ft. The bearing stiffeners are assumed to be identical in both examples. In Example 9.6, the web plate weight = (1/4)(56)(490/144) = 47.6 lb/ft. The weight of one intermediate stiffener (56 x 6 x 0.5 plate) is 48 lb. For a girder span of 80 ft, 12 intermediate stiffeners are required, which is 12(48/80) = 7.2 lb/ft due to these stiffeners in Example 9.6. According to a queried fabricator, due to fabrication labor this 7.2 lb/ft costs about 4.8 times as much as an equivalent weight in the thicker web of the member in Example 9.4. Thus, for the member in Example 9.6, the total effective web weight accounting for the difference in fabrication costs is about 47.6 + 4.8(7.2) = 82.2 lb/ft, which is 1.2 lb/ft lighter (1.4% lighter) than the web in Example 9.4.
3. The design bending strength for the member in Example 9.6 divided by the design bending strength for the member in Example 9.4 is (2960 ft-kips)/(3555 ft-kips) = 0.833. Therefore, the member in Example 9.6 is 17.7% weaker in bending than the member in Example 9.4. The member designed by the Tension Field Method is not the preferred choice for the comparison design examples.

Example 9.7

The built-up section used in Example 9.6 is chosen as a trial section and is investigated for the following conditions: L_b = 60 ft, C_b = 1, and $F_{yw} = F_{yf}$ = 36 ksi. Find ϕM_{nx} for compression flange buckling due to LTB.

Solution

$$\lambda_p = \frac{300}{\sqrt{F_{yf}}} = \frac{300}{\sqrt{36}} = 50.0$$

$$\left[\lambda = \frac{L_b}{r_T} = \frac{60(12)}{4.89} = 147\right] > \left(\lambda_r = \frac{756}{\sqrt{F_{yf}}} = \frac{756}{\sqrt{36}} = 126\right)$$

$$\left\{F_{cr} = C_b F_{yf}\left[1 - \frac{1}{2}\left(\frac{\lambda - \lambda_p}{\lambda_r - \lambda_p}\right)\right]\right\} \le F_{yf}$$

$$\left\{F_{cr} = 1.0(36)\left[1 - \frac{1}{2}\left(\frac{147.2 - 50.0}{126 - 50.0}\right)\right] = 13.0 \text{ ksi}\right\} \le \left(F_{yf} = 36 \text{ ksi}\right)$$

$$\left[R_{PG} = 1 - \frac{a_r}{1200 + 300a_r}\left(\frac{h_c}{t_w} - \frac{970}{\sqrt{F_{cr}}}\right)\right] \le 1.0$$

$$R_{PG} = \text{ smaller of } \begin{cases} 1 - \dfrac{0.778}{1200 + 300(0.778)}\left(224 - \dfrac{970}{\sqrt{13.0}}\right) = 1.024 \\ 1.0 \end{cases}$$

$$\phi M_{nx} = 0.9\left(S_{xc} R_{PG} R_e F_{cr}\right) = 0.9(1135)(1.0)(1.0)(13.0) = 13{,}280 \text{ in.-kips} = 1107 \text{ ft-kips}$$

Summary for the determination of ϕM_{nx} by the Tension Field Design Method

a. For tension flange yield:

$$\phi M_{nx} = 3064 \text{ ft-kips}$$

d. For compression-flange buckling:

$$\phi M_{nx} = 1107 \text{ ft-kips}$$

The governing value is $\phi M_{nx} = 1107$ ft-kips.

For comparison purposes, the summary from the solution in Example 9.5 for the determination of ϕM_{nx} by the Conventional Design Method is given here:

a. For FLB: $\phi M_{nx} = 3699$ ft-kips

b. For WLB: $\phi M_{nx} = 3555$ ft-kips

c. For LTB: $\phi M_{nx} = 1328$ ft-kips

The governing value is $\phi M_{nx} = 1328$ ft-kips. The member designed by the Tension Field Method is 16.6% weaker, more expensive to fabricate, and is not the preferred choice for the comparison design examples.

PROBLEMS

9.1 Is the section investigated in Example 9.1 satisfactory to use for the factored loading shown in Figure P9.1? If the section is satisfactory for shear and moment, design full-depth web stiffeners at the unframed member ends and at the concentrated load points. See Section 5.10 and Example 5.8 for the design of these stiffeners.

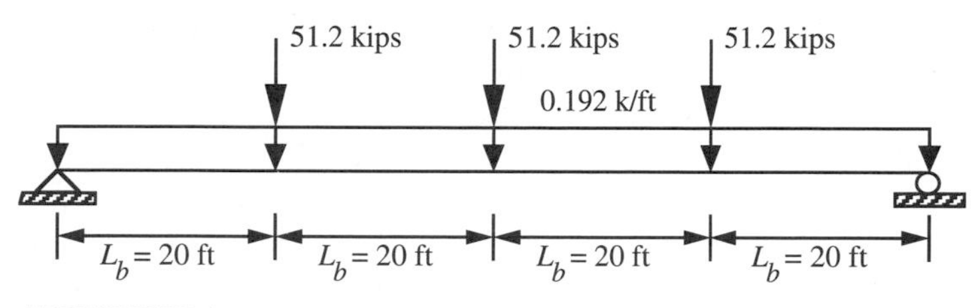

FIGURE P9.1

9.2 Is the section investigated in Example 9.2 satisfactory to use for the factored loading shown in Figure P9.2? If the section is satisfactory for shear and moment, design full-depth web stiffeners at the unframed member ends and at the concentrated load points. See Section 5.10 and Example 5.8 for the design of these stiffeners.

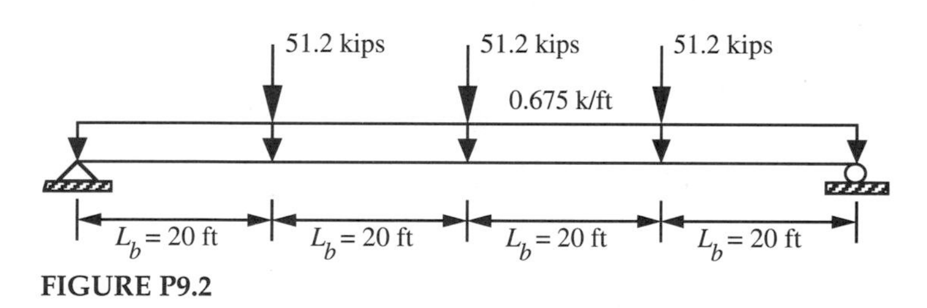

FIGURE P9.2

9.3 The section investigated in Example 9.4 is to be used for the factored loading shown in Figure P9.3. Find the maximum permissible value of the uniformly distributed load q_u.

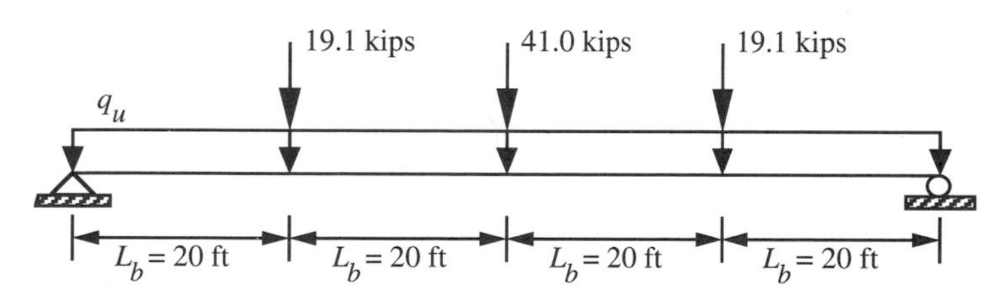

FIGURE P9.3

9.4 For LRFD Example 4-10 (p. 4-168), find the minimum web thickness that can be used in the Conventional Design Method without any intermediate web stiffeners. Make a comparison of these two solutions as we did at the end of Example 9.6 (see items 2 and 3).

9.5 For LRFD Example 4-13 (p. 4-180), try a girder section with 1.25×18 flange plates and a 0.25×50 web plate. Use the Tension Field Design Method wherever applicable. Design the intermediate stiffeners. Make a comparison of these two solutions as we did at the end of Example 9.6 (see items 2 and 3).

9.6 A plate girder that spans 60 ft between two columns in a multistory building is to be designed. This girder is one of a series over an assembly room where the columns were deleted within the assembly room. The overall girder depth cannot exceed 73.5 in. Service loads on the girder are as follows: live = 0.84 kips/ft; dead = 2.35 kips/ft (includes an estimate for the girder weight). Concentrated loads (from the columns above the girder) at 20 ft from each end of the girder are as follows: live = 110 kips; dead = 31 kips. Use the Conventional Design Method without any intermediate web stiffeners and $F_y = 36$ ksi for all steel. Assume that the governing LRFD loading combination is $1.2D + 1.6L$.

9.7 Solve Problem 9.6 using $F_y = 50$ ksi for the flanges and $F_y = 36$ ksi for the web.

9.8 Solve LRFD Example 4-12 (p. 4-176) by the Tension Field Design Method using $F_{yw} = F_{yf} = 50$ ksi and a web thickness of 7/16 in. Try a 28 x 2.5 in. flange plate. Try a = (30 in., 5 @ 42 in., 2 @ 120 in.). Design the intermediate web stiffeners using a single plate for each stiffener. Make a comparison of these two solutions as we did at the end of Example 9.6 (see items 2 and 3).

9.9 Solve Problem 9.6 by the Tension Field Design Method using a web plate thickness of 1/4 in.

9.10 Solve Problem 9.7 by the Tension Field Design Method using a web plate thickness of 1/4 in.

Composite Members

10.1 INTRODUCTION

In this chapter, a *composite member* is defined as consisting of a rolled or a built-up structural steel shape that is either filled with concrete, encased by reinforced concrete, or structurally connected to a reinforced concrete slab (see Figures 10.1 to 10.3). Composite members are constructed such that the structural steel shape and the concrete act together to resist axial compression and/or bending.

Most likely the reader has seen a highway bridge of composite steel–concrete construction at the various stages of construction. The girders are structural steel members with shear studs (steel connectors) welded to the top flange of each girder at discrete intervals along the length direction of the bridge, as shown in Figures 10.2(a) and 10.3. A shear stud is shaped like a bolt with a cylindrical head, but the shear stud does not have threads. After the steel girders have been erected, a concrete slab is poured on the top surface of the girders [see Figure 10.2(a)] or on a cold-formed steel deck supported by the girders (see Figure 10.3), and the concrete encases the shear studs. When the concrete has cured, the steel girders and the concrete slab are

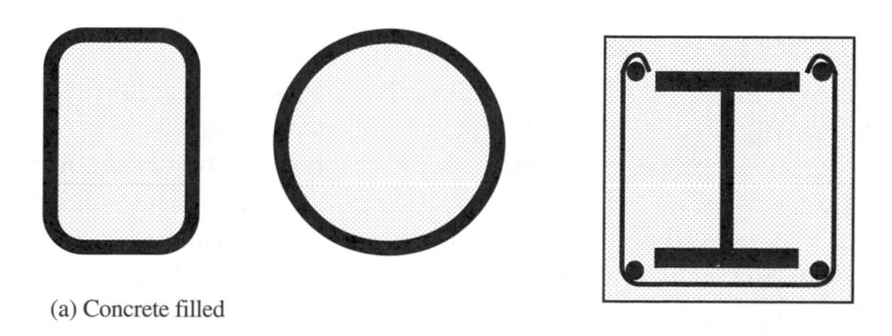

FIGURE 10.1 Composite column sections.

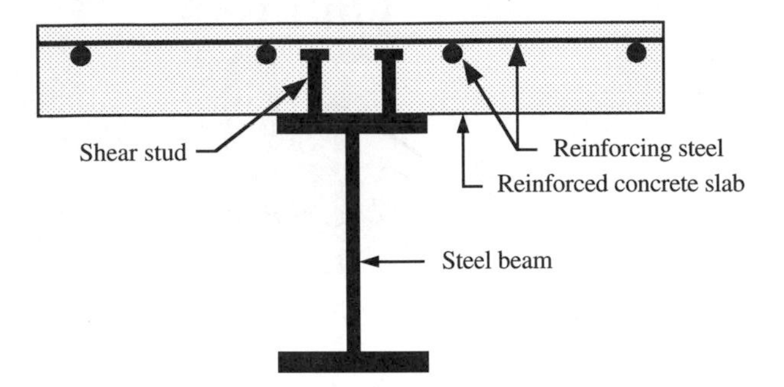

(a) Steel beam interactive with and supporting a concrete slab

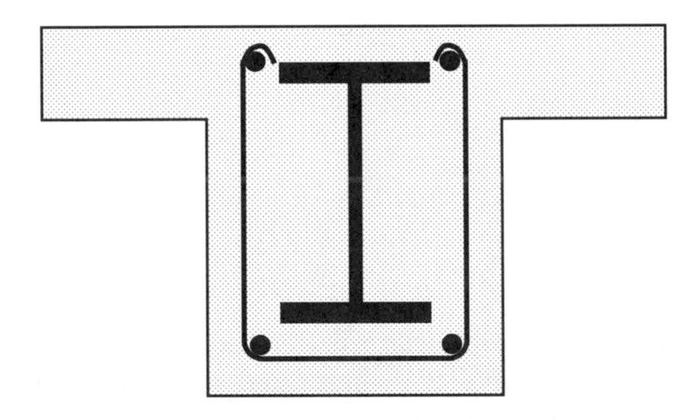

(b) Concrete-encased steel beam

FIGURE 10.2 Composite beam sections.

interconnected at discrete intervals by the shear studs. At the interconnected points, the concrete slab is prevented from slipping relative to the top surface of the steel girders. The steel connectors are subjected to shear at the concrete–steel construction interface, which explains why the steel connectors are called shear studs. Loads that occur on the bridge surface after the concrete slab has cured are resisted by flexural action of the composite steel–concrete construction.

There are more LRFD specifications for composite flexural members than composite compression members. Consequently, we chose the following discussion sequence: composite columns, composite beams, and composite beam columns.

10.2 COMPOSITE COLUMNS

As shown in Figure 10.1, composite columns consist of rolled or built-up steel shapes that are either filled with concrete or encased by reinforced concrete.

An axial compression load is applied at the top end of the column by whatever the column was designed to support. The bottom end of the column sits on a surface capable of providing the bearing forces from the steel and concrete in the composite column. Since the bottom end of the column is supported, the steel and concrete composing a composite column act together to resist the axial compressive load. Note that in most cases there is not any need for steel connectors in a composite column in order for the steel and concrete to act together in resisting an axial compressive load.

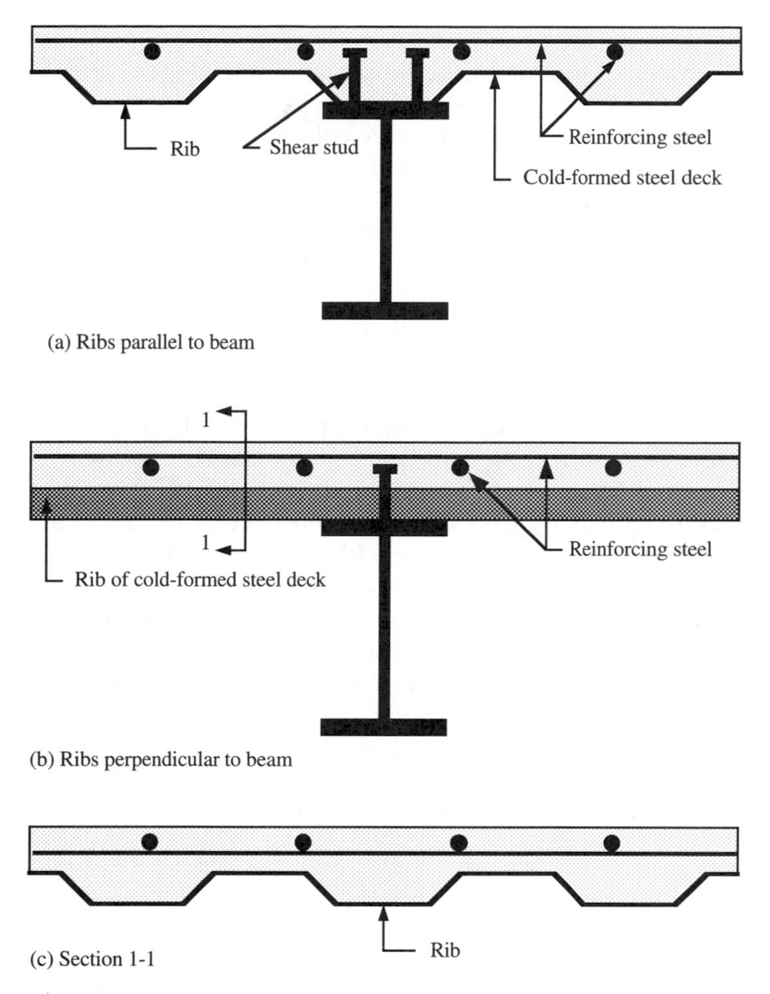

FIGURE 10.3 Cold-formed steel deck and composite beam sections.

A steel pipe or tube filled with concrete is the most efficient type of composite column. The perimeter steel provides stiffness and confinement of the concrete core, which resists compression and prevents local inward buckling of the steel encasement. The triaxally confined concrete core in some cases is capable of resisting a compressive strength in excess of f_c' which is the 28-day compressive strength of a concrete cylinder. This type of composite column has the toughness and ductility needed for earthquake-resistant structures.

When a structural steel shape is encased by concrete [see Figure 10.1(b)], a longitudinal steel reinforcing bar is located in each corner of the encasing concrete. Lateral ties are wrapped around the longitudinal bars at sufficiently close intervals along the member length. These U-shaped ties stabilize the longitudinal bars during construction and prevent outward local buckling of the longitudinal bars when an axial compression load is applied after the concrete has cured. Prior to failure of the composite column, the reinforced concrete encasing the steel shape prevents local buckling of the compression elements in the steel shape. The minimum acceptable amount of longitudinal and transverse reinforcement in the concrete encasement is approximately the same as the minimum amount specified for a tied reinforced concrete column. The behavior of this type of composite column is similar to the behavior of a tied reinforced conrete column. At a uniform compressive strain of 0.002, the encasing concrete starts to become unsound, and spalling is likely to occur. A section analysis for an eccentrically loaded composite column may be performed with the assumption that strains vary linearly across the section with a maximum acceptable compressive strain of 0.003 in the concrete.

Perhaps the most popular application of concrete-encased steel shapes is the perimeter columns in tube-type high-rise buildings. In high-rise construction, the steel shape is placed many stories prior to the placement of the encasing concrete. Therefore, the steel shape alone resists the axial compressive load until the encasing concrete has been placed and has cured sufficiently to resist some of the load. Temporary lateral bracing of the steel shapes may be required during construction until the encasing concrete has cured sufficiently [$0.75 f_c'$ according to LRFD Commentary I1 (p. 6-203)]. Concrete subjected to compression creeps (the shortening deformation continues to increase with respect to time without any increase in the compressive stress), but steel does not creep. When composite columns are used, a carefully controlled construction process must be maintained to ensure that the floors remain level (within the allowable tolerance limits). This is particularly true when some of the columns within a story are not composite columns since the rate of shortening is different for composite and noncomposite columns.

Chapter 10 in the *Guide to Stability Design Criteria for Metal Structures* [25] provides the reader with a discussion on the behavior of composite columns, laboratory test results, and analytical studies.

10.2.1 Limitations

LRFD I2 (p. 6-61) gives the following restrictions on composite columns:

1. $A_s \geq 0.04 A_g$ is required, where A_s is the cross-sectional area of structural steel and A_g is the gross cross-sectional area. Otherwise, the member is to be designed as a reinforced concrete column in accordance with ACI 318-92R [10].
2. Concrete encasing a steel core must be reinforced with longitudinal bars and lateral ties. Spacing of the lateral ties must not exceed two-thirds of the least dimension of the composite cross section. A minimum clear cover of 1.5 in. outside all reinforcement is required for fire and corrosion protection. The minimum cross-sectional area of the lateral ties and longitudinal bars is 0.007 in.^{2}/in. of bar spacing. Longitudinal bars provided only for the attachment of the lateral ties do not have to be continuous at the floor levels. However, any load-carrying longitudinal bars in the member must be continuous at the floor levels.
3. For normal weight concrete, 8 ksi $\geq f_c' \geq$ 3 ksi is required, where f_c' is the specified compressive strength. For lightweight concrete, $f_c' \geq$ 4 ksi is required. The reasons for these restrictions are explained in LRFD Commentary I2 (p. 6-203).
4. For the structural and reinforcing steel, $F_y \leq 55$ ksi is required in the strength calculations. This stress level corresponds to a limit compressive strain of

0.0018. The encasing concrete starts to become unsound at a uniform compressive strain of about 0.002 and subsequently spalling is likely to occur. Since $F_y = 60$ ksi reinforcing steel is commonly used, it should be noted that for this grade and higher grades of steel, the analyst must use $F_y = 55$ ksi in the strength calculations.

5. To ensure that structural steel tubes and pipes filled with concrete yield prior to the occurrence of local buckling, the following thicknesses are required:

   (a) $t \geq b\sqrt{F_y / 3E}$ is required for each face width b in a tube.

   (b) $t \geq D\sqrt{F_y / 8E}$ is required for a pipe with an outside diameter of D.

   These values are the same as those given in the ACI Code.

6. When concrete encases a steel core consisting of two or more steel shapes, lacing, tie plates, or batten plates must be used to interconnect the steel shapes to prevent buckling of each shape prior to the hardening of the concrete encasement.

7. At connections, the transfer of load to concrete must be by direct bearing to prevent overloading either the structural steel section or the concrete. When the supporting concrete area exceeds the loaded area on all sides, the design strength of the concrete is

$$\phi_c P_{nc} \leq 0.6\left(1.7 f_c' A_B\right)$$

   where A_B is the loaded area and the limit shown is due to bearing. *Note*: When the supporting concrete area is identical to the loaded area, the LRFD specifications do not give the limiting design strength of the concrete due to bearing. For this case, we would use

$$\phi_c P_{nc} \leq 0.6\left(0.85 f_c' A_B\right)$$

   which is based on Section 10.15.1 of the ACI Code [10].

10.2.2 Column Design Strength

The design compressive strength of a composite column is as defined in LRFD E2 (p. 6-47) with the following modifications:

1. A_g in LRFD E2 is replaced with A_s = gross area of structural steel shape.
2. r in LRFD E2 is replaced with $r_{m,}$ which is the larger of:
   (a) The radius of gyration of the steel shape about the flexural buckling axis.
   (b) 0.3 times the overall thickness of the composite cross section in the plane of buckling.
3. F_y and E in LRFD E2 are replaced, respectively, with

$$F_{my} = F_y + c_1 F_{yr}\left(A_r / A_s\right) + c_2\left(A_c / A_s\right)$$
$$E_m = E + c_3 E_c\left(A_c / A_s\right)$$

where

A_c = cross-sectional area of concrete, (in.[2])

A_r = cross-sectional area of longitudinal reinforcing bars, (in.^{2})

A_s = cross-sectional area of the structural steel shape, (in.^{2})

E = 29,000 ksi

$$E_c = w^{1.5}\sqrt{f'_c}$$

w = unit weight of concrete, (pcf)

f'_c = concrete strength, (ksi)

F_y = specified minimum yield stress of structural steel shape, (ksi)

F_{yr} = specified minimum yield stress of longitudinal reinforcing bars, (ksi)

$c_1 = 1.0, \quad c_2 = 0.85, \quad c_3 = 0.4$ for concrete-filled steel shapes

$c_1 = 0.7, \quad c_2 = 0.6, \quad c_3 = 0.2$ for concrete-encased steel shapes

Using these definitions, we find that the design compressive strength of a composite column is

$$\phi_c P_n = 0.85 A_s F_{cr}$$

where

$$F_{cr} = \left(0.658^{\lambda_c^2}\right) F_{my} \qquad \text{when } \lambda_c \le 1.5$$

$$F_{cr} = \left(\frac{0.877}{\lambda_c^2}\right) F_{my} \qquad \text{when } \lambda_c > 1.5$$

$$\lambda_c = \frac{KL}{r_m \pi}\sqrt{\frac{F_{my}}{E_m}}$$

Column design strength tables are provided as a design aid on LRFD pp. 5-73 to 5-142 for some configurations of composite columns. For other choices of structural steel shapes, or concrete grades, or longitudinal and lateral reinforcing bars and their arrangements, the preceding formulas are applicable.

Example 10.1

On LRFD p. 5-129 for a ST14 × 10 × 0.375 filled with concrete and $(KL)_y$ = 30 ft, verify the ϕP_n = 548 kips entry.

Solution

Check the wall thickness:

$$(t = 0.375 \text{ in.}) \ge \left[b\sqrt{\frac{F_y}{3E}} = 14\sqrt{\frac{46}{3(29{,}000)}} = 0.322 \text{ in.} \right] \quad \text{as required}$$

For the composite section:

$$A_g = A_s + A_c$$

A_c = area of concrete

From LRFD p. 1-127, ST14 x 10 x 0.375, $A_s = 17.1$ in.^{2} and $r_y = 4.08$ in. According to LRFD p. 1-120, the outside corner radius of a structural tube was assumed to be two times the tube thickness in computing the listed section properties. The area to be deducted for the concrete due to the four corners is

$$A_{\text{corners}} = (4-\pi)(2t)^2$$

$$= (4-\pi)[2(0.375)]^2 = 0.483 \text{ in.}^2$$

$$A_c = bh - A_{\text{corners}} - A_s$$

$$= 10(14) - 0.483 - 17.1 = 122.4 \text{ in.}^2$$

$$A_g = A_c + A_s = 122.4 + 17.1 = 139.5 \text{ in.}^2$$

$$(A_s/A_g = 17.1/139.5 = 0.123) \geq 0.04 \quad \text{as required}$$

Determine F_{my} and E_m:

$$c_1 = 1 \qquad c_2 = 0.85 \qquad c_3 = 0.4$$

$$\frac{A_c}{A_s} = \frac{122.4}{17.1} = 7.16$$

$$F_{my} = F_y + \frac{c_2 f'_c A_c}{A_s} = 46 + 0.85(3.5)(7.16) = 67.3 \text{ ksi}$$

$$E_c = w^{1.5}\sqrt{f'_c} = 145^{1.5}\sqrt{3.5} = 3267 \text{ ksi}$$

$$E_m = E + \frac{c_3 E_c A_c}{A_s} = 29000 + 0.4(3267)(7.16) = 38{,}357 \text{ ksi}$$

$$r_{my} = r_y\,(\text{ST14 x 10 x 0.375}) = 4.08 \text{ in.}$$

$$\lambda_{cy} = \left(\frac{KL/r_m}{\pi}\right)_y \sqrt{\frac{F_{my}}{E_m}} = \frac{360/4.08}{\pi}\sqrt{\frac{67.2}{38{,}357}} = 1.18 \qquad \lambda_{cy}^2 = 1.382$$

$$\phi P_{ny} = 0.85(0.658^{\lambda_{cy}^2} F_{my} A_s)$$

$$\phi P_{ny} = 0.85(0.658)^{1.382}(67.2)(17.1) = 548 \text{ kips}$$

This agrees with the $\phi P_n = 548$ kips entry on LRFD p. 5-129.

Example 10.2

For the composite ST14 x 10 x 0.375 column in Example 10.1, determine the minimum acceptable area of a bearing plate located on the concrete at the top of the column.

Solution

From LRFD p. 5-129:
Composite ST14 × 10 × 0.375

$$(KL)_y = 30 \text{ ft}, \ \phi P_n = 548 \text{ kips}$$

From LRFD p. 3-45,
Bare ST14 × 10 × 0.375

$$(KL)_y = 30 \text{ ft}, \ \phi P_{ns} = 396 \text{ kips}.$$

$$\phi P_{nc} = \phi P_n - \phi P_{ns} = 547 - 396 = 151 \text{ kips}$$

$$A_B \geq \left[\frac{\phi_c P_{nc}}{1.7 \phi_B f_c'} = \frac{151}{1.7(0.6)(3.5)} = 42.3 \text{ in.}^2 \right]$$

is required since ($A_B = 42.3$ in.^{2}) < ($A_c = 122.6$ in.^{2})

For a plate with $h/b = 1.4$ (ratio of tube dimensions), choose $b = 5.5$ in. and $h = 7.75$ in., which give ($A_B = 42.6$ in.^{2}) ≥ 42.3 in.^{2} as required.

Example 10.3

Verify the $\phi P_n = 1420$ kips entry at $(KL)_y = 30$ ft on LRFD p. 5-81 for a W12 x 120 (F_y = 36 ksi) encased by concrete (f_c' = 3.5 ksi) such that the composite column dimensions are $b = h = 20$ in.

Solution

Check the lateral reinforcement (no. 3 at 13 in. on center) requirements:

$$\text{Maximum tie spacing} = 0.667(20) = 13.3 \text{ in.}$$

$$(\text{Tie spacing} = 13 \text{ in.}) \leq 13.3 \text{ in.} \quad \text{as required.}$$

For one no. 3 bar,

$$(A_r = 0.11 \text{ in.}^2) \geq [0.007(13) = 0.091 \text{ in.}^2] \quad \text{as required}$$

Check the longitudinal reinforcement (four #9) requirements:

$$\text{Clear cover} = 1.5 \text{ in.} \quad \text{is required.}$$

$$\text{Bar spacing} = \text{thickness - 2 (clear cover} + d_t) - d_b$$

$$\text{Bar spacing} = 20 - 2\,(1.5 + 0.375) - 1.128 = 15.1 \text{ in.}$$

For one no. 9 bar,

$$(A_r = 1.00 \text{ in.}^2) \geq [0.007(15.1) = 0.106 \text{ in.}^2] \quad \text{as required}$$

For the composite section,

$$A_g = bh = 20(20) = 400 \text{ in.}^2$$

From LRFD p. 1-38 for a bare W12 x 120:

$$A_s = 35.3 \text{ in.}^2 \qquad \text{and} \qquad r_y = 3.13 \text{ in.}$$

For four no. 9 bars,

$$A_r = 4(1.00) = 4.00 \text{ in.}^2$$

$$A_c = A_g - (A_s + A_r)$$

$$= 400 - (35.3 + 4.00) = 361 \text{ in.}^2$$

$$(A_s/A_g = 35.3/400 = 0.0883) \geq 0.04 \quad \text{as required}$$

Determine F_{my} and E_m:

$$c_1 = 0.7 \qquad c_2 = 0.6 \qquad c_3 = 0.2$$

$$\frac{A_c}{A_s} = \frac{361}{35.3} = 10.2$$

$$\frac{A_r}{A_s} = \frac{4.00}{35.3} = 0.113$$

$$F_{my} = F_y + \frac{c_1 F_{yr} A_r}{A_s} + \frac{c_2 f_c' A_c}{A_s}$$

$$= 36 + 0.7(55)(0.113) + 0.6(3.5)(10.2) = 61.8 \text{ ksi}$$

$$E_c = w^{1.5}\sqrt{f_c'} = 145^{1.5}\sqrt{3.5} = 3267 \text{ ksi}$$

$$E_m = E + \frac{c_3 E_c A_c}{A_s} = 29000 + 0.2(3267)(10.2) = 35{,}665 \text{ ksi}$$

$$r_{my} = \text{larger of} \begin{cases} (r_y \text{ of W12} \times 120) = 3.13 \text{ in.} \\ 0.3b = 0.3(20) = 6.00 \text{ in.} \end{cases}$$

$$r_{my} = 6.00 \text{ in.}$$

$$\lambda_{cy} = \left(\frac{KL/r_m}{\pi}\right)_y \sqrt{\frac{F_{my}}{E_m}} = \frac{360/6.00}{\pi}\sqrt{\frac{61.8}{35{,}665}} = 0.795; \qquad \lambda_{cy}^2 = 0.632$$

$$\phi P_{ny} = 0.85(0.658^{\lambda_{cy}^2} F_{my} A_s)$$

$$\phi P_{ny} = 0.85(0.658)^{0.632}(61.8)(35.3) = 1423 \text{ kips}$$

This agrees very well with the $\phi P_n = 1420$ kips entry on LRFD p. 5-81.
Note: Before the encasing concrete hardens, $\phi P_n = 538$ kips (from LRFD, p. 3-23).

Example 10.4

On LRFD p. 5-78 for a W14 × 68 ($F_y = 36$ ksi) encased by concrete($f_c' = 3.5$ ksi) such that the composite column dimensions are $b = 18$ in. and $h = 20$ in.,

1. Verify $r_{mx}/r_{my} = 1.22$.
2. For $(KL)_x = 30$ ft and $(KL)_y = 15$ ft, determine ϕP_n.

Solution

$$r_{mx} = \text{larger of} \begin{cases} (r_x \text{ of W14 x 68}) = 6.01 \text{ in.} \\ 0.3h = 0.3(22) = 6.60 \text{ in.} \end{cases}$$

$$r_{mx} = 6.60 \text{ in.}$$

$$r_{my} = \text{larger of} \begin{cases} (r_y \text{ of W14 x 68}) = 2.46 \text{ in.} \\ 0.3b = 0.3(18) = 5.40 \text{ in.} \end{cases}$$

$$r_{my} = 5.40 \text{ in.}$$

$$\frac{r_{mx}}{r_{my}} = \frac{6.60}{5.40} = 1.22$$

$$\left[\frac{(KL)_x}{(r_{mx}/r_{my})} = \frac{30}{1.22} = 24.6 \text{ ft}\right] > (KL)_y = 15 \text{ ft}$$

Enter LRFD p. 5-78 at 24.6 ft for $F_y = 36$ ksi and find

$$\phi P_n = (\phi P_{nx} = 1088 \text{ kips})$$

10.3 COMPOSITE BEAMS WITH SHEAR CONNECTORS

The most common case of a composite beam is a steel shape that supports and interacts with a concrete slab as shown in Figures 10.2(a) and 10.3. Either a solid reinforced concrete slab [Figure 10.2(a)] or a concrete slab poured on a steel-ribbed metal deck (Figure 10.3) can be used. In either case, the shear studs shown or some other type of shear connectors are essential to ensure composite beam action. If the steel-ribbed metal deck in Figure 10.3 has embossments on the upper surface to bond the deck to the concrete slab, the steel deck is classified as a composite deck, and the steel deck can be used in the reinforcement requirement for the concrete slab.

10.3.1 Composite Construction

In *shored construction*, temporary shores are used to help the structural steel members support the poured concrete. Temporary shores are gravity direction supports located beneath the bottom flange of the steel shape at discrete intervals along the beam length and between the permanent beam supports. After the concrete has cured sufficiently, the temporary shores are removed. For shored construction, composite beam action supports all loads (dead weight of the structure and loads applied on the top surface of the concrete slab).

In *unshored construction*, temporary shores are not used, and only the structural steel members support the freshly poured concrete. After the concrete has cured sufficiently [0.75 f_c' is specified in LRFD I3.4 (p. 6-65)], composite beam action supports the loads applied on the top surface of the concrete slab. Therefore, the structural steel members must be adequately designed to support all factored loads that exist before the concrete has cured sufficiently. When an *elastic stress distribution* is required, superposition must be used in the member section analysis. In stage 1 the structural steel members support all factored permanent loads before the concrete has cured sufficiently. In stage 2, composite beam action supports the factored loads applied after the concrete has cured sufficiently. When a *plastic stress distribution* is permitted and used, load tests have shown that the factored loads can be assumed to be resisted by composite beam action for unshored construction.

10.3.2 Effective Concrete Flange Width

The behavior of a concrete slab compositely connected by shear studs to steel beams may be conceptually described as follows. A uniform gravity direction load on the top surface of the slab causes:

1. Compression forces in the gravity direction to occur between the slab and the steel beams.
2. Longitudinal shear forces to occur in the shear studs at the concrete–steel interface.

Along their length direction and at the top surface of their flanges, the steel beams are subjected to a gravity direction distributed load and to longitudinal shear forces at the shear stud locations. Therefore, each steel beam is subjected to an eccentrically applied axial compressive load and a lateral load. These loads are distributed along the member length.

The concrete slab transfers the gravity direction load via an interactive compression force to the steel beams. Also, the concrete slab is subjected to longitudinal shear forces at the bottom surface of the slab and distributed along the length direction centerlines of the steel beams. Due to the eccentric longitudinal shear forces, the in-plane compressive stress distribution in the concrete slab is not uniform. Along the lines where the eccentric longitudinal shear forces are applied, the compressive stress is maximum. Midway between these lines of eccentric longitudinal shear forces, the compressive stress is minimum. If the distance between these lines of eccentric longitudinal shear forces is increased, the maximum compressive stress divided by the minimum compressive stress also increases. To simplify the cross-sectional analysis of a composite beam, the compressive stress distribution described is assumed to be constant across an effective slab width attached at the shear studs to a steel beam. The cross section of each composite beam in Figures 10.2(a) and 10.3 consists of the structural steel shape and the effective flange width of the concrete slab, which are interactively connected by the shear studs, which transfer the longitudinal shear forces.

Three criteria are given in LRFD I3.1 for the determination of the effective concrete slab width. On each side of the beam centerline, the effective concrete slab width is the least of the following values:

1. $b_1 = L/8$, where L = beam span length.
2. $b_1 = S/2$, where S = distance to the centerline of the adjacent beam.
3. b_2 = distance to the edge of the slab.

For a composite beam on the interior of the concrete slab [see Figure 10.4(a)], the effective flange width is $b = b_{1L} + b_{1R}$.

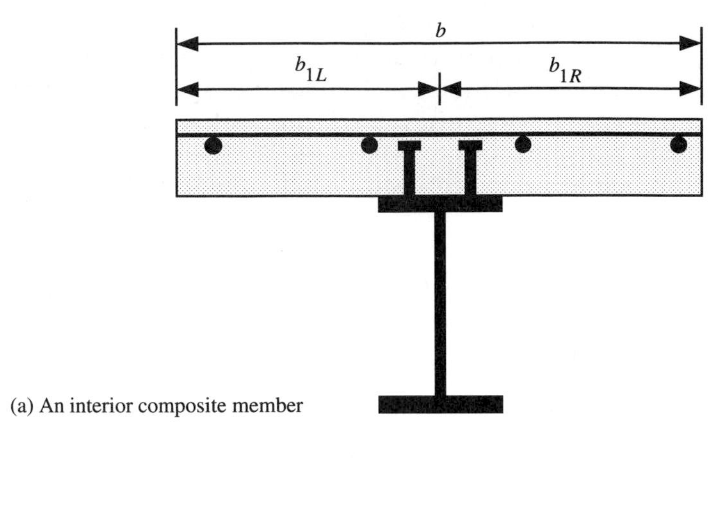

(a) An interior composite member

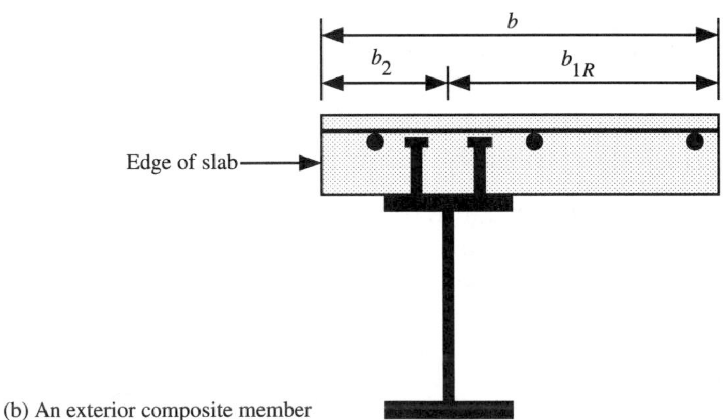

(b) An exterior composite member

FIGURE 10.4 Effective slab width.

For a composite beam on the exterior edge of the concrete slab [see Figure 10.4(b)], the effective flange width is $b = b_2 + b_{1R}$.

Examples involving the determination of the effective flange width are given at the end of Section 10.3.4.

10.3.3 Shear Design Strength

Only the shear design strength of the web for the structural steel shape is usable, and LRFD F2 (p. 6-56) is applicable for the determination of ϕV_n, which acts parallel to the gravity direction on a composite beam cross section. The shear perpendicular to the gravity direction is named horizontal shear in the LRFD Specification. In the behavioral discussion we used the terminology *longitudinal shear* instead of *horizontal shear*. Horizontal shear requirements are given in the next section.

10.3.4 Shear Connectors

LRFD I5.1 (p. 6-67) defines the material requirements for shear studs and channel shear connectors. The length of a shear stud must be at least four stud diameters. For other types of shear connectors, LRFD I6 is applicable.

The horizontal shear force at the interface between the steel beam and the concrete slab must be transferred by shear connectors and is defined as follows:

1. *In regions of positive moment* (top surface of the concrete slab is in compression), the *total horizontal shear* between the maximum moment point and the zero moment point is the least of:
   (a) $0.85\, f_c'\, A_c$, which is the maximum possible compressive force in the effective width of the concrete slab.
   (b) A_sF_y, which is the maximum possible tensile force in the structural steel shape. When a hybrid structural steel shape is used, ΣA_sF_y is applicable, and the sum is made for all elements in the cross section of the steel shape.
   (c) ΣQ_n, which is the sum of nominal strengths of the shear connectors in the indicated region.
2. *In regions of negative moment* (top surface of the concrete slab is in tension), the *total horizontal shear* between the maximum moment point and the zero moment point is the smaller of:
   (a) A_sF_{yr}, which is the maximum possible tensile force in the longitudinal reinforcing steel.
   (b) ΣQ_n, which is the sum of nominal strengths of the shear connectors in the indicated region.

When either item 1c or 2b governs the definition of the total horizontal shear, this behavior is classified as *partial composite action*. Otherwise, the behavior is classified as *full composite action*.

For a shear stud embedded in a solid concrete slab, the nominal strength of one shear connector is

$$\left(Q_n = 0.5 A_{sc} \sqrt{f_c' E_c}\right) \le A_{sc} F_u$$

where

$$A_{sc} = \text{cross-sectional area of a shear stud, (in.}^2\text{)}$$

$$F_u = \text{minimum specified tensile strength of a shear stud, (ksi)}$$

For a channel shear connector embedded in a solid concrete slab, the nominal strength of one shear connector is

$$Q_n = 0.3\left(t_f + 0.5 t_w\right) L_c \sqrt{f_c' E_c}$$

where

$$t_f = \text{flange thickness of channel shear connector, (in.)}$$

$$t_w = \text{web thickness of channel shear connector, (in.)}$$

$$L_c = \text{length of channel shear connector, (in.)}$$

When full composite action is desired, the minimum required number of shear connectors in each region between a maximum moment point and an adjacent zero moment point is the *total horizontal shear in each region* divided by Q_n.

When partial composite action is desired, let N = trial number of shear connectors in a region between a maximum moment point and an adjacent zero moment point. The *total horizontal shear transferred in the indicated region* = $\Sigma Q_n = NQ_n$. Then, as described later in Section 10.3.5 and as illustrated in Example 10.12, the design bending strength ϕM_n for partial composite action must be calculated to determine if the design requirement $\phi M_n \geq M_u$ is satisfied.

Shear connectors may be uniformly spaced within each region. However, the number of shear connectors placed between each concentrated load point and the nearest zero moment point must be sufficient to develop M_u needed at each concentrated load point. The following restrictions on the placement and spacing of shear connectors are imposed by LRFD I5.6:

1. Minimum lateral concrete cover is 1 in. except for shear connectors installed in the ribs of a steel deck.
2. $d \leq 2.5t_f$, where d = stud diameter and t_f = thickness of beam flange to which studs are welded. However, $d > 2.5t_f$ is permissible for studs located over the web in a solid slab.
3. Minimum longitudinal spacing of studs is $6d$ in solid slabs and $4d$ in the ribs of a steel deck.
4. Minimum transverse spacing of studs is $4d$ in all cases. However, see LRFD Figure C-I5.1, p. 6-215, for special provisions when the studs are staggered.
5. Maximum spacing of studs is eight times the total slab thickness.

There are special provisions for shear studs embedded in concrete on a formed steel deck (see LRFD Figure C-I3.3, p. 6-211):

1. The usual practice is to field-weld the studs through the steel deck to the beam flange. When the studs are shop-welded to the beam flange, holes must be made in the steel deck at the stud locations.
2. Maximum stud diameter is 0.75 in.
3. When the ribs in the steel deck are perpendicular to the steel beam:
   (a) The steel deck must be anchored to all supporting members at a spacing ≤16 in. Welds at the studs and puddle welds (or other devices) elsewhere are acceptable anchorages.
   (b) Concrete below the top of the steel deck must be ignored in calculating A_c.
   (c) Longitudinal spacing of shear studs ≤ 36 in.
   (d) Nominal strength of one shear connector is

$$\left(Q_n = 0.5 A_{sc} \sqrt{f_c' E_c}\right) \leq A_{sc} F_u$$

where

A_{sc} = cross-sectional area of a shear stud, (in.^{2})

F_u = minimum specified tensile strength of a shear stud, (ksi)

$$\left[R_{sc} = \frac{0.85\, w_r}{h_r \sqrt{N_r}} \left(\frac{H_s}{h_r} - 1.0\right)\right] \leq 1.0$$

N_r = number of shear studs in one rib at a beam intersection

($N_r \leq 3$ must be used in the R_{sc} formula; however, more than three studs may be installed.) w_r, h_r, and H_s are defined in LRFD Figure C-I3.3 p. 6-211. $H_s \leq (h_r + 3)$ must be used in the computations.

4. When the ribs in the metal deck are parallel to the steel beam:

(a) As shown in the last figure of LRFD Figure C-I3.3 p. 6-211, at a rib the deck may be cut longitudinally and separated to form a concrete haunch over the steel beam.

(b) Concrete below the top of the steel deck can be included in calculating A_c.

(c) When $h_r \geq 1.5$ in., $w_r \geq 2$ in. is required for the first stud in the transverse direction; for N studs where $N \geq 2$, $w_r \geq [2 \text{ in.} + 4(N\text{-}1)d]$ is required where d = stud diameter, (in.)

(d) When $w_r/h_r < 1.5$, the nominal strength of one shear connector is

$$\left(Q_n = 0.5 A_{sc} \sqrt{f_c' E_c}\right) \leq A_{sc} F_u$$

where

A_{sc} = cross-sectional area of a shear stud, (in.^{2})

F_u = minimum specified tensile strength of a shear stud, (ksi)

$$\left[R_{sc} = \frac{0.6 w_r}{h_r}\left(\frac{H_s}{h_r} - 1.0\right)\right] \leq 1.0$$

w_r, h_r, and H_s are defined in LRFD Figure C-I3.3 p. 6-211. $H_s \leq (h_r + 3)$ must be used in the computations.

Example 10.5

Figure 10.5 shows the cross section of a fully composite structural system consisting of a solid 5-in.-thick concrete slab connected via shear studs to the top flange of steel beams.

Beams:

W16 × 31 $\quad A_s = 9.12$ in.^{2} $\quad t_f = 0.440$ in. $\quad F_y = 36$ ksi.

$L = 30$ ft = simply supported span length,

$S = 10$ ft = transverse spacing,

Concrete:

$$f_c' = 3.5 \text{ ksi} \qquad w = 145 \text{ pcf}$$

$$E_c = w^{1.5}\sqrt{f_c'} = 145^{1.5}\sqrt{3.5} = 3267 \text{ ksi.}$$

Shear studs:

$$d = 0.75 \text{ in.} \qquad H_s = 4d = 3 \text{ in.} \qquad F_u = 60 \text{ ksi}$$

For full composite action, determine:

1. b = effective concrete slab width for an interior steel beam.

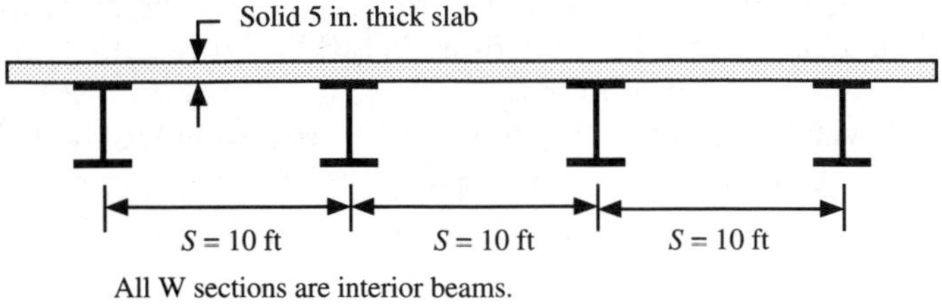

Solid 5 in. thick slab

$S = 10$ ft $S = 10$ ft $S = 10$ ft

All W sections are interior beams.

FIGURE 10.5 Example 10.5.

2. V_h = horizontal shear force that must be transferred. Assume that the interior composite beam is subjected to a uniformly distributed factored load.
3. Minimum required number of shear studs.

Solution 1

The effective slab width on each side of the beam centerline is the smaller of

$$\frac{L}{8} = \frac{30(12)}{8} = 45 \text{ in.}$$

$$\frac{S}{2} = \frac{10(12)}{2} = 60 \text{ in.}$$

Therefore, the effective slab width, $b = 2(45) = 90$ in.

Solution 2

In a positive moment region, the horizontal shear force that must be transferred for full composite action is

$$V_h = \text{smaller of } \begin{cases} 0.85 f_c' A_c = 0.85(3.5)(90)(5) = 1339 \text{ kips} \\ A_s f_y = 9.12(36) = 328 \text{ kips} \end{cases}$$

$$V_h = 328 \text{ kips}$$

Solution 3

Check the stud diameter. Unless each stud is welded directly over the beam web,

$$d \le [2.5t_f = 2.5(0.440) = 1.10 \text{ in.}] \quad \text{is required}$$

$$(d = 0.75 \text{ in.}) \le 1.10 \text{ in.} \quad \text{as required}$$

For one 0.75-in.-diameter shear stud, the nominal shear strength is

$$\left(Q_n = 0.5 A_{sc} \sqrt{f_c' E_c}\right) \le A_{sc} F_u$$

$$A_{sc} = \pi(d/2)^2 = \pi(0.75)^2 / 4 = 0.4418 \text{ in.}^2$$

$$A_{sc}F_u = 0.4418(60) = 26.5 \text{ kips}$$

$$\left[0.5(0.4418)\sqrt{3.5(3267)} = 23.6\right] \leq 26.5$$

$$Q_n = 23.6 \text{ kips/stud}$$

Between the maximum positive moment point and each adjacent zero moment point, the minimum required number of shear studs is

$$N = V_h/Q_n = 328/23.6 = 13.9$$

Use 14 shear studs between midspan and each simple support. The minimum total number of shear studs required is 2(14) = 28. For only one shear stud directly over the beam web at each location,

$$\text{Stud spacing} = 30(12)/28 = 12.9 \text{ in.}$$

$$\text{Minimum spacing} = 6d = 6(0.75) = 4.5 \text{ in.}$$

$$\text{Maximum spacing} = 8t_{\text{slab}} = 8(5) = 40 \text{ in.}$$

For placement convenience, use stud spacing = 12 in. and 30 studs.

Example 10.6

Solve Example 10.5 for the exterior beam configuration shown in Figure 10.6.

Solution 1

The effective slab width on the interior side of the beam centerline is the smaller of

$$\frac{L}{8} = \frac{30(12)}{8} = 45 \text{ in.}$$

$$\frac{S}{2} = \frac{10(12)}{2} = 60 \text{ in.}$$

The effective slab width on the exterior side of the beam centerline is the smaller of

$$\frac{L}{8} = \frac{30(12)}{8} = 45 \text{ in.}$$

$$\text{Edge distance} = 1 \text{ ft} = 12 \text{ in.}$$

Therefore, the effective slab width is

$$b = 12 + 45 = 57 \text{ in.}$$

Solution 2

In a positive moment region, the horizontal shear force which must be transferred for full composite action is

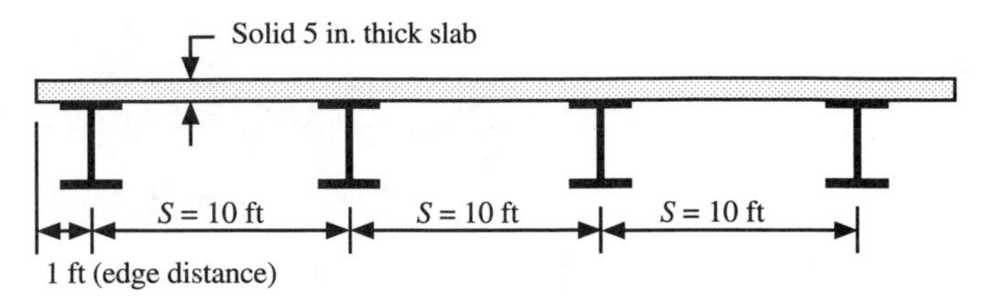

FIGURE 10.6 Example 10.6.

$$V_h = \text{smaller of} \begin{cases} 0.85 f_c' A_c = 0.85(3.5)(57)(5) = 848 \text{ kips} \\ A_s f_y = 9.12(36) = 328 \text{ kips} \end{cases}$$

$$V_h = 328 \text{ kips}$$

Solution 3

Same as in Example 10.5.

Example 10.7

Solve Example 10.5 with the following modifications: The 5-in. total slab thickness consists of a 2-in. concrete topping on a 3-in. formed steel deck, with the deck ribs spanning perpendicular to the steel beam as shown in Figure 10.7:

$$H_s = 4.5 \text{ in.} = \text{shear stud height}$$

$$h_r = 3 \text{ in.} = \text{deck rib height}$$

$$w_r = (4.5 + 7.5)/2 = 6.00 \text{ in.} = \text{average rib width}$$

$$t_c = 2 \text{ in.} = \text{thickness of concrete slab above the steel deck}$$

Solution 1

Check the special requirements when the concrete slab is poured on a formed steel deck:

$$(h_r = 3 \text{ in.}) \le 3 \text{ in.} \quad \text{as required}$$

$$(t_c = 2 \text{ in.}) \ge 2 \text{ in.} \quad \text{as required}$$

$$(w_r = 6.00 \text{ in.}) \ge 2.0 \text{ in.} \quad \text{as required}$$

$$(d = 0.75 \text{ in.}) \le 0.75 \text{ in.} \quad \text{as required}$$

$$(H_s = 4.5 \text{ in.}) \ge (h_r + 1.5 \text{ in.} = 3 + 1.5 = 4.5 \text{ in.}) \quad \text{as required}$$

The solution is the same as in Example 10.5.

Solution 2

The concrete below the top of the steel deck must be ignored in calculating A_c. In a positive moment region, the horizontal shear force that must be transferred for full

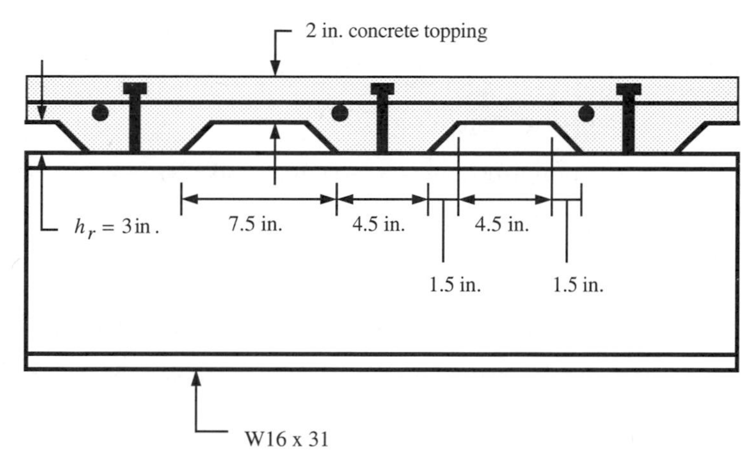

Ribs of metal deck are perpendicular to length centerline of W16 x 31.

FIGURE 10.7 Example 10.7.

composite action is

$$V_h = \text{smaller of } \begin{cases} 0.85 f_c' A_c = 0.85(3.5)(90)(2) = 459 \text{ kips} \\ A_s f_y = 9.12(36) = 328 \text{ kips} \end{cases}$$

$$V_h = 328 \text{ kips}$$

Solution 3

For one 0.75-in.-diameter shear stud, the nominal shear strength is

$$\left(Q_n = 0.5 A_{sc} \sqrt{f_c' E_c}\right) \le A_{sc} F_u$$

$$A_{sc} = \pi \left(\frac{d}{2}\right)^2 = \frac{\pi (0.75)^2}{4} = 0.4418 \text{ in.}^2$$

$$A_{sc} F_u = 0.4418(60) = 26.5 \text{ kips}$$

$$N_r = \text{number of shear studs located in each rib}$$

Try $N_r = 2$ since (from Example 10.5) stud spacing < 12 in. is probable for $N_r = 1$ and $R_{sc} < 1$:

$$\left[R_{sc} = \frac{0.85 w_r}{h_r \sqrt{N_r}} \left(\frac{H_s}{h_r} - 1.0\right) = \frac{0.85(6)}{3\sqrt{2}} \left(\frac{4.5}{3} - 1.0\right) = 0.601\right] \le 1.0$$

$$\left[Q_n = 0.601(0.5)(0.4418)\sqrt{3.5(3267)} = 14.2 \text{ kips}\right] \le 26.5 \text{ kips}$$

$$Q_n = 14.2 \text{ kips/stud}$$

Between the maximum positive moment point and each adjacent zero moment point, the minimum required number of shear studs is

$$N = V_h / Q_n = 328/14.2 = 23.1$$

Use 24 shear studs (12 pairs) between midspan and each simple support. The total number of shear studs required is 2(24) = 48:

$$\text{Longitudinal stud spacing} = 30(12)/24 = 12.9 \text{ in.}$$

$$\text{Minimum spacing} = 12 \text{ in.} \quad \text{(rib spacing)}$$

$$\text{Maximum spacing} = 8t_{\text{slab}} = 8(5) = 40 \text{ in.}$$

For placement convenience, use 60 studs and longitudinal spacing = 12 in:

$$\text{Minimum lateral spacing} = 4d = 4(0.75) = 3 \text{ in.}$$

For each pair of shear studs, use lateral spacing = 3 in.

Check the stud diameter. Since each pair of shear studs cannot be welded directly over the beam web,

$$(d = 0.75 \text{ in.}) \le [2.5t_f = 2.5(0.440) = 1.10 \text{ in.}] \quad \text{as required}$$

10.3.5 Flexural Design Strength

For a composite beam with shear connectors, there are two design bending strength definitions. One definition is for the positive moment region and the other definition is for the negative moment region.

Positive Moment Region

In the *positive moment region* of a fully composite beam with shear connectors, the design bending strength ϕM_n is determined as follows:

1. When $h/t_w \le 640/\sqrt{F_y}$, $\phi = 0.85$ and M_n is calculated from the plastic stress distribution on the composite section. For convenience, this calculation of M_n is named the plastic section analysis. The assumptions made in performing a fully composite plastic section analysis are [see LRFD Figure C-I3.1 (p. 6-207)]:

   (a) Concrete tensile strength is zero. A uniform compressive stress of $0.85f_c'$ is applicable for the concrete compression zone.
   (b) A uniform stress of F_y is applicable in the tension and compression zones of the structural steel shape.

   The resultant of the forces in the compression zone is equal to the resultant of the forces in the tension zone. This is an equilibrium requirement and is not an assumption.

2. When $h/t_w > 640/\sqrt{F_y}$, $\phi = 0.9$, and M_n is calculated from the superposition of elastic stress distributions on the composite section. For convenience, this calculation of M_n is named the elastic section analysis. The effects of shoring must be considered in computing M_n. The assumptions made in performing an elastic section analysis are:

(a) Strain is proportional to the distance from the neutral axis.
(b) Steel stress is E times steel strain, but cannot exceed F_y. For a hybrid structural steel shape, strain in the web may exceed the yield strain; however, at such locations, the stress that must be used in the calculations is F_{yw}.
(c) Concrete tensile strength is zero. Concrete compressive stress is E_c times concrete compressive strain, but cannot exceed $0.85f_c'$.

The resultant of the forces in the compression zone is equal to the resultant of the forces in the tension zone. This is an equilibrium requirement and is not an assumption.

Negative Moment Region

In the *negative moment region* of a fully composite beam with shear connectors, the design bending strength ϕM_n can be determined from either of the following definitions:

1. Only the structural steel shape can be used to determine ϕM_n in accordance with LRFD F1 (p. 6-52).
2. ϕM_n can be determined using $\phi = 0.85$ and M_n calculated for the composite section from a fully composite plastic section analysis [see LRFD Figure C-I3.1 (p. 6-207)], provided that:

(a) For the structural steel shape, $0.5b_f/t_f \le 65/\sqrt{F_y}$, $h/t_w \le 640/\sqrt{F_y}$, and $L_b \le L_p$ for the flange not in contact with the concrete slab or the metal deck.
(b) Adequately designed shear connectors exist in this region.
(c) Longitudinal reinforcement in the effective width of the slab is adequately anchored such that A_rF_{yr} can be fully developed.

Example 10.8

For the fully composite section in Example 10.5 and shored construction, determine ϕM_{nx}.

Solution

From Example 10.5, we have

Effective slab width,

$$b = 90 \text{ in.}$$

Solid slab thickness,

$$t = 5 \text{ in.}$$

$$f_c' = 3.5 \text{ ksi}$$

$$0.85f_c'A_c = 0.85\,(3.5)(90)(5) = \text{kips}$$

W16 x 31:

$A_s = 9.12 \text{ in.}^2$ $\quad t_w = 0.275$ in. $\quad t_f = 0.440$ in. $\quad d = 15.88$ in.

$$F_y = 36 \text{ ksi} \qquad A_sF_y = 9.12(36) = 328.32 \text{ kips}$$

$$\phi M_{px} = 146 \text{ ft-kips} \qquad L_p = 4.9 \text{ ft.}$$

$$\left(h/t_w = 51.6\right) \le \left(640/\sqrt{F_y} = 640/\sqrt{36} = 106.7\right)$$

When $h/t_w \le 640/\sqrt{F_y}$, $\phi = 0.85$ and M_{nx} is calculated from the plastic stress distribution on the composite section. From LRFD Eq. (C-I3-5), p. 6-207, we find that

$$\phi M_{nx} = 0.85\left[C\left(d_1 + d_2\right) + A_s F_y \left(d_3 - d_2\right)\right]$$

$$d_1 = \text{distance from top of the W section to compressive concrete force}$$

$$d_2 = \text{distance from top of the W section to compressive steel force}$$

$$d_2 = 0 \text{ when there is no compressive steel force}$$

$$d_3 = 0.5d \text{ of the W section}$$

These formulas are valid for positive bending of a W section compositely connected to the effective concrete slab width. Since

$$(C = 328.32 \text{ kips}) = (P_y = A_sF_y = 328.32 \text{ kips})$$

the plastic neutral axis (PNA) is located at or above the concrete–steel interface. The depth of the compressive stress block acting on the effective concrete slab width in the plastic section analysis is determined from horizontal force equilibrium:

$$a = \frac{A_s F_y}{0.85 f_c' b} = \frac{328.32}{0.85(3.5)(90)} = 1.226 \text{ in.}$$

Since (a = 1.226 in.) < (t = 5 in.), the PNA is located in the slab.

$$d_1 = t - a/2 = 5 - 1.226/2 = 4.387 \text{ in.}$$

$$d_2 = 0$$

$$d_3 = 15.88/2 = 7.94 \text{ in.}$$

$$\phi M_{nx} = 0.85(328.32)(4.387 + 7.94) = 3440 \text{ in.-kips} = 286.7 \text{ ft-kips}$$

Note that:

1. On LRFD p. 5-28 for a W16 x 31:

$$\text{PNA} = \text{top flange location (TFL)}$$

$$Y1 = 0$$

$$[2.5t_f = 2.5(0.440) = 1.10 \text{ in.}]\,Q_n = 328 \text{ kips}$$

$$Y2 = t - a/2 = 5 - 1.226/2 = 4.387 \text{ in.}$$

$$\phi M_{nx} = 286.5 \text{ ft-kips}$$

which is the same as the value computed here. See LRFD p. 5-6 for the definitions of the column headings for the table on LRFD p. 5-28.

2. (Composite ϕM_{nx})/ (steel ϕM_{px}) = 286.7/146 = 1.96; The composite section is 1.96 times stronger in bending than the steel section alone for $L_b \le L_p$.

Example 10.9

For the fully composite section in Example 10.7 and shored construction, determine ϕM_{nx}.

Solution

When the steel deck ribs are perpendicular to the steel beam, the concrete below the top of the deck must be neglected. Only the 2-in. concrete topping on the steel deck can be considered in the determination of ϕM_{nx}.

From the solution for Example 10.8, PNA was located in the slab and ($a = 1.226$ in.) < ($t_c = 2$ in.). For the current example problem, the steel deck is entirely within the tension zone of the concrete slab. Therefore, the solution for the current example problem is identically the same as the solution for Example 10.8 and $\phi M_{nx} = 286.7$ ft-kips.

Example 10.10

For the fully composite section in Example 10.8 and unshored construction, determine ϕM_{nx}.

Solution

The solution for Example 10.8 is also valid here since ϕM_{nx} is determined for a fully composite section and the plastic stress distribution. Therefore, $\phi M_{nx} = 286.7$ ft-kips. However, the steel beam alone must be checked for adequacy to resist all loads applied before the concrete strength in the slab becomes $0.75 f'_c$. Those loads are:

1. Dead loads

   Beam

$$0.031 \text{ kips/ft}$$

   Slab

$$\text{thickness} = 5 \text{ in.} \qquad \text{width} = 10 \text{ ft}$$

$$\text{weight} = (5/12)(10)(0.150) = 0.625 \text{ kips/ft}$$

$$D = 0.031 + 0.625 = 0.656 \text{ kips/ft}$$

2. Live loads

$$\text{Assume 20 psf:}$$

$$L = 0.02(10) = 0.2 \text{ kips/ft}$$

LRFD Commentary on *Strength During Construction* (p. 6-210) should be studied before dealing with the following solution. If we use the load factors in LRFD A4.1 (p. 6-30), the factored loading on an interior steel beam is the larger of

$$q_u = 1.4D = 1.4(0.656) = 0.918 \text{ k/ft}$$

$$q_u = 1.2D + 1.6L = 1.2(0.656) + 1.6(0.2) = 1.11 \text{ k/ft}$$

for which the required bending strength is

$$M_{ux} = q_u L^2/8 = 1.11\,(30)^2/8 = 124.9 \text{ ft-kips}$$

If the forms for the concrete slab are attached to the compression flange of the steel beam,

$$L_b = 0$$

$$[\phi M_{nx} = (\phi M_{px} = 146 \text{ ft-kips})] \geq (M_{ux} = 125 \text{ ft-kips}) \quad \text{as required}$$

If the forms for the concrete slab are not attached to the compression flange of the steel beam, try a lateral brace at midspan:

$$L_b = 15 \text{ ft} \qquad C_b = 1.75$$

$$\phi M_{nx} = \text{smaller of} \begin{cases} C_b \overline{M}_1 = 1.67(84.8) = 142 \text{ ft-kips} \\ \phi M_{px} = 9.12(36) = 146 \text{ ft-kips} \end{cases}$$

$$[\phi M_{nx} = 142 \text{ ft-kips})] \geq (M_{ux} = 125 \text{ ft-kips}) \quad \text{as required}$$

Also, the serviceability requirement for deflection should be checked using I_x of the W16 x 31, service condition loads (D and L), and the applicable deflection limitation according to the governing building code for the construction site.

In the design requirement for shear that is $\phi V_n \geq V_u$, note that ϕV_n is the design shear strength of the W16 × 31 web only for a composite beam regardless of the type of construction (shored or unshored). The live loading after the concrete has hardened usually is larger than the construction live loading. Therefore, the factored loading after the concrete has hardened will govern V_u unless there is a heavy construction load applied near the beam support.

Example 10.11

Solve Example 10.5 with the following modification: The interior composite beam is subjected to the loading shown in Figure 10.8.

Solution 1 and 2

These solutions are the same as in Example 10.5

Solution 3

The solution in Example 10.5 is applicable up to the statement that we need 14 shear studs between midspan and each simple support. Also, from the solution to Example 10.8 or from LRFD p. 5-28 for a W16 x 31,

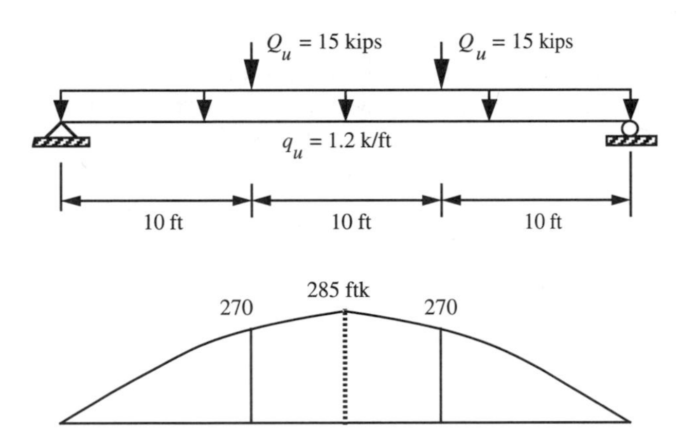

FIGURE 10.8 Example 10.11.

$$\text{PNA} = \text{TFL}$$
$$Y1 = 0$$
$$\Sigma Q_n = 328 \text{ kips}$$
$$Y2 = t - a/2 = 5 - 1.226/2 = 4.387 \text{ in.}$$
$$(\phi M_{nx} = 286.5 \text{ ft-kips}) \geq (M_u = 285 \text{ ft-kips}) \quad \text{as required at midspan}$$

Therefore, 14 shear studs between each simple support and midspan are satisfactory.

An additional requirement in this example is that the number of shear studs N_p placed between each concentrated load point and the nearest zero moment point must be sufficient to develop M_u needed at each concentrated load point. In order to satisfy the additional requirement, we can enter LRFD p. 5-28 for a W16 x 31:

$$Y2 = 4.5 \text{ in.}$$
$$(\phi M_{nx} = 275 \text{ ft-kips}) \geq (M_u = 270 \text{ ft-kips})$$
$$\Sigma Q_n = 285 \text{ kips}$$

which probably is adequate at the concentrated load point. Try

$$N_p \approx (\Sigma Q_n)/Q_n = 285/23.6 = 12.08$$

For 12 shear studs,

$$V_h = 12(23.6 \text{ kips/stud}) = 283.2 \text{ kips}$$

$$a = \frac{V_h}{0.85 f_c' b} = \frac{283.2}{0.85(3.5)(90)} = 1.058 \text{ in.}$$

$$Y2 = 5 - a/2 = 5 - 1.058/2 = 4.47 \text{ in.}$$
$$(\phi M_{nx} = 274 \text{ ft-kips}) \geq (M_n = 220 \text{ ft-kips})$$

as required at each concentrated load point. Therefore, we must use at least 12 shear studs between each support and the adjacent concentrated load point. If these studs are uniformly spaced, the stud spacing in the indicated regions is 10 ft = 120 in./12 = 10 in. Use 12 studs spaced 10 in. on center in the indicated regions.

In the region between the two concentrated load points, the maximum shear stud spacing = 40 in. can be used. For this spacing, at least (10 ft = 120 in.)/(40 in./stud) = 3 shear studs must be provided in the region between the two concentrated load points. However, for fully composite action 4 shear studs are required in this region since 14 -12 = 2 shear studs are required between the concentrated load point and midspan.

The minimum total number of studs required in this example is the larger of 4 + 2(12) = 28 and 2(14) = 28.

Example 10.12

Solve Example 10.11 with the following modification: For the moment diagram in Figure 10.9, use partial composite action and determine the number of shear studs.

Solution

We can enter LRFD p. 5-28 for a W16 x 31:

$$Y2 = 4.5 \text{ in.}$$

$$(\phi M_{nx} = 246 \text{ ft-kips}) \geq (M_u = 232.5 \text{ ft-kips})$$

$$\Sigma Q_n = 197 \text{ kips}$$

which probably is adequate at midspan. Try

$$N_p \approx (\Sigma Q_n)/Q_n = 197/23.6 = 8.35$$

For eight shear studs,

$$V_h = 8(23.6 \text{ kips/stud}) = 188.8 \text{ kips}$$

$$a = \frac{V_h}{0.85 f'_c b} = \frac{188.8}{0.85(3.5)(90)} = 0.705 \text{ in.}$$

$$Y2 = 5 - \frac{a}{2} = 5 - \frac{0.705}{2} = 4.65 \text{ in.}$$

$$(\phi M_{nx} = 245 \text{ ft-kips}) \geq (M_n = 232.5 \text{ ft-kips}) \quad \text{as required at midspan.}$$

We can enter LRFD p. 5-28 for a W16 x 31:

$$Y2 = 5 \text{ in.}$$

$$(\phi M_{nx} = 221 \text{ ft-kips}) \geq (M_u = 220 \text{ ft-kips})$$

$$\Sigma Q_n = 197 \text{ kips}$$

which may be adequate at midspan. Try

$$N_p \approx (\Sigma Q_n)/Q_n = 118 \text{ kips}/23.6 = 5.00$$

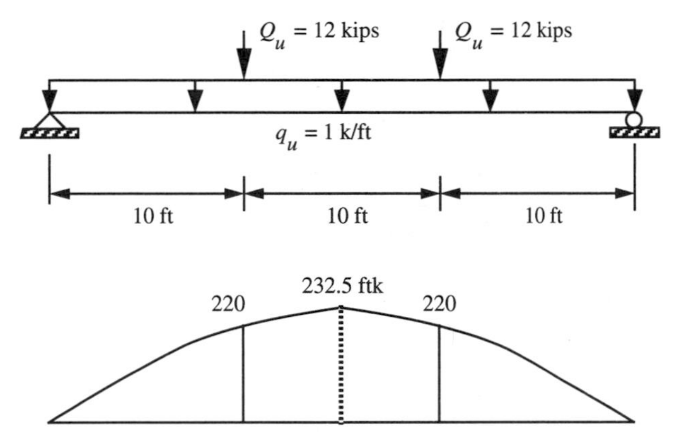

FIGURE 10.9 Example 10.12.

For five shear studs,

$$V_h = 5(23.6 \text{ kips/stud}) = 118.0 \text{ kips}$$

$$a = \frac{V_h}{0.85 f_c' b} = \frac{118}{0.85(3.5)(90)} = 0.441 \text{ in.}$$

$$Y2 = 5 - \frac{a}{2} = 5 - \frac{0.441}{2} = 4.78 \text{ in.}$$

$$(\phi M_{nx} = 219.2 \text{ ft-kips}) \approx (M_n = 220 \text{ ft-kips})$$

as required at each concentrated load point. Therefore, only five shear studs between each simple support and the adjacent concentrated load point are satisfactory.

For the 8 - 5 = 3 shear studs required between each concentrated load point and midspan, the spacing = 60/3 = 20 in. is less than the maximum spacing = 40 in. Therefore, the minimum total number of required shear studs is 2(8) = 16.

Let us verify that ϕM_{nx} = 245 ft-kips obtained by interpolation from the tables is the correct value at midspan for eight shear studs between each simple support and midspan.

$$C = [V_h = 8(23.6 \text{ kips/stud}) = 188.8 \text{ kips}]$$

$$P_y = A_s F_y = 9.12(36) = 328.32 \text{ kips}$$

$$\frac{P_y + C}{2} = \frac{328.32 + 188.8}{2} = 258.56 \text{ kips}$$

Location of PNA is in the flange at y_p from the top of the W section:

$$y_p = \frac{0.5(P_y + C) - C}{b_f F_y} = \frac{258.56 - 188.8}{5.525(36)} = 0.3507 \text{ in.}$$

$$\frac{y_p}{t_f} = \frac{0.3507}{0.440} = 0.797$$

$$d_1 = Y_2 = 4.65 \text{ in.}$$

$$d_2 = 0.5\, y_p = 0.5(0.3507) = 0.175 \text{ in.}$$

$$d_3 = 0.5d = 0.5(15.88) = 7.94 \text{ in.}$$

$$\phi M_{nx} = 0.85[C(d_1 + d_2) + A_s F_y (d_3 - d_2)]$$

$$= 0.85[188.8(4.65 + 0.175) + 3.28.32(7.94 - 0.175)]$$

$$= 2941.3 \text{ in.-kips} = 245.1 \text{ ft - kips}$$

This is the same as $\phi M_{nx} = 245$ ft-kips, which was obtained by interpolation from the table on LRFD p. 5-28.

Let us verify that $\phi M_{nx} = 219.2$ ft-kips obtained by interpolation from the tables is the correct value at the concentrated load point for five shear studs between each simple support and the adjacent concentrated load point:

$$C = [V_h = 5(23.6 \text{ kips/stud}) = 118.0 \text{ kips}]$$

$$\frac{P_y + C}{2} = \frac{328.32 + 118.0}{2} = 223.16 \text{ kips}$$

Location of PNA is in the web at y_p from the top of the W section:

Area of top flange plus the two fillets = 2.68 in.^{2}

$$y_p = t_f + \frac{0.5(P_y + C) - C - 2.68 F_y}{t_w F_y}$$

$$y_p = 0.440 + \frac{223.16 - 118.0 - 2.68(36)}{0.25(36)} = 1.404 \text{ in.}$$

$$d_1 = Y2 = 4.78 \text{ in.}$$

$$d_2 = \frac{2.68(36)(0.440/2) + 0.25(36)(1.404 - 0.440)^2/2}{2.68(36) + 0.25(1.404 - 0.440)(36)} = 0.242 \text{ in.}$$

$$d_3 = 0.5d = 0.5(15.88) = 7.94 \text{ in.}$$

$$\phi M_{nx} = 0.85[C(d_1 = d_2) + A_s F_y (d_3 - d_2)]$$

$$\phi M_{nx} = 0.85[118.0(4.78 + 0.242) + 328.32(7.94 - 0.242)]$$

$$\phi M_{nx} = 2652 \text{ in. - kips} = 221 \text{ ft - kips}$$

This is very nearly the same as $\phi M_{nx} = 219.2$ ft-kips, which was obtained by double interpolation from the table on LRFD p. 5-28.

Example 10.13

For the fully composite section in Figure 10.10, determine ϕM_{nx} in a positive moment region. The steel beams are the built-up W57 × 18 × 206 investigated in Example 5.9

for $F_y = 50$ ksi. Properties of the concrete flange are
Effective slab width,

$$b = 90 \text{ in.}$$

Solid slab thickness,

$$t = 5 \text{ in.}$$

$$f'_c = 3.5 \text{ ksi}$$

$$E_c = w^{1.5}\sqrt{f'_c} = 145^{1.5}\sqrt{3.5} = 3267 \text{ ksi}$$

Solution

From LRFD p. 4-184 and Example 5.9, W57 x 18 x 206, $F_y = 50$ ksi:

$$A_s = 60.5 \text{ in.}^2 \qquad I_s = I_x = 35600 \text{ in.}^4 \qquad d = 58 \text{ in.}$$

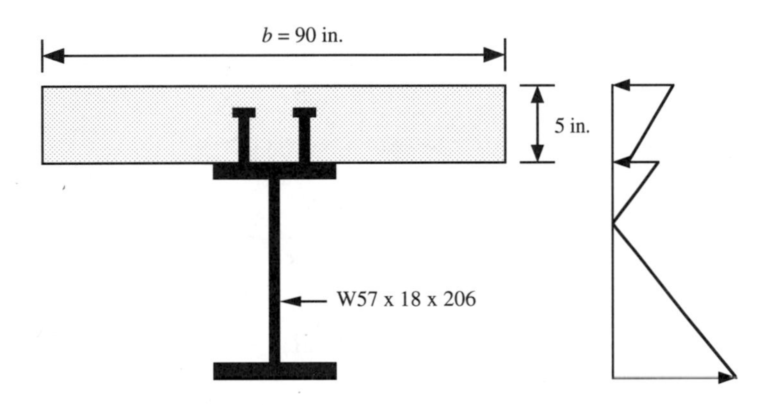

(a) An interior, fully composite section

(b) Actual stresses

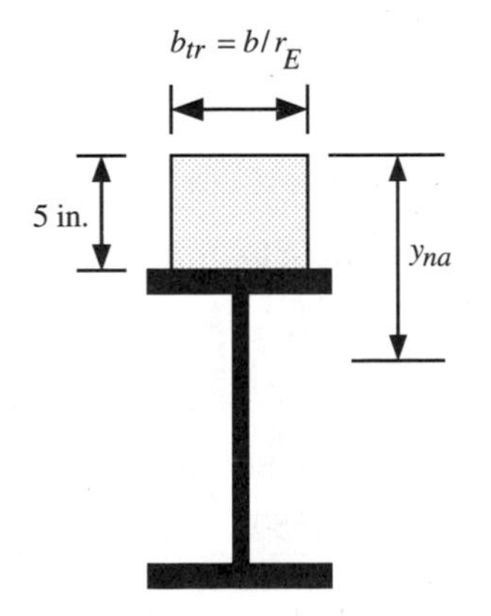

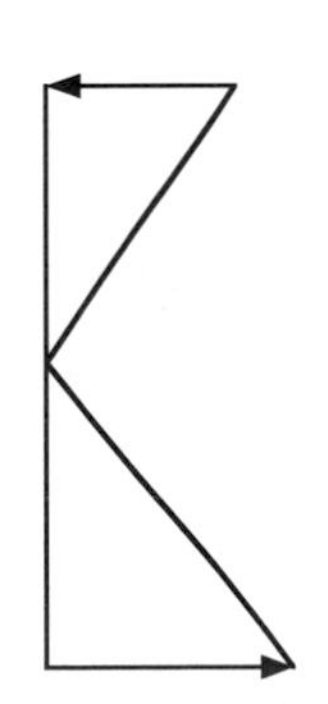

(c) Transformed composite section

(b) Stresses on transformed section

FIGURE 10.10 Example 10.13.

$$\left(h/t_w = 128\right) > \left(640/\sqrt{F_y} = 640/\sqrt{50} = 90.5\right)$$

$$\phi M'_{nx} = 2998 \text{ ft-kips}$$

which is valid for $L_b \leq \left(L'_p = 30.77 \text{ ft}\right)$ and the steel shape alone.

For the fully composite section, when $h/t_w > 640/\sqrt{F_y}$, $\phi = 0.9$ and M_{nx} is calculated from the elastic stress distribution on the composite section (see Figure 10.10). The concrete is transformed to steel and the flexural formula is used for computing the elastic stresses:

$$\rho_E = E/E_c = 29{,}000/3267 = 8.877$$

$$b_{tr} = b/\rho_E = 90/8.877 = 10.14 \text{ in.}$$

$$A_{tr} = b_{tr}t = 10.14(5) = 50.7 \text{ in.}^2$$

Locate the neutral axis:

$$y_{na} = \frac{A_{tr}t/2 + A_s\left(t + d/2\right)}{A_{tr} + A_s}$$

$$= \frac{50.7\left(2.5\right) + 60.5\left(5 + 29\right)}{50.7 + 60.5} = 19.6 \text{ in.}$$

The moment of inertia for the transformed section is obtained from the parallel axis theorem:

$$I_{tr} = \sum_{i=1}^{2}\left(I_{cg} + Ad^2\right)$$

$$I_{tr} = \frac{10.14\left(5\right)^3}{12} + 50.7\left(19.6 - 2.5\right)^2 + 35{,}600 + 60.5\left(34 - 19.6\right)^2$$

$$I_{tr} = 63{,}076 \text{ in.}^4$$

The section moduli for the extreme compression fiber and the extreme tension fiber, respectively, are

$$S_{xc} = I_{tr} \,/\, y_c = 63{,}076 \,/\, 19.6 = 3218 \text{ in.}^3$$

$$S_{xt} = I_{tr} \,/\, y_t = 63{,}076 \,/\, (5 + 58 - 19.6) = 1453 \text{ in.}^3$$

The stress in the extreme compressive fiber must be limited to $0.85f'_c$ and the stress in the extreme tension fiber must be limited to F_y:

$$0.85f'_c\, r_E S_{xc} = 0.85\ (3.5)\ (8.877)\ (3218) = 84{,}984 \text{ in.-kips}$$

$$M_{nx} = \text{smaller of} \begin{cases} 0.85f'_c\rho_E S_{xc} = 84{,}984 \text{ in.-kips} \\ S_{xt}F_y = 1453(50) = 72{,}650 \text{ in.-kips} \end{cases}$$

$$\phi M_{nx} = 0.9(72{,}650) = 65{,}385 \text{ in.-kips} = 5449 \text{ ft-kips}$$

Note: If partial composite beam action had been used for this example, the effects of slip should be accounted for (see LRFD p. 6-208) in computing the design bending strength.

10.4 CONCRETE-ENCASED BEAMS

Shear connectors are not required for this type of composite action. A steel beam totally encased in concrete cast with the slab [see Figure 10.2(b)] may be assumed to be bonded to the concrete, when:

1. Concrete cover below and on all edges of the steel shape is at least 2 in.
2. The top flange surface of the steel shape is at least 1.5 in. below the top of the concrete slab and at least 2 in. above the bottom of the concrete slab.
3. The concrete encasement is adequately reinforced with welded wire mesh or reinforcing bars to prevent spalling of the concrete.

Local buckling and lateral-torsional buckling do not have to be considered after the concrete encasement has hardened since the concrete encasement prevents the occurrence of these types of failure. Therefore, the design bending strength is ϕM_n, where $\phi = 0.90$ and M_n can be determined from either of the following methods:

1. The *plastic stress distribution* on the steel section alone ($M_{nx} = M_{px} = Z_x F_y$).
2. The *elastic stress distribution* on the composite section, considering the effects of shoring.

For method 2 and shored construction, bending stresses due to factored loads are based on the properties of the transformed composite section. Concrete on the tension side of the neutral axis is not included in the transformed section. For the concrete on the compression side of the neutral axis, each concrete width is divided by $\rho_E = E/E_c$, thereby transforming the concrete width to an equivalent width of steel. Bending stresses are computed by the flexural formula, $f = M_u y/I_x$, where f is the bending stress at a point of interest located at a distance y from the x axis and I_x is the moment of inertia of the transformed section. At the extreme fiber in either flange of the steel shape, $f \leq F_y$ is required. At the extreme compression fiber of the transformed section, $f = 0.85 f_c' \rho_E$ is required.

For method 2 and unshored construction, stresses are computed at two stages and superimposed:

1. Stage 1 (prior to hardening of the concrete)

   The steel section alone resists the loads until the concrete has hardened. The extreme fiber stresses in the steel section are

$$f = M_u/S_{xt} \qquad \text{and} \qquad f = M_u/S_{xc}$$

where

M_u = factored moment at section where stresses are to be superimposed

$$S_{xt} = I_x/y_t \qquad S_{xc} = I_x/y_c$$

I_x = moment of inertia for the steel section

y_t = distance from x-axis to the extreme tension fiber in the steel section

y_c = distance from x-axis to the extreme compression fiber in the steel section

At the extreme fiber in either flange of the steel shape, $f \leq F_y$ is required.

2. Stage 2 (after the concrete has hardened)

   Bending stresses due to factored loads applied after the concrete has hardened are based on the properties of the transformed composite section. Concrete on the tension side of the neutral axis is not included in the transformed section. For the concrete on the compression side of the neutral axis, each concrete width is divided by $\rho_E = E/E_c$, thereby transforming the concrete width to an equivalent width of steel. Bending stresses are computed by the flexural formula, $f = M_u y/I_x$, where f is the bending stress at a point of interest located at a distance y from the x-axis and I_x the moment of inertia of the transformed section.

For the stresses superimposed from stages 1 and 2:

1. At the extreme fiber in either flange of the steel shape, $f \leq F_y$ is required.
2. At the extreme compression fiber of the transformed section, $f = 0.85f'_c \rho_E$ is required.

Example 10.14

Determine ϕM_{nx} and ϕM_{ny} for the concrete-encased W14 x 68 (F_y = 36 ksi) shown on LRFD p. 5-78 when this composite section is used as a beam.

For the concrete,

$$f'_c = 3.5 \text{ ksi} \qquad b = 18 \text{ in.,} \qquad h = 20 \text{ in.}$$

Solution

According to LRFD I3.3 (p. 6-65), we can choose either of the two definitions given for the determination of the design bending strength about each principal axis. We choose to use the simpler definition, which is for the plastic stress distribution on the steel section alone. Local buckling and lateral-torsional buckling are prevented by the concrete encasement, when the concrete encasement satisfies the following minimum requirements:

1. Concrete cover below and on all edges of the steel shape is at least 2 in. On the edges,

   $$[\mathit{Cover} = (18 - b_f)/2 = (18 - 10.035)/2 = 3.98 \text{ in.}] \geq 2 \text{ in.}$$

   Beneath the bottom flange,

   $$[\mathit{Cover} = (22 - d)/2 = (22 - 14.04)/2 = 3.98 \text{ in.}] \geq 2 \text{ in.}$$

2. The top flange surface of the steel shape is at least 1.5 in. below the top of the concrete slab and at least 2 in. above the bottom of the concrete slab.

   The top flange surface is 3.98 in. below the top of the concrete. When the section being investigated is as shown in Figure 10.2(b) and when the concrete slab thickness $t \geq (2 + 3.98 = 5.98$ in.), this requirement is satisfied.

3. The concrete encasement is adequately reinforced with welded wire mesh or reinforcing bars to prevent spalling of the concrete.

From LRFD I2.1.b:

1. For the longitudinal steel:

$$\text{Minimum } A_s = 0.007(2)(11 - 0.375 - 1.128/2) = 0.141 \text{ in.}^2$$

$$[A_s = 4(1.00) = 4.00 \text{ in.}^2] \geq 0.141 \text{ in.}^2 \quad \text{as required}$$

2. For the stirrups,

$$\text{Minimum clear cover} = 1.5 \text{ in.}$$

$$\text{Minimum } A_s = 0.007(12) = 0.084 \text{ in.}^2$$

$$[A_s = 2(0.11) = 0.22 \text{ in.}^2] \geq 0.084 \text{ in.}^2 \quad \text{as required.}$$

Since the concrete encasement satisfies all minimum requirements, the design bending strengths are

$$\phi M_{nx} = 0.9Z_xF_y = 0.9(115)(36) = 3726 \text{ in.-kips} = 310.5 \text{ ft-kips}$$

$$\phi M_{ny} = 0.9Z_yF_y = 0.9(36.9)(36) = 1196 \text{ in.-kips} = 99.6 \text{ ft-kips}$$

10.5 DEFLECTIONS OF COMPOSITE BEAMS

As described on LRFD p. 5-9, deflections calculated on the basis of the lower bound moment of inertia I_{LB} may be satisfactory in checking serviceability requirements. For a concrete slab connected to the top flange of a steel shape and fully composite beam action, LRFD Figure 5-4 (p. 5-10) shows the definitions for the cross section to be used in calculating I_{LB} in a positive moment region. Formulas are given on LRFD p. 5-10 for the location of the elastic neutral axis Y_{ENA} and for I_{LB}. Values of I_{LB} are listed on LRFD pp. 5-50 to 5-65 for some plastic composite sections in a positive moment region. If there is a negative moment region, $I_{LB} = I_x$ of the steel shape. For partially composite beam action, LRFD Eq. (C-I3-6) on p. 6-208 is the recommended definition for the effective moment of inertia, which accounts for slip at the concrete–steel interface in a positive moment region.

For a composite beam consisting of a steel shape encased by concrete (see Figure 10.11), for simplicity we recommend the following approach for calculating a value of the moment of inertia in a positive moment region:

1. On the cracked section, transform the concrete in the compression region to steel. The transformed concrete width is

$$b_{tr} = b/\rho_E$$

where

$$E_c = w^{1.5}\sqrt{f_c'}$$

$$f_c' = \text{specified compressive concrete strength (ksi)}$$

$$w = \text{unit weight of concrete (lb/ft}^{3)}$$

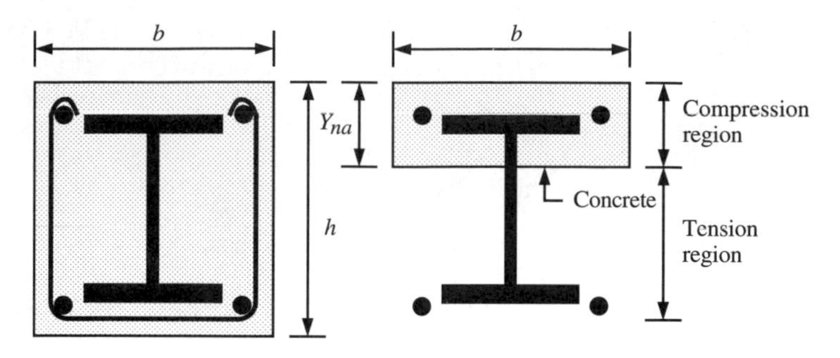

(a) Composite section

(b) Cracked cross section

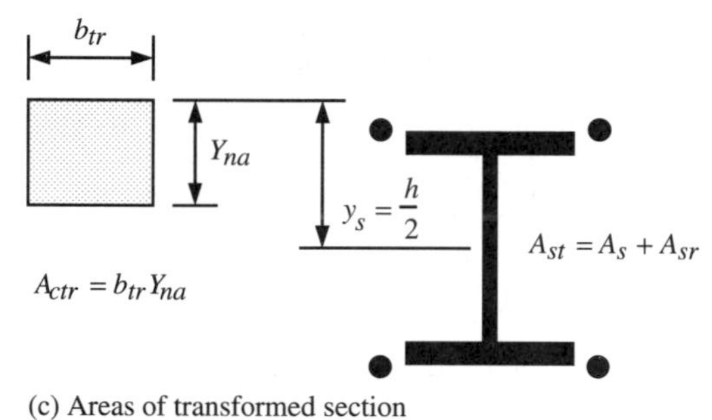

(c) Areas of transformed section

FIGURE 10.11 Assumptions for transformed moment of inertia.

2. Compute Y_{na}, which is the location of the neutral axis of the transformed section with respect to the top of the slab:

$$A_s = \text{area of the steel shape}$$

$$A_r = \text{area of the reinforcing steel}$$

$$A_{st} = A_s + A_r$$

$$A = A_{st} + b_{tr} Y_{na}$$

$$Y_{na} = \frac{A_{st} y_s + b_{tr} Y_{na}^2 / 2}{A}$$

$$y_s = \frac{h}{2}$$

$$Y_{na} = \frac{(b_{tr} - 2A_{st}) + \sqrt{(b_{tr} - 2A_{st})^2 + 8A_{st} y_s b_{tr}}}{2b_{tr}}$$

3. Calculate the moment of inertia of the transformed section:

$$I_{tr} = \frac{b_{tr} Y_{na}^3}{3} + I_s + A_{st}\left(y_s - Y_{na}\right)^2$$

where I_s = (I_x of the steel shape)

Example 10.15

Determine the moment of inertia for the concrete-encased W14 x 68 (F_y = 36 ksi) shown on LRFD p. 5-78 when this composite section is used as a simply supported beam.

Solution

$$E_c = w^{1.5}\sqrt{f_c'} = 145^{1.5}\sqrt{3.5} = 3267 \text{ ksi}$$

$$\rho_E = E/E_c = 29{,}000/3267 = 8.877$$

$$b_{tr} = b/\rho_E = 18/8.877 = 2.028 \text{ in.}$$

$$A_s = 20.0 \text{ in.}^2$$

$$A_r = 4(1.00) = 4.00 \text{ in.}^2$$

$$A_{st} = A_s + A_r = 20.0 + 4.00 = 24.0 \text{ in.}^2$$

$$y_s = \frac{h}{2} = \frac{22}{2} = 11 \text{ in.}$$

$$Y_{na} = \frac{(b_{tr} - 2A_{st}) + \sqrt{(b_{tr} - 2A_{st})^2 + 8A_{st}y_s b_{tr}}}{2b_{tr}}$$

$$(b_{tr} - 2A_{st}) = 2.028 - 2(24.0) = -45.97$$

$$Y_{na} = \frac{-45.97 + \sqrt{(-45.97)^2 + 8(24.0)(11.0)(2.028)}}{2 * 2.028} = 8.38 \text{ in.}$$

In the region where the section is cracked, the moment of inertia of the transformed section is

$$I_{tr} = \frac{b_{tr}Y_{na}^3}{3} + I_s + A_{st}(y_s - Y_{na})^2$$

$$= \frac{2.028(8.38)^3}{3} + 723 + 24.0(11.0 - 8.38)^2 = 1286 \text{ in.}^4$$

In the region where the section is not cracked, the moment of inertia of the transformed section is approximately obtained as follows:

$$A_{tr} = b_{tr}h - A_{st} = 2.028(22) - 24.0 = 20.62 \text{ in.}^2$$

$$\text{Equivalent } b_{tr} = A_{tr}/h = 20.62/22 = 0.937 \text{ in.}$$

$$\textit{Equivalent } I_g = \frac{(\text{Equivalent } b_{tr})h^3}{12} + I_s + A_r\left(\frac{d_r}{2}\right)^2$$

where

d_r = y-direction distance between the reinforcing steel bars

$$d_r = h - 2d_c$$

d_c = structural cover for the reinforcing steel bars

$$d_r = 22 - 2\,(1.5 + 0.375 + 1.128/2) = 17.12 \text{ in.}$$

$$\text{Equivalent } I_g = \frac{0.937(22)^3}{12} + 723 + 4.00\left(\frac{17.12}{2}\right)^2 = 1848 \text{ in.}^4$$

The *effective moment of inertia* is taken as the average of the inertia values in the uncracked and cracked regions. Therefore, in the deflection calculations use EI_{eff}, where

$$E = 29{,}000 \text{ ksi}$$

$$I_{eff} = (1848 + 1286)/2 = 1567 \text{ in.}^4$$

Example 10.16

Determine the moment of inertia for the fully composite section in Example 10.8 when this member is used as shown in Figure 10.12 for q_u = 2.95 kips/ft.

Solution

$$\text{Left reaction} = (30 - 19)(38)(2.95)/30 = 41.1 \text{ kips}$$

$$\text{Zero shear point occurs at } x = 41.1/2.95 = 13.93 \text{ ft}$$

$$\text{Maximum positive } M_u = 41.1(13.93)/2 = 286.3 \text{ ft-kips}$$

$$\text{Maximum negative } M_u = 2.95(8)^2/2 = 94.4 \text{ ft-kips}$$

Information applicable in the positive moment region is, from Example 10.8:

W16 x 31 PNA = TFL, $Y1 = 0$, $Y2 = 4.387$ in., $\phi M_{nx} = 286.7$ ft-kips

From LRFD p. 5-28, we find that I_{LB} = 1069 in.^{4}.

In the negative moment region, we can conservatively use

$$I_{LB} = I_x \text{ of the W16 x 31} = 375 \text{ in.}^4$$

$$(\phi M_{nx} = 128 \text{ ft-kips}) \geq (M_u = 94.4 \text{ ft-kips})$$

(ϕM_{nx} was obtained from the LRFD beam charts for W16 x 31, $C_b = 1$ and $L_b = 8$ ft.)

In the region between the supports, the average $I_{LB} = (1069 + 375)/2 = 722$ in.^{4} is appropriate. For the cantilevered region, I_{LB} = 375 in.^{4} is appropriate. In the negative moment region, we can utilize the reinforcing steel (no. 4 at 6 in. on center, A_r = 0.4 in.^{2}/ft) to obtain a larger value of I_{LB} and ϕM_{nx} when this steel is properly anchored and the number of shear studs provided is adequate. In this problem, partial composite action for a shear stud spacing approximately equal to the 40-in.

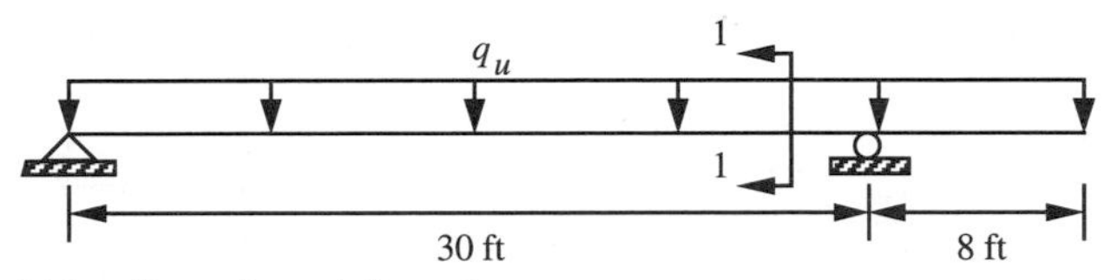

(a) Loading and span information

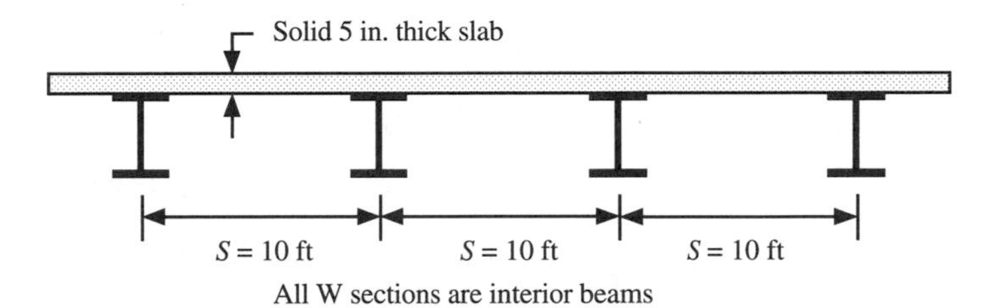

(b) Section 1-1

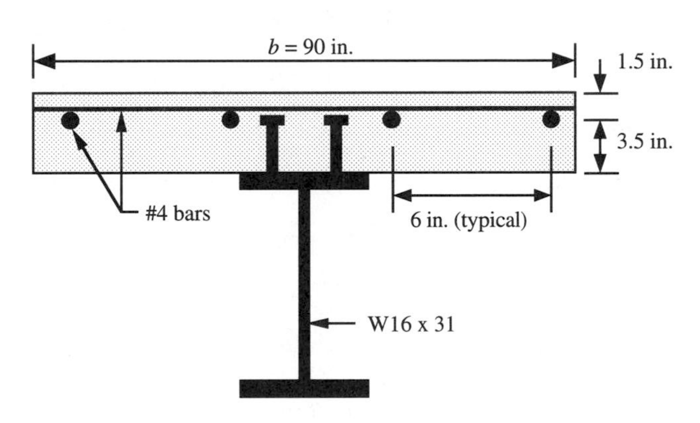

(c) An interior, fully composite member at Section 1-1

FIGURE 10.12 Example 10.16.

maximum stud spacing probably will be sufficient for deflection control purposes. LRFD Eq. (C-I3-6) (p. 6-208) can be used to compute an estimate of I_{LB} for the negative moment region when I_{LB} for the bare steel section is not adequate.

10.6 COMPOSITE BEAM-COLUMNS

Doubly and singly symmetric composite beam-columns must satisfy LRFD H1 (p. 6-59) with the following modifications:

1. In LRFD Eq. (H1-1a) and (H1-1b):

(a) ϕP_n is as defined for a composite column (see Section 10.2.2). Consequently, the seven limitations given in Section 10.2.1 must be satisfied.

(b) When $(P_u/\phi P_n) \geq 0.3$,

$$\phi_b M_n = 0.85 M_{nC}$$

where M_{nC} is the nominal bending strength determined from plastic stress distribution on the composite cross section.

An approximate formula for M_{nC} from LRFD Eq. (C-I4-1) on p. 6-213 is

$$M_{nC} = F_y Z + \tfrac{1}{3}(h_2 - 2c_r)A_r F_{yr} + \left(\frac{h_2}{2} - \frac{A_w F_y}{1.7 f_c' h_1}\right) A_w F_y$$

where

Z = plastic section modulus of the steel section (in.^{3})

F_y = yield strength of the steel shape (ksi)

A_w = web area of an encased steel shape (in.^{2})

$A_w = 0$ for a concrete-filled steel shape

A_r = total area of longitudinal reinforcing steel (in.^{2})

F_{yr} = yield strength of longitudinal reinforcing steel, (ksi)

$c_r = (c_{rc} + c_{rt})/2$

c_{rc} = distance from compression face to longitudinal reinforcing steel in that face (in.)

c_{rt} = distance from tension face to longitudinal reinforcing steel in that face (in.)

h_1 = width of composite cross section parallel to the plane of bending (in.)

h_2 = width of composite cross section perpendicular to the plane of bending (in.)

When $(P_u/\phi P_n) < 0.3$, $\phi_b M_n$ is linearly interpolated at $P_u/\phi P_n$ on the straight line joining point $C = [0.85M_{nC}, (P_u/\phi P_n) = 0.3]$ and point $B = [\phi_b M_{nB}, (P_u/\phi P_n) = 0]$, where $\phi_b M_{nB}$ is the design bending strength for a beam as defined in LRFD I3.2 for plastic stress distribution on the composite cross section. For convenience and conservatism, we may choose to use $\phi_b M_{nB} = 0.9ZF_y$.

If shear connectors are required for $P_u = 0$, they must be provided when $(P_u/\phi P_n) \le 0.3$.

2. In LRFD Eqn (H1-3) and Eqn (H1-6):

$$P_e = \frac{A_s F_{my}}{\lambda_c^2}$$

where A_s, F_{my}, and λ_c^2 are as defined in Section 10.2.2.

Example 10.17

On LRFD p. 5-78, we find a W14 × 68 (F_y = 36 ksi) encased by concrete (f_c' = 3.5 ksi) such that the composite column dimensions are b = 18 in. and h = 20 in. This section is to be used as a beam-column. Determine the design flexural strengths $\phi_b M_{nx}$ and $\phi_b M_{ny}$, needed in the interaction formulas.

Solution

When $(P_u/\phi P_n) \geq 0.3$,

$$\phi_b M_{nx} = 0.85 M_{nxC}$$

$$M_{nxC} = Z_x F_y + \frac{1}{3}(h - 2c_r) A_r F_{yr} + \left(\frac{h}{2} - \frac{A_w F_y}{1.7 f'_c b}\right) A_w F_y$$

$$c_r = 1.5 + 0.375 + 1.128/2 = 2.44 \text{ in.}$$

$$A_r = 4(1.00) = 4.00 \text{ in.}^2$$

$$F_{yr} = 60 \text{ ksi}$$

$$A_w = t_w d = 0.415(14.04) = 5.83 \text{ in.}^2$$

$$A_w F_y = 5.83(36) = 209.9 \text{ kips}$$

$$1.7 f'_c b = 1.7\,(3.5)\,(18) = 107.1 \text{ kips/in.}$$

$$M_{nxC} = 115(36) + \left[\frac{22 - 2(2.44)}{3}\right](4.00)(60) + \left(\frac{22}{2} - \frac{209.9}{107.1}\right)(209.9)$$

$$M_{nxC} = 7407 \text{ in.-kips}$$

$$\phi_b M_{nx} = 0.85 M_{nxC} = 0.85(7407) = 6296 \text{ in.-kips} = 525 \text{ ft-kips}$$

Note: On LRFD p. 5-78, the value listed for $\phi_b M_{nx} = 535$ ft-kips.

$$\phi_b M_{ny} = 0.85 M_{nyC}$$

$$M_{nyC} = Z_y F_y + \frac{1}{3}(b - 2c_r) A_r F_{yr} + \left(\frac{b}{2} - \frac{A_w F_y}{1.7 f'_c h}\right) A_w F_y$$

$$1.7 f'_c b = 1.7\,(3.5)\,(22) = 130.9 \text{ kips/in.}$$

$$M_{nyC} = 36.9(36) + \left[\frac{18 - 2(2.44)}{3}\right](4.00)(60) + \left(\frac{18}{2} - \frac{209.9}{130.9}\right)(209.9)$$

$$M_{nyC} = 3931 \text{ in. - kips}$$

$$\phi_b M_{ny} = 0.85 M_{nC} = 0.85(3931) = 3341 \text{ in.-kips} = 278 \text{ ft-kips}$$

Note: On LRFD p. 5-78, the value listed for $\phi_b M_{ny} = 279$ ft-kips.

When $(P_u/\phi P_n) < 0.3$, $\phi_b M_{nx}$ is obtained at $P_u/\phi P_n$ on the straight line joining point $C = [0.85 M_{nxC} = 525$ ft-kips, $(P_u/\phi P_n) = 0.3]$ and point $B = [\phi_b M_{nxB}, (P_u/\phi P_n) = 0]$, where $\phi_b M_{nxB} = 0.9 Z_x F_y = 0.9(115)(36) = 3726$ in.-kips $= 311$ ft-kips.

$\phi_b M_{ny}$ is obtained at $P_u/\phi P_n$ on the straight line joining point $C = [0.85 M_{nyC} = 278$ ft-kips, $(P_u/\phi P_n) = 0.3]$ and point $B = [\phi_b M_{nyB}, (P_u/\phi P_n) = 0]$, where $\phi_b M_{nyB} = 0.9 Z_y F_y = 0.9(36.9)(36) = 1196$ in.-kips $= 99.6$ ft-kips.

Note that the applicable value of $P_e = A_s F_{my} / \lambda_c^2$ in LRFD Eqs. (C1-2) (p.6-41) and (C1-5) (p. 6-42) can be obtained for this composite section from the values listed in lines 3 and 4 up from the bottom on LRFD p. 5-78. For example, if P_{ex} is desired for $(KL)_x = 30$ ft,

$$P_{ex} = 246\,(10^4) \,/\, (30)^2 = 2733 \text{ kips}$$

10.7 DESIGN EXAMPLES

LRFD pp. 5-12 to 5-17 contain design examples for a concrete flange compositely connected to steel beams. The tables in Part 5 of the LRFD Manual are used in these design examples.

LRFD pp. 5-68 to 5-71 contain design examples for a concrete-encased column, a concrete-filled column, and a concrete-encased beam-column. The tables in Part 5 of the LRFD Manual are used in these design examples.

PROBLEMS

10.1 On LRFD p. 5-136 for a structural steel tube ST14 × 10 × 0.375 filled with f_c' = 5 ksi concrete and $(KL)_y = 30$ ft, verify the $\phi P_n = 593$ kips entry.

10.2 For the composite column in Problem 10.1, determine the minimum acceptable area of a bearing plate located on the concrete at the top of the column.

10.3 On LRFD p. 5-89 for an $F_y = 50$ ksi, W14 x 120 encased by concrete and $(KL)_y$ = 30 ft, verify the $\phi P_n = 2070$ kips entry.

10.4 On LRFD p. 5-90 for an $F_y = 36$ ksi, W14 x 68 encased by concrete:

(a) Verify $r_{mx}/r_{my} = 1.22$.
(b) For $(KL)_x = 30$ ft and $(KL)_y = 15$ ft, determine ϕP_n.

10.5 A fully composite structural system consists of a solid 5-in.-thick concrete slab connected via shear studs to the top flange of steel beams:

Beams

W24 x 62 $\quad F_y = 36$ ksi

Simply supported span length

$L = 40$ ft

Transverse spacing

$S = 12$ ft

Concrete

$f_c' = 3.5$ ksi $\quad w = 145$ pcf $\quad E_c = w^{1.5}\sqrt{f_c'} = 145^{1.5}\sqrt{3.5} = 3267$ ksi

Shear studs

$d = 3/4$ in. $\quad H_s = 4d = 3$ in. $\quad F_u = 60$ ksi

For full composite action, determine

(a) Effective concrete slab width b for an interior steel beam.
(b) Horizontal shear force V_h, which must be transferred. Assume that the interior composite beam is subjected to a uniformly distributed factored load.
(c) Minimum required number of shear studs.

10.6 Solve Problem 10.5 with the following modification: The 5-in. total slab thickness consists of a 2-in. concrete topping on a 3-in. formed steel deck with the deck ribs spanning perpendicular to the steel beam. Shear stud height is $H_s = 4.5$ in. Deck rib height is $h_r = 3$ in. Average rib width is $w_r = (4.5 + 7.5)/2 = 6.00$ in. Thickness of concrete slab above the steel deck is $t_c = 2$ in.

10.7 For the fully composite section in Problem 10.5 and shored construction, determine ϕM_{nx}. The composite beam is subjected to a uniformly distributed load $q_u = 1.2$ kips/ft and a concentrated load Q_u at each third point of the span. Determine the maximum acceptable value of Q_u. Determine the minimum acceptable number of shear studs.

10.8 For the fully composite section in Problem 10.6 and shored construction, determine ϕM_{nx}. The composite beam is subjected to a uniformly distributed load $q_u = 1.2$ kips/ft and a concentrated load Q_u at each third point of the span. Determine the maximum acceptable value of Q_u. Determine the minimum acceptable number of shear studs.

10.9 Use the solution information from Examples 10.4 and 10.14 in the interaction equation for $P_u = 544$ kips, $C_{mx} = 0.85$, and $B_2 = 0$. Determine the maximum acceptable value of M_{ux}.

10.10 Use the solution information from Example 10.4 and Example 10.14 in the interaction equation for $P_u = 163$ kips, $C_{mx} = 0.85$, and $B_2 = 0$. Determine the maximum acceptable value of M_{ux}.

10.11 Simply supported beams spanning 30 ft and spaced at 7.5 ft on centers are to be made composite with a solid concrete slab whose properties are thickness = 5 in., weight = 145 pcf, and $f'_c = 3.5$ ksi. Shored construction is to be used. Select the $F_y = 36$ ksi steel beam required to support a service live load of 200 psf and a service dead load of 68 psf (does not include an estimate for the beam weight). Also determine the number of 0.75 -in.-diameter shear studs required and the service live-load deflection.

10.12 Solve Problem 10.11 for unshored construction.

10.13 Solve Problem 10.11 with the concrete slab being a 3-in. topping on a 3-in.-deep steel deck, with an average rib width of 6 in. The ribs are oriented perpendicular to the beam centerline. Lightweight concrete ($w = 115$ pcf) is to be used. Change the service dead load to 62 psf.

10.14 Solve Problem 10.13 for unshored construction.

Plastic Analysis and Design

11.1 INTRODUCTION

This chapter introduces the concepts and methods of plastic analysis for continuous beams and one-story frames in which the member ends are rigidly interconnected at the joints. A book written by Beedle [28] in 1958 was used as a textbook for plastic analysis and design for nearly 30 years. Beginning with Kazinczy of Hungary in 1914, Beedle gives an extensive list of references on the development of plastic analysis and design.

The concepts of plastic analysis and design are dependent on the great ductility of steel and the plastic plateau characteristic shown on the stress–strain curves for F_y = 36 and 50 ksi in Figure 1.2 (see Section 1.1.3). Members used in plastic analysis are primarily those W sections for which the plastic moment M_{px} is reached and maintained until bending deformations well into the plastic range occur before plastic buckling of the compression elements occurs (see Section 5.4).

In a continuous beam subjected to a uniformly distributed load, for example, yielding first occurs at the support where the maximum bending moment occurs. As the load is increased, the relative distribution of moment changes along the continuous member, and yielding at the first yield location continues to occur until the plastic moment is reached. If the load is further increased, plastic bending deformations increase without any increase in moment at the first yield location (a *plastic hinge* has formed), and the plastic moment is reached at other locations. When the plastic moment is reached at a sufficient number of locations, plastic bending deformations increase at each of these locations without any increase in load (a *plastic collapse mechanism* has formed).

At each plastic hinge location, except at the last plastic hinge to form, the compression flange must be laterally braced. For a W section subjected to bending about the major axis, except at the last plastic hinge to form, adjacent to each plastic hinge the laterally unbraced length L_b of the compression flange must not exceed

$$L_{pd} = \left(3600 + 2200\frac{M_1}{M_2}\right)\frac{r_y}{F_y}$$

where

F_y = specified minimum yield strength of the compression flange (ksi)

M_1 = smaller moment at the ends of an unbraced length (in.-kips)

M_2 = larger moment at the ends of an unbraced length (in.-kips)

r_y = radius of gyration about minor bending axis (in.)

(M_1/M_2) is positive when moments cause reverse curvature

For each L_b region with one end located at the last plastic hinge to form and in L_b regions not adjacent to a plastic hinge, $L_b > L_{pd}$ is permissible. For these L_b regions, the flexural design strength ϕM_{nx} must be determined as for beams analyzed by elastic methods (see Chapters 5, 6, and 9).

When design by plastic analysis is used, the following LRFD Specification requirements must be satisfied:

1. LRFD A5.1 (p. 6-31)—The steel must exhibit a plastic plateau on the stress–strain curve; consequently, $F_y \leq 65$ ksi must be used.
2. LRFD B5.2 (p. 6-36)—Compression elements in the section must have a width–thickness ratio $\leq \lambda_p$ [see LRFD Table B5.1 (p. 6-32)].
3. LRFD C2 (p. 6-42)—The axial force in a column caused by factored gravity plus factored horizontal loads shall not exceed $0.85\phi_c A_g F_y$ in a braced frame nor $0.75\phi_c A_g F_y$ in an unbraced frame.
4. LRFD E1.2 (p. 6-47)—The column design strength must be governed by inelastic column buckling; that is, the requirement is that

$$\left(\frac{\lambda_c}{K} = \frac{L/r}{\pi}\sqrt{\frac{F_y}{E}}\right) \leq 1.5$$

5. LRFD F1.2d (p. 6-55)—The compression flange must be laterally braced at each plastic hinge location, except at the last plastic hinge to form, and such that $L_b \leq L_{pd}$, where

L_b = laterally unbraced length of the compression flange

L_{pd} = limiting value of L_b for plastic design

For an I-shaped member with the compression flange larger than the tension flange and bending about the major axis,

$$L_{pd} = \left(3600 + 2200\frac{M_1}{M_2}\right)\frac{r_y}{F_y}$$

For solid rectangular bars and symmetric box beams bending about their major axis,

$$\left[L_{pd} = \left(5000 + 2200\frac{M_1}{M_2}\right)\frac{r_y}{F_y}\right] \geq 3000\frac{r_y}{F_y}$$

There is no limit on L_b for members with circular or square cross sections or for any member bending about its minor axis.

In the L_b region of the last plastic hinge to form, and regions not adjacent to a plastic hinge, ϕM_{nx} must be determined from LFRD F1.2 to F1.4.

6. LRFD H1.2 (p. 6-60)—M_u shall be determined in accordance with LRFD C1 which stipulates that second-order ($P\Delta$) effects shall be considered in the design of frames and the requirements stated in item 3 above must be satisfied.
7. LRFD I1 (p. 6-61)—ϕM_{nx} of composite members shall be determined from the plastic stress distributions specified in LRFD I3.

The following discussion pertains to item 6 in the preceding list of requirements. In our plastic analyses of one-story frames, we use a first-order plastic analysis since that has been the traditional approach. LRFD C1 requires second-order ($P\Delta$) effects to be considered in the design of frames, but does not give an approximate procedure to determine M_u based on a first-order plastic analysis. For one-story frames, we recommend the following approach to account for second-order effects.

Case 1

For a member subjected to axial compression and bending in a PCM for which sidesway does not occur

$$M_u = B_1 M_{pcm}$$

where

M_{pcm} = maximum moment in each L_b region from the plastic collapse mechanism

$$\left(B_1 = \frac{C_m}{1 - P_u / P_{e1}}\right) \geq 1$$

$C_m = 0.85$ for a member subjected to transverse loads

Otherwise, $C_m = 0.6 - 0.4M_1/M_2$

M_1/M_2 = ratio of smaller to larger moments at the ends of an L_b region

M_1/M_2 is negative when the member is bent in single curvature in an L_b region

Otherwise, M_1/M_2 is positive

P_u = axial compression force obtained from the plastic collapse mechanism

$$P_{e1} = \pi^2 EI/(KL)^2$$

I and KL are for the axis of bending

$K \leq 1$ for braced frame members

B_1 accounts for the local second-order $P\delta$ effect. There is not any global second-order $P\Delta$ effect when sidesway does not occur.

Case 2

For a member subjected to axial compression and bending in a PCM for which sidesway occurs,

$$M_u = B_2 M_{pcm}$$

where

M_{pcm} = maximum moment in each L_b region from the plastic collapse mechanism

$$B_2 = \frac{1}{1 - \Sigma P_u / \Sigma P_{e2}}$$

ΣP_u = required axial strength of all columns in a story

$\Sigma P_{e2} = \Sigma\, [\pi^2 EI/(KL)^2]$ of all columns in a story

I and KL are for the axis of bending

$K > 1$ for unbraced frame members

B_2 accounts for the $P\Delta$ effect when sidesway occurs. Since we cannot separately determine the portion of M_{pcm} due only to the lateral loads and the portion of M_{pcm} due only to the gravity loads, M_{pcm} due to all loads is amplified by B_2. This approach overcompensates for the $P\Delta$ effect, but **Case 2** usually does not govern for one-story frames. For multistory frames, we recommend that M_u be obtained from a second-order plastic analysis that directly accounts for the $P\Delta$ effect.

11.2 PLASTIC HINGE

In Figure 11.1, the simply supported beam is subjected to a concentrated load at midspan. For convenience in the following discussion, assume that the member weight is negligible. For strong-axis bending, the load W_{ux} causes the plastic moment M_{px} to occur at the maximum moment location. When the load W_{ux} has been applied, the beam has no more resistance to bending at the maximum moment point until strain-hardening starts to occur (see Figure 5.4). Accounting for strain-hardening is not easy or necessary. It is conservative to ignore the extra bending strength due to strain-hardening. Thus, in plastic analysis, the plastic moment is accepted as being the maximum possible moment. Also, for a simple plastic theory, the moment–curvature relationship (see Figure 5.4) is idealized as linearly elastic up to M_p and plastic thereafter. Using this idealized moment–curvature relationship is permissible for a plastic analysis to determine the required bending strength since the presence of residual stresses does not affect the plastic moment (see Figure 5.6).

A *plastic hinge* is a zone of yielding due to flexure in a structural member. The plastic hinge length L_{ph} is the beam region length in which the moment exceeds the yield moment. As shown in the following discussion, L_{ph} is dependent on the geometry of the cross section and the loading configuration. In Figure 11.1, the moment diagram due to a single concentrated load is linear. Since

$$M_{px} = F_y Z_x$$

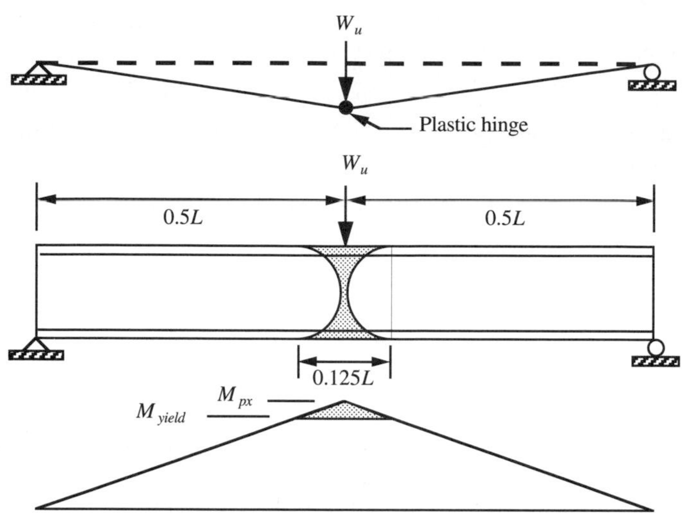

(a) X-axis bending of a W section beam

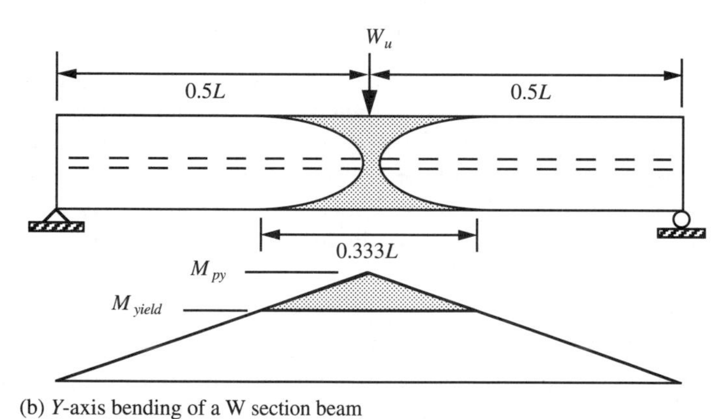

(b) Y-axis bending of a W section beam

FIGURE 11.1 Plastic hinge.

$$M_{yield\ (x)} = F_y S_x$$

then

$$M_{px}/M_{yield\ (x)} = Z_x / S_x$$

where

$$Z_x = \text{plastic section modulus}$$

$$S_x = \text{elastic section modulus}$$

For W sections used as beams, the shape factor Z_x / S_x ranges from 1.10 to 1.18 and the average is 1.14. For some W sections used as columns, Z_x / S_x ranges as high as 1.23. From similar triangles on the moment diagram in Figure 11.1,

$$\frac{M_{px}}{L} = \frac{M_{px} - F_y S_x}{L_{ph}}$$

$$L_{ph} = \left(\frac{M_{px} - F_y S_x}{M_{px}} \right) L = \left(1 - \frac{S_x}{Z_x} \right) L$$

For $Z_x/S_x = 1.14$,

$$L_{ph} = 0.1228L \approx L/8$$

Similarly, for weak-axis bending,

$$L_{ph} = \left(1 - \frac{S_y}{Z_y} \right) L \approx \left(1 - \frac{1}{1.5} \right) L = \frac{L}{3}$$

For a uniformly distributed load,

$$L_{ph} = 0.35L \text{ for strong-axis bending}$$

$$L_{ph} = 0.577L \text{ for weak-axis bending}$$

For a simple plastic theory, the moment–curvature relationship is idealized as linearly elastic up to M_p and plastic thereafter, all of the *plastic rotation* is assumed to occur at the plastic hinges, and the length of each plastic hinge is assumed to be zero. This means that the idealized load–deflection curve for the member and loading shown in Figure 11.1 is linearly elastic up to W_u and is plastic thereafter. At a plastic hinge location, the member behaves as though it were hinged with a constant restraining moment M_p. For the member and loading shown in Figure 11.1, a mechanism (geometrically unstable structure) consisting of the real hinges at the member ends and the plastic hinge at midspan forms when W_u causes M_p to occur at midspan. Since the real hinges cannot resist any moment and the plastic hinge cannot resist any additional moment after M_p occurs, the structure does not have any bending strength to resist any load after W_u has been applied. Therefore, for the member and loading shown in Figure 11.1, W_u is the plastic collapse load.

11.3 PLASTIC COLLAPSE MECHANISM

In a continuous beam, the maximum moment points occur at the supports and at a point of zero shear between the supports. In regard to the strength concepts of plastic analysis, the behavior of an interior span of a continuous beam is the same as for a fixed-ended beam. Thus, for graphical convenience, we choose to discuss the behavior of a fixed-ended beam (see Figure 11.2). For the idealized moment–curvature relationship (linearly elastic up to M_p and plastic thereafter), plastic hinges form at each support when the applied load value is w_{ph}, which is obtained from $w_{ph}L^2/12 = M_{px}$. As shown in Figure 11.3, when the applied load is increased from w_{ph} to the plastic collapse load w_u, the member behavior is elastic between the supports, and the member-end moments remain constant at their maximum value of M_{px}. We see that the first plastic hinge forms at the supports (maximum moment point as defined by elastic behavior). Then, redistribution of moment occurs until a plastic hinge forms at the zero shear point between the supports when the applied load value is w_u. Each beam segment between plastic hinges is able to move without any increase in load. A system of such beam segments is called a *plastic collapse mechanism*, which is an unstable structure until strain hardening occurs at one or more of the plastic hinge locations. In Figure 11.2(b), the deflected shape of the PCM (*plastic*

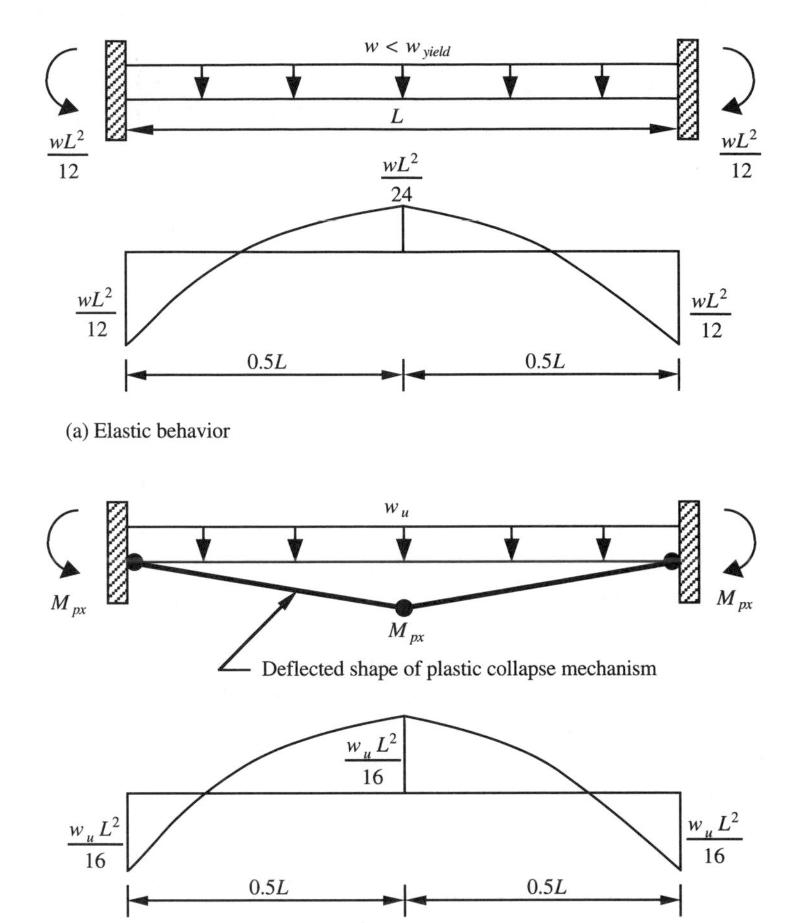

FIGURE 11.2 Moment redistribution.

collapse mechanism) is shown as straight beam segments between the plastic hinges. When the PCM forms, each beam segment is bent due to M_{px}, but the further behavior (motion) of the PCM does not involve any additional bending of any beam segment, and the PCM behavior is identical to a lin-kipsage of straight bars connecting real hinges. This fact is very useful in performing a plastic analysis by a method that involves the motion of the PCM due to an imposed virtual displacement.

The types of *independent mechanisms* are shown in Figure 11.4 and discussed here:

1. *Beam Mechanism* This mechanism can form in any span of a continuous beam and, as shown in Figure 11.4, in any member of a frame. For a member subjected to a uniformly distributed load, there is only one possible beam mechanism and only three possible plastic hinge locations (only two possible plastic hinge locations if one member end is a real hinge). For a beam subjected to concentrated loads at n locations between the member ends,

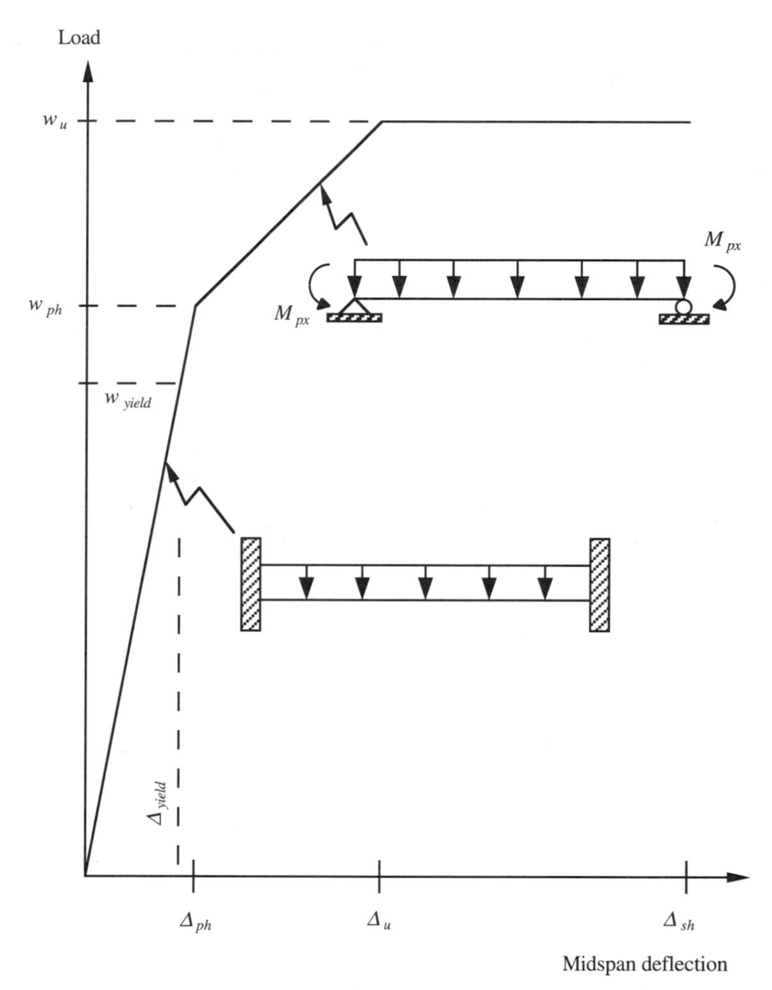

FIGURE 11.3 Idealized load-deflection curve for beam in Figure 11.2.

there are n possible beam mechanisms and $n + 2$ possible plastic hinge locations (only $n + 1$ possible plastic hinge locations if one member end is a real hinge).

2. *Panel Mechanism* This mechanism can occur due to lateral loads.
3. *Joint Mechanism* This mechanism can occur at the junction of three or more members.
4. *Gable Mechanism* This mechanism is characteristic of gabled frames and involves spreading of the column tops with respect to the column bases. Various combinations of the independent mechanisms can be made to form a *composite mechanism*. Examples of some composite mechanisms are shown in Figure 11.5.

For a structure that is indeterminate to the nth degree, let the number of

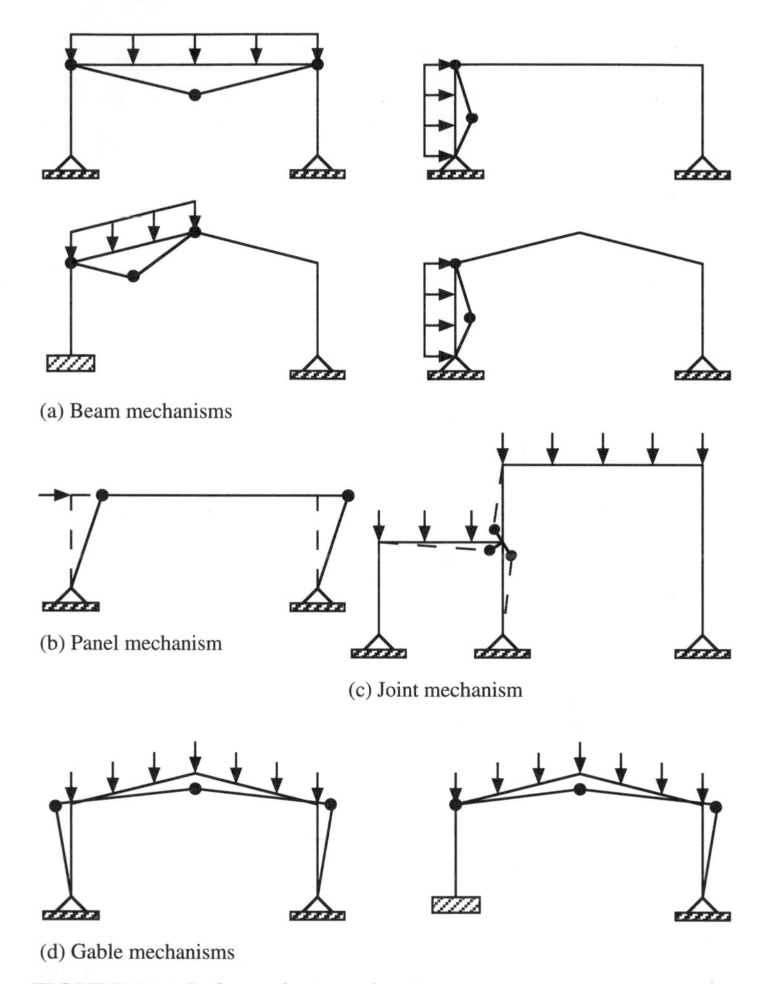

FIGURE 11.4 Independent mechanisms.

redundants be denoted by $NR = n$. The number of independent mechanisms NIM for such a structure is

$$NIM = NPPH - NR$$

where $NPPH$ = number of possible plastic hinge locations.

11.4 EQUILIBRIUM METHOD OF ANALYSIS

The *equilibrium method* of plastic analysis is useful for solving beam mechanisms in continuous beams and frames. This method is also useful for solving a plane frame for which there is only one redundant. In this method the objective is to find an equilibrium moment diagram in which $M_u \leq M_{up}$ and $M_u = M_{up}$ at a sufficient number of locations to produce a PCM. In the preceding notation, M_u is a moment diagram value due to factored loads and M_{up} is the required plastic hinge strength. The steps in the analysis procedure are:

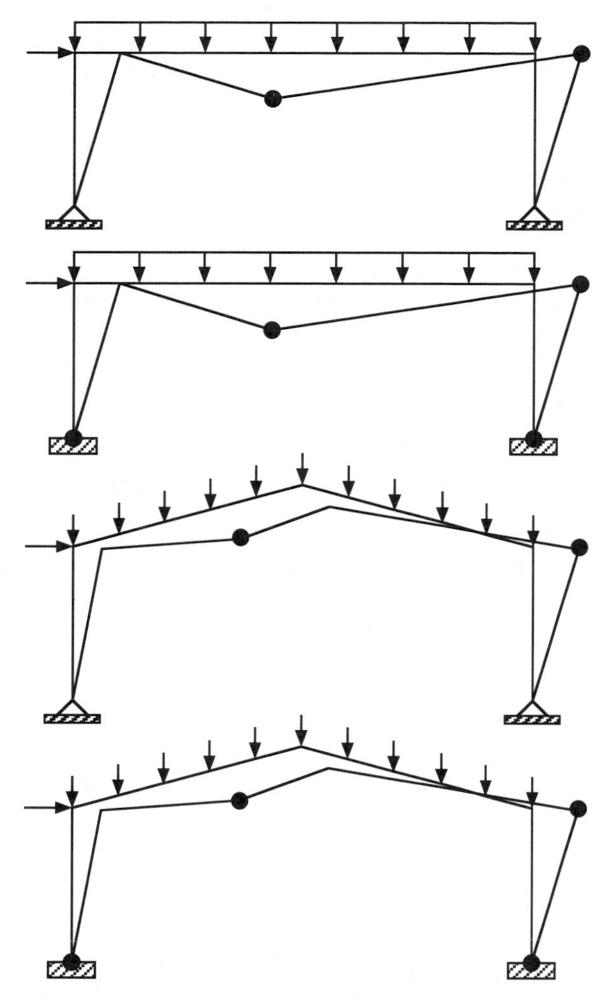

FIGURE 11.5 Composite mechanisms.

1. Select the redundant(s). For a continuous beam, the moments at the supports are chosen as the redundants.
2. Draw the moment diagram for the factored loads.
3. Draw the moment diagram for the redundant(s).
4. Assume that a plastic hinge forms at a sufficient number of locations to produce a PCM.
5. Solve the moment equilibrium equation for M_{up}.
6. Check to see if $M_u \leq M_{up}$.

Our purpose in the analysis is to determine M_{up} for the governing plastic collapse mechanism. A value of M_{up} obtained by the equilibrium method is a lower bound value. Therefore, the governing value of M_{up} is the maximum of the required strength values.

A beam mechanism is simpler than a composite mechanism for a frame. Therefore, we begin our discussion for the equibilibrium method and continuous beams.

Example 11.1

Use $F_y = 36$ ksi and the equilibrium method of plastic analysis. In Figure 11.6 the same W section is to be used for both spans. Select the lightest acceptable W section assuming lateral braces can only be provided as stated in Figure 11.6 for the following conditions:

Dead load

$$W_2 = 7.10 \text{ kips} \qquad W_3 = 11.9 \text{ kips} \qquad w = 0.533 \text{ kips/ft}$$

(W_2, W_3, and w contain an estimate accounting for the member weight.)

Live load

$$W_2 = 22.8 \text{ kips} \qquad W_3 = 34.2 \text{ kips} \qquad w = 2.10 \text{ kips/ft}$$

The governing LRFD loading combination is $1.2D + 1.6L$.

Solution

The factored loads are

$$W_{u2} = 1.2(7.10) + 1.6(22.8) = 45.0 \text{ kips}$$

$$W_{u3} = 1.2(11.9) + 1.6(34.2) = 69.0 \text{ kips}$$

$$w_u = 1.2(0.533) + 1.6(2.10) = 4.00 \text{ kips/ft}$$

In Figure 11.7(a), each real hinge location and each possible plastic hinge location is numbered for convenience in the following discussion. The number of independent mechanisms *NIM* are

$$NIM = NPPH - NR = 5 - 2 = 3$$

The three independent mechanisms are shown in Figure 11.7(b–d).

The redundants are chosen as the support moments at points 4 and 6. For span 1, there is only one redundant M_4, which is the moment at point 4. As shown in Figure 11.8(a), M_4 causes tension in the top fiber of the section at the support and this is a

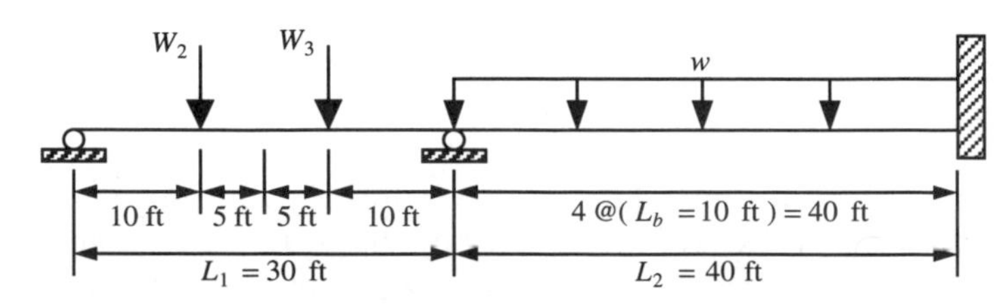

FIGURE 11.6 Example 11.1: structure and loading.

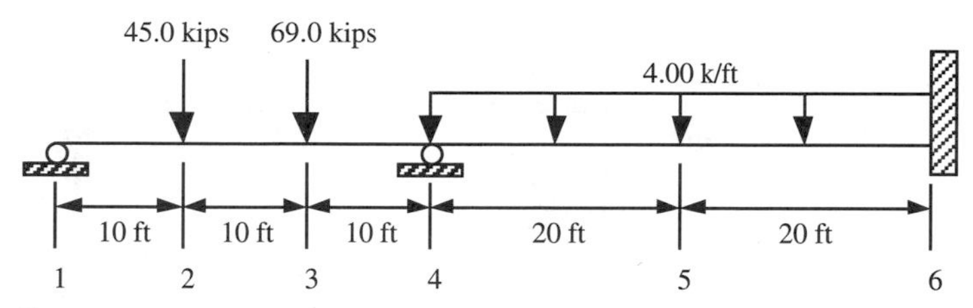

Comments:
1. Point 1 is a real hinge location.
2. Points 2 to 6 are possible plastic hinge locations.

(a) Governing factored loading: $1.2D + 1.6L$

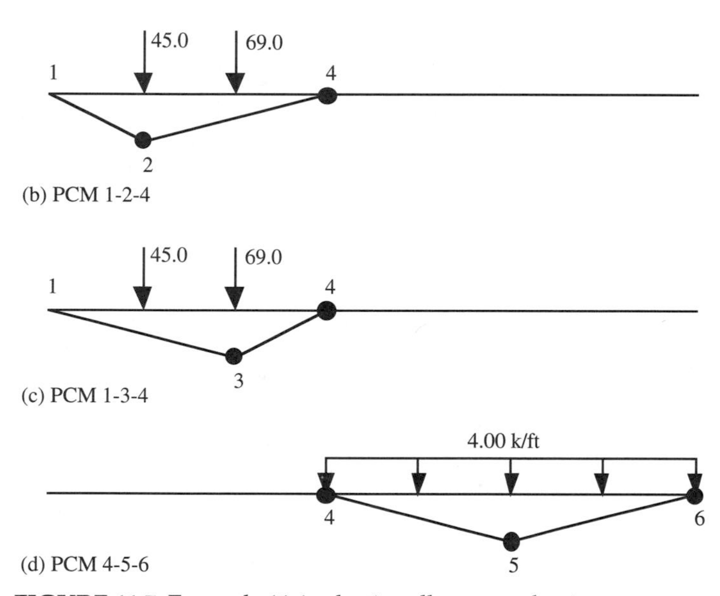

FIGURE 11.7 Example 11.1: plastic collapse mechanisms.

negative moment. For convenience, we let the moment vector show the correct direction and M_4 is the magnitude of this vector. Hence, M_4 is an absolute value.

As shown in Figure 11.8(b), the moment diagram is drawn by parts for the factored loads and the redundant applied on a simply supported beam. The factored loads cause a positive moment and the redundant causes a negative moment.

For PCM 1-2-4, the equilibrium requirements at the possible plastic hinge locations are:

1. $-M_4 = -M_{up}$
where M_{up} = required plastic hinge strength.

2. $\left(M_2 = 530 - \frac{1}{3}M_4\right) = M_{up}$

Since $M_4 = M_{up}$, we obtain

$$\frac{4}{3}M_{up} = 530 \text{ ft-kips}$$

$$M_{up} = 397.5 \text{ ft-kips}$$

45.0 kips 69.0 kips

10 ft 10 ft 10 ft M_4

2 3

$V_1 = 53 - \frac{M_4}{30}$ $V_4 = 61 + \frac{M_4}{30}$

Note:
M_4 is the magnitude of the indicated moment vector.

(a) Loading and member end forces

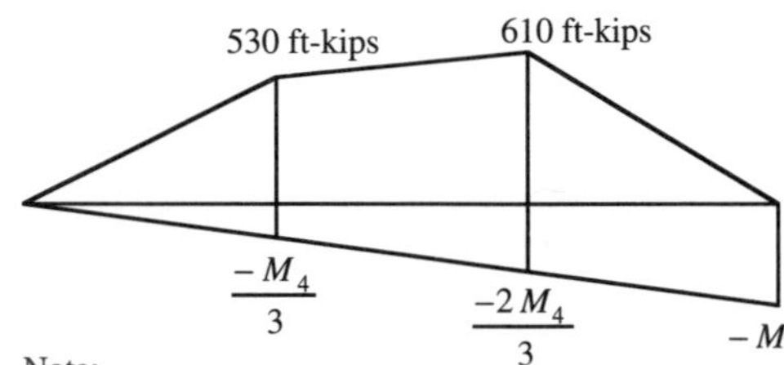

Note:
M_4 is an absolute value.

(b) Factored moment diagram

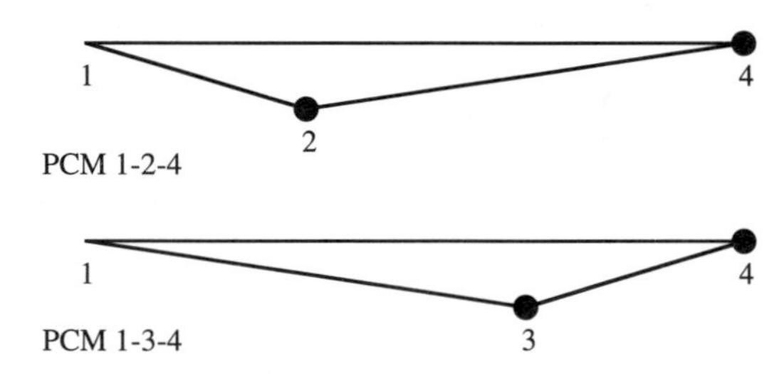

(c) Possible plastic collapse mechanisms

FIGURE 11.8 Example 11.1: span 1

3. $M_3 = 610 - \frac{2}{3}M_4 = 610 - \frac{2}{3}(397.5) = 345$ ft-kips

   Since

$$(M_3 = 345 \text{ ft-kips}) \leq (M_{up} = 397.5 \text{ ft-kips})$$

   no moment ordinate exceeds $M_{up} = 397.5$ ft-kips and PCM 1-2-4 governs for span 1.

We could have started the solution for span 1 by assuming that PCM 1-3-4 governs and for illustration purposes this is done. For PCM 1-3-4, the equilibrium requirements at the possible plastic hinge locations are:

1. $-M_4 = -M_{up}$
2. $\left(M_3 = 610 - \frac{2}{3}M_4\right) = M_{up}$

Since $M_4 = M_{up}$, we obtain

$$\frac{5}{3} M_{up} = 610 \text{ ft-kips}$$

$$M_{up} = 366 \text{ ft-kips}$$

3. $M_2 = 530 - \frac{1}{3} M_4 = 530 - \frac{1}{3}(366) = 408$ ft-kips

Since

$$(M_2 = 408 \text{ ft-kips}) > (M_{up} = 366 \text{ ft-kips})$$

the governing PCM must contain a plastic hinge at point 2.

M_{up} values obtained by the equilibrium method are lower bound values. Therefore, the required plastic hinge strength in span 1 cannot be less than the maximum computed value, which is $M_{up} = 397.5$ ft-kips. Also we know that for the correct value of M_{up}, all ordinates on the M diagram are less than or equal to M_{up}. Both these important facts were illustrated in the above calculations.

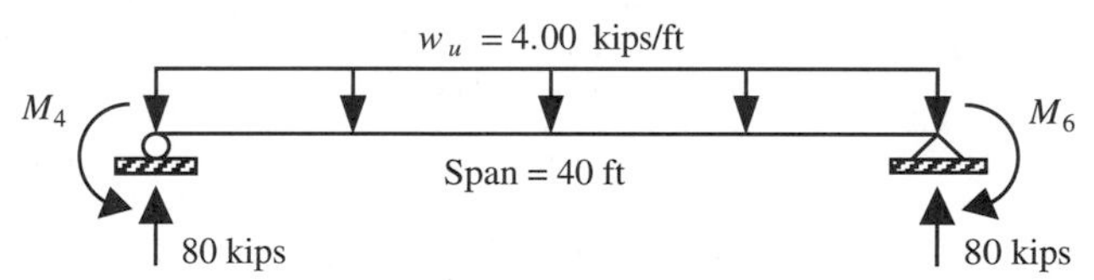

Note:
$M_4 = M_6 =$ is the magnitude of the indicated moment vector.

(a) Loading and member end forces

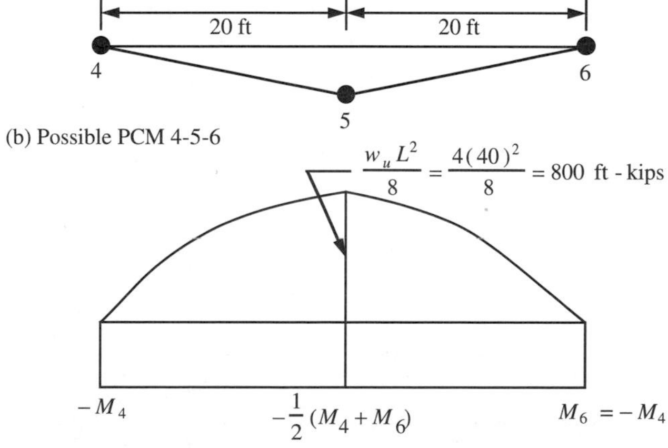

(b) Possible PCM 4-5-6

(c) Factored moment diagram

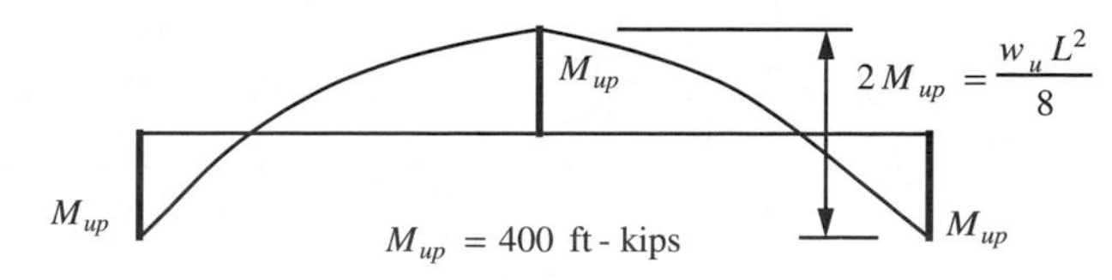

(d) Composite factored moment diagram

FIGURE 11.9 Example 11.1: span 2

As shown in Figure 11.9(a), for span 2 there is only one possible PCM, and there are two redundants (M_4 and M_6). However, since the structure and loading for this span are symmetric with respect to midspan, $M_4 = M_6$, and as illustrated in Figure 11.9(d), we can easily find M_{up} = 400 ft-kips.

Since the same W section is to be used in both spans, span 2 governs and M_{up} = 400 ft-kips. Therefore, the M_u diagram is shown in Figure 11.10. At each ordinate on the M_u diagram, the LRFD design requirement for strong-axis bending is

$$\phi M_{nx} \geq M_u$$

where ϕM_{nx} is the design bending strength for strong-axis bending. Therefore, at the plastic hinge locations, we must require

$$\phi M_{px} \geq M_{up}$$

where

$$\phi M_{px} = 0.9F_y Z_x = \text{design plastic hinge strength}$$

$$M_{up} = \text{required plastic hinge strength}$$

Using LRFD p. 4-18 for F_y = 36 ksi, we find that

$$\text{W24 x 62} \qquad (\phi M_{px} = 413 \text{ ft-kips}) \geq (M_{up} = 400 \text{ ft-kips})$$

is the lightest W section that satisfies the design bending requirement at the plastic hinge locations, when the lateral bracing requirements are satisfied.

Using the information given in Figure 11.10, check the lateral bracing requirements. Note that Figure 11.10 shows the M_u diagram associated with the governing plastic collapse mechanism. This is the moment diagram that should be used in checking the lateral bracing requirements. Therefore, in the following L_{pd} formula, we have shown the definitions as they apply to the M_u diagram associated with the governing plastic collapse mechanism. On the left side of the plastic hinge at the center support, we must require that

$$L_{pd} \geq (L_b = 10 \text{ ft})$$

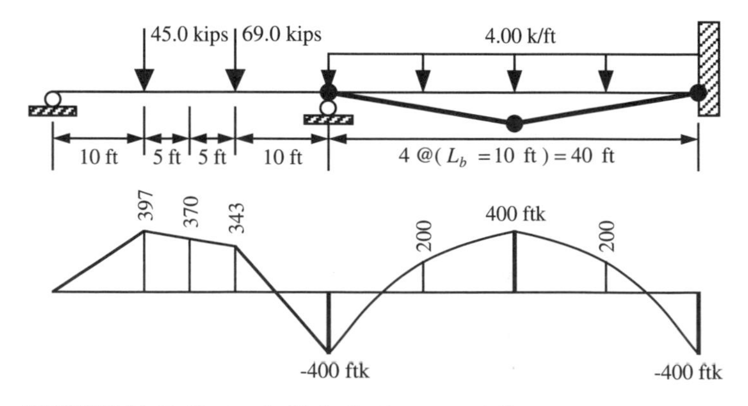

FIGURE 11.10 Example 11.1: final moment diagram.

where

$$\left[L_{pd} = \left(3600 + 2200\frac{M_1}{M_2} \right)\frac{r_y}{F_y} \right] \equiv \left(3600 + 2200\frac{M_1}{M_{up}} \right)\frac{r_y}{F_y}$$

F_y = specified minimum yield strength of the compression flange (ksi)

M_1 = smaller moment at the ends of the L_b region (ft-kips)

M_2 = larger moment at the ends of the L_b region (ft-kips)

r_y = radius of gyration about weak axis (ft.)

(M_1/M_2) is positive when moments cause reverse curvature

M_{up} = required plastic hinge strength (ft-kips)

Therefore,

$$F_y = 36 \text{ ksi}$$

$$M_1 = 343 \text{ ft-kips}$$

$$M_{up} = 400 \text{ ft-kips}$$

$$r_y = 1.38 \text{ in.} = 0.115 \text{ ft}$$

$$M_1/M_{up} = 343/400 = 0.8575$$

$$L_{pd} = [3600 + 2200(0.8575)]\frac{(0.115)}{(36)} = 17.53 \text{ ft}$$

$$(L_{pd} = 17.53 \text{ ft}) \geq (L_b = 10 \text{ ft}) \quad \text{as required}$$

On the right side of the plastic hinge at the center support, we must require that

$$L_{pd} \geq (L_b = 10 \text{ ft})$$

Since $M_1/M_{up} = 200/400 = 0.5$,

$$L_{pd} = [3600 + 2200(0.5)]\frac{(0.115)}{(36)} = 15.0 \text{ ft}$$

$$(L_{pd} = 15.0 \text{ ft}) \geq (L_b = 10 \text{ ft}) \quad \text{as required}$$

On the left side of the plastic hinge at the right support, we must require that

$$L_{pd} \geq (L_b = 10 \text{ ft})$$

Since $M_1/M_{up} = 200/400 = 0.5$,

$$L_{pd} = [3600 + 2200(0.5)]\frac{(0.115)}{(36)} = 15.0 \text{ ft}$$

$$(L_{pd} = 15.0 \text{ ft}) \geq (L_b = 10 \text{ ft}) \quad \text{as required}$$

For the other L_b regions, the design bending requirement is

$$\phi M_{nx} \geq M_u$$

where

$$\phi M_{nx} = \text{smaller of} \begin{cases} \phi M_{px} \\ C_b \overline{M}_1 \end{cases}$$

$$\overline{M}_1 = (\phi M_{nx} \text{ for } C_b = 1)$$

$$C_b = \frac{12.5 M_{\max}}{2.5 M_{\max} + 3 M_A + 4 M_B + 3 M_C}$$

where

M_{max} = absolute value of maximum M in the L_b region
M_A = absolute value of M at quarter point in the L_b region
M_B = absolute value of M at middle point in the L_b region
M_C = absolute value of M at three-quarter point in the L_b region

For the L_b = 10 ft region at the left support,

$$C_b = \frac{12.5(397)}{2.5(397) + 3(99.25) + 4(198.5) + 3(297.75)} = 1.67$$

and from LRFD p. 4-128 we find that

$$\left[C_b \overline{M}_1 = 1.67(355) = 592 \text{ ft-kips}\right] > \left(\phi M_{px} = 413 \text{ ft-kips}\right)$$

Therefore, (ϕM_{nx} = 413 ft-kips) ≥ (M_u = 397 ft-kips) as required.

For the L_b = 5 ft region at the 45.0 kip load,

$$(L_b = 5 \text{ ft}) \leq (L_p = 5.8 \text{ ft})$$

Therefore, (ϕM_{nx} = 413 ft-kips) ≥ (M_u = 397 ft-kips) as required.

For each L_b = 10 ft region at the last plastic hinge to form (at midspan of span 2),

$$C_b = \frac{12.5(400)}{2.5(400) + 3(287.5) + 4(350) + 3(387.5)} = 1.13$$

From LRFD p. 4-128, we find that

$$\left[C_b \overline{M}_1 = 1.13(355) = 401.2 \text{ ft-kips}\right] < \left(\phi M_{px} = 413 \text{ ft-kips}\right)$$

Therefore, (ϕM_{nx} = 401 ft-kips) ≥ (M_u = 400 ft-kips) as required.

All design bending requirements are satisfied for a W24 × 62, which is the lightest acceptable W section for this example problem.

Example 11.2

The objective of this example is to perform an equilibrium method of plastic analysis to determine the loading that causes the PCM to form in the structure of Example 11.1 when a W24 x 62 section is used. Therefore, the loads in Figure 11.11 are unknown

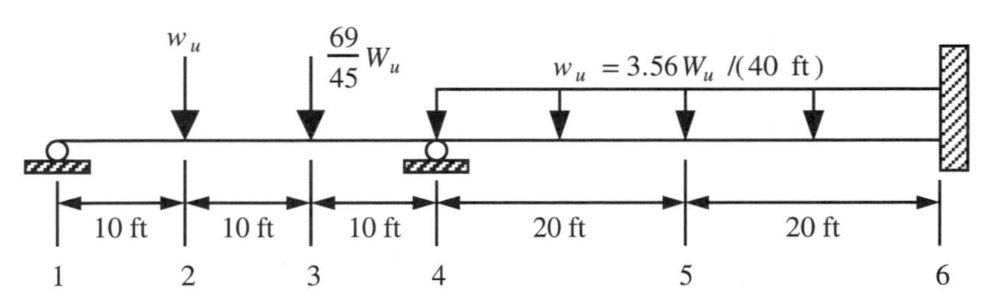

FIGURE 11.11 Example 11.2: structure and loading.

and the moment at each possible plastic hinge location is ϕM_{px} = 413 ft-kips. We assume that the loads shown in Figure 11.7 increase proportionately to cause the PCM for the indicated W section. Consequently, the 45.0-kip load is replaced by W_u, the 69.0-kip load is replaced by $(69/45)W_u$, and the uniform load is replaced by (4 kips/ft)(40 ft)/(45 kips)W_u /(40 ft) = 0.0889 kips/ft.

Solution

For span 1, the loads in Figure 11.12 are unknown and the moment at each possible plastic hinge location is ϕM_{px} = 413 ft-kips. Therefore, M_4 = 413 ft-kips. When a plastic hinge forms at point 2, the moment equilibrium requirement is

$$(M_2 = 11.78W_u - 138) = 413$$

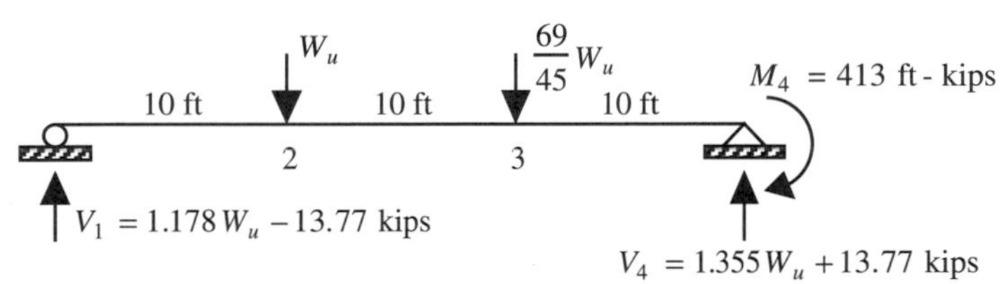

(a) Loading and member end forces

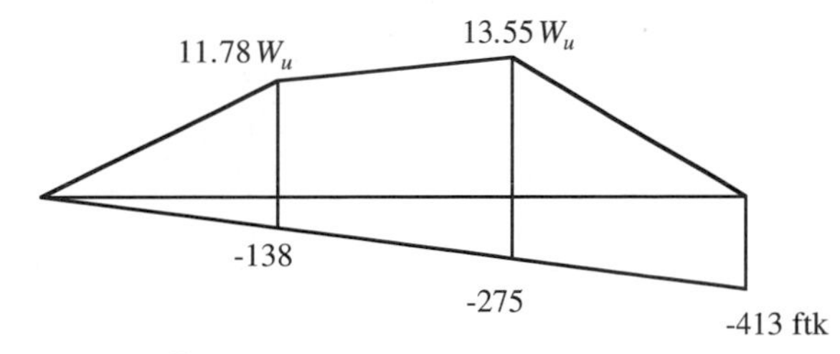

(b) Moment diagram

FIGURE 11.12 Example 11.2: span 1.

which gives

$$W_u = (413 + 138)/11.78 = 46.8 \text{ kips}$$

When a plastic hinge forms at point 3, the moment equilibrium requirement is

$$(M_3 = 13.55W_u - 275) = 413$$

which gives

$$W_u = (413 + 275)/13.55 = 50.8 \text{ kips}$$

Plastic collapse loads obtained by the equilibrium method are lower bound values. Therefore, the loads that produce plastic collapse in span 1 cannot be less than the least computed value of W_u and $(69/45)W_u$. The governing PCM for span 1 has a plastic hinge at point 2 and $W_u = 46.8$ kips and $(69/45)W_u = 71.8$ kips are required to produce this PCM. If the PCM in span 2 forms first, the equilibrium method requires

$$2\left(\phi M_{px}\right) = \frac{w_u L^2}{8}$$

which gives

$$(w_u = 0.0889W_u) = 8(2)(413)/\ (40)^2 = 4.13 \text{ k/ft}$$

$$W_u = 46.5 \text{ kips}$$

Therefore, the governing PCM for the entire structure is the PCM in span 2. The loads that produce this PCM and the corresponding moment diagram are shown in Figure 11.13.

Example 11.3

Modify Example 11.1, for the reasons stated, to obtain the conditions shown in Figure 11.14:

1. Change w_u from 4.00 kips/ft to 3.60 kips/ft to cause M_{up} to be different for each span.

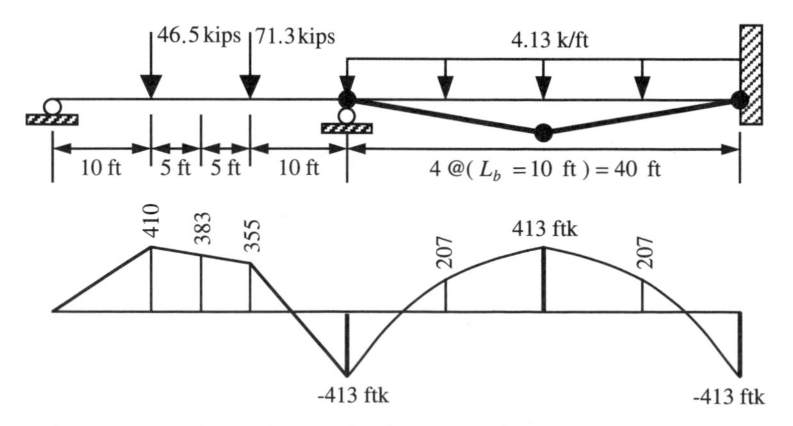

FIGURE 11.13 Example 11.2: final moment diagram.

2. Choose $L_b = 10$ ft for all regions to illustrate that the L_{pd} requirement may not dictate the choice of the section.

In Figure 11.14, a different W section is to be used in each span. Lateral braces can only be provided at intervals of $L_b = 10$ ft. Use $F_y = 36$ ksi and the equilibrium method of plastic analysis. Select the lightest acceptable W section for the span in which the governing PCM occurs. Use the actual ϕM_{px} of the selected section in the determination of M_{up} for the other span.

Solution

In Figure 11.15, the information for span 1 is the same (except for the L_b values) as in Example 11.1 from which we find that $M_{up} = 397.5$ ft-kips. For span 2,

$$M_{up} = \frac{w_u L^2}{16} = \frac{3.60(40)^2}{16} = 360 \text{ ft-kips}$$

Select the lightest available W section for span 1. $M_{up} = 397.5$ ft-kips and $L_b = 10$ ft. Using LRFD p. 4-18 for $F_y = 36$ ksi, we find that

$$\text{W}24 \times 62, \qquad (\phi M_{px} = 413 \text{ ft-kips}) \geq (M_{up} = 397.5 \text{ ft-kips})$$

is the lightest W section that satisfies the design bending requirement at the plastic hinge locations in span 1 when the lateral bracing requirements are satisfied.

At the right support of span 1,

$$M_1 = 345 \text{ ft-kips}$$
$$M_{up} = 397.5 \text{ ft-kips}$$
$$r_y = 1.38 \text{ in.} = 0.115 \text{ ft}$$
$$M_1/M_{up} = 345/397.5 = 0.8679$$

$$L_{pd} = [3600 + 2200(0.115)]\frac{(0.115)}{(36)} = 17.6 \text{ ft}$$

$$(L_{pd} = 17.6 \text{ ft}) \geq (L_b = 10 \text{ ft}) \quad \text{as required}$$

Between the two concentrated loads, $L_b = 10$ ft:

$$C_b = \frac{12.5(397)}{2.5(397) + 3(358) + 4(371) + 3(384)} = 1.0553$$

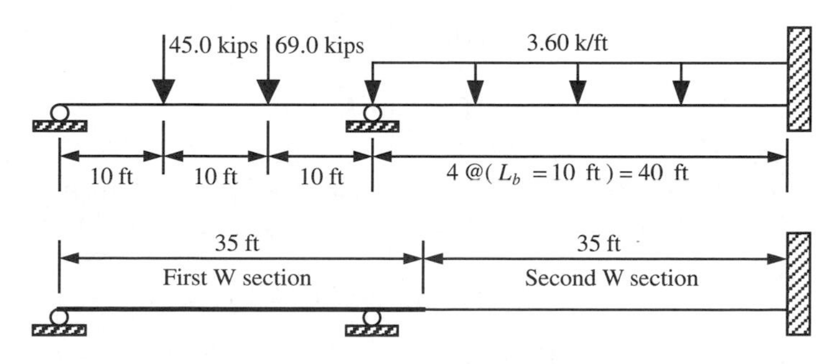

FIGURE 11.14 Example 11.3: structure and loading.

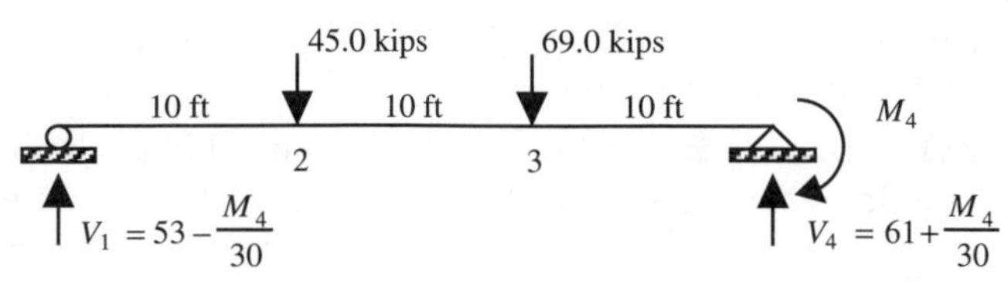

(a) Factored loading and member end forces

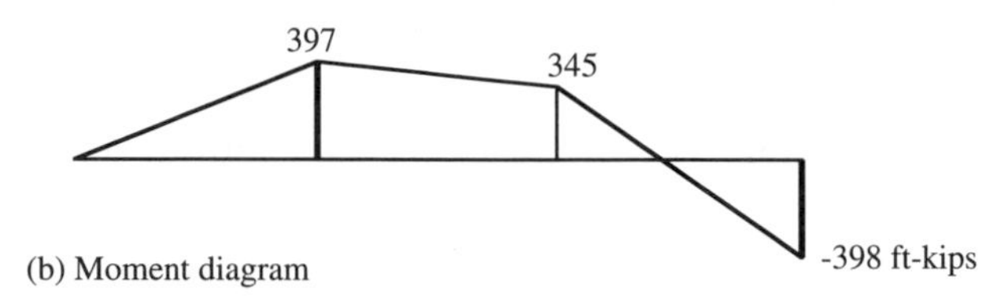

(b) Moment diagram

FIGURE 11.15 Example 11.3: span 1.

From LRFD p. 4-128 we find that

$$\left[C_b\overline{M}_1 = 1.0553(355) = 374.6\right] < \left(\phi M_{px} = 413 \text{ ft-kips}\right)$$

Therefore, (ϕM_{nx} = 375 ft-kips) < (M_u = 397 ft-kips), which is not satisfactory. Enter LRFD p. 4-128 with L_b = 10 ft and M_u/C_b = 397.5/1.0553 = 377 ft-kips. We find there are two sections of the same weight that may be satisfactory:

1. W21 × 68 (ϕM_{px} = 432 ft-kips) ≥ (M_{up} = 398 ft-kips) and r_y = 1.80 in.:

$$\left[C_b\overline{M}_1 = 1.0553(406) = 428\right] < \left(\phi M_{px} = 432 \text{ ft-kips}\right)$$

$$(\phi M_{nx} = 430 \text{ ft-kips}) \geq (M_u = 397 \text{ ft-kips}) \quad \text{as required}$$

$$L_{pd} = \left[3600 + 2200(0.8679)\right]\frac{(1.80/12)}{(36)} = 22.96 \text{ ft} > (L_b = 10 \text{ ft}) \quad \text{as required}$$

W21 × 68 with L_b = 10 ft is satisfactory.

2. W24 × 68, (ϕM_{px} = 478 ft-kips) ≥ (M_{up} = 398 ft-kips) and r_y = 1.87 in.:

$$[C_bM_1 = 1.0553(450) = 475] < (\phi M_{px} = 478 \text{ ft-kips})$$

$$(\phi M_{nx} = 475 \text{ ft-kips}) \geq (M_u = 397 \text{ ft-kips}) \quad \text{as required}$$

$$L_{pd} = \left[3600 + 2200(0.8679)\right]\frac{(1.87/12)}{(36)} = 23.85 \text{ ft} > (L_b = 10 \text{ ft}) \quad \text{as required}$$

W24 x 68 with L_b = 10 ft is satisfactory.

Suppose we choose to use the W24 x 68 section in span 1 since, as shown in Figure 11.16, this gives us the maximum benefit for span 2. Assume that a plastic hinge forms at x = 20 ft from the left support. The moment equilibrium requirement is

$$720 - \frac{1}{2}\left(478 + M_{up}\right) = M_{up}$$

which gives M_{up} = 321 ft-kips.

In Figure 11.16, using $M_6 = M_{up}$ = 321 ft-kips and $\Sigma M = 0$ at point 6, we find that the reaction at the left support is 75.9 kips. The zero shear point occurs at $x = 75.9/3.60 = 21.08$ ft from the left support, and the maximum positive moment is 323 ft-kips. Therefore, $321 < M_{up} < 323$ and we can conservatively use $M_{up} = 323$ ft-kips. The final moment diagram is shown in Figure 11.17.

For span 2, try

$$\text{W24 x 55} \qquad (\phi M_{px} = 362 \text{ ft-kips}) \geq (M_{up} = 323 \text{ ft-kips})$$

At the right support,

$$M_1 = 180 \text{ ft-kips}$$
$$M_{up} = 323 \text{ ft-kips}$$
$$r_y = 1.34 \text{ in.} = 0.112 \text{ ft}$$
$$M_1/M_{up} = 180/323 = 0.5573$$
$$L_{pd} = [3600 + 2200(0.5573)]\frac{(0.112)}{(36)} = 15.0 \text{ ft}$$
$$(L_{pd} = 15.0 \text{ ft}) \geq (L_b = 10 \text{ ft}) \quad \text{as required}$$

For the $L_b = 10$ ft region on the right side of midspan:

$$C_b = \frac{12.5(323)}{2.5(323) + 3(314) + 4(287) + 3(243)} = 1.113$$

From LRFD p. 4-128, we find that

$$\left[C_b \overline{M}_1 = 1.113(305) = 340 \text{ ft-kips}\right] < \left(\phi M_{px} = 362 \text{ ft-kips}\right)$$
$$(\phi M_{nx} = 340 \text{ ft-kips}) \geq (M_u = 323 \text{ ft-kips}) \quad \text{as required}$$

W24 × 55 with $L_b = 10$ ft is satisfactory for span 2.

For illustration purposes in span 2, suppose that lateral bracing can only be provided such that $L_b = 20$ ft. As in the preceding solution, we might try

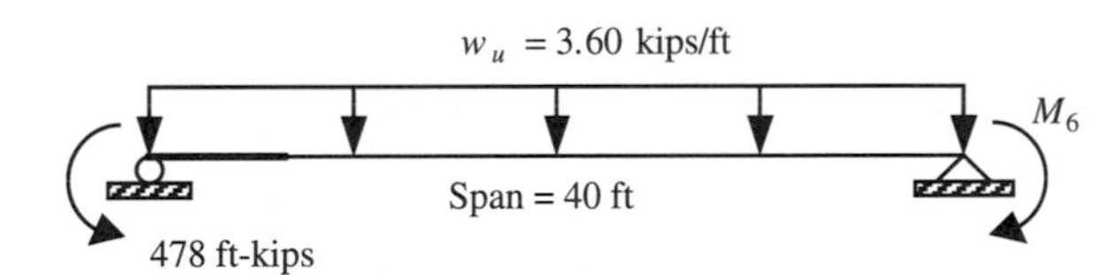

(a) Loading and member end moments using span 1 plastic design strength

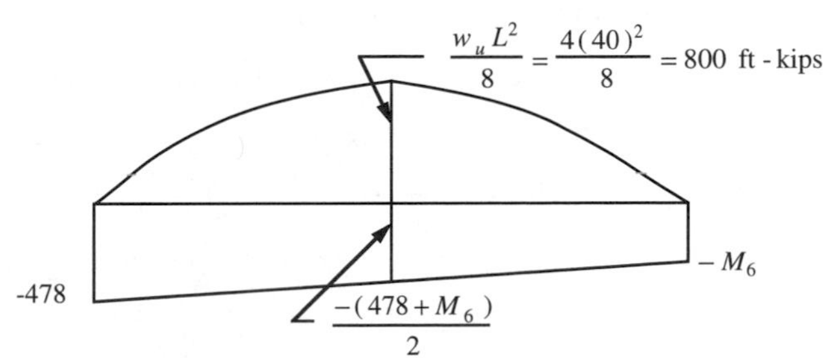

(b) Moment diagram

FIGURE 11.16 Example 11.3: span 2.

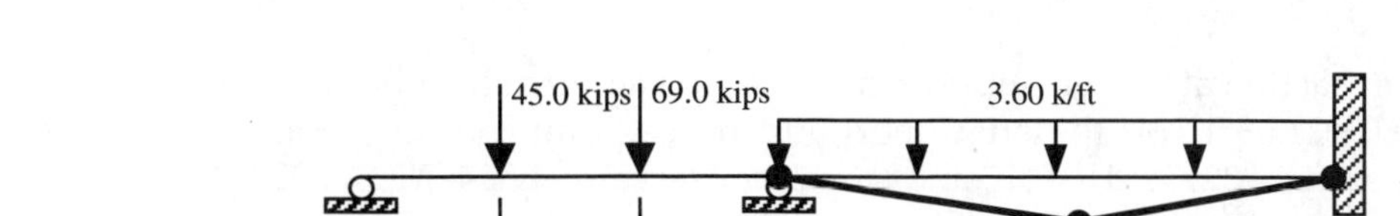

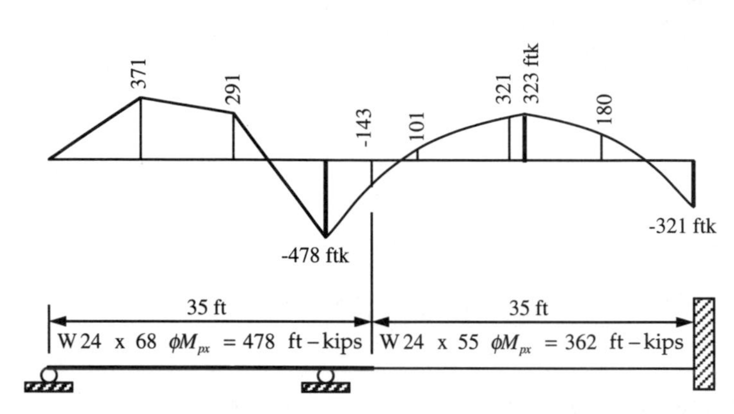

FIGURE 11.17 Example 11.3: final moment diagram.

$$\text{W24 x 55,} \qquad (\phi M_{px} = 362 \text{ ft-kips}) \geq (M_{up} = 323 \text{ ft-kips})$$

At the right support, from the previous solution we know that

$$(L_{pd} = 15.0 \text{ ft}) < (L_b = 20 \text{ ft})$$

This section is not satisfactory. To obtain a satisfactory section, we need:

1. $\phi M_{px} \geq (M_{up} = 323 \text{ ft-kips})$
2. $L_{pd} = [3600 + 2200(0.5573)](r_y/12)/(36) \geq 20.0 \text{ ft}$

which requires $r_y \geq 1.79$ in.

For the W24 series, the lightest choice is

$$\text{W24 x 68} \qquad (r_y = 1.87 \text{ in.}) \geq 1.79 \text{ in.}$$

$$(\phi M_{px} = 478 \text{ ft-kips}) \geq (M_{up} = 323 \text{ ft-kips})$$

For the W21 series, the lightest choice is

$$\text{W21 x 68,} \qquad (r_y = 1.80 \text{ in. }) \geq 1.79 \text{ in.}$$

$$(\phi M_{px} = 432 \text{ ft-kips}) \geq (M_{up} = 323 \text{ ft-kips})$$

For the W18 series, the lightest choice is

$$\text{W18 x 76,} \qquad (r_y = 2.61 \text{ in. }) \geq 1.79 \text{ in.}$$

$$(\phi M_{px} = 440 \text{ ft-kips}) \geq (M_{up} = 323 \text{ ft-kips})$$

Therefore, when $L_b = 20$ ft in span 2, the lightest acceptable choice is either W24 x 68 or W21 × 68.

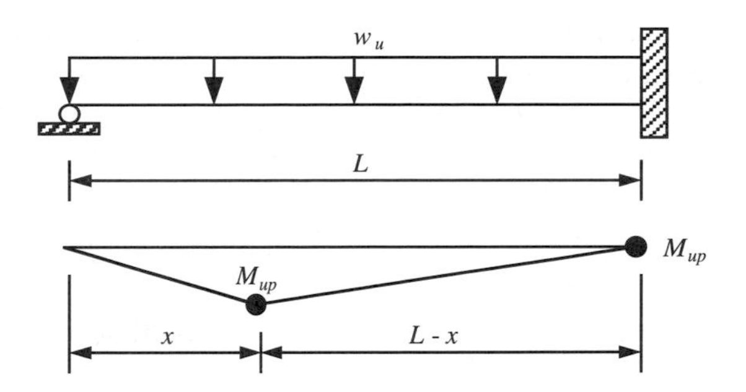

FIGURE 11.18 Example 11.4: structure and loading.

Example 11.4

Use $F_y = 36$ ksi and the equilibrium method of plastic analysis. For the structure and loading shown in Figure 11.18, determine x, which locates the last plastic hinge to form in the governing PCM. Also determine M_{up}.

Solution

As shown on the first free-body diagram in Figure 11.19, the shear is zero at x, and we can determine the left reaction R_1 as a function of x. This enables us to determine

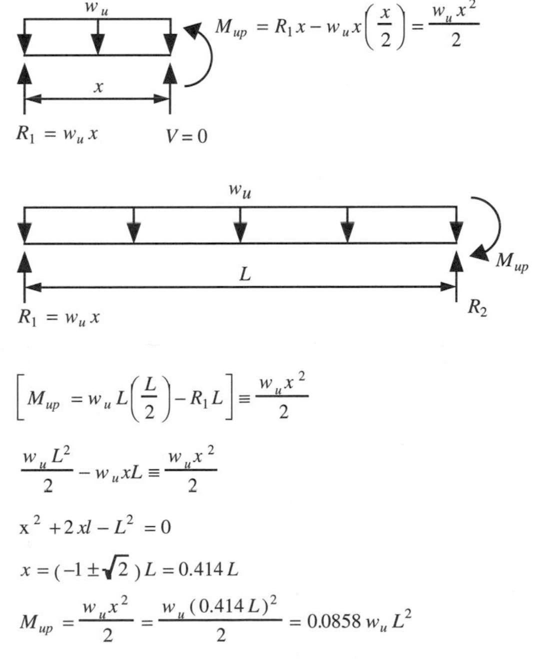

FIGURE 11.19 Example 11.4: locations and computations of M_{up}.

M_{up} as a function of w_u and x. The second free-body diagram enables us to determine M_{up} as a function of w_u and L. Equating these two requirements of M_{up} enables us to solve for

$$x = 0.414L$$

$$M_{up} = \frac{w_u x^2}{2} = 0.0858\, w_u L^2$$

As shown in Figures D.3 to D.9 of Appendix D, the solution for this example is useful in the determination of plastic analysis formulas for continuous beams subjected to a uniformly distributed load.

As shown in Figure 11.20, we can replace a uniformly distributed load with an equivalent set of concentrated loads, which can be used to obtain an approximate value of M_{up}. For the last two sets of concentrated loads, the approximate value obtained for M_{up} is only 2.75% less than the correct value. The indicated procedure can be used to replace the factored member weight by an equivalent set of concentrated loads when all other loads are concentrated loads.

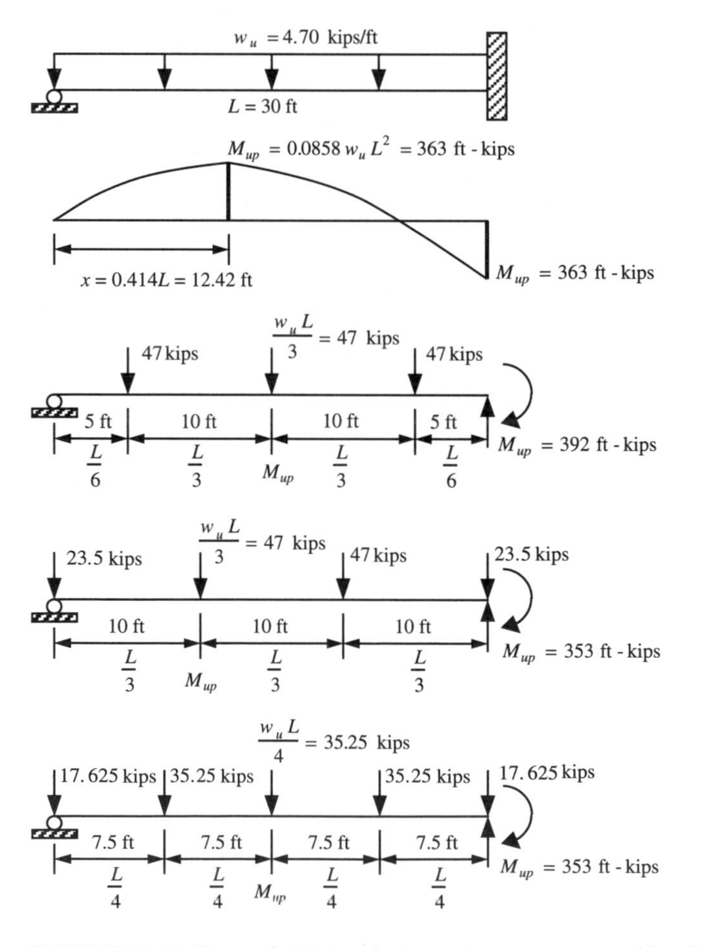

FIGURE 11.20 Example 11.4: solution using concentrated loads.

Example 11.5

Use $F_y = 36$ ksi and the equilibrium method of plastic analysis. For the structure and loadings shown in Figure 11.21, determine M_{up} for the governing PCM. For member 3, $(KL)_y = L_b = 5$ ft. For members 1 and 2, $(KL)_y = L_b = 15$ ft. Use the recommended design procedures in Section 6.6 for beam-columns. Select the lightest acceptable W18 for member 3 and the lightest acceptable W14 for members 1 and 2.

Solution

Our recommended approach given in Section 11.1 is used. In this approach, a tentative member size is needed in order to compute B_1 and B_2. Therefore, we assume that $B_1 = 1$, select trial sections for loading 1, and check these trial sections for loadings 2 and 3.

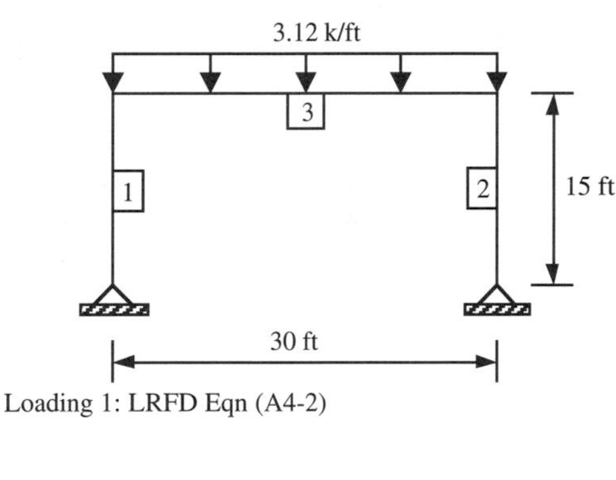

Loading 1: LRFD Eqn (A4-2)

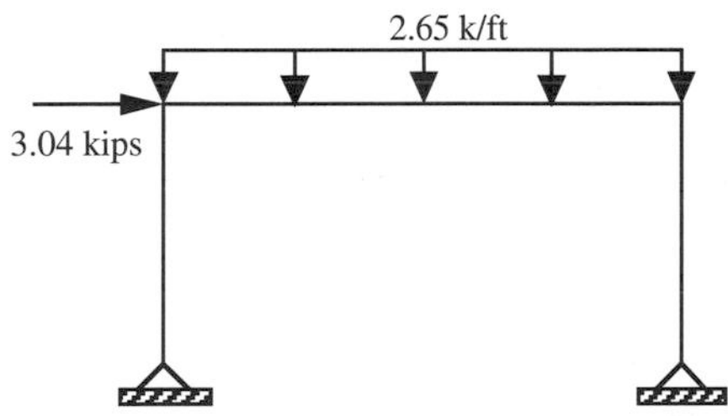

Loading 2: LRFD Eqn (A4-3)

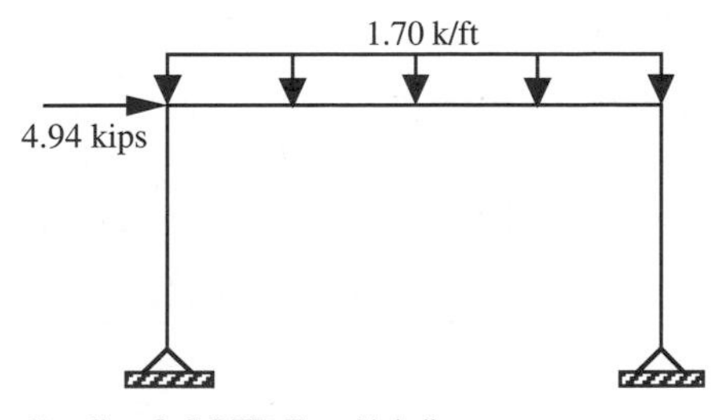

Loading 3: LRFD Eqn (A4-4)

FIGURE 11.21 Example 11.5: structure and loadings.

For loading 1 in Figure 11.21 and the equilibrium method of plastic analysis, we find

$$2M_{up} = \left[\frac{w_u L^2}{8} = \frac{3.12(30)^2}{8} = 351 \text{ ft-kips} \right]$$

$$M_{up} = 175.5 \text{ ft-kips}$$

The governing PCM and the corresponding moment diagram are shown in Figure 11.22.

Using recommended **Procedure 1** in Section 6.6 to account for the axial compression force, select a trial section for which

$$\phi M_{px} \geq (M_{up} + P_u L/40)$$

For member 3, try

$$\phi M_{px} \geq [176 + 11.7(30)/40 = 185 \text{ ft-kips}]$$

$$\text{W18} \times 40, \qquad (\phi M_{px} = 212 \text{ ft-kips}) \geq 185 \text{ ft-kips}$$

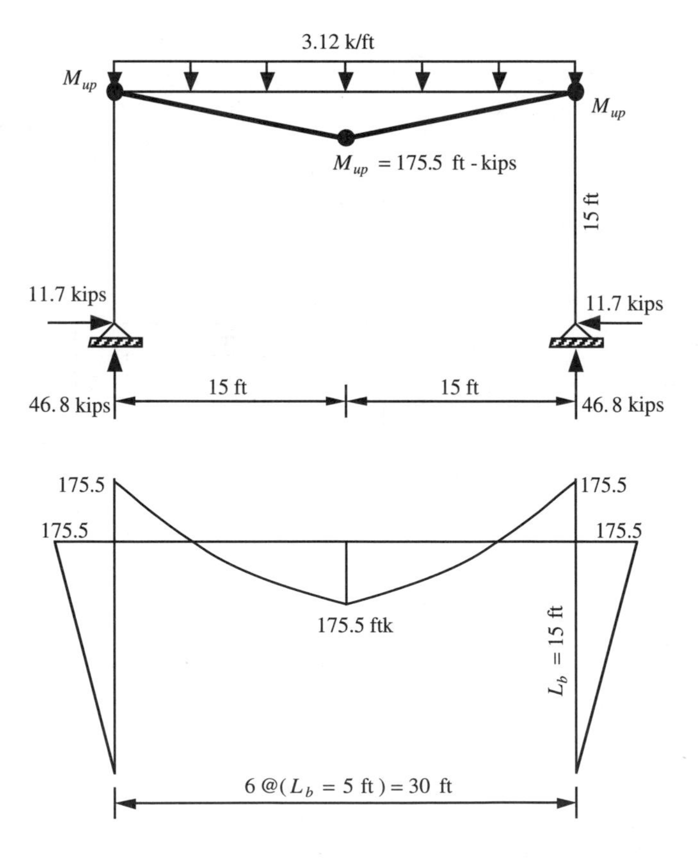

FIGURE 11.22 Example 11.5: PCM for loading 1.

For members 1 and 2, try

$$\phi M_{px} \geq [176 + 46.8(15)/40 = 194 \text{ ft-kips}]$$

$$\text{W14 x 48,} \qquad (\phi M_{px} = 212 \text{ ft-kips}) \geq 194 \text{ ft-kips}$$

For loading 2 in Figure 11.21, we need to use the information shown in Figure 11.23 where the redundant is chosen as the horizontal reaction at the right support. Note that w_u = 2.65 kips/ft on the girder for loading 2 is less than w_u = 3.12 kips/ft on the girder for loading 1. Therefore, we know that the beam mechanism solution for loading 2 does not govern.

Assume that the composite PCM shown in Figure 11.23(d) governs for loading 2. Let the moment ordinates that lie inside the frame be positive moments. At point 4, the moment equilibrium requirement is

$$(-M_4 = -15H) = -M_{up}$$

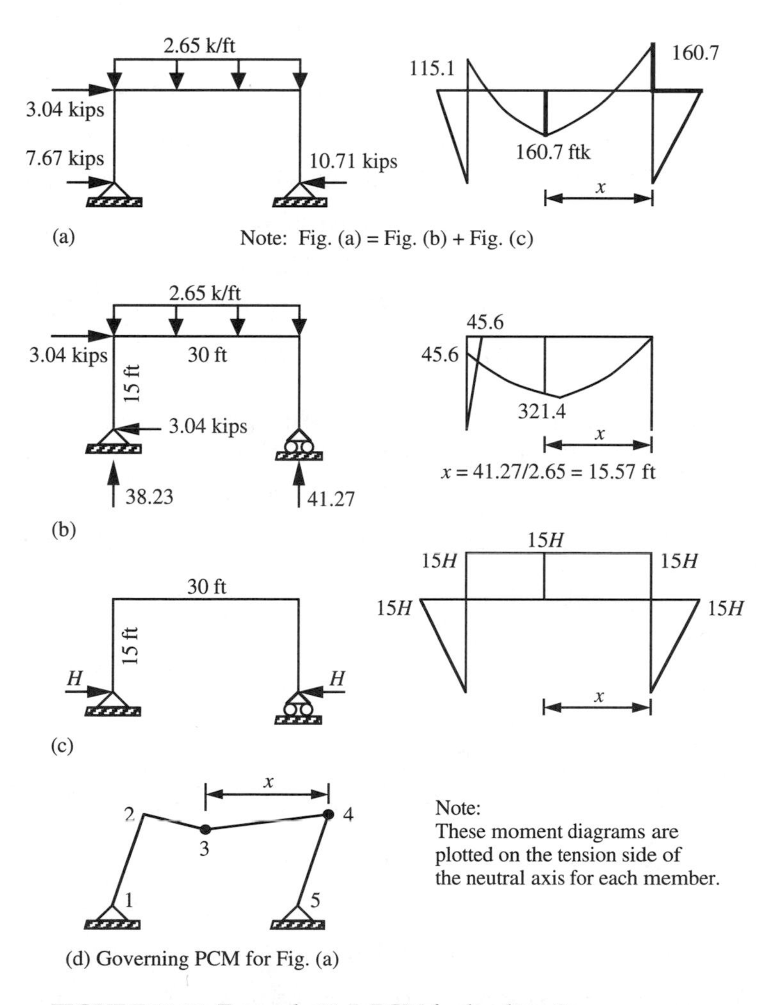

FIGURE 11.23 Example 11.5: PCM for loading 2.

Due only to the redundant H, the vertical reaction at each support is zero. Consequently, before we determine H, we can find the location of M_3, which is at the zero shear point in the girder for all loads. At point 3, the moment equilibrium requirement is

$$(M_3 = 321.4 - 15H) = M_{up}$$

Substitution of $(-15H) = -M_{up}$ gives

$$2M_{up} = 321.36 \text{ ft-kips}$$

$$M_{up} = 160.7 \text{ ft-kips}$$

$$H = 160.7/15 = 10.71 \text{ kips}$$

Since $M \leq (M_{up} = 161$ ft-kips) everywhere along the moment diagram in Figure 11.23(a), the assumed composite PCM is the governing PCM for loading 2.

To account for $P\Delta$ effects, $B_2 M_{up}$ is the required plastic hinge strength. Therefore, we must compute $B_2. = 1/(1 - \Sigma P_u/\Sigma P_{e2})$. For sidesway frame buckling of members 1 and 2,

$$G_{\text{bottom}} = 10$$

$$G_{\text{top}} = (485/15)/(612/30) = 1.58$$

$$(KL)_x = 2.01(15)(12) = 361.8 \text{ in.}$$

$$\pi^2 EI/(KL)^2 = \pi^2(29{,}000)(485)/(361.8)^2 = 1060.5 \text{ kips}$$

$$\Sigma P_{e2} = 2(1060.5) = 2121 \text{ kips}$$

$$\Sigma P_u = 38.23 + 41.27 = 79.5 \text{ kips}$$

$$B_2 = 1/(1 - \Sigma P_u/\Sigma P_{e2}) = 1/(1 - 79.5/2121) = 1.039$$

$$[B_2 M_{up} = 1.039(161) = 167 \text{ ft-kips}] < [M_{up} = 175.5 \text{ ft-kips (for loading 1)}]$$

Loading 2 does not govern.

For loading 3 in Figure 11.21, we use the process described in loading 2. Figure 11.24 shows that the results of this analysis is $M_{up} = 115.1$ ft-kips:

$$\Sigma P_u = 23.0 + 28.0 = 51.0 \text{ kips}$$

$$B_2 = 1/(1 - \Sigma P_u/\Sigma P_{e2}) = 1/(1 - 51.0/2121) = 1.025$$

$$[B_2 M_{up} = 1.025(115) = 118 \text{ ft-kips}] < [M_{up} = 175.5 \text{ ft-kips (for loading 1)}]$$

Loading 3 does not govern.

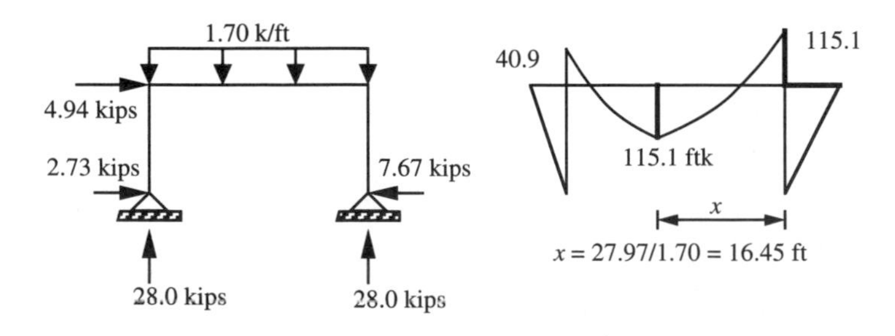

FIGURE 11.24 Example 11.5: PCM for loading 3.

Loading 1 governs. Perform the design checks. See Figure 11.22. No appreciable sidesway occurs due to this loading. However, sidesway frame buckling is not prevented and ϕP_{nx} due to sidesway frame buckling must be considered.

Check members 1 and 2: W14 × 48 $F_y = 36$ ksi.

As a column,

Check LRFD C2.2 (p. 6-43):

$$0.75A_gF_y = 0.75(14.1)(36) = 381 \text{ kips}$$

$$(P_u = 46.8 \text{ kips}) \le (0.75A_gF_y = 381 \text{ kips}) \quad \text{as required}$$

For sidesway frame buckling,

$$K_x = 2.01 \quad \text{(from loading 2 calculations)}$$

$$(KL)_x = 2.01(15) = 30.2 \text{ ft}$$

See LRFD p. 3-21 for W14 x 48, $F_y = 36$ ksi. The section has no flag on it. Therefore, local buckling does not govern ϕP_n and, when we get to the beam-column check, web local buckling does not govern ϕM_{nx}:

$$[(KL)_y = 15 \text{ ft}] > [(KL)_x/(r_x/r_y) = 30.2/3.06 = 9.87 \text{ ft}]$$

$$\phi P_n = (\phi P_{ny} = 270 \text{ kips})$$

$$\left[\frac{\lambda_c}{K} = \frac{(L/r)_y}{\pi}\sqrt{\frac{F_y}{E}} = \frac{(15)(12)/1.91}{\pi}\sqrt{\frac{36}{29{,}000}} = 1.057\right] < 1.5 \quad \text{as required}$$

As a beam,

$$\phi M_{px} = 212 \text{ ft-kips}$$

$$r_y = 1.91 \text{ in.} = 0.159 \text{ ft}$$

$$\frac{M_1}{M_{up}} = \frac{0}{176} = 0$$

$$L_{pd} = \frac{[3600 + 2200(0)](0.159)}{36} = 15.9 \text{ ft}$$

$$(L_{pd} = 15.9 \text{ ft}) \ge (L_b = 15 \text{ ft}) \quad \text{as required}$$

As a beam-column,

$$M_{ux} = B_1M_{up} = B_1(176 \text{ ft-kips})$$

$$[B_1 = C_{mx}/(1 - P_u/P_{ex})] \ge 1 \quad \text{is required}$$

$$P_{ex} = \frac{\pi^2 EI_x}{(KL)_x^2} = \frac{\pi^2(29{,}000)(485)}{(180)^2} = 4284 \text{ kips}$$

$$C_{mx} = 0.6 - 0.4(0/165) = 0.6$$

$$\frac{C_{mx}}{(1 - P_u / P_{ex})} = \frac{0.6}{(1 - 46.8/4284)} = 0.607$$

$$B_1 = \text{larger of} \begin{cases} 0.607 \\ 1 \end{cases} \quad \text{is required}$$

$$M_{ux} = B_1(176 \text{ ft-kips}) = 1.00(176) = 176 \text{ ft-kips}$$

$$[P_u/(\phi P_n) = 46.8/270 = 0.173] < 0.2$$

$$\left(\frac{0.173}{2} + \frac{176}{212} = 0.917 \right) \leq 1.00$$

W14 x 48, $F_y = 36$ ksi is acceptable for members 1 and 2.

Check member 3, W18 x 40, $F_y = 36$ ksi:

$$\left(\frac{0.5 b_f}{t_f} = 5.7 \right) \leq \left(\lambda_p = \frac{65}{\sqrt{36}} = 10.8 \right) \quad \text{as required}$$

$$\left[\frac{P_u}{\phi P_y} = \frac{11.7}{0.9 A F_y} = \frac{11.7}{0.9(11.8)(36)} = 0.0306 \right] \leq 0.125$$

$$\left(\frac{h}{t_w} = 51.0 \right) \leq \left[\lambda_p = [1 - 2.75(0.0306)] \frac{640}{\sqrt{36}} = 97.7 \right] \quad \text{as required}$$

As a beam,

$$r_y = 1.27 \text{ in.} = 0.10583 \text{ ft}$$

At the member ends,

$$M_1 = 176 - 3.12\,[30 - 2(5)]^2/8 = 20 \text{ ft-kips}$$

$$\frac{M_1}{M_{up}} = \frac{20}{176} = 0.102$$

$$L_{pd} = [3600 + 2200(0.102)] \frac{(0.10583)}{(36)} = 11.2 \text{ ft}$$

$$(L_{pd} = 11.2 \text{ ft}) \geq (L_b = 5 \text{ ft}) \quad \text{as required}$$

Elsewhere along the member length,

$$(L_p = 5.3 \text{ ft}) \geq (L_b = 5 \text{ ft})$$

Therefore, everywhere along the member length,

$$\phi M_{nx} = (\phi M_{px} = 212 \text{ ft-kips}) \quad \text{as required}$$

As a column,

Since the axial compression force in members 1 and 2 is not negligible, the braced frame nomograph on LRFD p. 6-186 cannot be used to obtain K_x for member 3. Be conservative and use $K_x = 1$. $(KL)_y = 5$ ft = 60 in. (given in the problem statement):

$$(KL/r)_y = 60/1.27 = 47.2$$

$$(KL/r)_x = 360/7.21 = 49.9 \quad (\text{governs } \phi P_n)$$

$$\left(\lambda_c = \frac{KL/r}{\pi}\sqrt{\frac{F_y}{E}} = \frac{49.9}{\pi}\sqrt{\frac{36}{29{,}000}} = 0.560\right) < 1.5$$

$$\lambda_c^2 = 0.313$$

$$\phi P_n = 0.85(11.8)(0.658)^{0.313}(36) = 317 \text{ kips}$$

$$[P_u/(\phi P_n) = 11.7/317 = 0.0369] < 0.2$$

As a beam-column,

$$P_{ex} = \frac{\pi^2(29{,}000)(612)}{(360)^2} = 1352 \text{ kips}$$

$$\frac{C_{mx}}{(1 - P_u/P_{ex})} = \frac{0.85}{(1 - 11.0/1352)} = 0.86$$

$$B_1 = \text{larger of } 0.86 \text{ and } 1.00$$

$$M_{ux} = B_1(176) = 1.00(176) = 176 \text{ ft-kips}$$

$$\left(\frac{0.0369}{2} + \frac{176}{212} = 0.849\right) \leq 1.00$$

W18 x 40, $F_y = 36$ ksi is acceptable for Member 3.

Serviceability checks need to be performed for deflection and drift at service loads, as we did in Example 6.4, which is for very nearly the same structure and loadings.

11.5 VIRTUAL WORK METHOD OF ANALYSIS

The *virtual work method* of plastic analysis is useful for solving all types of mechanisms. This method is particularly useful for frames in which the geometry is complicated and/or for which there is more than one redundant.

Our purpose in the analysis is to determine M_{up} for the governing plastic collapse mechanism. A value of M_{up} obtained by the virtual work method is an upper bound value. Therefore, the governing value of M_{up} is the least of the required strength values. The objective is to find a mechanism (independent or composite) such that $M_u \leq M_{up}$ for each member in the structure.

Immediately after the assumed plastic hinges have formed in an assumed PCM, we imagine that the structure only moves through a small additional displacement (an increment of the actual displacement that would occur). Since we wish to limit this increment to a small displacement, we choose to say that this is a virtual displacement. Let θ denote an angle in radians. For small displacement theory, sin θ = tan $\theta = \theta$. For the imposed virtual displacement on the assumed mechanism, the internal work done by the plastic hinges is equated to the external work done by the loads.

The steps in the analysis procedure are:

1. Identify the location of each possible plastic hinge, the number of possible plastic hinges *NPPH*, the number of redundants *NR*, and the number of independent mechanisms, *NIM* = *NPPH* - *NR*.
2. Identify the independent mechanisms and, whenever applicable, at least one possible composite mechanism.
3. For each mechanism, solve the virtual work equation of equilibrium for M_{up}.
4. Check to see if $M_u \leq M_{up}$. If this is true, we have found the correct PCM. Otherwise, we repeat steps 3 and 4.

Example 11.6

In Figure 11.25 the same W section is to be used for both spans. Use F_y = 36 ksi and the virtual work method of plastic analysis. Determine the governing PCM and M_{up}.

Solution

The number of independent mechanisms is

$$NIM = NPPH - NR = 5 - 2 = 3$$

These independent mechanisms are shown in Figure 11.25. No other mechanisms are applicable.

For span 1, try PCM 1-2-4:

The internal work done at each plastic hinge is the product of M_{up} and the angle through which the plastic hinge moves. The external work done at each concentrated load is the product of the concentrated load and the displacement through which the load moves. For mathematical convenience, we choose to say that the imposed virtual displacement is the smallest angle θ in the displaced configuration of the assumed PCM. All of the other virtual displacements must be expressed in terms of θ. Equating internal work and external work gives

$$M_{up}\,(3\theta + \theta) = 45\,(20\theta) + 69\,(10\theta)$$

Since $\theta \neq 0$, we can divide both sides of the equation by θ. For PCM 1-2-4, we obtain

$$M_{up} = 397.5 \text{ ft-kips}$$

We could perform the moment check and prove that this is the correct PCM for span 1. Instead, we will compute the M_{up} for each possible PCM to illustrate that a solution obtained by the virtual work method is a lower bound solution.

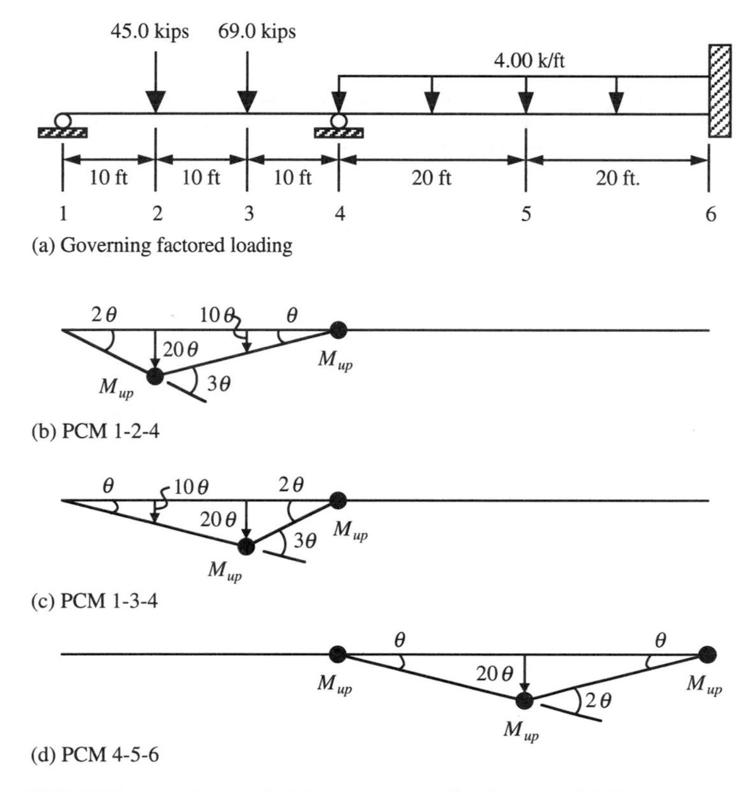

FIGURE 11.25 Example 11.6: structure, loading, and PCM.

For span 1, try PCM 1-3-4. Equating internal work and external work gives

$$M_{up}\,(3\theta + 2\theta) = 45\,(10\theta) + 69\,(20\theta)$$

$$M_{up} = 366 \text{ ft-kips}$$

For span 1, the maximum value obtained for $M_{up} = 398$ ft-kips is the governing value for this span. However, the governing value of M_{up} may be dictated by span 2 since the same W section is to be used in both spans.

For span 2, PCM 4-5-6 is the only possibility:

The external work is $w_u = 4.00$ kips/ft times the triangular area whose boundaries are the original beam line and the mechanism lines. Equating internal work and external work gives

$$M_{up}\,(\theta + 2\theta + \theta) = 4.00\left[\frac{1}{2}(40)(20\theta)\right]$$

$$M_{up} = 400 \text{ ft-kips}$$

PCM 4-5-6 governs and $M_{up} = 400$ ft-kips. This is verified in Figure 11.26 by plotting the M_u diagram and showing that $M_u \leq (M_{up} = 400$ ft-kips).

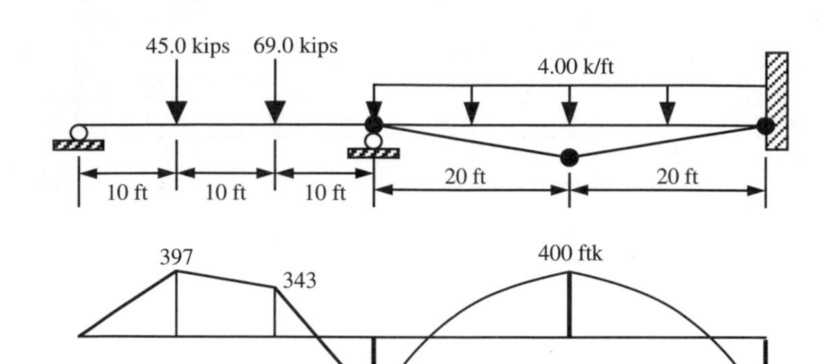

FIGURE 11.26 Example 11.6: final moment diagram.

Example 11.7

In Figure 11.27, a W24 x 62, $\phi M_{px} = 413$ ft-kips is used in both spans. Use the virtual work method of plastic analysis. Determine the governing PCM and the plastic collapse loads.

Solution

For span 1, try PCM 1-2-4. Equating external work and internal work gives

$$W_u\,(20\theta) + (69/45)W_u\,(10\theta) = 413\,(3\theta + \theta)$$
$$W_u = 46.8 \text{ kips}$$

We could perform the moment check and prove that this is the correct PCM for span 1. Instead, we will compute W_u for each possible PCM to illustrate that a solution obtained by the virtual work method is a lower bound solution.

For span 1, try PCM 1-3-4

$$W_u\,(10\theta) + (69/45)W_u\,(20\theta) = 413\,(3\theta + 2\theta)$$
$$W_u = 50.8 \text{ kips}$$

For span 1, the minimum value obtained for $W_u = 46.8$ kips is the governing value for this span. However, the governing value of W_u may be dictated by span 2 since the same W section is to be used in both spans.

For span 2, PCM 4-5-6 is the only possibility. Equating external work and internal work gives

$$0.0889W_u\left[\frac{1}{2}(40)(20\theta)\right] = 413(\theta + 2\theta + \theta)$$

$$W_u = 46.5 \text{ kips}$$

PCM 4-5-6 governs and $W_u = 46.5$ kips. This is verified in Figure 11.28 by plotting the M_u diagram and showing that $M_u \leq (M_{up} = 413$ ft-kips).

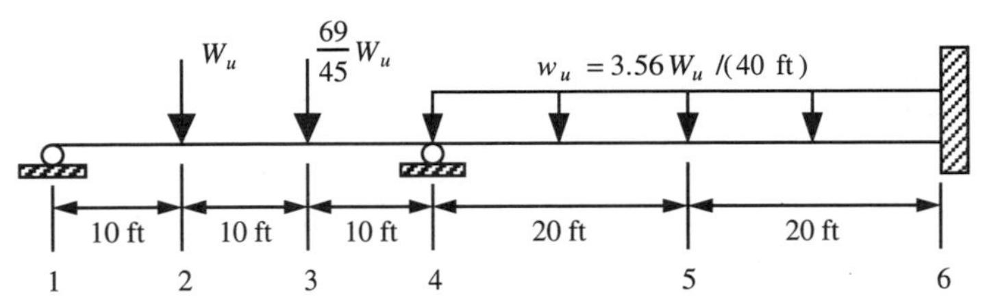

In both spans, the member is W24 x 62 ϕM_{px} = 413 ft - kips.

(a) Governing factored loading

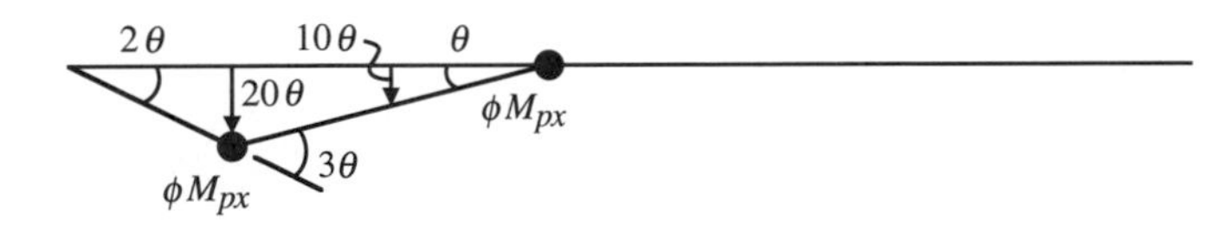

(b) PCM 1-2-4

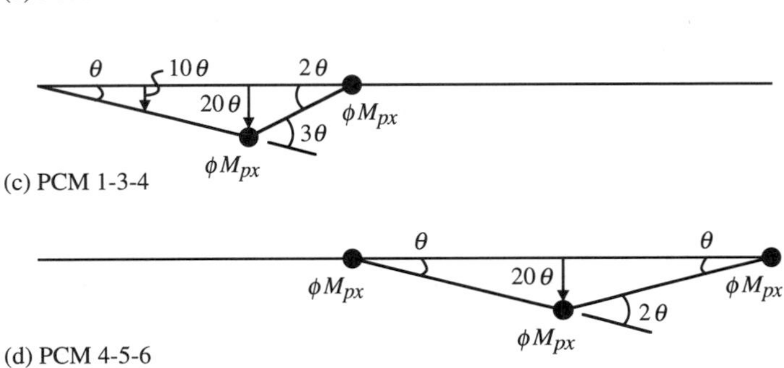

(c) PCM 1-3-4

(d) PCM 4-5-6

FIGURE 11.27 Example 11.7: structure, loading, and PCM.

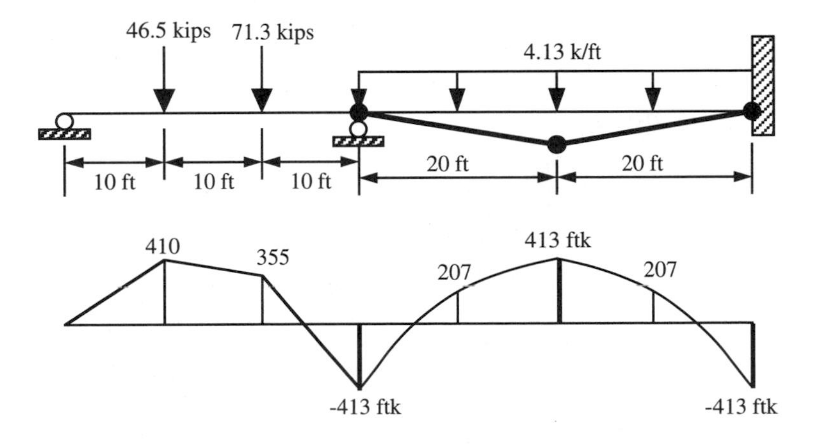

FIGURE 11.28 Example 11.7: final moment diagram.

Example 11.8

Use F_y = 36 ksi and the virtual work method of plastic analysis. In Figure 11.29, a different W section is to be used in each span. In both spans, $L_b = 0$ for the top flange and L_b = 10 ft for the bottom flange wherever lateral braces are needed. Select the lightest acceptable W section for span 1. Use the actual ϕM_{px} of the selected section for span 1 in the determination of M_{up} for span 2.

Solution

For span 1, try PCM 1-2-4:

$$M_{up}\ (3\theta + \theta) = 45\ (20\theta) + 69\ (10\theta)$$

$$M_{up} = 397.5 \text{ ft-kips}$$

As shown in Figure 11.30, the moment check reveals that this is the correct PCM for span 1.

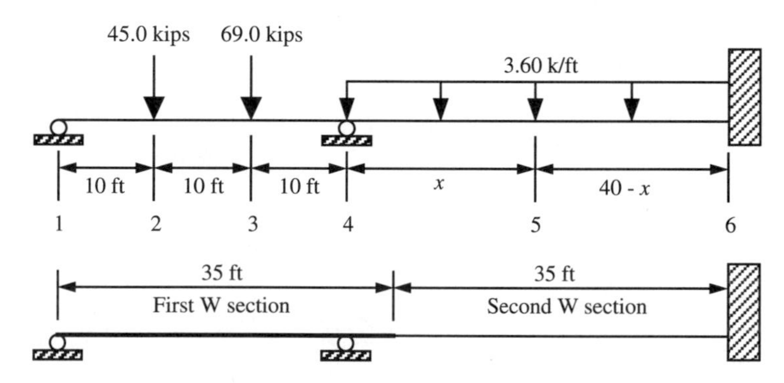

(a) Governing factored loading and beam layout

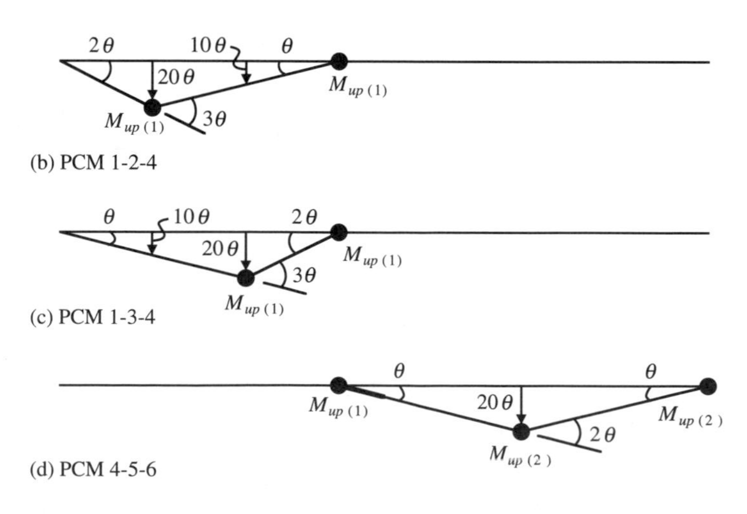

FIGURE 11.29 Example 11.8: structure, loading, and PCM.

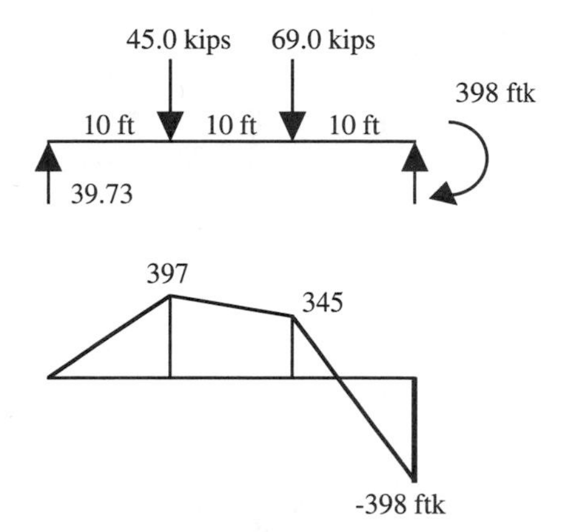

FIGURE 11.30 Example 11.8: span 1.

Using LRFD p. 4-18 for $F_y = 36$ ksi, we find that

$$\text{W24} \times 62, (\phi M_{px} = 413 \text{ ft-kips}) \geq (M_{up} = 398 \text{ ft-kips})$$

is the lightest W section that satisfies the design bending requirement at the plastic hinge locations in span 1, whenthe lateral bracing requirements are satisfied.

At the right support of span 1,

$$M_1 = 345 \text{ ft-kips}$$

$$M_{up} = 398 \text{ ft-kips}$$

$$r_y = 1.38 \text{ in.} = 0.115 \text{ ft}$$

$$\frac{0.9 M_1}{M_{up}} = \frac{0.9(345)}{398} = 0.780$$

$$L_{pd} = [3600 + 2200(0.780)]\frac{(0.115)}{(36)} = 17.0 \text{ ft}$$

$$(L_{pd} = 17.0 \text{ ft}) \geq (L_b = 10 \text{ ft}) \quad \text{as required}$$

For the top flange in span 1, $L_b = 0$; therefore, where the top flange of the W24 x 62 is in compression, we find that ($\phi M_{nx} = \phi M_{px} = 413$ ft-kips) ≥ ($M_u = 398$ ft-kips). Use W24 x 62 for span 1.

For span 2, assume that the plastic hinge at point 5 in PCM 4-5-6 occurs at midspan:

$$M_{up}(\theta + 2\theta + \theta) = 4.00\left[\frac{1}{2}(40)(20\theta)\right]$$

$$M_{up\,(2)} = 342 \text{ ft-kips}$$

As shown in Figure 11.31, the moment check reveals that the correct $M_{up\,(2)}$ cannot be more than 343 ft-kips.

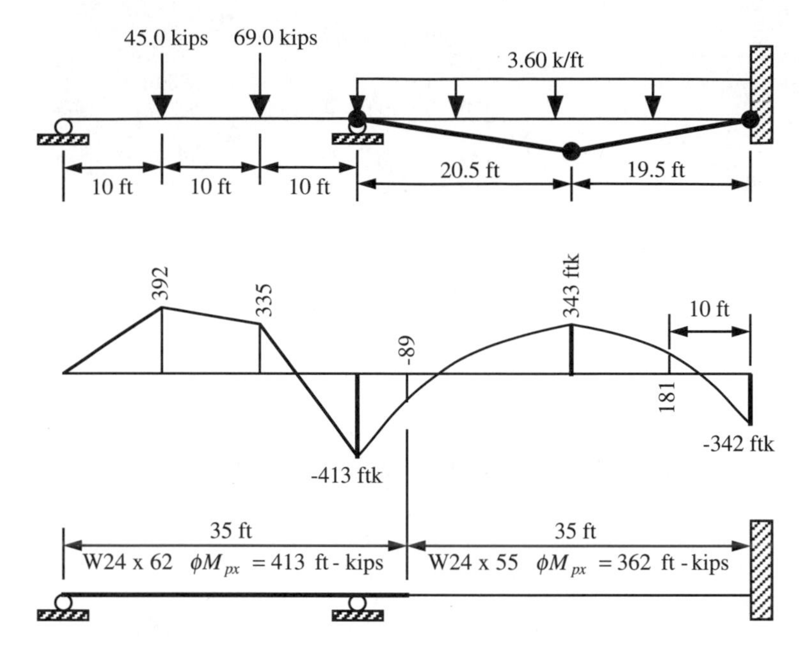

FIGURE 11.31 Example 11.8: final moment diagram.

For span 2, try

$$\text{W24 x 55}, (\phi M_{px} = 362 \text{ ft-kips}) \geq (M_{up} = 343 \text{ ft-kips})$$

At the right support, $L_b = 10$ ft for the bottom flange:

$$M_1 = 180 \text{ ft-kips}$$

$$M_{up\,(2)} = 343 \text{ ft-kips}$$

$$r_y = 1.34 \text{ in.} = 0.112 \text{ ft}$$

$$\frac{0.9\,M_1}{M_{up}} = \frac{0.9(181)}{343} = 0.475$$

$$L_{pd} = [3600 + 2200(0.475)]\frac{(0.112)}{(36)} = 14.5 \text{ ft}$$

$$(L_{pd} = 14.5 \text{ ft}) \geq (L_b = 10 \text{ ft}) \quad \text{as required}$$

For the top flange in span 2, $L_b = 0$; therefore, where the top flange of the W24 x 55 is in compression, we find that

$$(\phi M_{nx} = \phi M_{px} = 362 \text{ ft-kips}) \geq (M_u = 343 \text{ ft-kips}).$$

Use W24 x 55 for span 2.

Example 11.9

Use $F_y = 36$ ksi and the virtual work method of plastic analysis. For the structure and loadings shown in Figure 11.32, determine M_{up} for the governing PCM.

Solution

The beam mechanism shown in Figure 11.33 governs for loading 1, which is symmetric and gives

$$M_{up}(\theta + 2\theta + \theta) = 3.12\left[\frac{1}{2}(30)(15\theta)\right]$$

$$M_{up} = 175.5 \text{ ft-kips}$$

For loading 2, the panel mechanism shown in Figure 11.34(a) is assumed to govern for illustrative purposes and we obtain

$$M_{up}(\theta + \theta) = 3.04(15\theta)$$

$$M_{up} = 22.8 \text{ ft-kips}$$

From the moment check [see Figure 11.34(b)], we find that:

1. This is not the correct PCM since $M > (M_{up} = 22.8$ ft-kips).
2. The correct value of M_{up} lies in the range of 298.6 ft-kips $> M_{up} >$ 22.8 ft-kips.
3. We should investigate a composite mechanism containing a plastic hinge at approximately 14.43 ft from the left end of the girder.

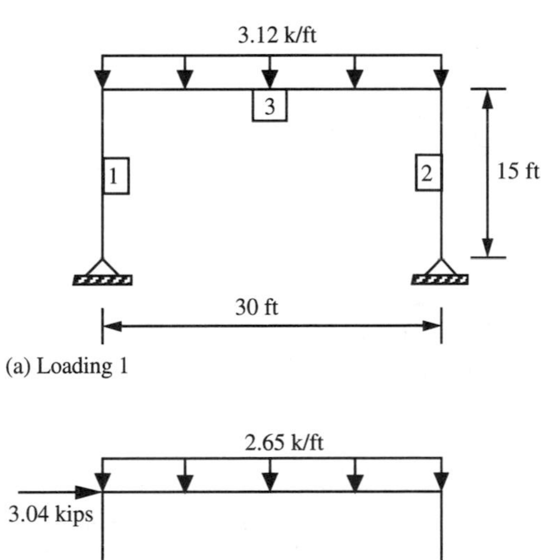

FIGURE 11.32 Example 11.9: structure and loadings.

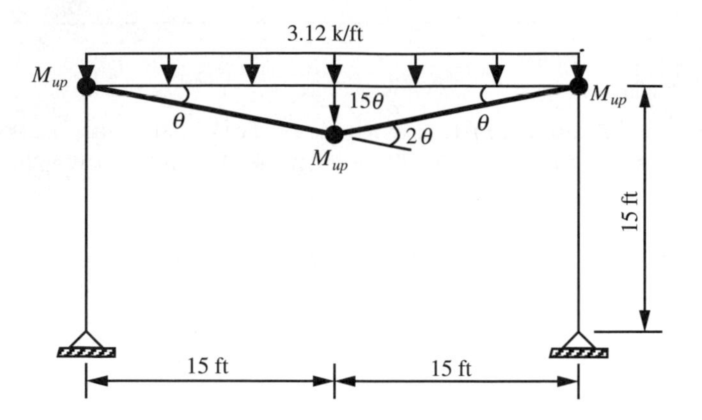

FIGURE 11.33 Example 11.9: no-sway mechanism.

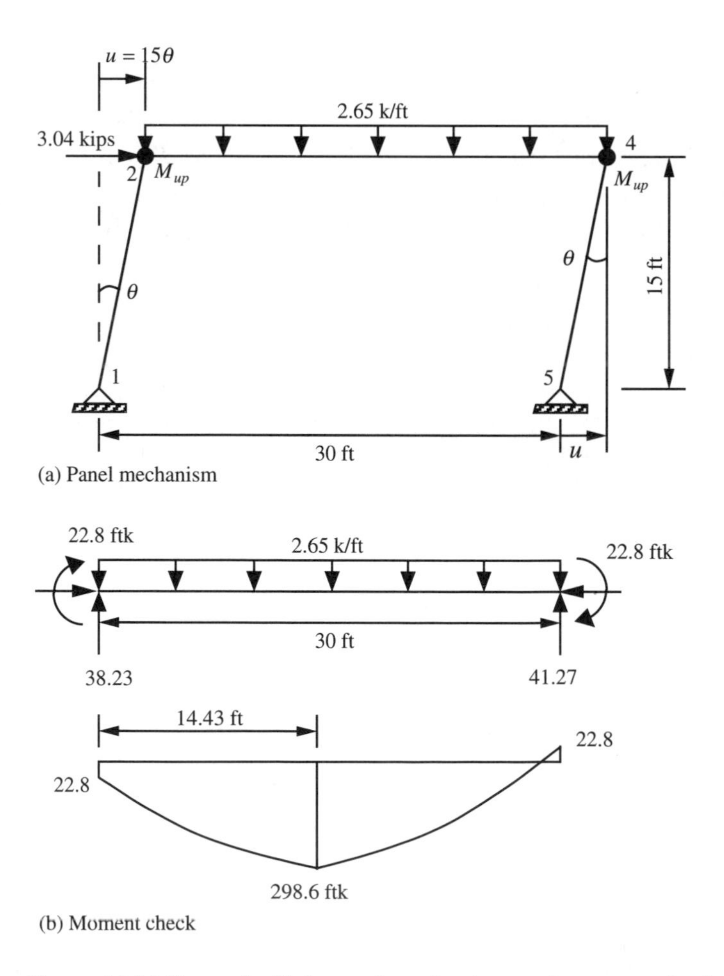

Figure 11.34 Example 11.9: panel mechanism and moment check.

For gravity loads plus wind on a one-story frame, as we have shown in the preceding calculations, the panel mechanism is not likely to be the governing PCM. Therefore, we usually assume that the composite mechanism [see Figure 11.35(a)] governs and perform the moment check to verify our assumption. Note that the plastic hinge in the girder is assumed to occur at 14 ft from the left end of the girder. Since supports 1 and 5 are real hinges for our structure, we can find the vertical reactions and the correct point of zero shear at which there is a plastic hinge in the composite mechanism. However, if both (or either one) of supports 1 and 2 are fixed instead of being hinged, we cannot find the correct point of zero shear. Therefore, we chose to illustrate the general procedure which involves an assumed location of the zero shear point. For the assumed PCM in Figure 11.35(a), we obtain

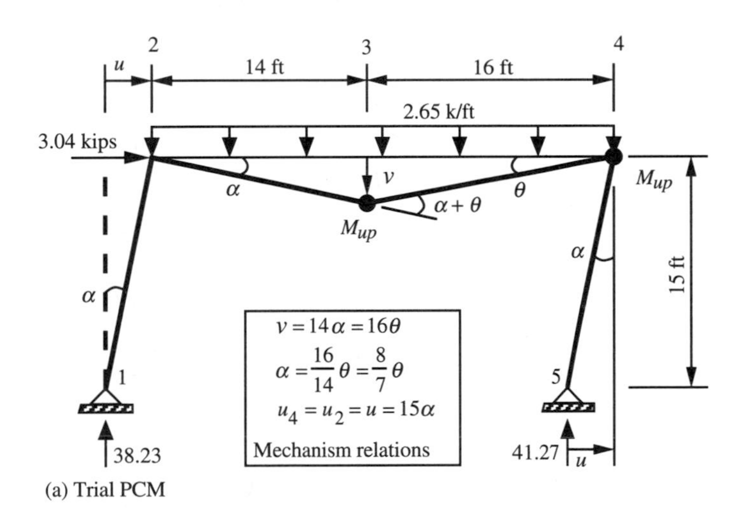

(a) Trial PCM

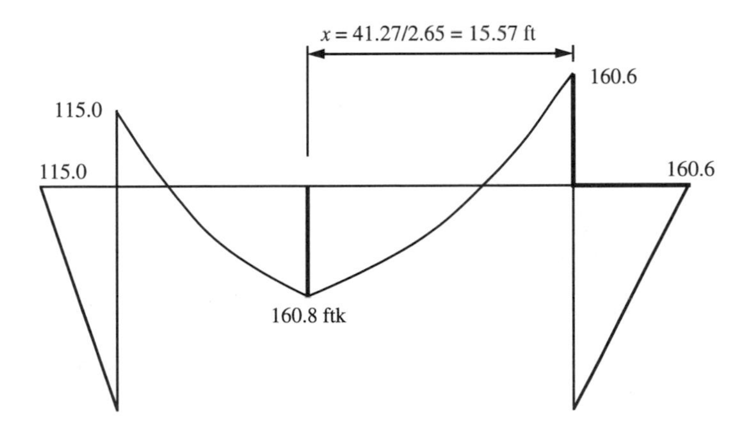

(b) Moment check

FIGURE 11.35 Example 11.9: composite mechanism and moment check.

$$M_{up}\left[2(\alpha+\theta)\right]=3.04(15\alpha)+2.65\left[\frac{1}{2}(30v)\right]$$

From the mechanism relations, we find and substitute $a=\left(\frac{8}{7}\right)\theta$ and $v=16\theta$ to obtain

$$M_{up} = 160.6 \text{ ft-kips}$$

From the moment check [see Figure 11.35(b)], we find that the correct value of M_{up} lies in the range of 160.8 ft-kips > M_{up} > 160.6 ft-kips. We can conservatively use M_{up} = 160.8 ft-kips as the correct value of M_{up}. Note that this correct value of M_{up} = 161 ft-kips does lie in the range of 298.6 ft-kips > M_{up} > 22.8 ft-kips found in the panel mechanism investigation.

The instantaneous center method illustrated in Figure 11.36 can be used to obtain the mechanism relations. When the roof is not flat, the instantaneous center method is the most direct procedure for finding the mechanism relations.

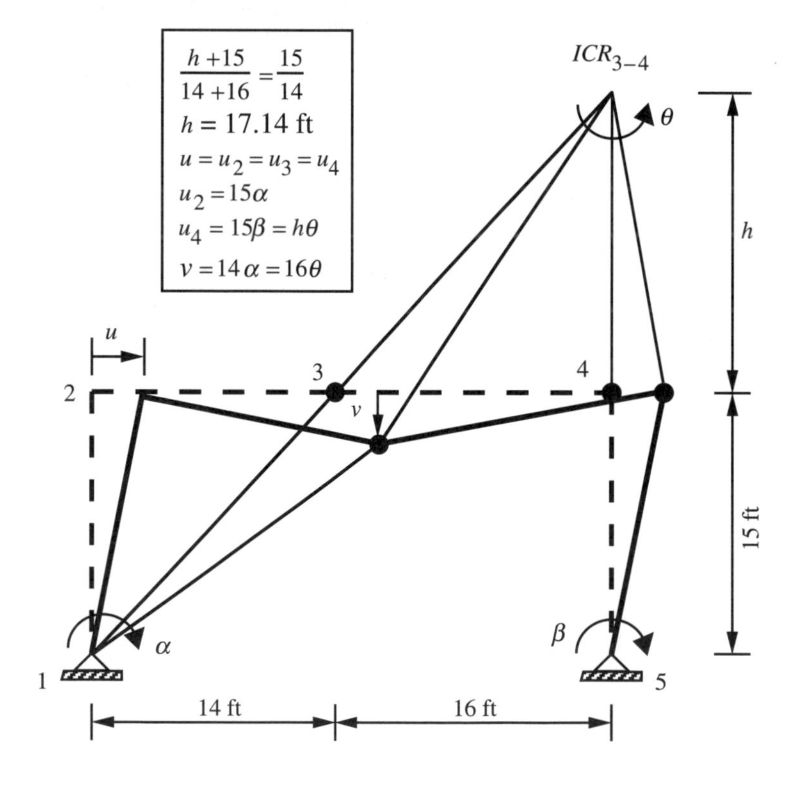

FIGURE 11.36 Example 11.9: composite PCM.

Example 11.10

Use $F_y = 36$ ksi and the virtual work method of plastic analysis to determine M_{up} for the structure and loadings shown in Figure 11.37. Assume that the two roof girders are identical and all plastic hinges form in the roof girders.

Solution

For illustration purposes, find M_{up} due to loading 1 for the gable mechanism shown in Figure 11.38(a). The mechanism relation is obtained by observing that the horizontal deflection at point 2 is

$$u_2 = 15\alpha = 20\theta$$

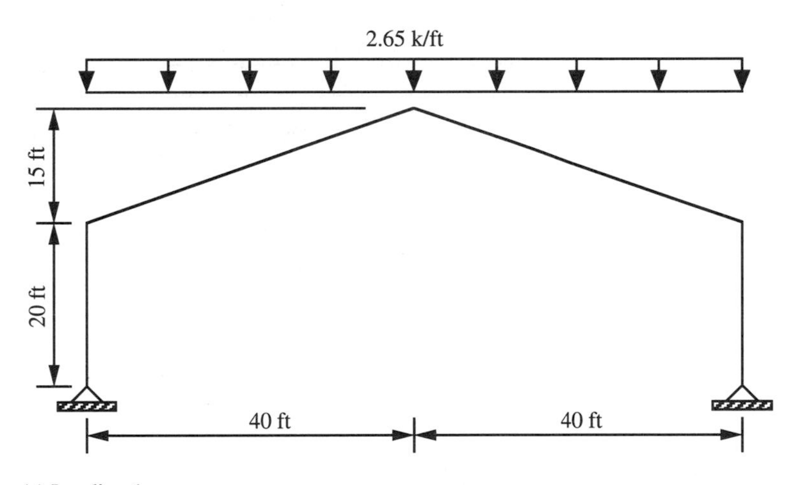

(a) Loading 1

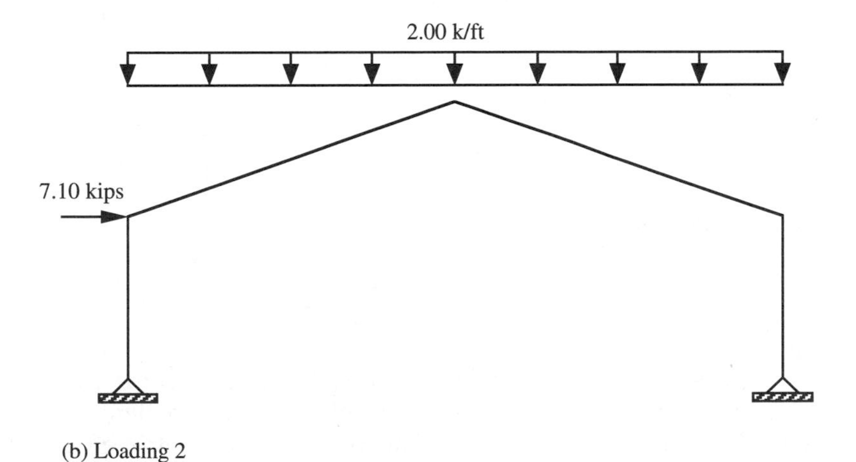

(b) Loading 2

FIGURE 11.37 Example 11.10: structure and loadings

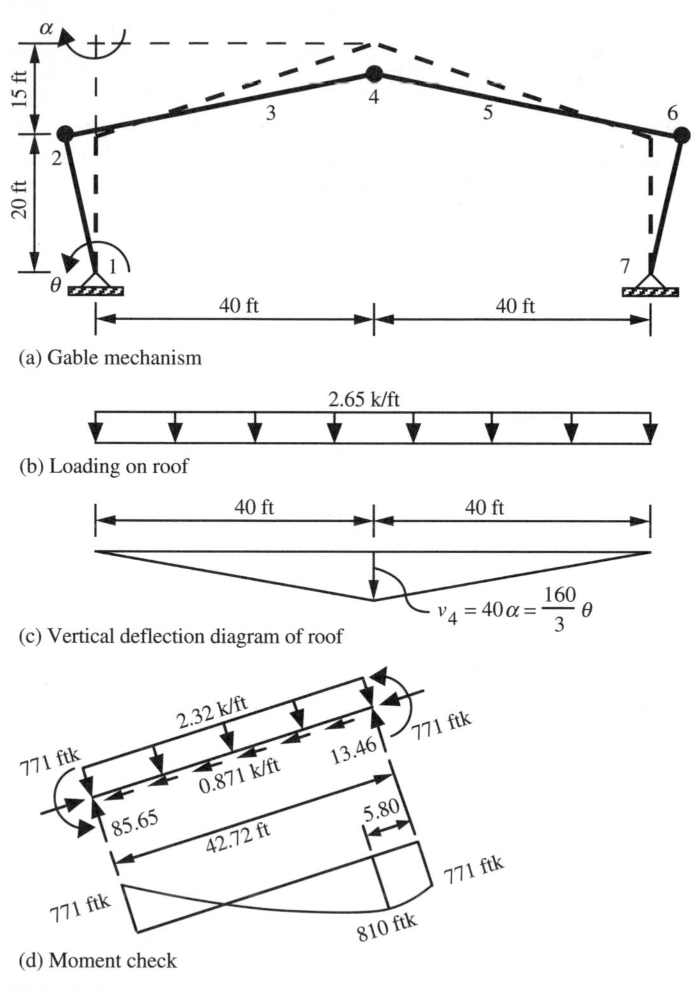

FIGURE 11.38 Example 11.10: panel mechanism for loading 1.

The virtual work equation is

$$M_{up}\left[2(\theta+\alpha)+(\alpha+\alpha)\right]=2.65\left[\frac{1}{2}(80v_4)\right]$$

Substitution of

$$\alpha=\frac{4}{3}\theta \qquad \text{and} \qquad v_4=40\alpha=\frac{160}{3}\theta$$

gives $M_{up} = 770.9$ ft-kips.

To perform the moment check shown in Figure 11.38(d), we proceed as follows. The vertical loading of 2.65 kips/ft is resolved into components perpendicular to and parallel to the roof girder. From moment equilibrium at the left end of the girder, we find that the shear at the right end of the girder is 13.46 kips. The zero shear point is located at (13.46 kips)/(2.32 kips/ft) = 5.80 ft from the right end of the girder and the moment at this point is 771 + 13.46 (5.80)/2 = 810 ft-kips, which exceeds our computed value of M_{up} = 771 ft-kips. Therefore, we know that the correct value of M_{up} lies in the range of 810 ft-kips > M_{up} > 771 ft-kips. Since 810/771 = 1.05, if we choose to say that M_{up} = 810 ft-kips, our M_{up} is not more than 5% larger than the correct M_{up}. In addition to having found a very good estimate of M_{up}, we also have found that plastic hinges numbered 3 and 4 in Figure 11.39(a) (the correct PCM for loading 1) are located $x \approx 40$ - 5.80 (40/42.72) = 34.57 ft from the eaves.

Assume that x = 34.5 ft in Figure 11.39(a) and compute M_{up}. The mechanism relation is

$$u_2 = (34.5/40)15\alpha = 20\theta$$

The virtual work equation is

$$M_{up}\left[2(\theta+\alpha)+2(\alpha)\right] = 2.65\left[(80-69.0)+\frac{1}{2}(69.0)\right]v_4$$

Substitution of $\alpha = 1.546\theta$ and $v_4 = 34.5\alpha$ gives M_{up} = 786 ft-kips, which lies in the previously established range of

$$(810 \text{ ft-kips} > M_{up} > 771 \text{ ft-kips})$$

The preceding calculations for a uniformly distributed load take less time to perform than the case of concentrated loads at the purlin locations on the roof girder. Purlins are spaced at about 5 or 6 ft along the roof girder axis. We have shown for a

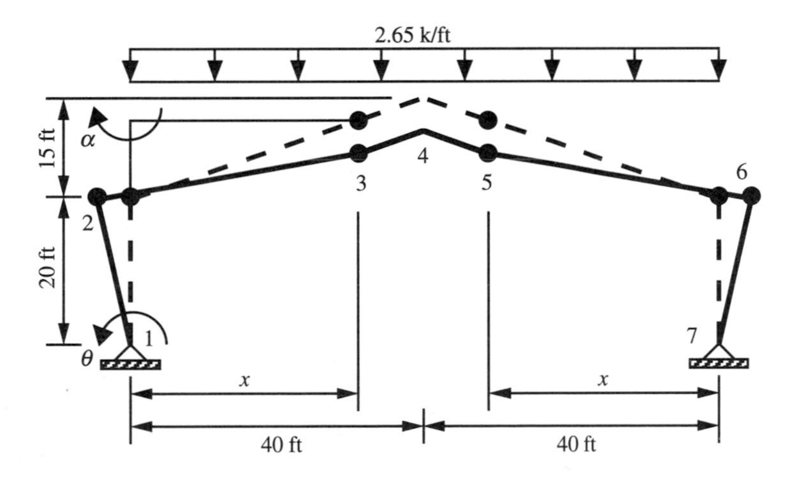

(a) Governing PCM for loading 1

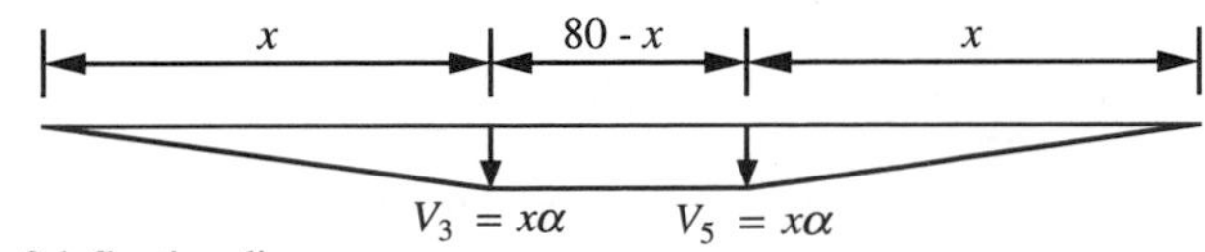

(b) Roof deflection diagram

FIGURE 11.39 Example 11.10: composite PCM for loading 1.

uniformly distributed load that the correct locations of the plastic hinges, numbered 3 and 4 in Figure 11.39(a), are approximately at a purlin location. Therefore, the solution for a uniformly distributed load is a very good approximation of the solution for the concentrated loads case. For preliminary design purposes, gabled frame solutions based on a uniformly distributed roof loading can be used.

We have determined that $M_{up} = 786$ ft-kips for loading 1 in Figure 11.37. Usually, the loading combination that includes wind does not govern M_{up} for a gabled frame. Now we need to show that loading 2 does not govern M_{up} or we must compute the governing value of M_{up}. We prefer to compute M_{up} for the generalized version of loading 2 and the composite mechanism shown in Figure 11.40. As we will show, the deletion of H_u from this solution gives us the exact solution for only a uniformly distributed load.

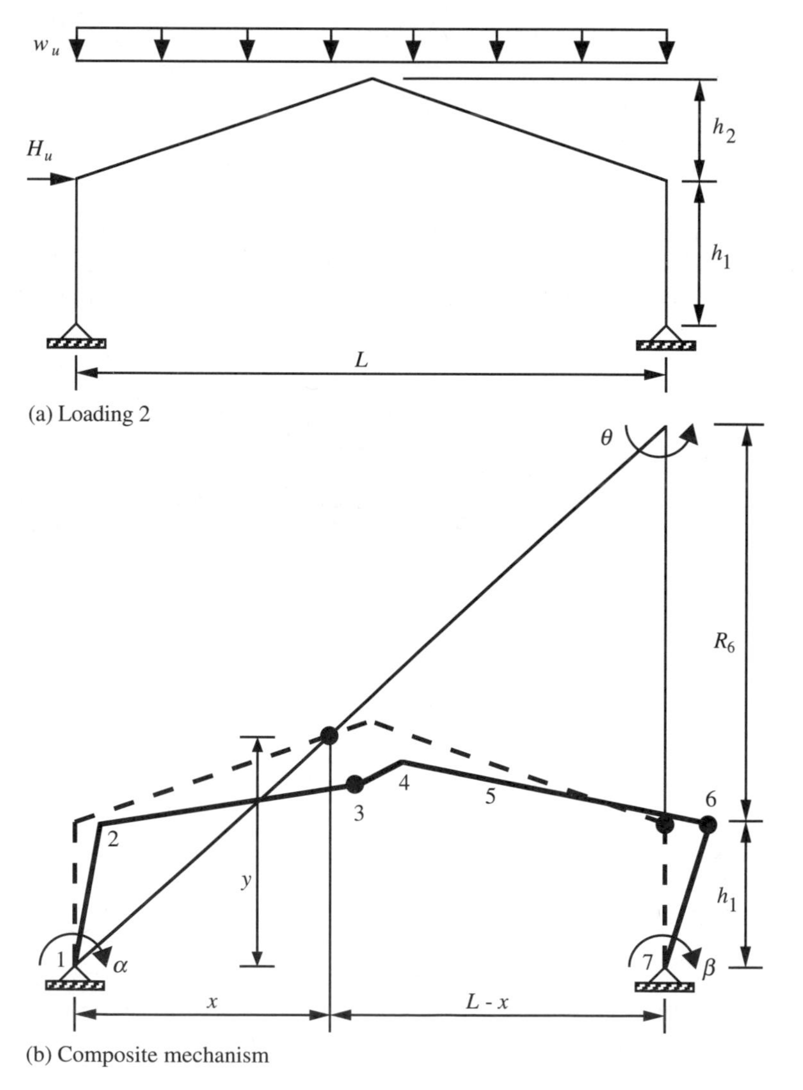

FIGURE 11.40 Example 11.10: composite PCM for loading 2.

We need to locate the instantaneous center of rotation (ICR) point for the structural segment connecting points 3 and 6 in Figure 11.40. From similar triangles, we find that

$$\frac{y}{x} = \frac{h_1 + R_6}{L}$$

where

$$y = h_1 + \left(\frac{x}{L/2}\right) h_2 = h_1 + \frac{2xh_2}{L}$$

Solving for R_6, we obtain

$$R_6 = \frac{h_1 L}{x} + 2h_2 - h_1$$

The horizontal displacement at point 6 is

$$u_6 = h_1 \beta = R_6 \theta$$

which gives the following mechanism relation:

$$\beta = \frac{R_6}{h_1}\theta = \left(\frac{L}{x} + \frac{2h_2}{h_1} - 1\right)\theta$$

The vertical displacement at point 3 is

$$v_3 = x\alpha = (L - x)\theta$$

which gives the following mechanism relation:

$$\alpha = \left(\frac{L - x}{x}\right)\theta$$

The virtual work equation is

$$M_{up}\left[(\alpha + \theta) + (\theta + \beta)\right] = \frac{1}{2} w_u L v_3 + H_u h_1 \alpha$$

Substitution of the mechanism relations gives

$$M_{up} = \frac{\left(\frac{1}{2} w_u L x + H_u h_1\right)(L - x)}{2\left(L + \frac{h_2 x}{h_1}\right)}$$

Differentiating M_{up} with respect to x, setting this result to zero, and solving for x give

$$x = \frac{h_1 L}{h_2}\left\{-1 + \sqrt{1 + \frac{h_2}{h_1}\left[1 - \frac{2H_u(h_1 + h_2)}{w_u L^2}\right]}\right\}$$

For our loading 2 (see Figure 11.37), $H_u = 7.1$ kips, $w_u = 2.00$ kips/ft, $L = 80$ ft, $h_1 = 20$ ft, and $h_2 = 15$ ft. Substitution of these values into the preceding formulas gives us x

= 33.26 ft and M_{up} = 624 ft-kips. Therefore, loading 1 governs and M_{up} = 786 ft-kips. Note that if we set $H_u = 0$ and w_u = 2.65 kips/ft in the formulas obtained for the composite mechanism of loading 2, we obtain x = 34.44 ft and M_{up} = 786 ft-kips. Therefore, the formulas for x and M_{up} are valid for the case when $H_u = 0$.

Example 11.11

See the structure and loadings shown in Figure 11.41. Use F_y = 36 ksi. Assume that the simultaneous occurrence of the beam mechanisms for loading 1 is the governing PCM and determine M_{up} for each member by the equilibrium method. Use the virtual work method and the relative M_{up} values obtained from loading 1 to verify that loading 2 does not govern the M_{up} values.

Solution

For loading 1 and the beam mechanisms [see Figure 11.42(a)], we find, for member 1,

$$M_{up} = \frac{w_u L^2}{16} = \frac{3.12(60)^2}{16} = 702 \text{ ft-kips}$$

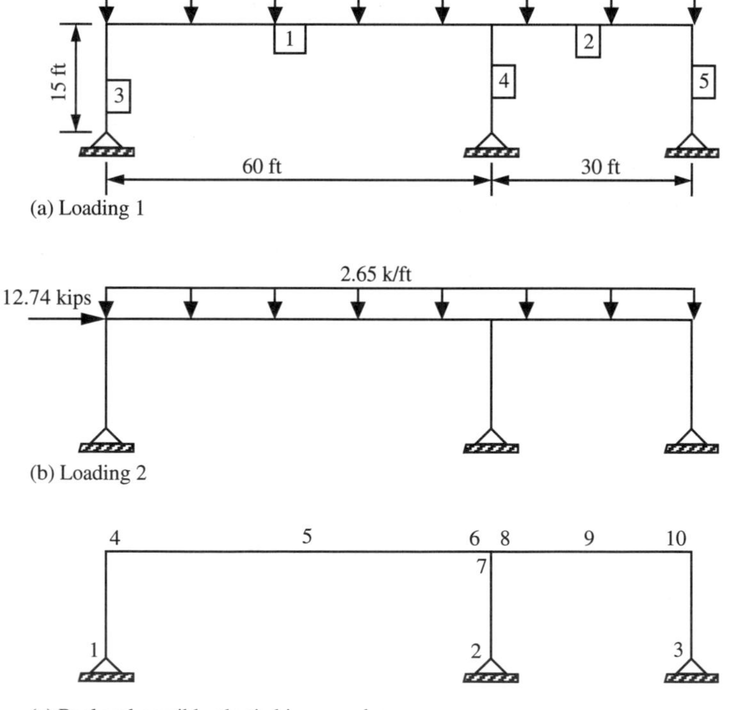

FIGURE 11.41 Example 11.11: structure and loadings.

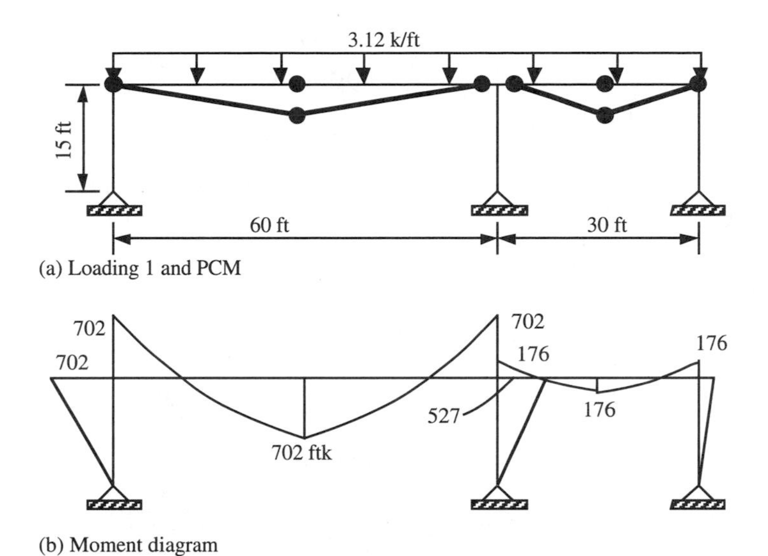

FIGURE 11.42 Example 11.11: composite PCM for loading 1.

For member 2,

$$M_{up} = \frac{w_u L^2}{16} = \frac{3.12(30)^2}{16} = 176 \text{ ft-kips}$$

The moment check [see Figure 11.42(b)], establishes the indicated M_{up} for the columns. If loading 1 (loading combination that does not involve wind) governs, we have established that M_{up} = 176 ft-kips for members 2 and 5, M_{up} = 527 ft-kips for member 4, and M_{up} = 702 ft-kips for members 1 and 3. If we choose the least of these values as the base value and express the other values as a function of this base value, we find that the relative M_{up} values are M_{up} for members 2 and 5, $3M_{up}$ for member 4, and $4M_{up}$ for members 1 and 3.

Now, we choose to prove that loading 2 does not govern the M_{up} values. Any assumed composite mechanism solution for which the required moment diagram ordinates do not exceed the M_{up} values obtained for loading 1 is proof that loading 2 does not govern the M_{up} values. We choose to investigate the trial PCM shown in Figure 11.43(b). The total internal virtual work is

$$W_{int} = 4M_{up}(2\theta) + 4M_{up}\theta + 3M_{up}\theta + M_{up}\theta = 16M_{up}\theta$$

and the total external virtual work is

$$W_{ext} = 12.74(15\theta) + 2.65\left[\frac{1}{2}(60)(30\theta)\right] = 2576.1\theta$$

From $W_{int} = W_{ext}$, we find that M_{up} = 161 ft-kips, $3M_{up}$ = 483 ft-kips, and $4M_{up}$ = 644 ft-kips. When we perform the moment check [see Figure 11.43(c)], we find that our trial PCM has enabled us to prove that loading 2 does not govern the M_{up} values.

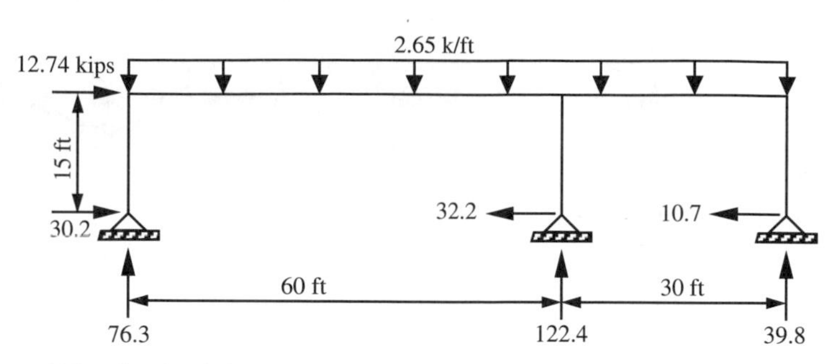

(a) Loading 2 and the reactions for the trial PCM

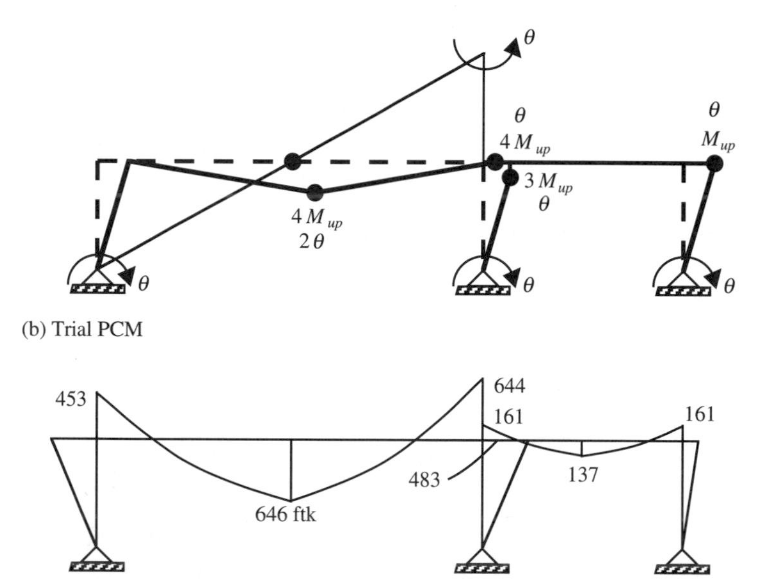

(b) Trial PCM

(c) Moment diagram

FIGURE 11.43 Example 11.11: composite PCM for loading 2.

We have established that loading 1 governs and M_{up} = 176 ft-kips for members 2 and 5, M_{up} = 527 ft-kips for member 4, and M_{up} = 702 ft-kips for members 1 and 3.

We can use the approach illustrated for two bays and a one-story frame when there are two or more bays. As the number of bays increases, a loading combination that involves wind is less likely to govern the relative M_{up} values.

11.6 JOINT SIZE

In a frame, the plastic hinges do not form at the intersection of the member centerlines as we assumed in the preceding example problems. Consider a one-story rectangular frame (see Figure 11.44). If we use the same W section for members 1 and 3, due to a larger axial force in member 3, the plastic hinge at the intersection of members 1 and 3 will form at the end of member 3. If the W section chosen for member 3 is not the same as the one chosen for member 1, the plastic hinge will form at the end of the weaker member. At the intersection of members 1, 2, and 4, there are three possible plastic hinge locations (one at the end of each

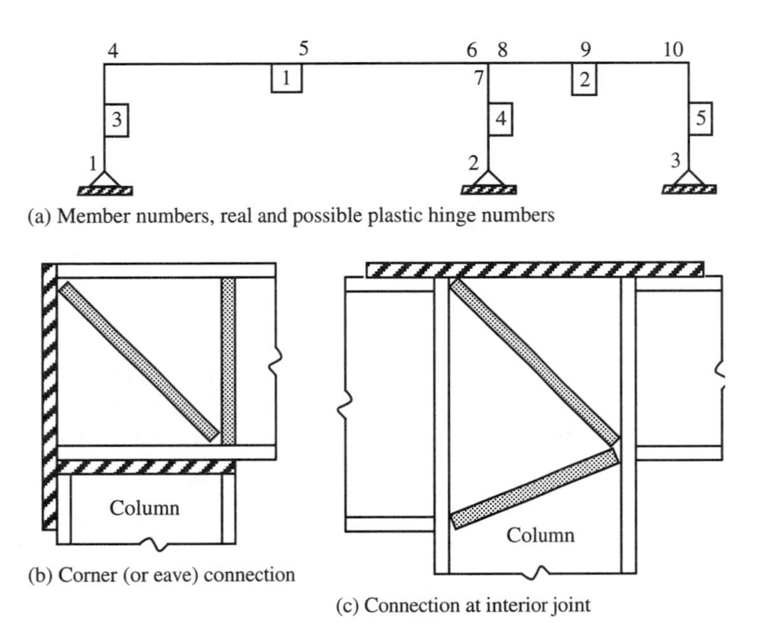

(a) Member numbers, real and possible plastic hinge numbers

(b) Corner (or eave) connection

(c) Connection at interior joint

FIGURE 11.44 Connection details at corner and interior joints.

of the intersecting members). Actually, as we have described, the plastic hinges form at the interface of the beam and column members. This interface is at the connection locations. Hence, connections are coincident with plastic hinge locations. Connections must be strong enough to develop the required plastic hinge strength for an assumed inelastic rotation capacity of 3 (see footnote c on LRFD p. 6-39) at the plastic hinge in a member. According to LRFD p. 6-171, the "target" reliability index is $\beta = 3.0$ for members and $\beta = 4.5$ for connections. The larger value of $\beta = 4.5$ means that the connections are expected to be stronger than the members they connect. This means that a plastic hinge at a joint is expected to form outside the connection in one of the interconnected member ends. Therefore, the plastic collapse strength of the frame is somewhat larger than that predicted on the basis of centerline dimensions. Thus, if we ignore the joint size in the plastic analysis, the results obtained are conservative.

In a gabled frame (see Figure 11.45), the eave connection usually is haunched in some manner. Sometimes, the crown connection is also haunched. In design, we assume that a plastic hinge will not form within a haunch. Therefore, in the plastic analysis made for the final design considerations, we should assume plastic hinge locations on each end of the haunch. This complicates the plastic analysis computations somewhat, but our final analysis assumptions should be a reasonably good representation of how we intend for the structure to behave.

If a beam mechanism governs in a plastic analysis of a frame, we can use the clear span of the beam when we compute M_{up}. The clear span of the beam is the distance between the column flanges. If the column has not already been designed, we use an assumed depth of the column section in computing the clear span and check our assumed depth.

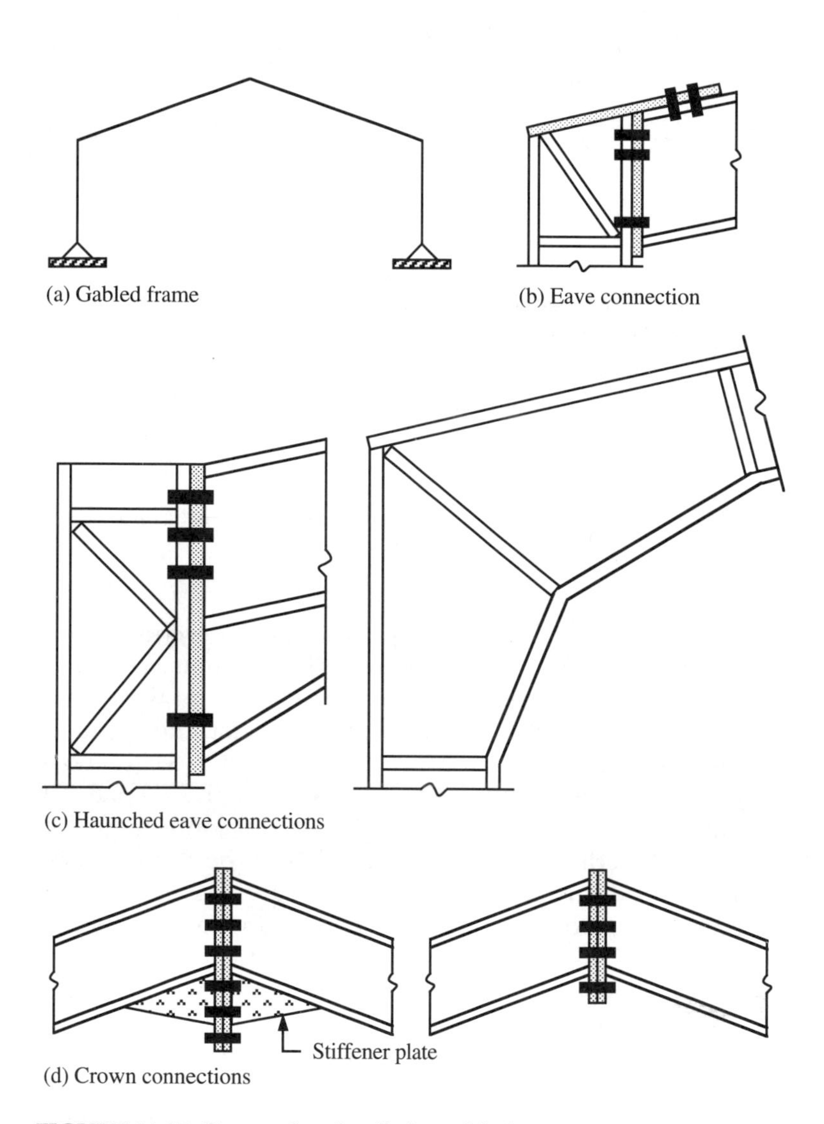

FIGURE 11.45 Connection details for gable frames.

PROBLEMS

11.1 For the structure and loading in Figure P11.1, find the correct PCM and M_{up}.

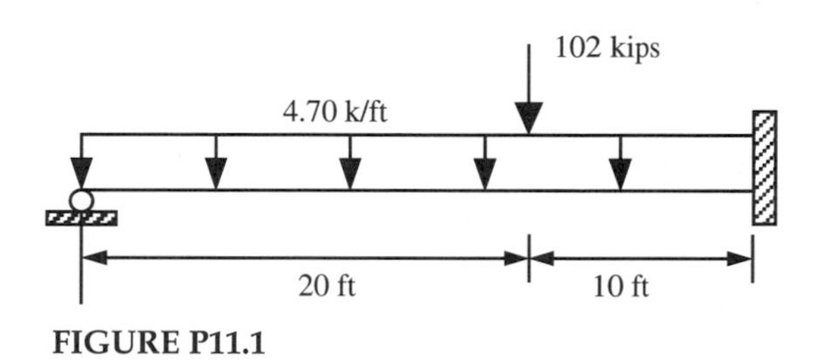

FIGURE P11.1

11.2 Solve Problem 11.1 with the uniform load replaced by 23.5 kips at each support and 47 kips at the third span points.

11.3 See Appendix D, Figures D.8 and D.9. For five spans of $L = 30$ ft, $w_u = 0.832$ kips/ft, $L_b = 0$ for the top flange, $L_b = 6.0$ ft wherever needed for the bottom flange, $F_y = 36$ ksi, select the lightest acceptable W sections for each case and compare the total weights.

11.4 Solve Problem 11.3 for $F_y = 50$ ksi and $L_b = 4.0$ ft for bottom flange.

11.5 See Figure P11.5. For the 30-ft span, $L_b = 10$ ft. For the 36-ft span, $L_b = 9$ ft. Use $F_y = 36$ ksi and the same W section in both spans. Select the lightest acceptable W section.

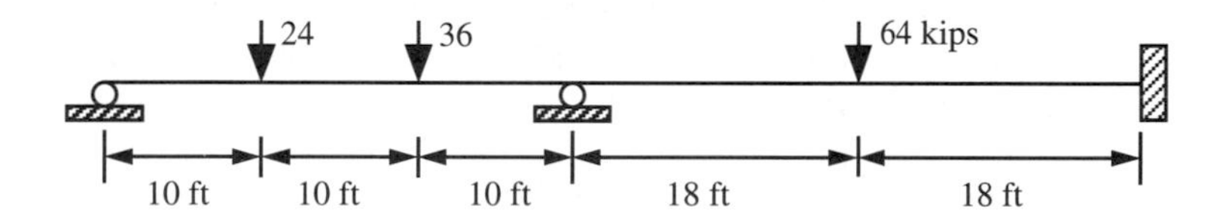

FIGURE P11.5

11.6 See Figure P11.6. For the 30-ft span, $L_b = 10$ ft. For the 36-ft span, $L_b = 9$ ft. Use $F_y = 36$ ksi. Select the lightest W section that is satisfactory for the 30-ft span. Use this W section in the other span and design a pair of cover plates only in the region indicated by a heavy line.

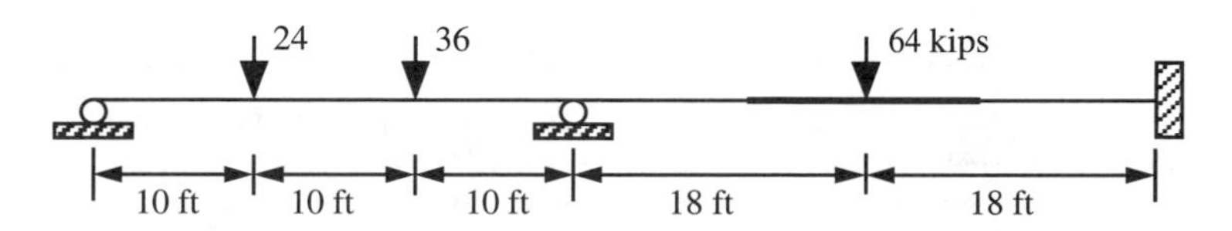

FIGURE P11.6

11.7 See Figure P11.7. For the 30-ft span, L_b = 10 ft. For the 36-ft span, L_b = 9 ft. Use F_y = 36 ksi. First, select the lightest W section that is satisfactory for the 30-ft span. Use ϕM_{px} for this W section in the PCM for the other span and select the lightest acceptable W section for the 36-ft span.

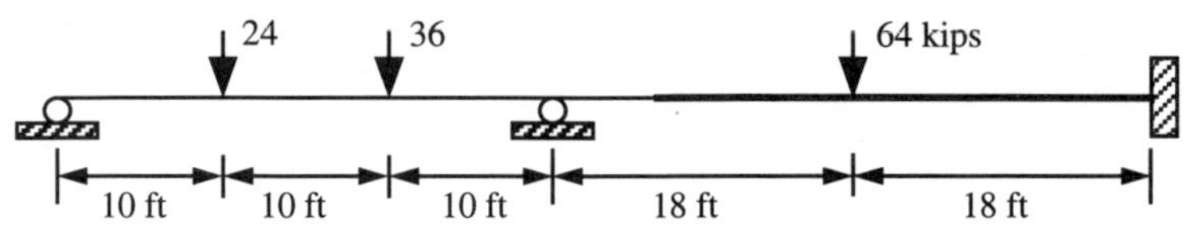

FIGURE P11.7

11.8 See Figure P11.8. For the 30-ft span, L_b = 10 ft. For the 36-ft span, L_b = 12 ft. Use F_y = 36 ksi. First, select the lightest W section that is satisfactory for the 30-ft span. Use ϕM_{px} for this W section in the PCM for the other span and select the lightest acceptable W section for the 36-ft span.

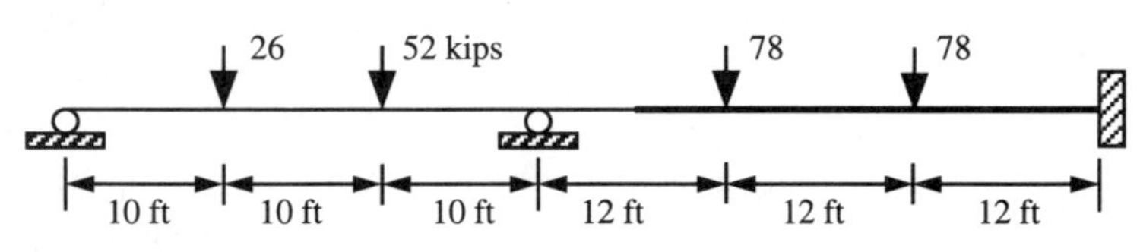

FIGURE P11.8

11.9 See Figure P11.9. For the 30-ft span, L_b = 10 ft. For the 36-ft span, L_b = 12 ft. Use F_y = 36 ksi. First, select the lightest W section that is satisfactory for the 36-ft span. Use ϕM_{px} for this W section in the PCM for the other span and select the lightest acceptable W section for the 30-ft span.

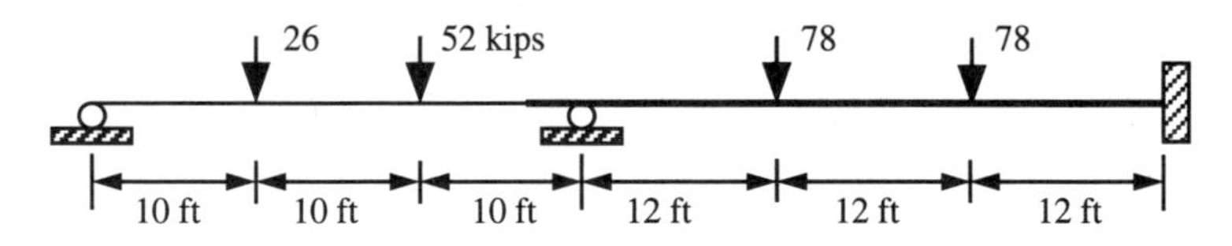

FIGURE P11.9

11.10 In Figure P11.10, use L_b = 7.5 ft and F_y = 36 ksi. Select the lightest acceptable W section.

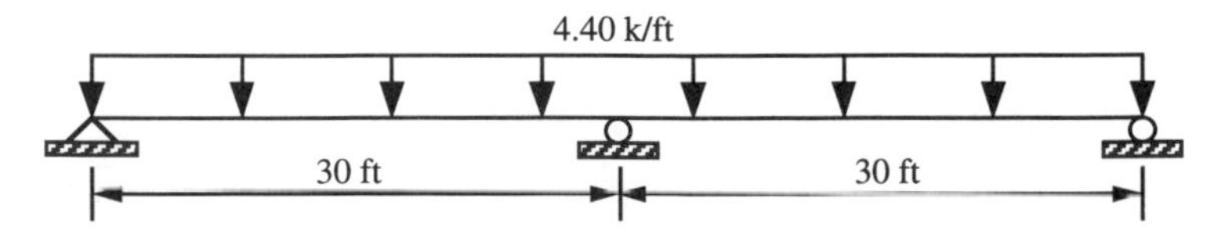

FIGURE P11.10

11.11 In Figure P11.11, use $L_b = 7.5$ ft and $F_y = 36$ ksi. Select the lightest acceptable W section.

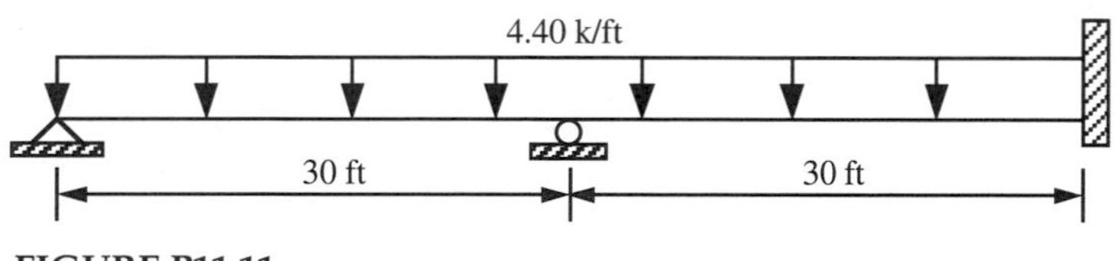

FIGURE P11.11

11.12 In Figure P11.12, use $L_b = L/4$ in each span, the same W section in all spans, and $F_y = 36$ ksi. Select the lightest acceptable W section that is valid for all spans.

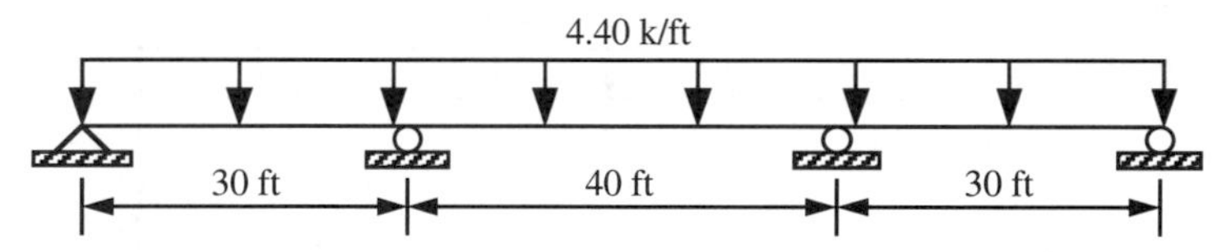

FIGURE P11.12

11.13 In Figure P11.13, use $L_b = L/4$ in each span and $F_y = 36$ ksi. First, select the lightest acceptable W section that is valid for the exterior spans. Use ϕM_{px} for this W section in the PCM for the interior span and select the lightest acceptable W section for the 40-ft span.

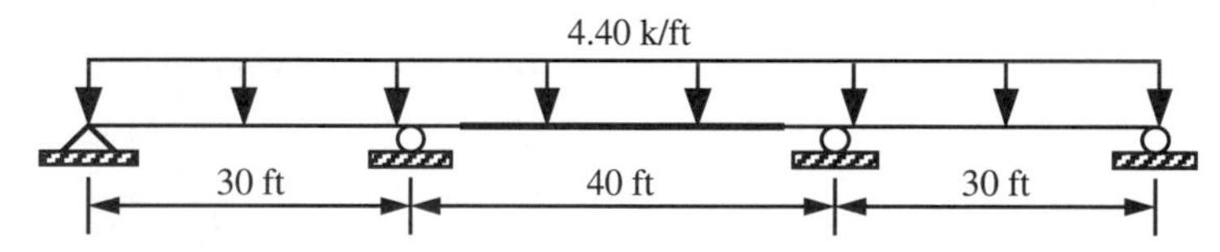

FIGURE P11.13

11.14 In Figure P11.14, use $L_b = L/4$ in each span and $F_y = 36$ ksi. First, select the lightest acceptable W section that is valid for the interior span. Use ϕM_{px} for this W section in the PCM for the exterior span and select the lightest acceptable W section for the 30-ft spans.

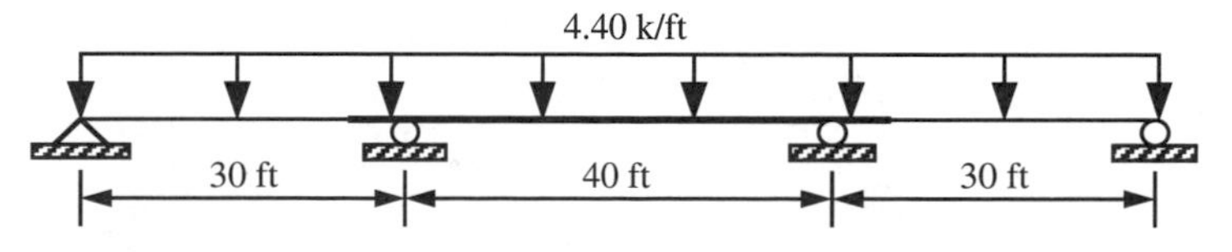

FIGURE P11.14

11.15 In Figure P11.15, use $F_y = 36$ ksi, $L_b = 6$ ft for the girder, and $L_b = (KL)_y = 20$ ft for the column. Select the lightest acceptable W section for the girder. Select the lightest acceptable W12 section for the column. Also, select the lightest acceptable W14 section for the column. Use the lighter of the two column selections.

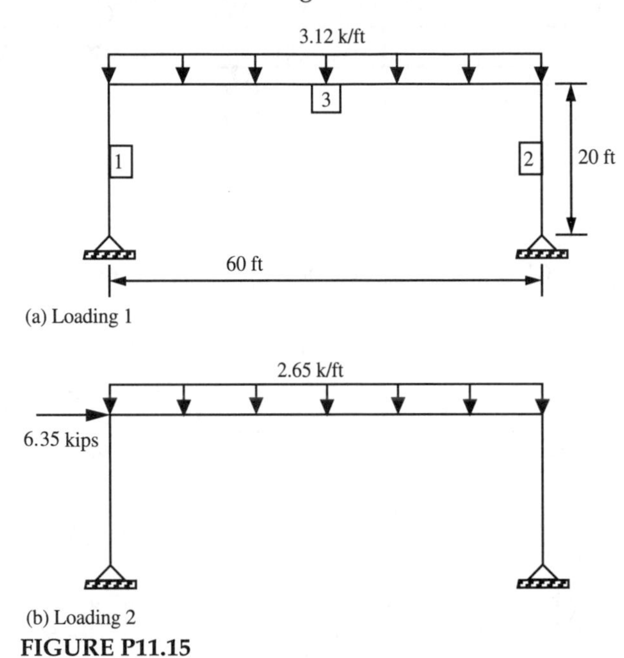

FIGURE P11.15

11.16 In Figure P11.16, use $F_y = 36$ ksi, $L_b = 7.5$ ft for the girder, and $L_b = (KL)_y = 12.5$ ft for the column. Select the lightest acceptable W section for the girder. Select

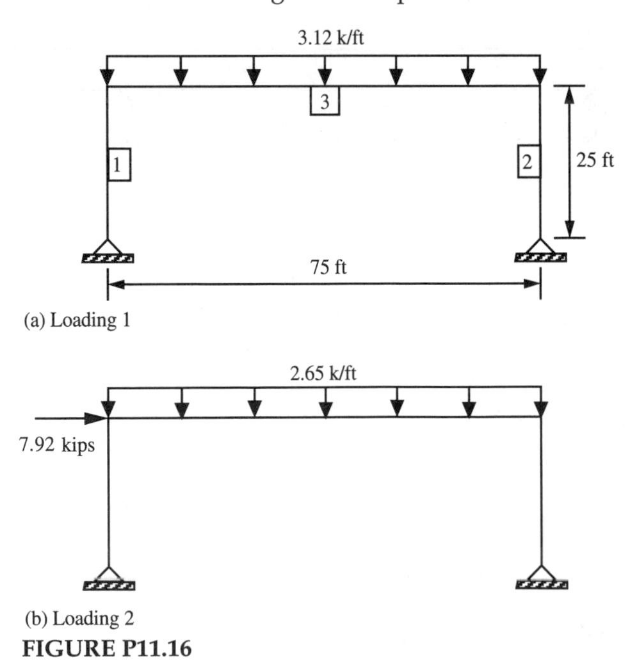

FIGURE P11.16

the lightest acceptable W12 section for the column. Also, select the lightest acceptable W14 section for the column. Use the lighter of the two column selections.

11.17 Finish Example 11.10. Use $F_y = 36$ ksi, eight equal spaces of L_b for the girder, and $L_b = (KL)_y = 20$ ft for the column. Select the lightest acceptable W section for the girder. Select the lightest acceptable W12 section for the column. Also, select the lightest acceptable W14 section for the column. Use the lighter of the two column selections.

11.18 Finish Example 11.11. Use $F_y = 36$ ksi, $L_b = 6$ ft for the girders, and $L_b = (KL)_y = 15$ ft for the columns. Select the lightest acceptable W section for each girder. Select the lightest acceptable W12 section for each column. Also, select the lightest acceptable W14 section for each column. Use the lighter column selections.

11.19 In Figure P11.19, use $F_y = 36$ ksi, five equal spaces of L_b for the girder, and $L_b = (KL)_y = 20$ ft for the column. Assume that the plastic hinges form only in the girders. Select the lightest acceptable W section for the girders. Select the lightest acceptable W12 section for the column. Also, select the lightest acceptable W14 section for the column. Use the lighter of the two column selections.

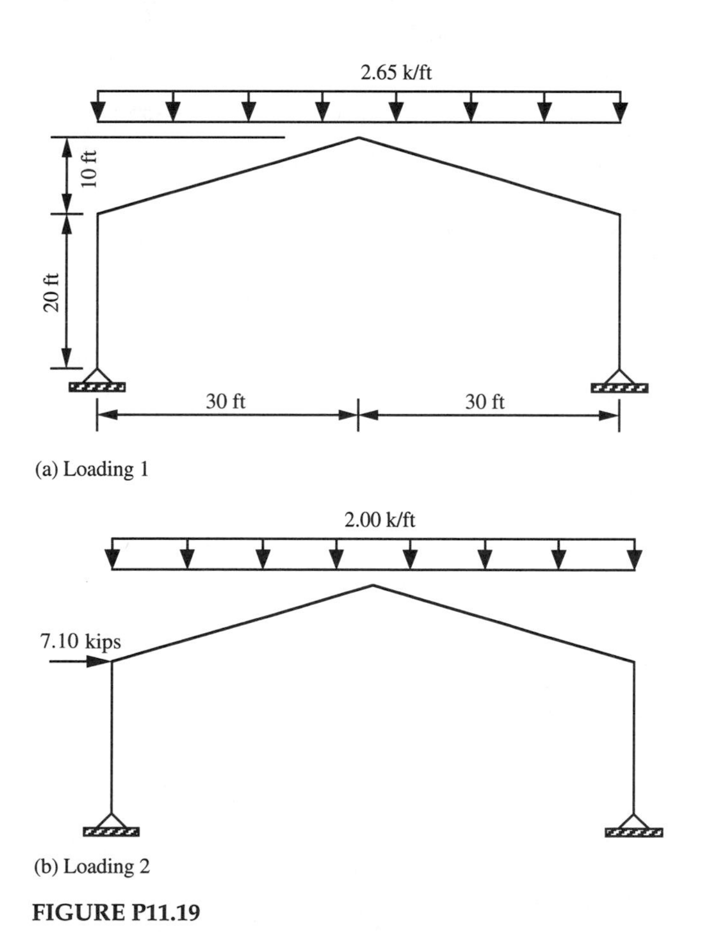

(a) Loading 1

(b) Loading 2

FIGURE P11.19

11.20 In Figure P11.20, determine the relative values of M_{up} for the trial PCM shown. Compare these values with those that were obtained in Example 11.11 when the side-wind load was applied in the opposite direction.

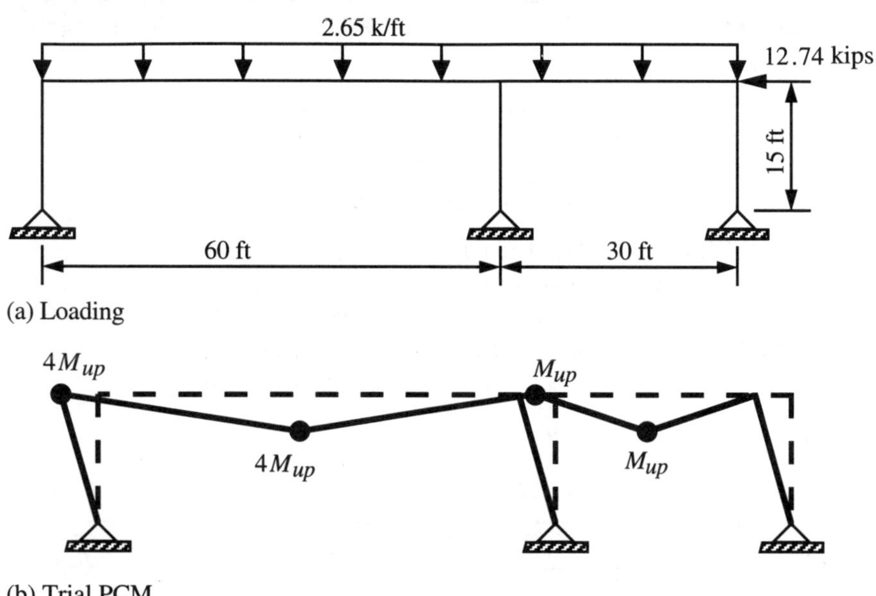

FIGURE P11.20

Computer Output for an Example of an Elastic, Factored Load Analysis

STAAD (structural analysis and design), a proprietary computer software program of Research Engineers, Inc., Yorba Linda, CA, was used to obtain a P-DELTA analysis of a plane frame (Figures 1.14 and 1.15) from the structure shown in Figure 1.16.

The output from STAAD begins immediately after Figure A.1 in which the STAAD sign conventions for input and output are given. For the user's convenience, STAAD places a number in front of each line of input and prints these line numbers in the output. The elastic P-DELTA analysis was performed for applicable LRFD Load Combinations (A4-1) to (A4-6) (p. 6-30) and accounts for the second-order (PΔ) effects due to joint displacements. Therefore, we do not have to use LRFD Eq.(C1-1) (p. 6-41) in designing the members of this structure. To minimize the number of output pages, member-end forces are listed for only the governing members and a few other members to indicate the range of axial force and bending moments.

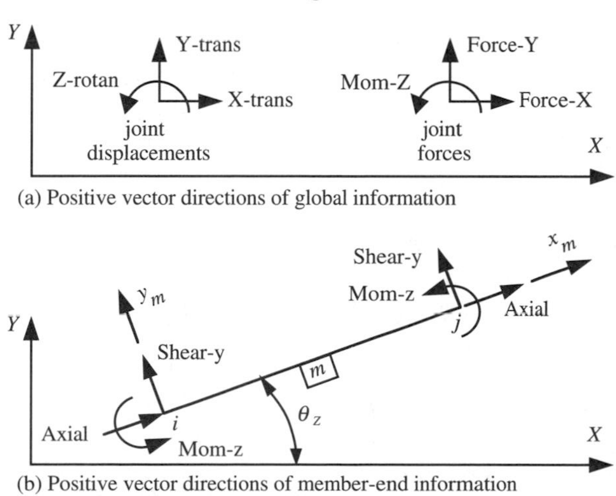

FIGURE A.1 Positive directions of global and local forces.

```
 1. STAAD PLANE  (SEE FIGURES 1.14 - 1.17)
 2. UNIT KIP FEET
 3. JOINT COORDINATES
 4.  1  0.  0.;  2  0. 21.;  3  0. 25.5
 5.  4  6. 21.;  5  6. 26.;  6 12. 21.;  7 12. 26.5
 6.  8 18. 21.;  9 18. 27.; 10 24. 21.; 11 24. 27.5
 7. 12 30. 21.; 13 30. 28.; 14 36. 21.; 15 36. 27.5
 8. 16 42. 21.; 17 42. 27.; 18 48. 21.; 19 48. 26.5
 9. 20 54. 21.; 21 54. 26.; 22 60. 21.; 23 60. 25.5
    24 60. 0.
10. MEMBER INCIDENCES
11.  1  1  2;  2  2  3;  3 24 22;  4 22 23
12.  5  2  4;  6  4  6;  7  6  8;  8  8 10;  9 10 12
13. 10 12 14; 11 14 16; 12 16 18; 13 18 20; 14 20 22
14. 15  3  5; 16  5  7; 17  7  9; 18  9 11; 19 11 13
15. 20 13 15; 21 15 17; 22 17 19; 23 19 21; 24 21 23
16. 25  4  5; 26  6  7; 27  8  9; 28 10 11; 29 12 13
17. 30 14 15; 31 16 17; 32 18 19; 33 20 21
18. 34  3  4; 35  5  6; 36  7  8; 37  9 10; 38 11 12
19. 39 12 15; 40 14 17; 41 16 19; 42 18 21; 43 20 23
20. UNIT INCHES
21. MEMBER PROPERTIES
22.  1 TO  4 PRISMATIC  AX 11.8   IZ 310.
23.  5 TO 14 PRISMATIC  AX  5.    IZ  20.9
24. 15 TO 24 PRISMATIC  AX  5.89  IZ  23.3
25. 25 33    PRISMATIC  AX  3.13  IZ  3.83
26. 26 TO 32 PRISMATIC  AX  2.38  IZ  2.17
27. 34 TO 43 PRISMATIC  AX  2.38  IZ  2.17
28. CONSTANTS
29. E 29000. ALL
30. UNIT FEET
31. SUPPORTS
32. 1 24  FIXED BUT KMZ 140.6
33. *               UNITS ARE FT-KIPS/DEGREE
34. *               VALUE SHOWN IS FOR G = 2 AT THE SUPPORT
35. LOADING 1
36. *       DEAD
37. JOINT LOADS
38. 5 7 9 11 15 17 19 21 FY -2.20;  3 23 FY -1.10
                         13 FY -2.90
39. 4 6 8 10 12 14 16 18 20 FY -0.20;  2 22 FY -10.90
40. *
41. LOADING 2
42. *       LIVE LOAD (CRANES)
43. JOINT LOADS
44. 6 18 FY -8.;  12 FY -16.
45. *
```

```
46. LOADING 3
47. *       ROOF LOAD (SNOW)
48. JOINT LOADS
49. 5 7 9 11 13 15 17 19 21 FY -3.60;  3 23 FY -1.80
50. *
51. LOADING 4
52. *       WIND LOAD
53. JOINT LOADS
54. 3        FX -0.085  FY 1.;  23          FX 0.085  FY 1.
55. 5 7 9 11 FX -0.170  FY 2.;  15 17 19 21 FX 0.170  FY 2.
           13 FY 2.
56. MEMBER LOADS
57. 1 2 UNIFORM GX 0.240;  3 4 UNIFORM GX 0.150
58. *
59. LOADING 5
60. *       0.9D + W - - TO CHECK DRIFT
61. REPEAT LOAD
62. 1 0.9  4 1.
63. *
64. LOADING 6
65. *       1.4D
66. REPEAT LOAD
67. 1 1.4
68. *
69. LOADING 7
70. *       1.2D + 1.6L +0.5R
71. REPEAT LOAD
72. 1 1.2  2 1.6  3 0.5
73. *
74. LOADING 8
75. *       1.2D + 0.5L +1.6R
76. REPEAT LOAD
77. 1 1.2  2 0.5  3 1.6
78. *
79. LOADING 9
80. *       1.2D + 1.6R + 0.8W
81. REPEAT LOAD
82. 1 1.2  3 0.5  4 0.8
83. *
84. LOADING 10
85. *       1.2D + 1.3W + 0.5L + 0.5R
86. REPEAT LOAD
87. 1 1.2  4 1.3  2 0.5  3 0.5
88. *
89. LOADING 11
90. *       0.9D + 1.3W
91. REPEAT LOAD
```

```
92. 1 0.9  4 1.3
93. *
94. PDELTA ANALYSIS
95. *
96. *  OUTPUT FOR DRIFT & DEFLECTION CHECKS
97. LOAD LIST 1 TO 5
98. PRINT JOINT DISPLACEMENTS LIST 2 3 12

JOINT DISPLACEMENTS - (INCHES  RADIANS)
---------

JT  LOAD  X-TRANS   Y-TRANS   Z-ROTAN

 2    1  -0.03821  -0.01705  -0.00083
      2  -0.05922  -0.01178  -0.00130
      3  -0.05508  -0.01326  -0.00119
      4   1.04229   0.00818  -0.00023
      5   1.03581  -0.00714  -0.00100

 3    1   0.01814  -0.01891  -0.00118
      2   0.02927  -0.01421  -0.00185
      3   0.02610  -0.01599  -0.00170
      4   1.00881   0.00980   0.00106
      5   1.05344  -0.00718   0.00001

12    1   0.00000  -0.35005   0.00000
      2   0.00000  -0.55002   0.00000
      3   0.00000  -0.49100   0.00000
      4   1.01927   0.27932   0.00006
      5   1.04743  -0.03551   0.00007

 99. *
100. *  OUTPUT FOR STRENGTH CHECKS
101. LOAD LIST 6 TO 11
102. PRINT REACTIONS ALL

SUPPORT REACTIONS - (KIPS  FEET)
--------

JOINT  LOAD  FORCE-X  FORCE-Y   MOM Z

   1     6      0.88    32.41   -4.31
         7      2.77    62.38  -13.63
         8      2.69    64.58  -13.22
         9     -3.53    27.84   21.41
        10     -6.01    30.25   36.21
        11     -7.15     6.36   40.68

  24     6     -0.88    32.41    4.31
```

```
         7       -2.77      62.38    13.63
         8       -2.69      64.58    13.22
         9       -4.42      29.72    27.85
        10       -6.92      33.31    44.10
        11       -5.78       9.31    37.48

103. PRINT MEMBER FORCES LIST    1 TO 5  10 14  15 20 24
                               25 29 33  34 39 43

MEMBER END FORCES - (KIPS  FEET)

MEMB  LOAD  JT   AXIAL  SHEAR-Y  MOM-Z

   1     6   1   32.41    -0.88   -4.31
             2  -32.41     0.88  -14.38
         7   1   62.38    -2.77  -13.63
             2  -62.38     2.77  -45.49
         8   1   64.58    -2.69  -13.22
             2  -64.58     2.69  -44.11
         9   1   27.84     3.53   21.41
             2  -27.84     0.50   12.28
        10   1   30.25     6.01   36.21
             2  -30.25     0.54   24.45
        11   1    6.36     7.15   40.68
             2   -6.36    -0.60   41.37

   2     6   2   16.48     2.09   11.59
             3  -16.48    -2.09   -2.06
         7   2   47.39     6.68   37.12
             3  -47.39    -6.68   -6.06
         8   2   49.54     6.32   35.67
             3  -49.54    -6.32   -6.21
         9   2   14.49    -3.06  -13.18
             3  -14.49     3.92   -2.44
        10   2   16.99    -5.58  -24.91
             3  -16.99     6.98   -3.19
        11   2   -2.82    -7.99  -38.48
             3    2.82     9.39   -0.64

   3     6  24   32.41     0.88    4.31
            22  -32.41    -0.88   14.38
         7  24   62.38     2.77   13.63
            22  -62.38    -2.77   45.49
         8  24   64.58     2.69   13.22
            22  -64.58    -2.69   44.11
         9  24   29.72     4.42   27.85
```

		22	-29.72	-1.90	40.67
	10	24	33.31	6.92	44.10
		22	-33.31	-2.82	62.00
	11	24	9.31	5.78	37.48
		22	-9.31	-1.68	41.91
4	6	22	16.48	-2.09	-11.59
		23	-16.48	2.09	2.06
	7	22	47.39	-6.68	-37.12
		23	-47.39	6.68	6.06
	8	22	49.54	-6.32	-35.67
		23	-49.54	6.32	6.21
	9	22	15.66	-7.50	-36.33
		23	-15.66	8.04	1.31
	10	22	18.90	-11.66	-55.92
		23	-18.90	12.54	1.36
	11	22	-1.03	-8.56	-39.38
		23	1.03	9.44	-1.13
5	6	2	2.98	0.67	2.79
		4	-2.98	-0.67	1.26
	7	2	9.46	1.91	8.37
		4	-9.46	-1.91	3.46
	8	2	9.01	1.96	8.44
		4	-9.01	-1.96	3.71
	9	2	-2.56	0.27	0.90
		4	2.56	-0.27	0.70
	10	2	-5.04	0.19	0.46
		4	5.04	-0.19	0.59
	11	2	-8.59	-0.62	-2.90
		4	8.59	0.62	-0.84
10	6	12	-35.78	0.04	-0.17
		14	35.78	-0.04	0.48
	7	12	-114.24	-0.22	-1.62
		14	114.24	0.22	0.95
	8	12	-109.64	0.01	-0.81
		14	109.64	-0.01	1.31
	9	12	-31.27	0.03	-0.17
		14	31.27	-0.03	0.40
	10	12	-40.60	-0.06	-0.59
		14	40.60	0.06	0.33
	11	12	5.66	-0.02	-0.04
		14	-5.66	0.02	-0.11
14	6	20	2.98	-0.67	-1.26
		22	-2.98	0.67	-2.79

```
        7   20     9.46    -1.91    -3.46
            22    -9.46     1.91    -8.37
        8   20     9.01    -1.96    -3.71
            22    -9.01     1.96    -8.44
        9   20     9.40    -0.98    -1.63
            22    -9.40     0.98    -4.34
       10   20    14.48    -1.33    -2.10
            22   -14.48     1.33    -6.08
       11   20    10.25    -0.53    -0.62
            22   -10.25     0.53    -2.52

15      6    3    15.48     0.50     2.04
             5   -15.48    -0.50     1.20
        7    3    46.41     1.23     6.02
             5   -46.41    -1.23     3.36
        8    3    46.87     1.29     6.15
             5   -46.87    -1.29     3.57
        9    3    18.52     0.56     2.35
             5   -18.52    -0.56     1.28
       10    3    24.64     0.70     3.06
             5   -24.64    -0.70     1.58
       11    3     5.18     0.11     0.53
             5    -5.18    -0.11     0.16

20      6   13    36.16     0.11     0.04
            15   -36.16    -0.11     0.61
        7   13   122.47     0.33    -0.30
            15  -122.47    -0.33     1.91
        8   13   112.48     0.38     0.10
            15  -112.48    -0.38     1.92
        9   13    32.43     0.10     0.03
            15   -32.43    -0.10     0.56
       10   13    44.75     0.11    -0.13
            15   -44.75    -0.11     0.70
       11   13    -4.14    -0.03    -0.07
            15     4.14     0.03    -0.09

24      6   21    15.48    -0.50    -1.20
            23   -15.48     0.50    -2.04
        7   21    46.41    -1.23    -3.36
            23   -46.41     1.23    -6.02
        8   21    46.87    -1.29    -3.57
            23   -46.87     1.29    -6.15
        9   21     9.41    -0.36    -0.89
            23    -9.41     0.36    -1.36
       10   21     9.77    -0.37    -0.95
            23    -9.77     0.37    -1.46
```

	11	21	-9.17	0.25	0.46
		23	9.17	-0.25	1.02
25	6	4	13.53	-0.30	-0.72
		5	-13.53	0.30	-0.70
	7	4	41.33	-1.09	-2.28
		5	-41.33	1.09	-2.20
	8	4	41.59	-1.06	-2.21
		5	-41.59	1.06	-2.14
	9	4	10.93	-0.24	-0.53
		5	-10.93	0.24	-0.57
	10	4	13.07	-0.30	-0.65
		5	-13.07	0.30	-0.72
	11	4	-3.96	0.09	0.27
		5	3.96	-0.09	0.18
29	6	12	-1.72	0.00	0.00
		13	1.72	0.00	0.00
	7	12	-14.40	0.00	0.00
		13	14.40	0.00	0.00
	8	12	-8.69	0.00	0.00
		13	8.69	0.00	0.00
	9	12	-1.51	0.00	0.02
		13	1.51	0.00	0.01
	10	12	-4.55	0.01	0.03
		13	4.55	-0.01	0.02
	11	12	0.64	0.01	0.02
		13	-0.64	-0.01	0.02
33	6	20	13.53	0.30	0.72
		21	-13.53	-0.30	0.70
	7	20	41.33	1.09	2.28
		21	-41.33	-1.09	2.20
	8	20	41.59	1.06	2.21
		21	-41.59	-1.06	2.14
	9	20	13.90	0.32	0.79
		21	-13.90	-0.32	0.71
	10	20	17.92	0.43	1.07
		21	-17.92	-0.43	0.94
	11	20	0.73	0.04	0.14
		21	-0.73	-0.04	0.04
34	6	3	-21.88	0.04	0.02
		4	21.88	-0.04	-0.05
	7	3	-66.32	0.38	0.04
		4	66.32	-0.38	-0.18
	8	3	-66.43	0.37	0.06

```
              4   66.43    -0.37    -0.16
         9    3  -18.23     0.05     0.09
              4   18.23    -0.05     0.03
        10    3  -22.09     0.08     0.13
              4   22.09    -0.08     0.05
        11    3    5.14     0.03     0.11
              4   -5.14    -0.03     0.12

  39     6   12    0.92     0.00    -0.02
             15   -0.92     0.00     0.02
         7   12   -7.47    -0.01    -0.07
             15    7.47     0.01     0.07
         8   12    0.30     0.00    -0.06
             15   -0.30     0.00     0.06
         9   12   -0.46     0.00     0.00
             15    0.46     0.00     0.03
        10   12   -4.51     0.00    -0.01
             15    4.51     0.00     0.04
        11   12   -2.53     0.00     0.02
             15    2.53     0.00     0.01

  43     6   20  -21.88    -0.04     0.05
             23   21.88     0.04    -0.02
         7   20  -66.32    -0.38     0.18
             23   66.32     0.38    -0.04
         8   20  -66.43    -0.37     0.16
             23   66.43     0.37    -0.06
         9   20  -21.82    -0.01     0.12
             23   21.82     0.01     0.05
        10   20  -27.96    -0.02     0.19
             23   27.96     0.02     0.10
        11   20   -0.52     0.03     0.11
             23    0.52    -0.03     0.12

104. FINISH
```

Cross-Sectional Properties and Flexure

B.1 NOTATION

In Figure B.1

1. x and y are orthogonal *reference axes*.
2. X and Y are *centroidal axes* parallel to the x-and y-axes.

From similar triangles, for Figure B.2,

$$\frac{w}{b} = \frac{h-y}{h} \qquad w = \frac{b(h-y)}{h}$$

For Figure B.3,

$$\frac{Z}{h} = \frac{b-x}{b} \qquad Z = \frac{h(b-x)}{b}$$

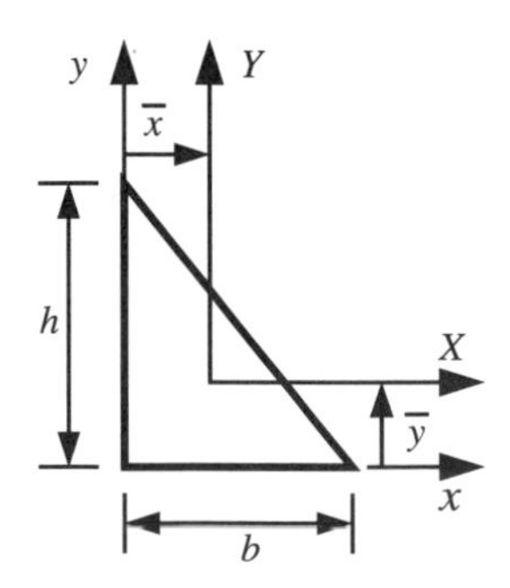

FIGURE B.1

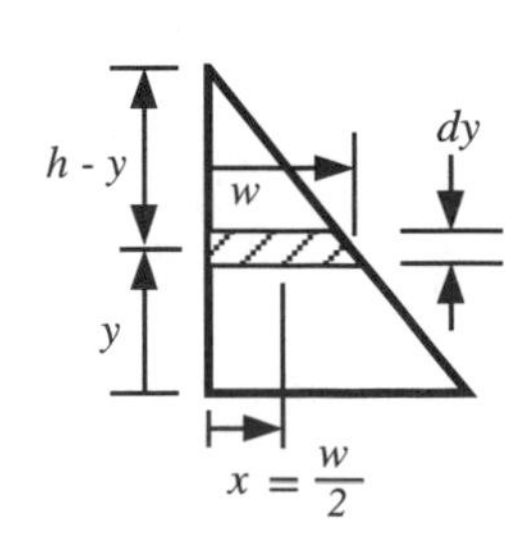

FIGURE B.2

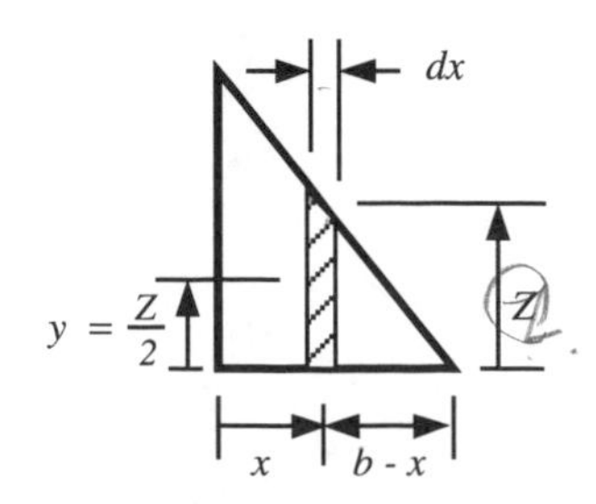

FIGURE B.3

B.2 CENTROIDAL AXES

In Figure B.1, the origin of the *centroidal axes* (X, Y) is at $x = \overline{x}$ and $y = \overline{y}$ where $\overline{x}$ and $\overline{y}$ are defined as follows:

$$A = \int dA = \int_0^h wdY = \frac{b}{h}\int_0^h (h-y)\ dy = \frac{bh}{2}$$

$$\overline{x} = \frac{\int xdA}{A} = \frac{\int_0^b xZdx}{A} = \frac{\frac{h}{b}\int_0^b x(b-x)\ dx}{0.5bh} = \frac{b}{3}$$

$$\overline{y} = \frac{\int ydA}{A} = \frac{\int_0^h ywdy}{A} = \frac{\frac{h}{b}\int_0^h y(h-y)\ dy}{0.5bh} = \frac{h}{3}$$

B.3 MOMENTS AND PRODUCT OF INERTIA

For the *reference axes* (x, y), the definition of the *moments of inertia* (I_x and I_y) and the *product of inertia* (I_{xy}) are as follows. Using Figure B.2,

$$I_x = \int y^2 dA = \int_0^h y^2 w\,dy = \frac{b}{h}\int_0^h y^2\ (h-y)\ dy = \frac{bh^3}{12}$$

$$I_{xy} = \int xy dA = \int_0^h xy\ \ w\,dy = \frac{1}{2}\int_0^h w^2 y\ dy$$

$$I_{xy} = \frac{b^2}{2h^2}\int_0^h y(h-y)^2\ dy = \frac{b^2h^2}{24}$$

Using Figure B3,

$$I_y = \int y^2\ dA = \int_0^b x^2 Z\ dx = \frac{h}{b}\int_0^b x^2\ (h-x)\ dx = \frac{hb^3}{12}$$

The preceding properties for the reference axes are easier to obtain by integration of the calculus definitions than the properties for the centroidal axes by integration. The properties for the centroidal axes are needed, however, and they are most easily obtained by using the transfer axes formulas defined in the next section.

Hot-rolled steel sections (L-shaped and I-shaped sections, e.g.) are composed of two or more rectangles with some rounded in-filled corners (fillets) where the rectangles intersect in the steel section. The L section also has some rounded exterior corners (toe radii). The properties of a rectangle and the fillets are obained by integration of the calculus definitions, and the summation equation definitions given below are used to find the properties of a steel-rolled section.

B.4 TRANSFER AXES FORMULAS

Definitions are as follows:

$$I_x = I_X + A\overline{y}^2$$

$$I_y = I_Y + A\overline{x}^2$$

$$I_{xy} = I_{XY} + A\bar{x}\bar{y}$$

Note that $\bar{x}$ and $\bar{y}$ have signs and are positive in the arrow directions in Figure B1.

For Figure B.1

$$I_X = I_x - A\bar{y}^2 = \frac{bh^3}{12} - \left(\frac{bh}{2}\right)\left(\frac{h}{3}\right)^2 = \frac{bh^3}{36}$$

$$I_Y = I_y - A\bar{x}^2 = \frac{hb^3}{12} - \left(\frac{bh}{2}\right)\left(\frac{b}{3}\right)^2 = \frac{hb^3}{36}$$

$$I_{XY} = I_{xy} - A\bar{x}\bar{y} = \frac{h^2b^2}{24} - \left(\frac{bh}{2}\right)\left(\frac{b}{3}\right)\left(\frac{h}{3}\right) = -\frac{b^2h^2}{72}$$

Note that the definitions assume the properties for the centroidal axes are known and the properties for the reference axes are to be obtained. That is, the transfer is made from the centroidal axes to the reference axes. Then, using algebra, we obtain the last three equations from the definitions. With the last three equations, the properties for the centroidal axes are easily obtained from the properties for the reference axes.

B.5 SUMMATION FORMULAS

Properties for the centroidal axes and reference axes are given for a rectangle in Figure B.4. We will use these properties in an example problem. Properties for many other geometric sections are given on LRFD pp. 7-17 to 7-23.

As shown in Figure B.5, an angle section is L-shaped and has some rounded in-filled corners (fillets) and some rounded exterior corners (toe radii). Other hot-rolled sections have fillets and some of them have toe radii. For all hot-rolled sections except L sections, the fillets and toe radii are accounted for in finding the section properties. There is no standardized requirement for the fillet and toe radii sizes in each L section, and all rolling mills do not use the same fillet and toe radii sizes. Consequently, as illustrated in Figure B.5 for an L section, in the LRFD Manual the fillets and toe radii are ignored in finding the section properties and the L section can be treated as two rectangles.

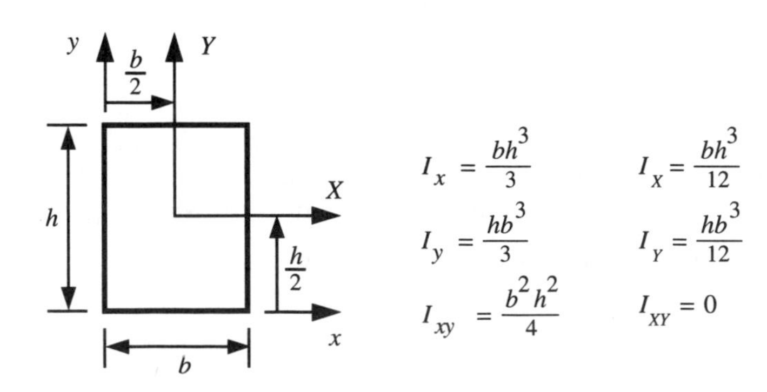

FIGURE B.4

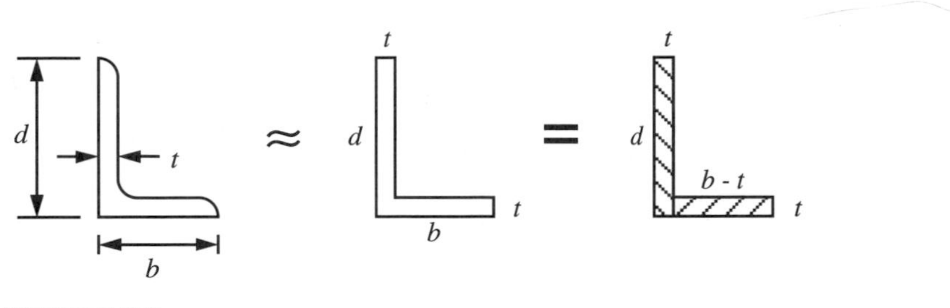

FIGURE B.5

Example B.1

Use the centroidal axes properties given in Figure B.4 to obtain via summation the centroidal axes properties for an L8 × 6 × 1 section subdivided into the two rectangles shown in Figure B.6. Note in Figure B.6 that we have omitted the usual superscript bar symbol on the coordinates of the centroid for each of the individual areas since a subscript must be used and makes the definition unique. However, for the coordinates of the original or composite area, we must retain the superscript bar symbol.

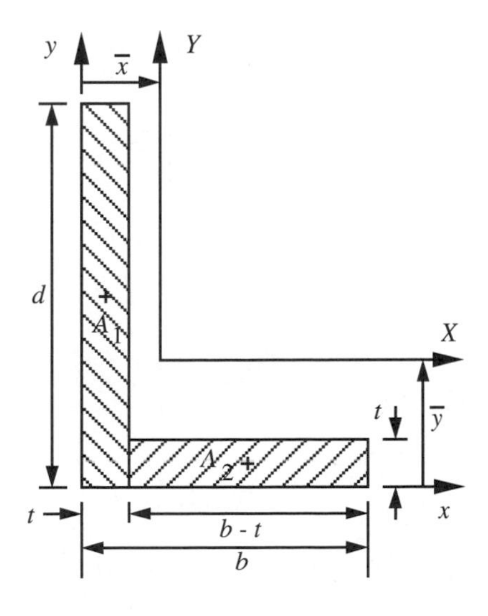

FIGURE B.6

Solution

$$A_1 = dt = (8)(1) = 8 \text{ in.}^2$$

$$x_1 = 0.5t = (0.5)(1) = 0.5 \text{ in.}$$

$$y_1 = 0.5d = (0.5)(8) = 4 \text{ in.}$$

$$A_2 = (b - t)t = (6 - 1)(1) = 5 \text{ in.}^2$$

$$x_2 = t + 0.5(b - t) = 1 + (0.5)(6 - 1) = 3.5 \text{ in.}$$

$$y_2 = 0.5t = (0.5)(1) = 0.5 \text{ in.}$$

$$A = \sum_{i=1}^{2} A_i = 8 + 5 = 13 \text{ in.}^2$$

$$\bar{x} = \frac{\sum_{i=1}^{2} x_i A_i}{A} = \frac{0.5(8) + 3.5(5)}{13} = 1.65 \text{ in.}$$

$$\bar{y} = \frac{\sum_{i=1}^{2} y_i A_i}{A} = \frac{4(8) + 0.5(5)}{13} = 2.65 \text{ in.}$$

$$X_i = x_i - \bar{x}$$

$$Y_i = y_i - \bar{y}$$

$$I_X = \sum_{i=1}^{2} \left[(I_X)_i + A_i Y_i^2 \right]$$

$$I_X = \left[\frac{1(8)^3}{12} + 8(4 - 2.65)^2 \right] + \left[\frac{5(1)^3}{12} + 5(0.5 - 2.65)^2 \right] = 80.78 \text{ in.}^4$$

$$I_Y = \sum_{i=1}^{2} \left[(I_Y)_i + A_i X_i^2 \right]$$

$$I_Y = \left[\frac{8(1)^3}{12} + 8(0.5 - 1.65)^2 \right] + \left[\frac{1(5)^3}{12} + 5(3.5 - 1.65)^2 \right] = 38.78 \text{ in.}^4$$

$$I_{XY} = \sum_{i=1}^{2} \left[(I_{XY})_i + A_i X_i Y_i \right]$$

$$I_{XY} = \left[0 + 8(0.5 - 1.65)(4 - 2.65) \right] + \left[0 + 5(3.5 - 1.65)(0.5 - 2.65) \right] = -32.31 \text{ in.}^4$$

Note: Our computed values of I_x and I_y agree with those given for an L8 x 6 x 1 on LRFD pp. 1-56 and 1-57.

B.6 PRINCIPAL AXES

The *principal axes* are those centroidal axes for which the product of inertia is zero.

Let

W denote the *major principal axis* (maximum inertia axis), and

Z denote the *minor principal axis* (minimum inertia axis).

Then, by definition, $I_{WZ} = 0$.

An axis of symmetry is a principal axis. In many steel sections, the X-and/or Y-axis is an axis of symmetry, in which case the X, Y axes are also the principal axes. As shown in Figure B.7, the principal axes for an L8 x 6 x 1 section are not the X, Y-axes.

B.7 USING MOHR'S CIRCLE TO FIND THE PRINCIPAL AXES

Example B.2

For an L8 × 6 × 1 section and the notation shown in Figure B.7, use Mohr's circle to find:

1. α which locates the principal axes (W, Z)
2. I_W and I_Z

Solution

For an L8 x 6 x 1, from Section B.5 we know that

$$I_X = 80.78 \text{ in.}^4 \qquad I_Y = 38.78 \text{ in.}^4 \qquad I_{XY} = -32.31 \text{ in.}^4$$

The following steps are illustrated in Figures B.8 and B.9:

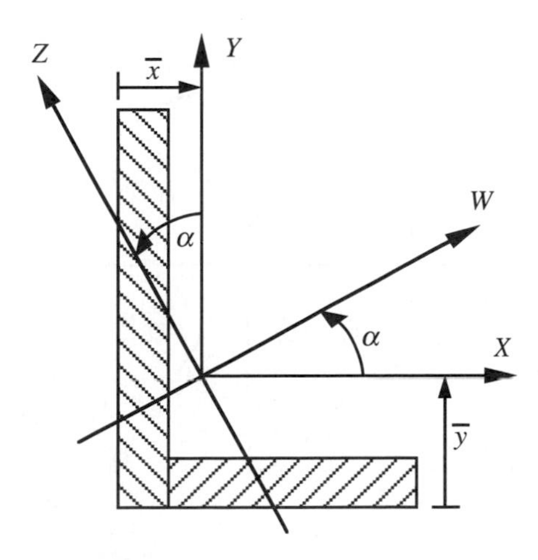

FIGURE B.7

Step 1

Plot point:

$$(I_X, I_{XY}) = (80.78, -32.31)$$

Plot point:

$$(I_Y, -I_{XY}) = (38.78, 32.31)$$

Step 2

Locate the center of the circle:

$$C = (I_X + I_Y)/2 = (80.78 + 38.78)/2 = 59.78$$

Step 3

Draw the circle (see Figure B.8). The center is at *C*; the circle passes through points plotted in Step 1.

Step 4

See Figure B.9. From point (I_X, I_{XY}), erect a perpendicular to the inertias axis. Calculate:

$$b = I_X - C = 80.78 - 59.78 = 21.0$$

b is the base of the shaded triangle

triangle height h = (absolute value of I_{xy})

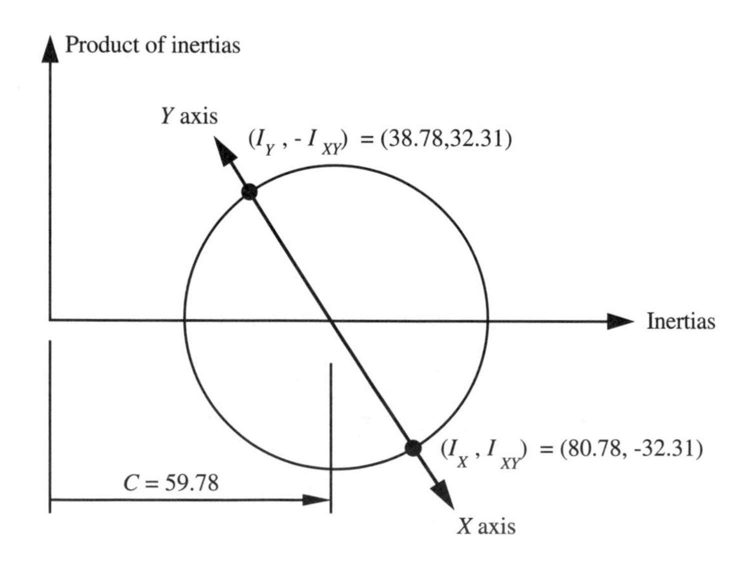

FIGURE B.8

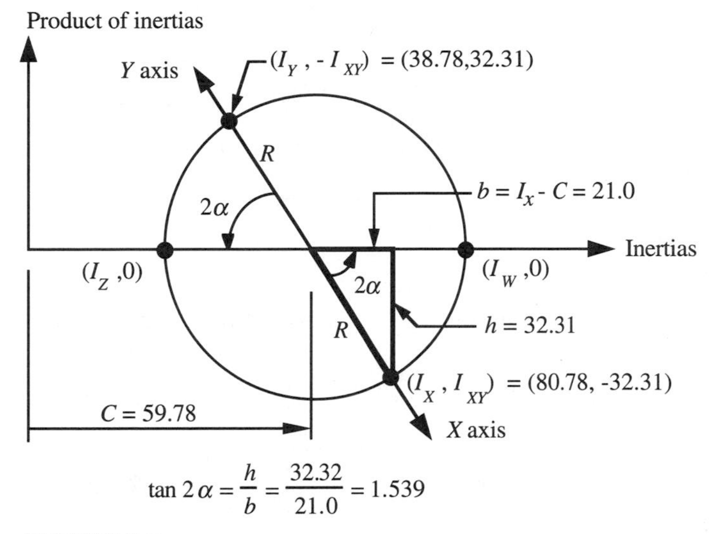

FIGURE B.9

$$R = \sqrt{b^2 + h^2} = \sqrt{(21.0)^2 + (32.31)^2} = 38.53$$

$$\tan\ 2\alpha = h/b = 32.31/21.0 = 1.539$$

$$2\alpha = 57.0^\circ \qquad \text{(on Mohr's circle)}$$

$$\alpha = 28.5^\circ \qquad \text{(on Figure B.7)}$$

$$\tan\ \alpha = 0.543$$

Note: Each angle in Figure B.7 is *doubled* on Mohr's circle.

Step 5

In Figure B.9, note that 2α on Mohr's circle is measured counterclockwise (CCW) from the *X*-axis to the *W*-axis (maximum inertia axis). Also, 2α is measured CCW from the *Y*-axis to the *Z*-axis (minimum inertia axis).

In Figure B.7

(a) At $\alpha = 28.5^\circ$ CCW from the *X*-axis, draw the *W*-axis.

(b) At $\alpha = 28.5^\circ$ CCW from the *Y*-axis, draw the *Z*-axis.

Calculate:

$$I_W = C + R = 59.78 + 38.53 = 98.31 \text{ in.}^4$$

$$I_Z = C - R = 59.78 - 38.53 = 21.25 \text{ in.}^4$$

B.8 RADIUS OF GYRATION

If the cross section for which I_W and I_Z were found by using Mohr's circle is used as a column, the radius of gyration is needed for each principal axis.

Definition:

$$r_i = \sqrt{\frac{I_i}{A}}$$

where r_i is the radius of gyration for i, any axis of interest.

Example B.3

Using A, I_W, and I_Z from Section B.7, find the radii for the principal axes of an L8 × 6 × 1 section.

Solution

$$r_W = \sqrt{\frac{I_W}{A}} = \sqrt{\frac{98.31 \text{ in.}^4}{13 \text{ in.}^2}} = 2.75 \text{ in.}$$

$$r_Z = \sqrt{\frac{I_Z}{A}} = \sqrt{\frac{21.25 \text{ in.}^4}{13 \text{ in.}^2}} = 1.28 \text{ in.}$$

Note: Our computed values of $\tan\ \alpha = 0.543$ (in Section B.8) and r_Z agree with those given for an L8 x 6 x 1 section on LRFD p. 1-57.

B.9 PROPERTIES OF A STEEL L SECTION

Example B.4

See LRFD pp. 1-56 and 1-57 for the properties of an L8 x 6 x 1. Note that all properties are given for the centroidal axes, but only $\tan \alpha$ and r_Z are given for the principal axes. Find I_Z and I_W.

Solution

$$I_Z = Ar_Z^2 = 13.0(1.28)^2 = 21.3 \text{ in.}^4$$

$$(I_W + I_Z) = (I_x + I_y)$$

$$I_W = (I_x + I_y) - I_Z = 80.8 + 38.8 - 21.3 = 98.3 \text{ in.}^4$$

$$r_W = \sqrt{\frac{I_W}{A}} = \sqrt{\frac{98.3}{13}} = 2.75 \text{ in.}$$

Example B.5

If the cross section of a member is a single angle, only the properties for the principal axes are needed and I_{xy} is not needed. However, if a single angle is fillet-welded to some other rolled section to create a combined section, I_{xy} of the single angle will be needed if the properties for the principal axes of the combined section are needed. Find I_{xy} and I_W of an L8 × 6 × 1 section.

Solution

From LRFD p. 1-57 for an L8 x 6 x 1 :

$$\tan\alpha = 0.543 \qquad I_X = 80.8 \text{ in.}^4 \qquad I_Y = 38.8 \text{ in.}^4$$

From Example B.3,

$$I_Z = 21.3 \text{ in.}^4$$

$$\alpha = \tan^{-1}(0.543) = 28.5^\circ$$

$$2\alpha = 57.0^\circ$$

Using Mohr's circle (see Figure B10):

$$C = (I_X + I_Y)/2 = (80.8 + 38.8)/2 = 59.8$$

$$R = C - I_Z = 59.8 - 21.3 = 38.5$$

$$I_W = C + R = 59.8 + 38.5 = 98.3 \text{ in.}^4$$

On LRFD p. 1-57, note that the Z-axis is located at $\alpha = 28.5^\circ$ CCW from the Y-axis. The W-axis is not shown, but W is located at $\alpha = 28.5^\circ$ CCW from the X-axis. On Mohr's

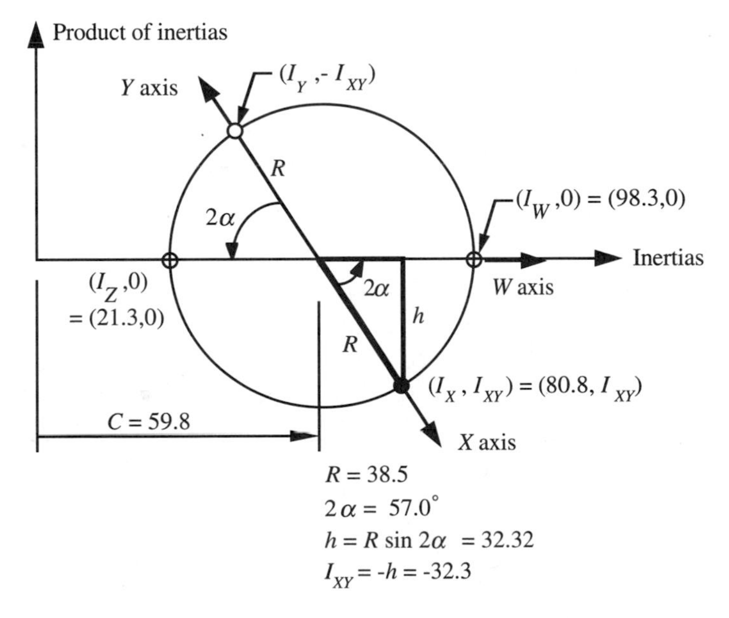

FIGURE B.10

circle (see Figure B.10), we need to locate the *X*-axis in order to find I_{xy} from the point whose coordinates are (I_x, I_{xy}). We know that the *W*-axis passes through the maximum inertia point on Mohr's circle and that the *X*-axis must be properly located such that the *W*-axis is at $2\alpha = 57.0°$ CCW from the X-axis. Since $h = R \sin 2\alpha = 32.3$ and since the point (I_x, I_{xy}) lies in the region where the product of inertias is negative, $I_{xy} = -32.3$. Alternatively, I_{xy} can be found by using the summation formulas and the figure in the LRFD Manual decomposed into two rectangles, or by using the formula on LRFD p. 7-23, but we prefer to use Mohr's circle.

B.10 FLEXURE FORMULA

Figure B.11 shows a W18 × 50 used as a simply supported beam (Case 1 on LRFD p. 4-190) with $L = 30$ ft and $w = 1.5$ kips/ft, which causes bending about only the *X*-axis.

The maximum bending moment is

$$M_X = \frac{wL^2}{8} = \frac{(1.5 \text{ kips/ft})(30 \text{ ft})^2}{8} = 168.75 \text{ ft-kips} = 2025 \text{ in.-kips}$$

The *X*-and *Y*-axes are axes of symmetry. Therefore, *X* and *Y* are principal axes. The flexure formula

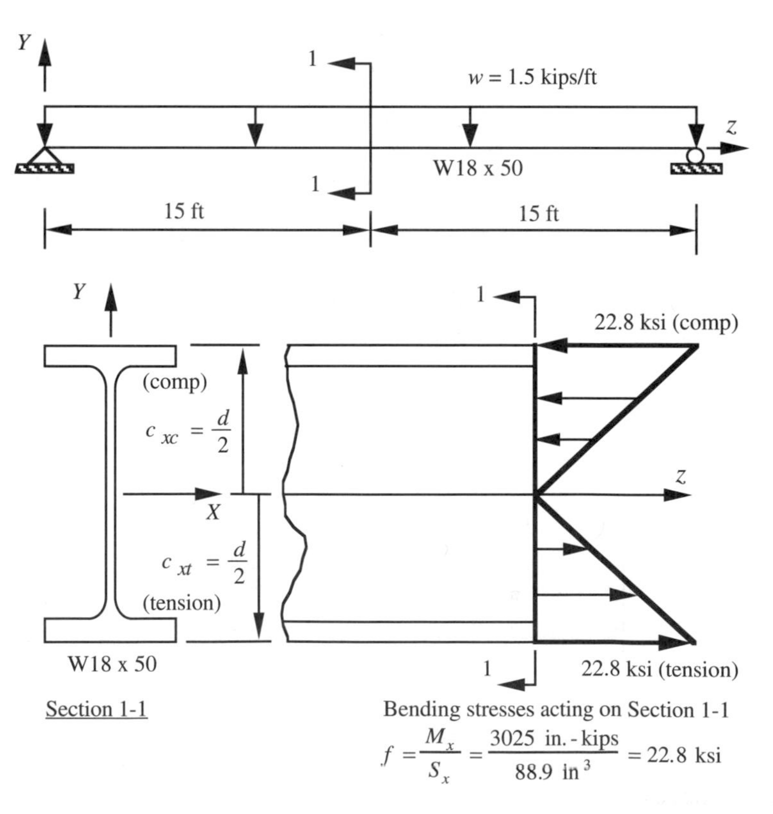

FIGURE B.11

$$f = \frac{Mc}{I}$$

is only valid

1. In the linearly-elastic region of the stress–strain curve
2. For each principal axis

The parameters in the flexure formula are

f = extreme fiber bending stress (either compression or tension)

M = bending moment about a principal axis

c = perpendicular distance from the bending axis to an extreme fiber

I = moment of inertia for the bending axis

Suppose that a section is bending about the *major principal axis* W; then, the flexure formula is

$$f = \frac{M_W c_W}{I_W}$$

Note that M, c, and I are for the same principal axis, W. A similar formula can be written for the other principal axis Z, by substituting Z for the subscript W in the preceding formula. The bending stress diagram (see Figure B.11) varies linearly from zero at the bending axis to f at the extreme fiber. On one side of the bending axis, f is a compressive bending stress, and on the other side of the bending axis, f is a tensile bending stress. For a W section subjected to bending about the major principal axis X, the LRFD Manual defines S_x as the elastic section modulus for X-axis bending:

$$S_X = \frac{I_X}{c_X}$$

where $c_x = d/2$; then,

$$f = \frac{M_X}{S_X}$$

On LRFD p. 1-32, we find for a W18 x 50 that $S_x = 88.9\text{ in.}^3$

$$f = \frac{M_X}{S_X} = \frac{2025\text{ in - kips}}{88.9\text{ in.}^3} = 22.8\text{ ksi}$$

B.11 BIAXIAL BENDING

Since we show only one example of biaxial bending of a beam, the example entails a completely general case of a single angle, L8 x 6 x 1, loaded through the shear center (see Figure B.12) to prevent twisting of the cross section, which avoids the need for computing torsional stresses.

The most convenient manner to fillet-weld a single angle to another steel section having flat exterior surfaces is to lay the longer leg of the single angle on a flat region of the other steel section. Then fillet welds are made on each edge of the longer leg

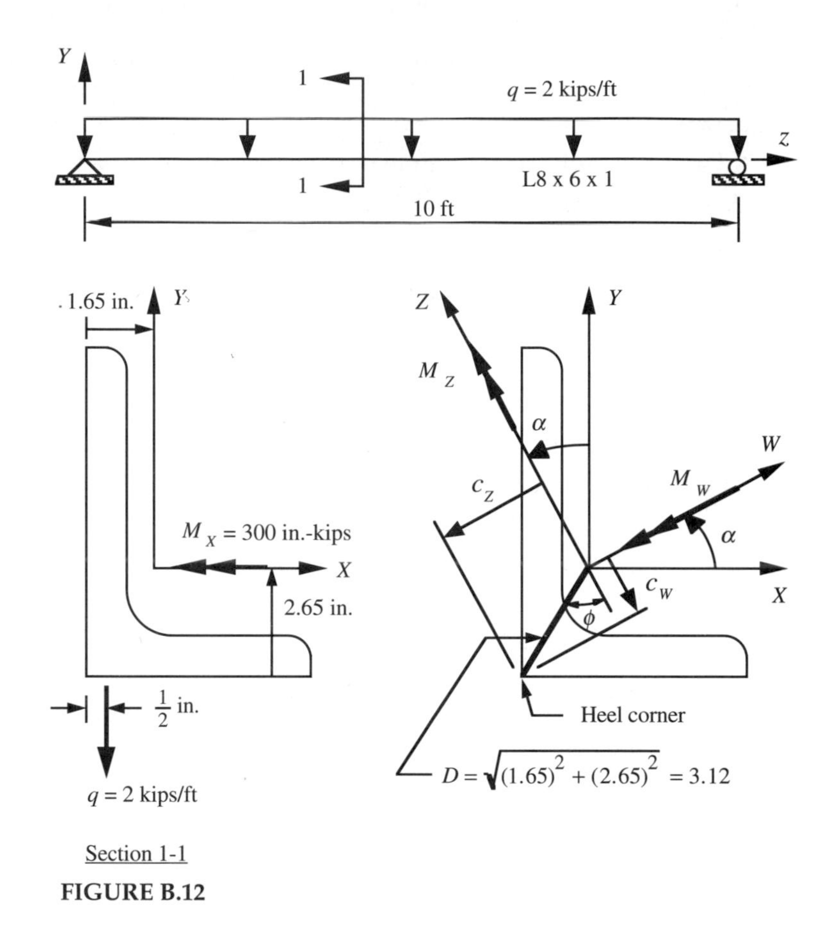

Section 1-1

FIGURE B.12

of the single angle. An example of biaxial bending of a single angle used as a compression member in a truss is given on LRFD p. 3-104. Such a truss member is classified as a beam-column.

Example B.6

Use the flexure formula to compute the total bending stress at the heel corner of the L8 × 6 × 1 subjected only to biaxial bending as shown in Figure B.12.

Solution

Since the flexure formula is only valid for each principal axis (see Section B.6), we need:

1. $I_W = 98.3$ in.^{4} and $I_Z = 21.3$ in.^{4} (from Example B.2)
2. M_W and M_Z
3. c_W and c_Z for the heel corner in Figure B.12

$$M_x = \frac{wL^2}{8} = \frac{(2\,\text{kips/ft})(10\,\text{ft})^2}{8} = 25\ \text{ft-kips} = 300\ \text{in.-kips}$$

$$\alpha = 28.5°$$

$$M_W = M_x \cos\alpha = 236.6\ \text{in.-kips}$$

$$M_Z = M_x \sin\alpha = 143.17\ \text{in.-kips}$$

$$D = \sqrt{(1.65)^2 + (2.65)^2} = 3.12\ \text{in.}$$

$$\tan\theta = \frac{1.65}{2.65} = 0.62264$$

$$\theta = 31.908°$$

$$\phi = \theta + \alpha = 60.408°$$

At the heel corner of the L8 x 6 x 1:

1. M_W produces a tensile stress.

2. M_Z produces a tensile stress.

$$f = \frac{M_W c_W}{I_W} + \frac{M_Z c_Z}{I_Z} = \frac{263.6(1.54)}{98.3} + \frac{143.1(2.71)}{21.3}$$

$$f = 4.13 + 18.21 = 22.34\ \text{ksi} \quad \text{(tension)}$$

PROBLEMS

B.1 The built-up box section in Figure B.13 is a pair of C15 × 40 and a pair of 1 x 10 plates. Each plate is 1 in. thick and 10 in. wide. See LRFD p. 1-50 for the properties of a C15 x 40 section:

(a) Locate the centroidal axes, *X* and *Y*.
(b) Compute the principal moments of inertia, I_X and I_Y.
(c) Compute the radius of gyration for each principal axis.

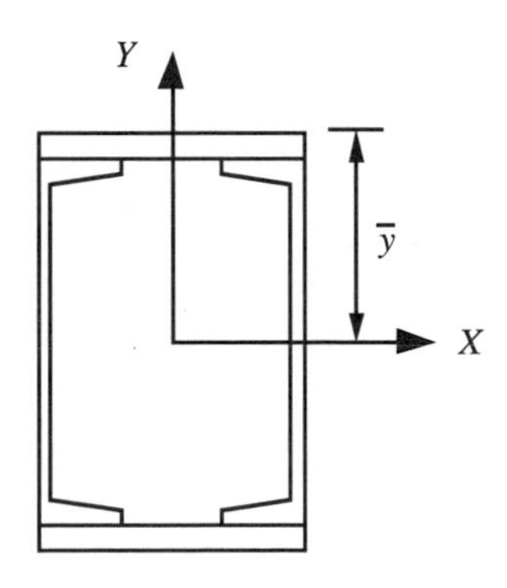

FIGURE B.13

B.2 The built-up box section in Figure B.13 is a pair of C15 x 33.9 and a pair of 1 x 10 plates. Each plate is 1 in. thick and 10 in. wide. See LRFD p. 1-50 for the properties of a C15 x 33.9 section:

(a) Locate the centroidal axes, X and Y.
(b) Compute the principal moments of inertia, I_X and I_Y.
(c) Compute the radius of gyration for each principal axis.

B.3 The built-up crane-girder section in Figure B.14 is a C10 x 15.3 cap on a W16 x 26. See LRFD p. 1-50 for the properties of a C10 x 15.3 section and LRFD p. 1-34 for the properties of a W16 x 26 section:

(a) Locate the centroidal axes, X and Y.
(b) Compute the principal moments of inertia, I_X and I_Y.
(c) Compute the radius of gyration for each principal axis.

B.4 The built-up crane-girder section in Figure B.14 is a C15 x 33.9 cap on a W24 x 55. See LRFD p. 1-50 for the properties of a C15 x 33.9 section and LRFD p. 1-30 for the properties of a W24 x 55 section:

(a) Locate the centroidal axes, X and Y.
(b) Compute the principal moments of inertia, I_X and I_Y.
(c) Compute the radius of gyration for each principal axis.

B.5 The built-up column section in Figure B.15 is a pair of L8 x 6 x 7/8 with long legs back to back and separated by s = 1 in. See LRFD p. 1-56 for properties of one L8 x 6 x 7/8 section:

(a) Locate the centroidal axes, X and Y.
(b) Compute the principal moments of inertia, I_X and I_Y.
(c) Compute the radius of gyration for each principal axis.

B.6 The built-up column section in Figure B.15 is a pair of L6 x 4 x 7/8 with long legs back to back and separated by s = 1 in. See LRFD p. 1-58 for properties of one L6 x 4 x 7/8 section:

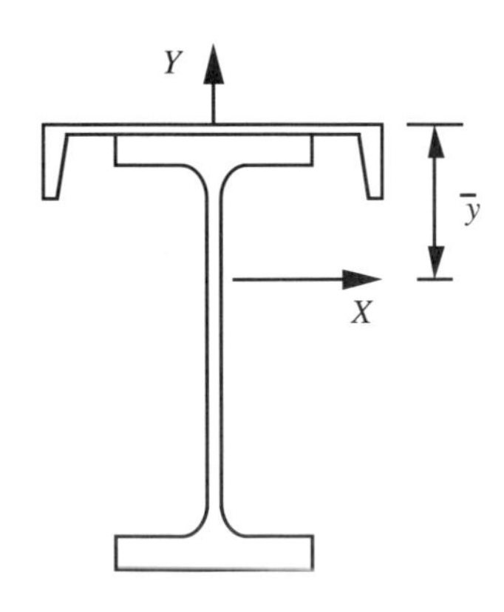

FIGURE B.14

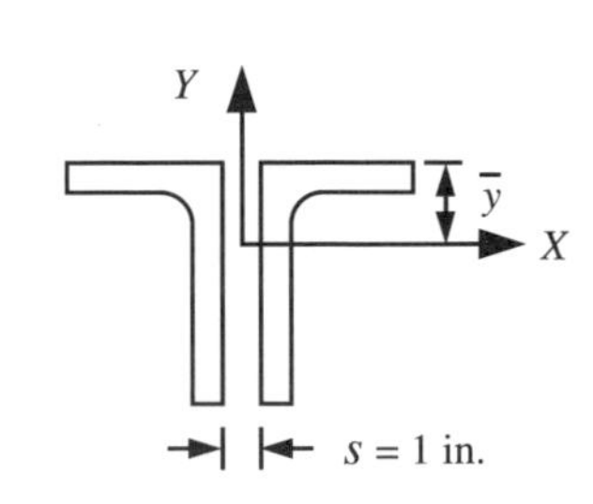

FIGURE B.15

a) Locate the centroidal axes, *X* and *Y*.
b) Compute the principal moments of inertia, I_X and I_Y.
c) Compute the radius of gyration for each principal axis.

B.7 The built-up crane-girder section in Figure B.16 is a C15 x 33.9, two web plates each of which is 0.25 in. x 36 in., and a 0.625 in. x 18 in. bottom flange plate. See LRFD p. 1-50 for properties of a C15 x 33.9 section:

(a) Locate the centroidal axes, *X* and *Y*
(b) Compute the principal moments of inertia, I_X and I_Y.
(c) Compute the radius of gyration for each principal axis.

B.8 See Figure B.17. On the bottom line of LRFD p. 1-118 , information is given for a built-up section composed of a C15 x 33.9 (see LRFD p.1-50) and a L6 x 3.5 x 5/16 (see LRFD p.1-58). Use the information given on the bottom line of LRFD p.1-118 only to check your solution:

(a) Locate the centroidal axes, *X* and *Y*.
(b) Locate the principal axes, *W* and *Z*.
(c) Compute the principal moments of inertia, I_W and I_Z.

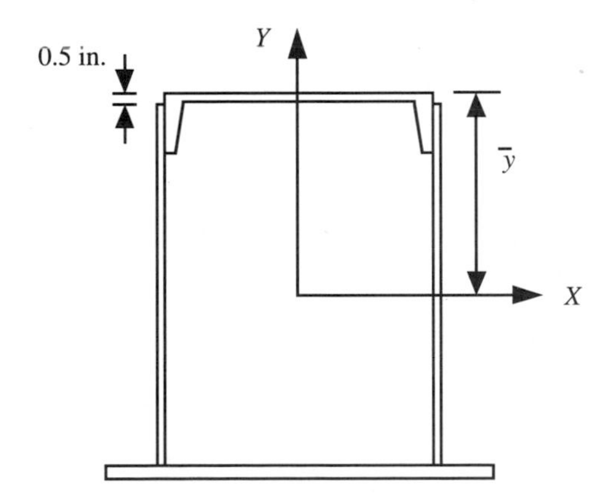

FIGURE B.16

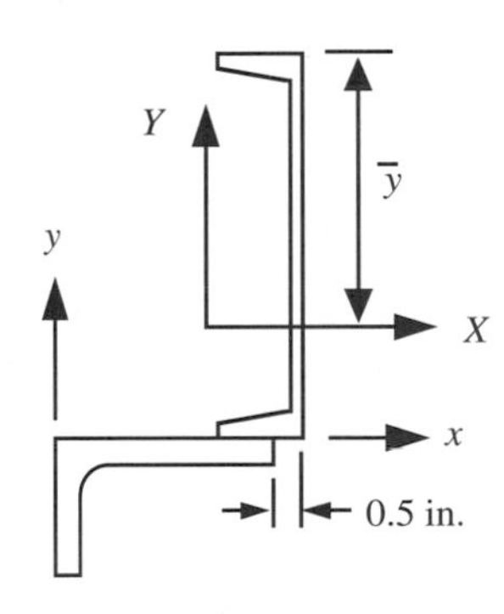

FIGURE B.17

Torsional Properties

NOTATION

Location of centroid is denoted by an addition symbol.
Location of shear center is denoted by a large black dot.
C_w = warping torsional constant
J = nonwarping torsional constant

For a cross section composed of thin rectangular elements of length b and thickness t,

$$J = \frac{1}{3}\sum bt^3$$

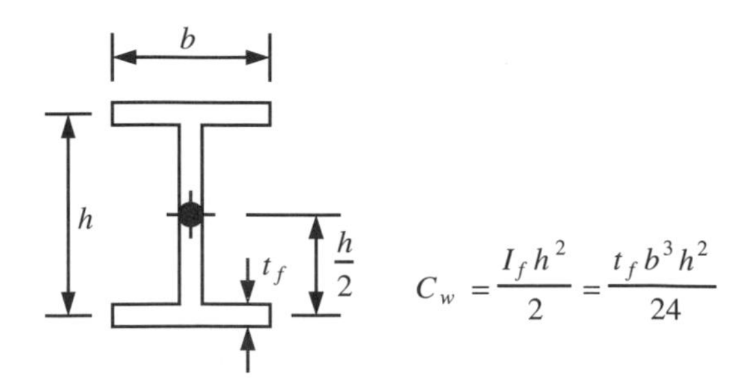

FIGURE C.1

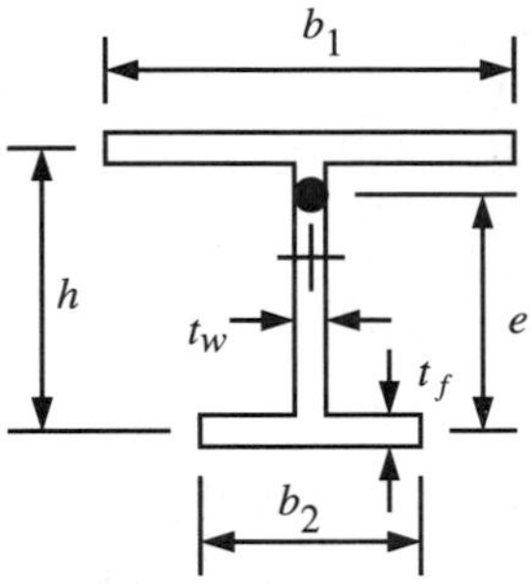

$$e = \frac{b_1^3 h}{b_1^3 + b_2^3}$$

$$C_w = \frac{I_f h^2}{12}\left(\frac{b_1^3 b_2^3}{b_1^3 + b_2^3}\right)$$

FIGURE C.2

$$e = \frac{3b^2 t_f}{6t_f b + t_w h}$$

$$C_w = \frac{t_f b^3 h^2}{12}\left(\frac{3t_f b + 2t_w h}{6t_f b + t_w h}\right)$$

FIGURE C.3

$$C_w = \frac{b^3 h^2 \left[2t_f(b^2 + bh + h^2) + 3t_w bh\right]}{12(2b + h)^2}$$

FIGURE C.4

$$C_w = \frac{1}{36}\left(\frac{b^3 t_f^3}{4} + h^3 t_w^3\right)$$

FIGURE C.5

$$C_w = \frac{b^3 t_1^3 + h^3 t_2^3}{36}$$

FIGURE C.6

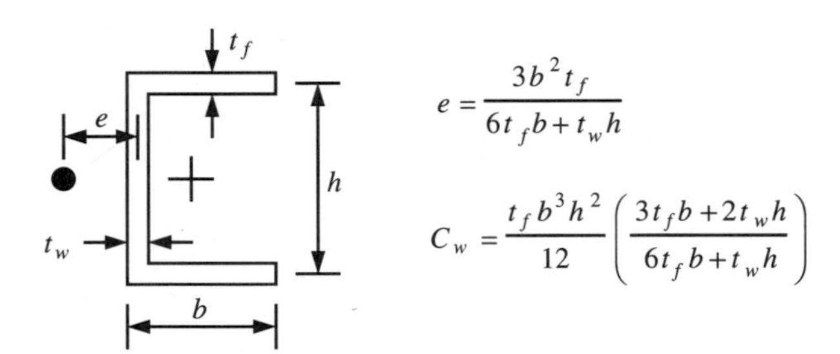

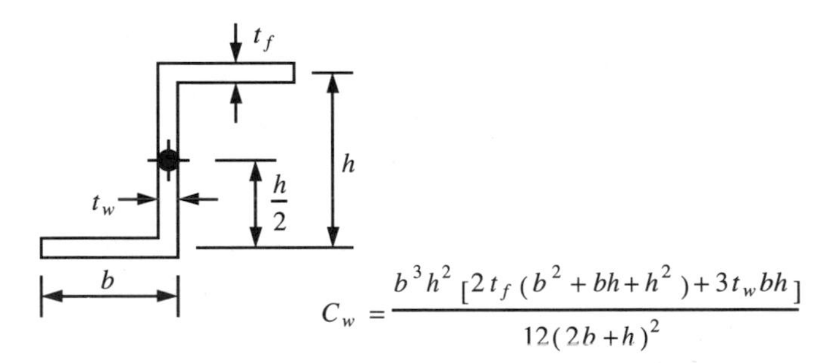

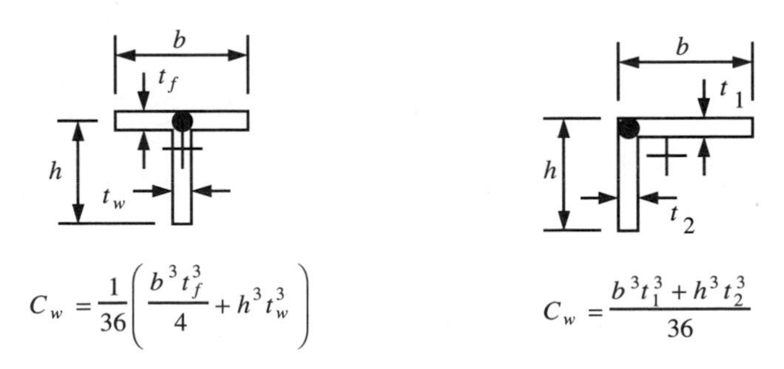

Appendix D

Plastic Analysis Formulas

NOTES

1. Except as otherwise noted, one W section is assumed throughout each structure.
2. M_{up} = required plastic hinge strength for a LRFD loading combination.
3. At a plastic hinge location, $\phi M_{px} \geq M_{up}$ is the LRFD design requirement for bending about the major axis.
4. Information is given to assist the designer in choosing the splice point locations of a continuous beam in regions where the required moment is much less than M_{up}.
5. For the cases that involve more than one span, the given information should not be used when a checkboard pattern of live loading can occur.

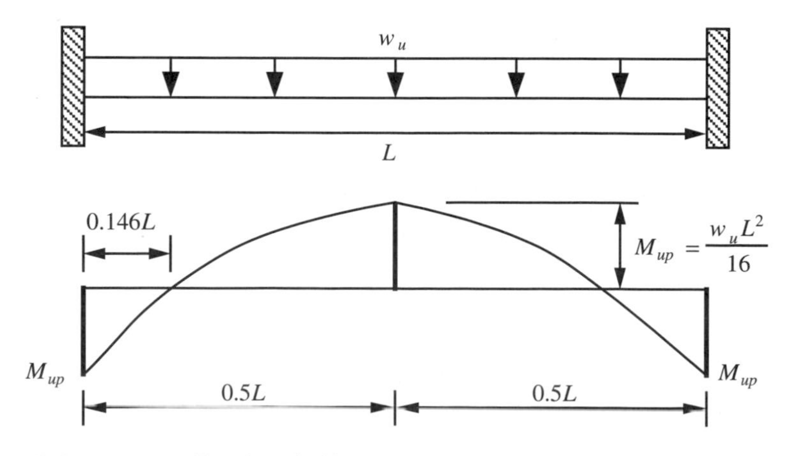

FIGURE D.1 Fixed-ended beam.

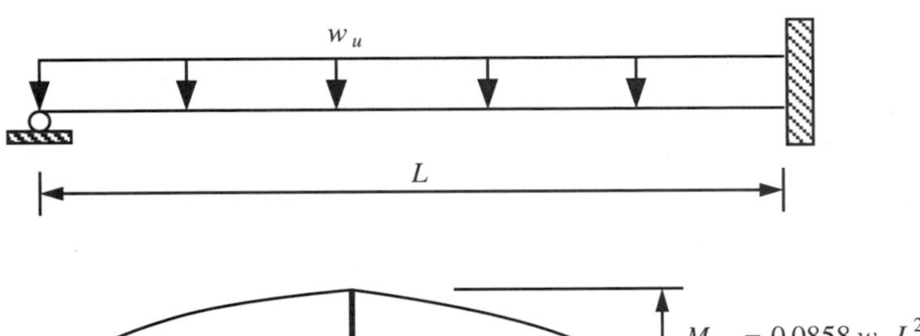

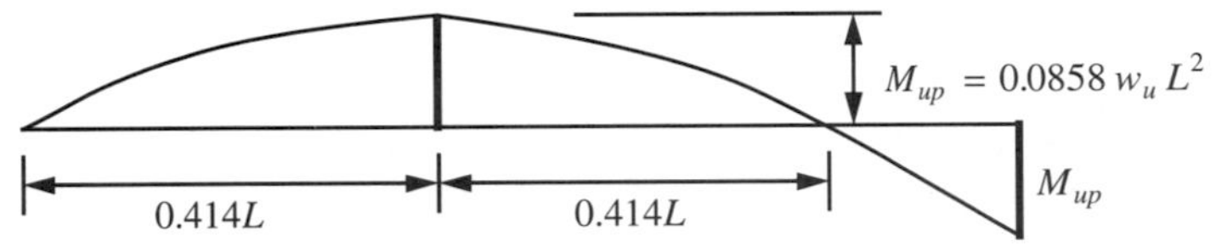

FIGURE D.2 Propped-cantilever beam.

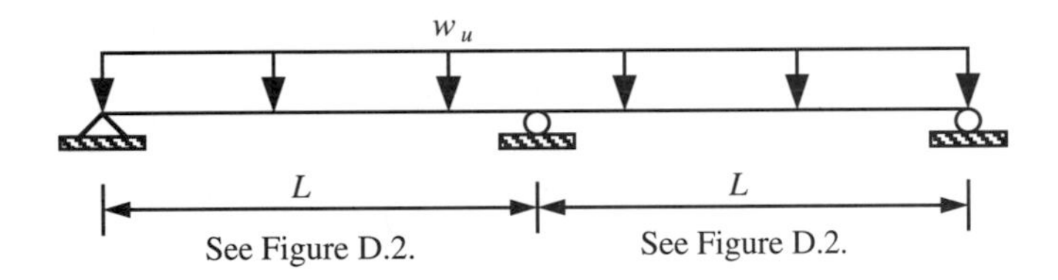

FIGURE D.3 Two equal span lengths.

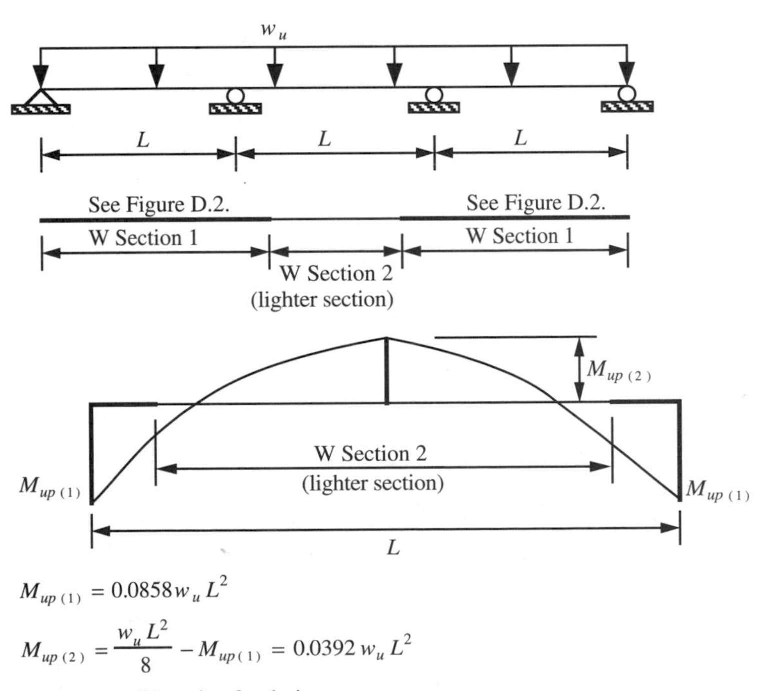

$$M_{up\,(1)} = 0.0858 w_u L^2$$

$$M_{up\,(2)} = \frac{w_u L^2}{8} - M_{up\,(1)} = 0.0392\, w_u L^2$$

Center span W section 2 solution

FIGURE D.4 Three equal span lengths—Case 1.

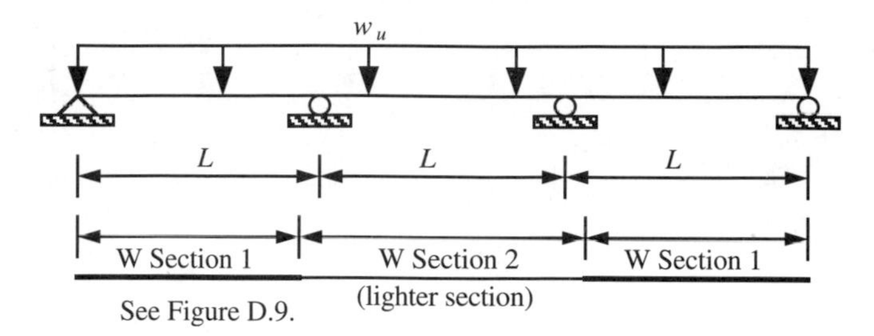

FIGURE D.5 Three equal span lengths—Case 2.

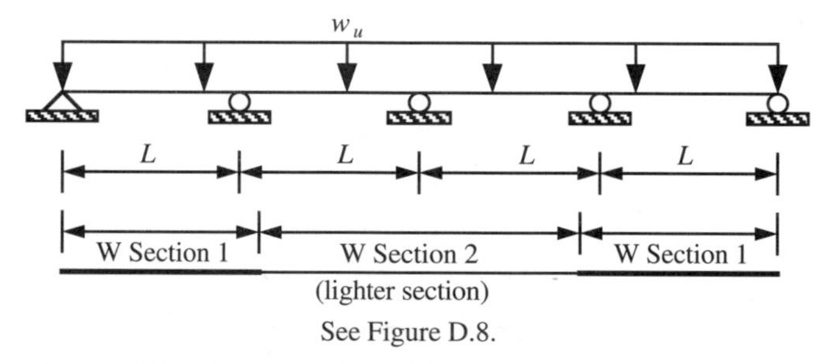

FIGURE D.6 Four equal span lengths.

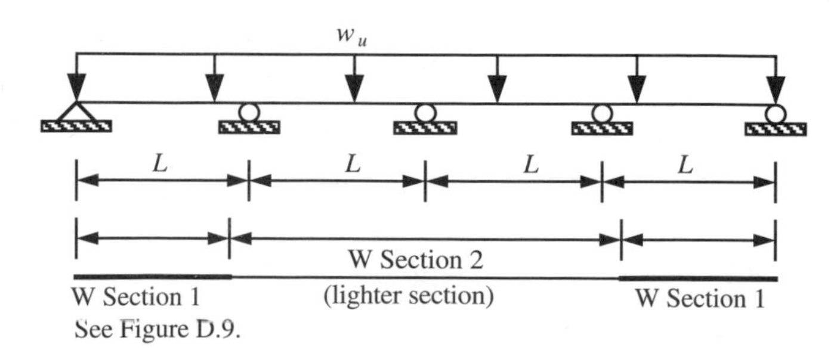

FIGURE D.7 Five or more equal span lengths—Case 1

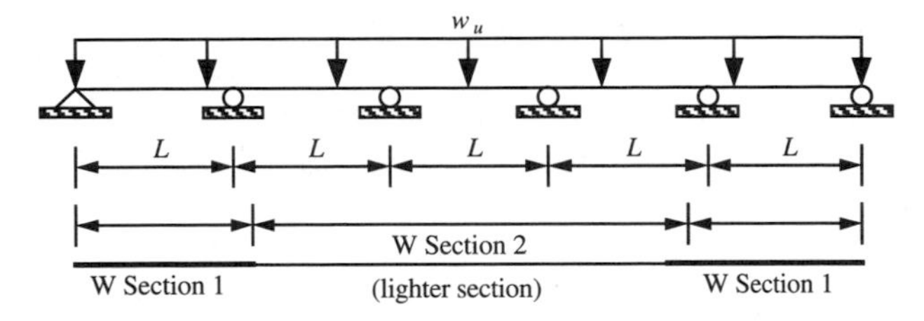

See Figure D.1 for the middle span and similar spans.

See Figure D.2 for each exterior span.

The following information is applicable for the first interior span:

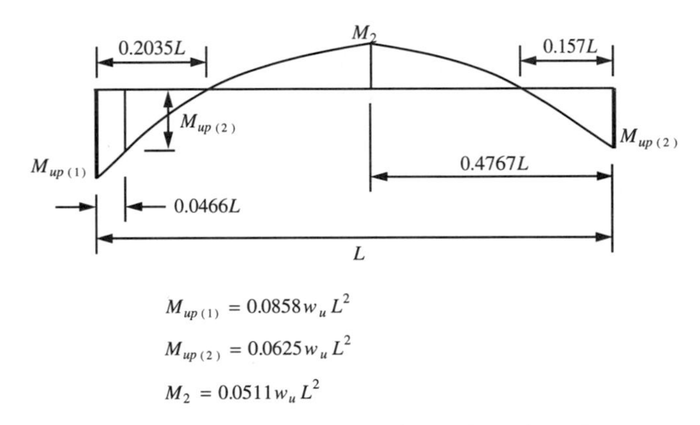

$$M_{up(1)} = 0.0858 w_u L^2$$

$$M_{up(2)} = 0.0625 w_u L^2$$

$$M_2 = 0.0511 w_u L^2$$

FIGURE D.8 Five or more equal span lengths—Case2

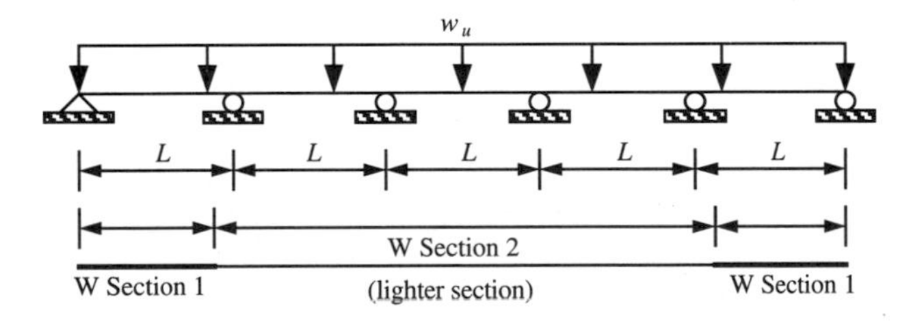

See Figure D.1 for each interior span.

The following information is applicable for exterior span:

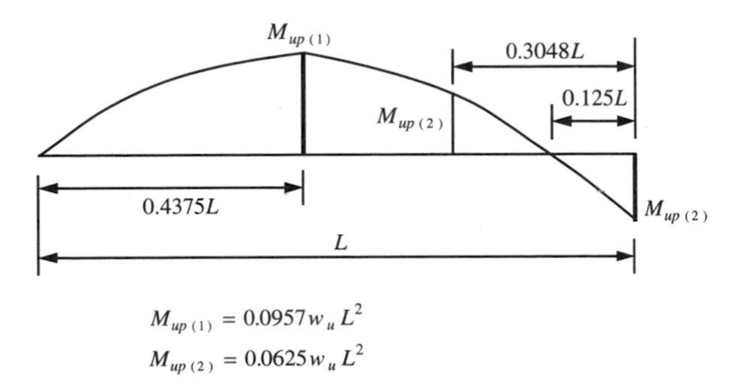

$$M_{up(1)} = 0.0957 w_u L^2$$

$$M_{up(2)} = 0.0625 w_u L^2$$

FIGURE D.9 Combined mechanism relations for a gable frame.

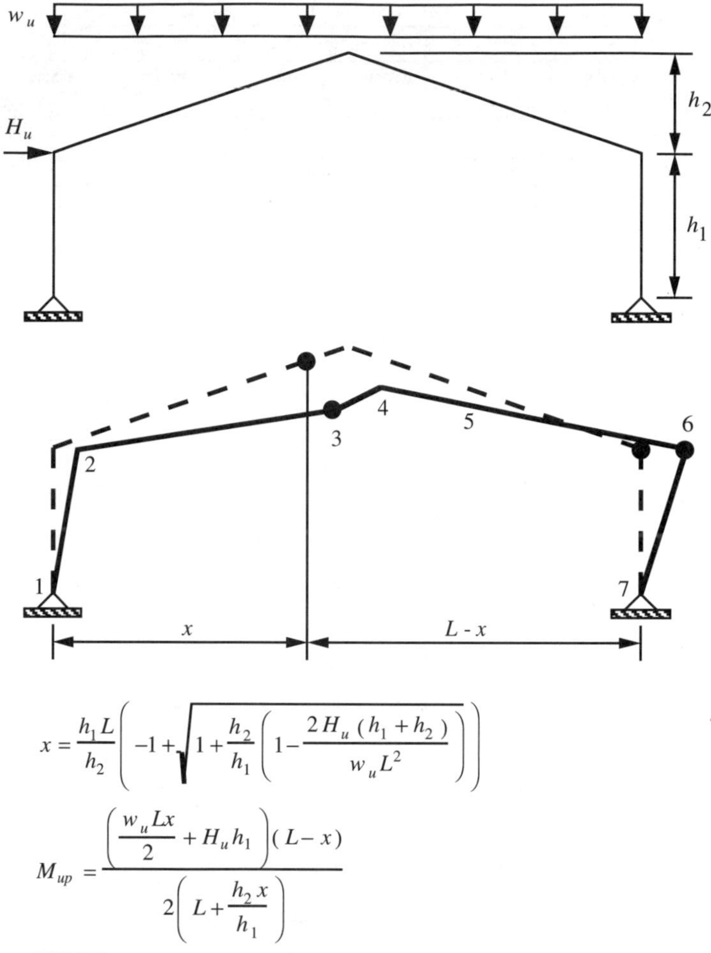

$$x = \frac{h_1 L}{h_2}\left(-1+\sqrt{1+\frac{h_2}{h_1}\left(1-\frac{2H_u(h_1+h_2)}{w_u L^2}\right)}\right)$$

$$M_{up} = \frac{\left(\frac{w_u L x}{2}+H_u h_1\right)(L-x)}{2\left(L+\frac{h_2 x}{h_1}\right)}$$

NOTES:
1. The required plastic hinge strength is for the girders.
2. The formulas are valid also for the case where the lateral load is zero.

FIGURE D.10

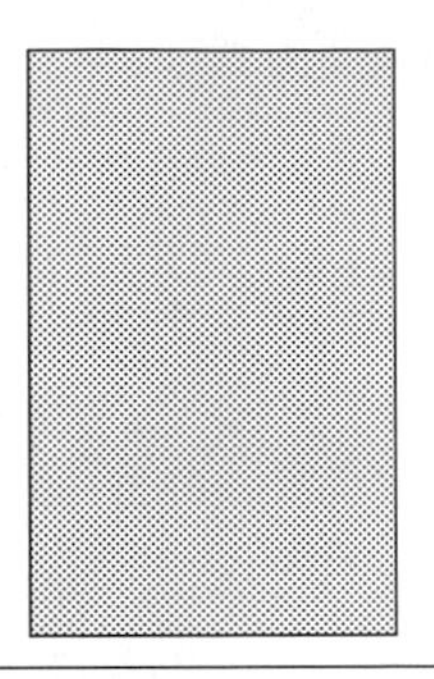

References

1. Smith, J.C., *Structural Analysis*, New York: Harper & Row, 1988.

2. AISC Manual of Steel Construction, *Load & Resistance Factor Design*, Volume I: Structural Members, Specifications, & Codes, 2nd ed.; Volume II: Connections, 2nd ed., Chicago, IL: American Institute of Steel Construction, 1994.

3. Galambos, T.V., and Ellingwood, B., "Serviceability Limit States: Deflection," *Journal of the Structural Division*, ASCE, Vol. 112, No. 1, Jan. 1986, pp. 68–84.

4. Euler, L., *Methodus Inveniendi Lineas Curvas Maximi Minimive Proprietate Gaudentes*, Appendix: *De Curvis Elasticis*, Lausanne and Geneva, 1744.

5. Euler, L., "Sur la force de colonnes," *Memoires de l'Academie de Berlin*, 1759.

6. Timoshenko, S.P., and Gere, J.M., *Theory of Elastic Stability*, New York: McGraw-Hill, 1961.

7. Bleich, F., *Buckling Strength of Metal Structures*, New York: McGraw-Hill, 1952.

8. Galambos, T.V., *Structural Members and Frames*, Englewood Cliffs, NJ: Prentice-Hall, 1968.

9. Wood, B.R., Beaulieu, D., and Adams, P.F., "Column Design by P-Delta Method," *Journal of the Structural Division*, ASCE, Vol. 102, No. ST2, Feb. 1976, pp. 411–427.

10. American Concrete Institute, *Building Code Requirements for Reinforced Concrete*, ACI 318–92R, Detroit, MI: 1992.

11. Kavanagh, T.C., "Effective Length of Framed Columns," *Transactions ASCE*, Vol. 127, Part II, 1962, pp. 81–101.

12. *Standard Specifications for Highway Bridges*, Washington, DC : American Association of State Highway and Transportation Officials (AASHTO), 1989.

13. *Specifications for Steel Railway Bridges*, Chicago, IL: American Railway Engineering Association, 1992.

14. Cochrane, V.H., "Rules for Riveted Hole Deductions in Tension Members," *Engineering News-Record* (New York), Nov. 16, 1922, pp. 847–848.

15. Munse, W.H., and Chesson, Jr., E., "Riveted and Bolted Joints," *Journal of the Structural Division*, ASCE, Vol. 89, No. ST1, February 1963.

16. Easterling, W.S. and Giroux, L. G., "Shear Lag Effects in Steel Tension Members," *Engineering Journal*, AISC, Vol. 30, No. 3 (3rd quarter), 1993, pp. 77-89.

17. Yura, J.A., "The Effective Length of Columns in Unbraced Frames," *Engineering*

Journal, AISC, Vol. 8, No. 2 (2nd quarter), 1971, pp. 37–42.

18. Winter, G., "Lateral Bracing of Columns and Beams," *Transactions ASCE*, Vol. 125, Part I, 1960, pp. 808–845.

19. McGuire, W., *Steel Structures*, Englewood Cliffs, NJ: Prentice-Hall, 1968.

20. Galambos, T.V., "Lateral Support for Tier Building Frames," *Engineering Journal*, AISC, Vol. 1, No. 1 (1st quarter), 1964, pp. 16–19; Discussion 1, No. 4 (4th quarter), 1964, p. 141.

21. Crawford, S.F., and Kulak, G.L., "Eccentrically Loaded Bolted Connections," *Journal of the Structural Division*, ASCE, Vol. 97, No. ST3, March 1971, pp. 765–783.

22. Butler, L. J., Pal, S., and Kulak, G.L., "Eccentrically Loaded Weld Connections," *Journal of the Structural Division*, ASCE, Vol. 98, No. ST3, May 1972, pp. 989–1005.

23. Shipp, J.G., and Haninger, E.R., "Design of Headed Anchor Bolts," *Engineering Journal*, AISC, Vol. 20, No. 2 (2nd quarter), 1983, pp. 58–69.

24. Curtis, L.E., and Murray, T.M., "Column Flange Strength at Moment End-Plate Connections," *Engineering Journal*, AISC, Vol. 26, No. 2 (2nd quarter), 1989, pp. 41–50.

25. Galambos, T.V., ed., *Guide to Stability Design Criteria for Metal Structures*, 4th ed., Structural Stability Research Council, New York: John Wiley & Sons, 1988.

26. Ravindra, M. K., and Galambos, T. V., "Load and Resistance Factor Design for Steel," *Journal of the Structural Division*, ASCE, Vol. 104, No. ST9, Sept. 1978, pp. 1337–1353.

27. Ellingwood, B., et al., *Development of a Probability Based Load Criterion for American National Standard A58 Building Code Requirements for Minimum Design Loads in Buildings and Other Structures*, Special Publication 577, Washington, DC: National Bureau of Standards, June 1980.

28. Beedle, L. S., *Plastic Design of Steel Frames*, New York: John Wiley & Sons, 1958.

29. Tall, L., "Residual Stresses in Welded Plates — A Theoretical Study," *Welding Journal*, Vol. 43, Jan. 1964.

30. American Welding Society (AWS), *Structural Welding Code—Steel(D1.1)*, AWS, Miami, FL, 1992.

31. Julian, O. G., and Lawrence, L. S., Notes on J and L Nomgrams for Determination of Effective Lengths, 1959 (unpublished).

32. Dumonteil, P., "Simple Equations for Effective Length Factors," *Engineering Journal*, AISC, Vol. 29, No. 3 (3rd quarter), 1992, pp. 111–115.

33. Lesik, D. F., and Kennedy, D. J. L., "Ultimate Strength of Fillet-Welded Connections Loaded in Plane," *Canadian Journal of Civil Engineering*, Vol. 17, No. 1, National Research Council of Canada, Ottawa, Canada, 1990.

34. Chen, W. F., and Lui, E. M., *Stability Design of Steel Frames*, Boca Raton, FL: CRC Press, Inc., 1991.

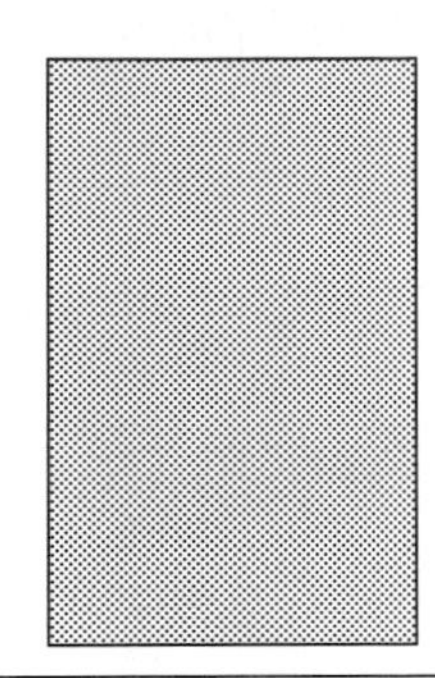

Index

C

D

Y